PROBABILITY
Theory and Examples

The Wadsworth & Brooks/Cole Statistics/Probability Series

R. Becker, J. Chambers, A. Wilks, *The New S Language: A Programming Environment for Data Analysis and Graphics*

P. Bickel, K. Doksum, J. Hodges, Jr., *A Festschrift for Erich L. Lehmann*

G. Box, *The Collected Works of George E. P. Box, Volumes I and II*, G. Tiao, editor-in-chief

L. Breiman, J. Friedman, R. Olshen, C. Stone, *Classification and Regression Trees*

G. Casella, R. Berger, *Statistical Inference*

W. S. Cleveland, M. McGill, *Dynamic Graphics for Statistics*

K. Dehnad, *Quality Control, Robust Design, and the Taguchi Method*

R. Durrett, *Lecture Notes on Particle Systems and Percolation*
   Probability: Theory and Examples

F. Graybill, *Matrices with Applications in Statistics, Second Edition*

L. Le Cam, R. Olshen, *Proceedings of the Berkeley Conference in Honor of Jerzy Neyman and Jack Kiefer, Volumes I and II*

P. Lewis, E. Orav, *Simulation Methodology for Statisticians, Operations Analysts, and Engineers*

H. J. Newton, *TIMESLAB*

J. Rawlings, *Applied Regression Analysis*

J. Rice, *Mathematical Statistics and Data Analysis*

J. Romano, A. Siegel, *Counterexamples in Probability and Statistics*

J. Tanur, F. Mosteller, W. Kruskal, E. Lehmann, R. Link, R. Pieters, G. Rising, *Statistics: A Guide to the Unknown, Third Edition*

J. Tukey, *The Collected Works of J. W. Tukey*, W. S. Cleveland, editor-in-chief
   Volume I: *Time Series: 1949–1964*, edited by David Brillinger
   Volume II: *Time Series: 1965–1984*, edited by David Brillinger
   Volume III: *Philosophy and Principles of Data Analysis: 1949–1964*, edited by L. Jones
   Volume IV: *Philosophy and Principles of Data Analysis: 1965–1986*, edited by L. Jones
   Volume V: *Graphics: 1965–1985*, edited by W. S. Cleveland
   Volume VI: *More Mathematical: 1938–1984*, edited by Colin L. Mallows

PROBABILITY
Theory and Examples

Richard Durrett

Cornell University

Wadsworth & Brooks/Cole Advanced Books & Software
Pacific Grove, California

Brooks/Cole Publishing Company
A Division of Wadsworth, Inc.

Printed in the United States of America

10   9   8   7   6   5   4   3   2   1

Library of Congress Cataloging-in-Publication Data
Durrett, Richard
     Probability : theories and examples / Richard Durrett.
          p.   cm.
     Includes bibliographical references and index.
     ISBN 0-534-13206-5
     1. Probabilities.   I. Title.
QA273.D865   1991
519.2—dc20   90-37802
                                                         CIP

Sponsoring Editor: *John Kimmel*
Editorial Assistant: *Jennifer Kehr*
Production Editor: *Penelope Sky*
Production Assistant: *Dorothy Bell*
Manuscript Editor: *Carol Reitz*
Interior and Cover Design: *Roy R. Neuhaus*
Cover Illustration: *Roy R. Neuhaus*
Art Coordinator: *Cloyce J. Wall*
Interior Illustration: *Cloyce J. Wall*
Typesetting: *Asco Trade Typesetting Ltd.*
Printing and Binding: *Arcata Graphics/Martinsburg*

Preface

The first and most obvious use for this book is as a textbook for a one-year graduate course in probability taught to students who are familiar with measure theory. An appendix gives complete proofs of the necessary results from measure theory, so that the book can be used whether or not students are familiar with measure theory.

The title indicates that as we develop the theory, we focus our attention on examples. So the book will be a useful reference for people who apply probability in their work, I have emphasized results that can be used to solve problems. However, as Chung said in the preface to *Markov Chains*, "In general, the practical man in search of ready-made solutions to his own problems will discover in this book, as elsewhere, that mathematicians are more inclined to build fire stations than put out fires."

Exercises are integrated into the text because they are an integral part of it. They are divided into two categories: (i) those at the end of sections and (ii) those that appear elsewhere. In general, exercises of type (ii) are consequences of the material just presented and readers should do these "finger exercises" to check their understanding. We use those results freely in later proofs. Exercises of type (i) involve extensions of the results and various complements.

This text developed from notes I wrote while teaching the first-year probability course three times at UCLA and once at Cornell. Of course, in preparing those notes, I borrowed heavily from existing books in the field. Chung (1974), Billingsley (1979), Feller, Vol. II (1971), and Breiman (1968) contributed to the entire book. For special topics, sources were Royden (1988) on measure theory; Neveu (1975) on martingales; Hoel, Port, and Stone (1972), and Revuz (1984) on Markov chains; Halmos (1956), and Krengel (1985) on ergodic theory; and Durrett (1984) on Brownian motion.

Other contributors are mentioned at appropriate places in the text. In the book, as in the paragraph above, citations give credit to my sources and indicate where the reader can learn more about the subject at hand. I have not tried to sort out who did what and when. This is not a history book! Its conflicting aims are to treat the subject as simply and as completely as possible.

Several people read versions alpha and beta of the book and pointed out numerous typos, obscurities, and outright errors. In this connection I would like to thank David Aldous, Alexandra Bellow, Ted Cox, Tom Mountford, Claudia Neuhauser, Robin Pemantle, Mark Pinsky, Jeff Steif, and Murad Taqqu. Many of the exercises came from problem sets and exams provided by David Aldous, Tom Liggett, and Harry Kesten.

Version alpha was written during the 1987–1988 academic year while I was teaching the graduate probability course. Support from the Guggenheim Foundation and the Army Research Office through the Mathematical Sciences Institute at Cornell gave me 18 months free from teaching to improve and polish the first version.

Closer to home, I would like to thank my wife, Susan, for her patience, understanding, and encouragement during the two and a quarter years it took to write this book. The same cannot be said for my two sons, Gregory (now 1) and David (a little over 3), but they have enriched my life in other ways. I also thank Marie Fredriksson, Per Gessele, and W. Axl Rose for their special assistance in putting the finishing touches on this manuscript.

Rick Durrett

Contents

*Asterisks indicate leaves on the tree of knowledge. These sections are not prerequisites for the ones that follow.

Before you begin,

You should know three things.

1. When we have a sequence of objects X_n or θ^n and we let $n = N_k$, then we will often write $X(N_k)$ or $\theta^{N(k)}$ for typographical reasons.
2. $A \equiv \{x: f(x) > 1\}$ is read "A is defined to be the set of x...."
3. Other notation we use is explained in the section on page 451.

Introductory Lecture

As Breiman (1968) should have said in his preface: "Probability theory has a right and a left hand. On the left is the rigorous foundational work using the tools of measure theory. The right hand 'thinks probabilistically,' reduces problems to gambling situations, coin-tossing, and motions of a physical particle." We have interchanged Breiman's hands in the quote because we learned in a high school English class that the left hand is sinister and the right is dextrous. While measure theory does not "threaten harm, evil or misfortune," it is an unfortunate fact that we will need four sections of definitions before we come to the first interesting result. To motivate the reader for this necessary foundational work, we will now give some previews of coming attractions.

For a large part of the first two chapters, we will be concerned with the laws of large numbers and the central limit theorem. To introduce these theorems and to illustrate their use, we will begin by giving their interpretation for a person playing roulette. In doing this we will use some terms (e.g., independent, mean, variance) without explaining them. If some of the words that we use are unfamiliar, don't worry. There will be more than enough definitions when the time comes.

A roulette wheel has 38 slots—18 red, 18 black, and 2 green ones that are numbered 0 and 00—so if our gambler bets $1 on red coming up, he wins $1 with probability 18/38 and loses $1 with probability 20/38. Let X_1, X_2, ... be the outcomes of the first, second, and subsequent bets. If the house and gambler are honest, X_1, X_2, ... are independent random variables and each has the same distribution—namely, $P(X_1 = 1) = 9/19$ and $P(X_1 = -1) = 10/19$. One of the first things we will have to do is to construct a probability space and define on it a sequence of independent random variables X_1, X_2, ... with this distribution, but our friend the gambler doesn't care about this technicality. He wants to know what we can tell him about the amount he has won at time n: $S_n = X_1 + \cdots + X_n$.

The first facts we can tell him are that (i) the average amount of money he will win on one play ($=$ the mean of X_1 and denoted EX_1) is

$$(9/19) \cdot \$1 + (10/19) \cdot (-\$1) = -\$1/19 = -\$.05263,$$

and (ii) on the average after n plays his winnings will be $ES_n = nEX_1 = -\$n/19$.

1

For most values of n the probability of having lost exactly $n/19$ dollars is zero, so the next question to be answered is: How close will his experience be to the average? The first answer is provided by:

The weak law of large numbers   If X_1, X_2, ... are independent and identically distributed random variables with mean $EX_1 = \mu$, then for all $\varepsilon > 0$,

$$P(|S_n/n - \mu| > \varepsilon) \to 0 \quad \text{as } n \to \infty.$$

Less formally, if n is large, S_n/n is close to μ with high probability.

This result provides some information but leaves several questions unanswered. The first one is: If our gambler was statistically minded and wrote down the values of S_n/n, would the resulting sequence of numbers converge to $-1/19$? The answer to the question is given by:

The strong law of large numbers   If X_1, X_2, ... are independent and identically distributed random variables with mean $EX_1 = \mu$, then with probability 1, S_n/n converges to μ.

An immediate consequence of the last result of interest to our gambler is that with probability one, $S_n \to -\infty$ as $n \to \infty$. That is, the gambler will eventually go bankrupt no matter how much money he starts with.

The laws of large numbers tell us what happens in the long run but do not provide much information about what happens over the short run. That gap is filled by:

The central limit theorem   If X_1, X_2, ... are independent and identically distributed random variables with mean $EX_i = \mu$ and variance $\sigma^2 = E(X_i - \mu)^2$, then for any y,

$$P((S_n - n\mu)/\sigma n^{1/2} \le y) \to \mathcal{N}(y)$$

where

$$\mathcal{N}(y) = \int_{-\infty}^{y} (2\pi)^{-1/2} e^{-x^2/2} dx$$

is the (standard) normal distribution.

If we let χ denote a random variable with a normal distribution, then the last conclusion can be written informally as

$$S_n \approx n\mu + \sigma n^{1/2}\chi.$$

In the example we have been considering, $\mu = -1/19$ and

$$\sigma^2 = (9/19)(1 + 1/19)^2 + (10/19)(-1 + 1/19)^2 = 1 - (1/19)^2 = .9973.$$

If we use $\sigma^2 \approx 1$ to simplify the arithmetic, then the central limit theorem tells us

$$S_n \approx -n/19 + n^{1/2}\chi,$$

or when $n = 100$,

$$S_{100} \approx -5.26 + 10\chi.$$

If we are interested in the probability $S_{100} \geq 0$, this is

$$P(-5.26 + 10\chi \geq 0) = P(\chi \geq -.526) \approx .30$$

from the table of the normal distribution in Section 4 of Chapter 2.

The last result shows that after 100 plays the negative drift is not too noticeable. The gambler has lost \$5.26 on the average and has a probability .3 of being ahead. To see why casinos make money, suppose there are 100 gamblers playing 100 times and set $n = 10,000$ to get

$$S_{10,000} \approx -526 + 100\chi.$$

Now $P(\chi \leq 2.3) = .99$, so with that probability $S_{10,000} \leq -296$; that is, the casino is slowly but surely making money.

$$100\chi \leq 230$$

$$\chi \leq 2.3 \checkmark$$

1 Laws of Large Numbers

In the first three sections we will recall some definitions and results from measure theory. Our purpose is not only to review that material but also to introduce the terminology of probability theory, which differs slightly from that of measure theory. In Section 4 we introduce the crucial concept of independence and explore its properties. In Section 5 we prove the weak law of large numbers and give several applications. In Sections 6–8 we develop the strong law of large numbers by proving some Borel–Cantelli lemmas and investigating the convergence of random series. Finally in Section 9 we show that in nice situations convergence in the weak law occurs exponentially rapidly.

1 Basic Definitions

Here and throughout the book terms being defined are set in *italic*. We begin with the most basic quantity. A *probability space* is a triple $(\Omega, \mathscr{F}, P)$ where Ω is a set of "outcomes," $\mathscr{F}$ is a set of "events," and $P: \mathscr{F} \to [0, 1]$ is a function that assigns probabilities to events. We assume that $\mathscr{F}$ is a *σ-field* (or *σ-algebra*)—that is, a (nonempty) collection of subsets of Ω that satisfy:

(i) if $A, B \in \mathscr{F}$, then $A \cup B$ and $A^c \in \mathscr{F}$, and
(ii) if $A_i \in \mathscr{F}$ for $i \geq 1$, then $\bigcup_{i=1}^{\infty} A_i \in \mathscr{F}$.

Since $\bigcap_i A_i = (\bigcup_i A_i^c)^c$, it follows that a σ-field is closed under finite and countable intersections. We omit the last property from the definition to make it easier to check.

Without P, $(\Omega, \mathscr{F})$ is called a *measurable space*; that is, it is a space on which we can put a measure. A *measure* is a nonnegative countably additive set function— that is, a function $\mu: \mathscr{F} \to \mathbb{R}$ with

(i) $\mu(A) \geq \mu(\phi) = 0$ for all $A \in \mathscr{F}$, and
(ii) if $A_i \in \mathscr{F}$ is a finite or countable sequence of disjoint sets, then

$$P\left(\bigcup_i A_i\right) = \sum_i P(A_i).$$

If $\mu(\Omega) = 1$, we call μ a *probability measure*. Probability measures are usually denoted by P.

The next three exercises give some consequences of the definition that we will need later. In all cases we assume that the $A_i \in \mathcal{F}$.

Exercise 1.1   Let P be a probability measure on $(\Omega, \mathcal{F})$.

(i) *Monotonicity.* If $A \subset B$, then $P(A) \leq P(B)$.
(ii) *Subadditivity.* If $A_i \in \mathcal{F}$ are arbitrary and $A \subset \bigcup_i A_i$, then $P(A) \leq \sum_i P(A_i)$.
(iii) *Continuity from below.* If $A_i \uparrow A$ (i.e., $A_1 \subset A_2 \subset \cdots$ and $\bigcup_i A_i = A$), then $P(A_i) \uparrow P(A)$.
(iv) *Continuity from above.* If $A_i \downarrow A$ (i.e., $A_1 \supset A_2 \supset \cdots$ and $\bigcap_i A_i = A$), then $P(A_i) \downarrow P(A)$.

In the next two exercises we suppose that P is a probability measure.

Exercise 1.2   *Inclusion–exclusion formula.* If $A_1, A_2, \ldots$ are any events,

$$P\left(\bigcup_{i=1}^n A_i\right) = \sum_{i=1}^n P(A_i) - \sum_{i<j} P(A_i \cap A_j)$$

$$+ \sum_{i<j<k} P(A_i \cap A_j \cap A_k) - \cdots + (-1)^{n-1} P\left(\bigcap_{i=1}^n A_i\right).$$

Exercise 1.3   *Bonferroni inequalities*

$$P\left(\bigcup_{i=1}^n A_i\right) \leq \sum_{i=1}^n P(A_i).$$

$$P\left(\bigcup_{i=1}^n A_i\right) \geq \sum_{i=1}^n P(A_i) - \sum_{i<j} P(A_i \cap A_j).$$

$$P\left(\bigcup_{i=1}^n A_i\right) \leq \sum_{i=1}^n P(A_i) - \sum_{i<j} P(A_i \cap A_j) + \sum_{i<j<k} P(A_i \cap A_j \cap A_k).$$

In general if we stop the inclusion exclusion formula after an even (odd) number of sums, we get a lower (upper) bound.

Some examples of probability measures should help to clarify the concept. Here, and in what follows, we leave it to the reader to check that they are examples.

Example 1.1 $\Omega =$ a countable set. $\mathcal{F} =$ the set of all subsets of Ω.

$$P(A) = \sum_{\omega \in A} p(\omega) \quad \text{where } p(\omega) \geq 0 \text{ and } \sum_{\omega \in \Omega} p(\omega) = 1.$$

A little thought reveals that this is the most general probability measure on this space. In many cases when Ω is a finite set, we have $p(\omega) = 1/|\Omega|$ where $|\Omega| =$ the number of points in Ω. Concrete examples in this category are:

(i) *flipping a fair coin*: $\Omega = \{$Heads, Tails$\}$, and
(ii) *rolling a die*: $\Omega = \{1, 2, 3, 4, 5, 6\}$.

Example 1.2   A trivial example but useful for counterexamples is: $\Omega = \mathbb{R}$, $\mathscr{F} = $ all subsets so that A or A^c is countable, $P(A) = 0$ in the first case and $= 1$ in the second.

Example 1.3   Let $\mathbb{R} = $ the real line, $\mathscr{R} = $ the *Borel sets* $= $ the smallest σ-field containing the open sets, $\lambda = Lebesgue\ measure = $ the only measure on $\mathscr{R}$ with $\lambda((a, b]) = b - a$ for all $a < b$. $\lambda(\mathbb{R}) = \infty$. To get a probability space let $\Omega = (0, 1)$, $\mathscr{F} = \{A \cap (0, 1) : A \in \mathscr{F}\}$, and $P(B) = \lambda(B)$ for $B \in \mathscr{F}$. P is Lebesgue measure restricted to the Borel subsets of $(0, 1)$. To check that there is a smallest σ-field containing any collection of sets, we take the intersection of all the σ-fields containing the collection and use:

Exercise 1.4   If $\mathscr{F}_i$, $i \in I$, are σ-fields, then $\bigcap_{i \in I} \mathscr{F}_i$ is. Here $I \neq \varnothing$ is an arbitrary index set (i.e., possibly uncountable).

The collection of all subsets of Ω is a σ-field, so the index set $I \neq \varnothing$. The construction of Lebesgue measure is carried out in Section 1 of the Appendix.

Example 1.4   *Product Spaces*. If $(\Omega_i, \mathscr{F}_i, P_i)$, $i = 1, \ldots, n$, are probability spaces, we can let $\Omega = \Omega_1 \times \cdots \times \Omega_n = \{(\omega_1, \ldots, \omega_n) : \omega_i \in \Omega_i\}$. $\mathscr{F} = \mathscr{F}_1 \times \cdots \times \mathscr{F}_n = $ the σ-field *generated* by sets $A_1 \times \cdots \times A_n$ with $A_i \in \mathscr{F}_i$—that is, the smallest σ-field containing these sets. $P = P_1 \times \cdots \times P_n = $ the measure on $\mathscr{F}$ that has

$$P(A_1 \times \cdots \times A_n) = P_1(A_1) \cdot P_2(A_2) \cdots P_n(A_n).$$

For more details see Section 6 of the Appendix. Concrete examples of product spaces are:
(i) $\Omega = \{1, 2, 3, 4, 5, 6\} \times \{1, 2, 3, 4, 5, 6\}$, $\mathscr{F} = $ all subsets of Ω, $P(A) = |A|/36$. This space corresponds to rolling two dice.
(ii) If $\Omega_i = (0, 1)$, $\mathscr{F}_i = $ the Borel sets, and $P_i = $ Lebesgue measure, then the product space defined above is the unit cube $\Omega = (0, 1)^n$, $\mathscr{F} = $ the Borel subsets of Ω, and P is n-dimensional Lebesgue measure restricted to $\mathscr{F}$.

Exercise 1.5   Let $\mathbb{R}^n = \{(x_1, \ldots, x_n) : x_i \in \mathbb{R}\}$. $\mathscr{R}^n = $ the Borel subsets of $\mathbb{R}^n$ is defined to be the σ-field generated by open subsets of $\mathbb{R}^n$. Prove this is the same as $\mathscr{R} \times \cdots \times \mathscr{R} = $ the σ-field generated by sets of the form $A_1 \times \cdots \times A_n$.

Hint   Show that both σ-fields coincide with the one generated by $(a_1, b_1) \times \cdots \times (a_n, b_n)$.

Exercise 1.6   A σ-field $\mathscr{F}$ is said to be *countably generated* if there is a countable collection $\mathscr{C} \subset \mathscr{F}$ so that $\sigma(\mathscr{C}) = \mathscr{F}$. Show that $\mathscr{R}^d$ is countably generated.

Probability spaces become a little more interesting when we define random variables on them. A real-valued function X defined on Ω is said to be a *random*

$= \{X \in B\}$

variable ("r.v." for short) if for every Borel set $B \subset \mathbb{R}$ we have $X^{-1}(B) = \{\omega : X(\omega) \in B\} \in \mathscr{F}$. When we need to emphasize the σ-field we will say that X is $\mathscr{F}$-*measurable* or write $X \in \mathscr{F}$. If Ω is a *discrete probability space*—that is, Ω is a finite or countable set and $\mathscr{F}$ is the set of all subsets of Ω, then any function $X : \Omega \to \mathbb{R}$ is a random variable. A second trivial, but useful, type of example of a random variable is the *indicator function* of a set $A \in \mathscr{F}$:

$$1_A(\omega) = \begin{cases} 1 & \omega \in A \\ 0 & \omega \notin A \end{cases}.$$

The notation is supposed to remind you that this function is 1 on A. Analysts call this object the characteristic function of A. In probability that term is used for something quite different. (See Section 3 of Chapter 2.)

Exercise 1.7 Suppose X and Y are random variables on $(\Omega, \mathscr{F}, P)$ and let $A \in \mathscr{F}$. Show that if we let $Z(\omega) = X(\omega)$ for $\omega \in A$ and $Z(\omega) = Y(\omega)$ for $\omega \in A^c$, Z is a random variable.

If X is a random variable, then X induces a probability measure on $\mathbb{R}$ called its *distribution* by setting $\mu(A) = P(X \in A)$ for Borel sets A. To check that this is a probability measure, we observe that if the A_i are disjoint,

$$\mu\left(\bigcup_i A_i\right) = P\left(X \in \bigcup_i A_i\right) = P\left(\bigcup_i \{X \in A_i\}\right) = \sum_i P(X \in A_i) = \sum_i \mu(A_i),$$

the next to last equality following from the fact that the events $\{X \in A_i\}$ are disjoint. The distribution of X is usually described by giving its *distribution function* (or *d.f.*) $F(x) = P(X \le x)$. F has three obvious properties:

(i) F is nondecreasing,
(ii) $\lim_{x \to \infty} F(x) = 1$, $\lim_{x \to -\infty} F(x) = 0$, and
(iii) F is right continuous; that is, $\lim_{y \downarrow x} F(y) = F(x)$.

To prove (iii) we recall that if $A_n \downarrow A$, then $P(A_n) \downarrow P(A)$ and observe that part (iv) of Exercise 1.1 implies

$$\lim_{y \downarrow x} F(y) = \lim_{y \downarrow x} P(X \in (-\infty, y]) = P(X \in (-\infty, x]) = F(x).$$

Exercise 1.8 Let $F(x-) = \lim_{y \uparrow x} F(y)$. $P(X \in (a, b)) = F(b-) - F(a)$ and $P(X = c) = F(c) - F(c-)$.

Notation If X and Y have the same distribution function—that is, $P(X \le z) = P(Y \le z)$ for all z, then we say X and Y are *equal in distribution* and write $X \overset{d}{=} Y$.

(1.1) THEOREM If F satisfies (i), (ii), and (iii), then it is the distribution function of some random variable.

Proof   Let $\Omega = (0, 1)$, $\mathcal{F}$ = the Borel sets, and P = Lebesgue measure. If $\omega \in (0, 1)$, let

$$X(\omega) = \sup\{y : F(y) < \omega\}.$$

Roughly speaking, X is the inverse of F. Now I claim

$$\{\omega : X(\omega) \le x\} = \{\omega : \omega \le F(x)\}, \tag{*}$$

from which the result follows immediately, since $P(\omega : \omega \le F(x)) = F(x)$. (Recall P is Lebesgue measure.) To check (*), we observe that if $\omega \le F(x)$, then $X(\omega) \le x$, since $x \notin \{y : F(y) < \omega\}$. On the other hand, if $\omega > F(x)$, then since F is right continuous, there is an $\varepsilon > 0$ so that $F(x + \varepsilon) < \omega$ and $X(\omega) \ge x + \varepsilon > x$.   □

An immediate consequence of (1.1) is:

(1.2) COROLLARY   If F satisfies (i), (ii), and (iii), there is a unique probability measure μ on $(\mathbb{R}, \mathcal{R})$ that has $\mu((a, b]) = F(b) - F(a)$ for all a, b.

Proof   (1.1) gives the existence of μ. It is unique because the sets $(a, b]$ are closed under intersection and generate the σ-field. (See (2.2) in the Appendix.)   □

When the distribution function $F(x) = P(X \le x)$ has the form

$$F(x) = \int_{-\infty}^{x} f(y)\, dy, \tag{*}$$

we say that X has *density function f*. It is useful to think of $f(x)$ as being $P(X = x)$, although

$$P(X = x) = \lim_{\varepsilon \to 0} \int_{x-\varepsilon}^{x+\varepsilon} f(y)\, dy = 0.$$

We can start with f and use (*) to define F. In other to end up with a distribution function, it is necessary and sufficient that $f(x) \ge 0$ and $\int f(x)\, dx = 1$. Three examples that will be important in what follows are:

Example 1.5   *Uniform distribution on (0, 1).* $f(x) = 1$ for $x \in (0, 1)$, 0 otherwise.

Example 1.6   *Exponential distribution.* $f(x) = e^{-x}$ for $x \ge 0$, 0 otherwise.

Example 1.7   *The (standard) normal distribution.* $f(x) = (2\pi)^{-1/2} \exp(-x^2/2)$.

In the first two cases it is easy to calculate the distribution function. In the third there is no closed-form expression, but we have the following useful asymptotic formula:

(1.3) THEOREM   For $x > 0$,

$$(x^{-1} - x^{-3})\exp(-x^2/2) \le \int_x^\infty \exp(-y^2/2)\,dy \le x^{-1}\exp(-x^2/2).$$

In particular,

$$\int_x^\infty \exp(-y^2/2)\,dy \sim x^{-1}\exp(-x^2/2) \quad \text{as } x \to \infty,$$

where $f(x) \sim g(x)$ means $f(x)/g(x) \to 1$.

Proof   Changing variables $y = x + z$ gives

$$\int_x^\infty \exp(-y^2/2)\,dy \le \exp(-x^2/2)\int_0^\infty \exp(-xz)\,dz = x^{-1}\exp(-x^2/2).$$

For the other direction we observe

$$\int_x^\infty (1 - 3y^{-4})\exp(-y^2/2)\,dy = (x^{-1} - x^{-3})\exp(-x^2/2). \qquad \square$$

Exercise 1.9   Let χ have the standard normal distribution. Use (1.3) to estimate $P(\chi \ge 4)$ and $P(\chi \ge 10)$.

Remark   The scheme in the proof of (1.1) is useful in generating random variables on a computer. Standard algorithms generate random variables U_n with a uniform distribution. To get an exponential distribution we let $X_n = -\log U_n$. This recipe does not work well for the normal distribution, since there is no formula for its distribution function. See Exercise 4.8 for an alternative approach to this case.

A distribution function on $\mathbb{R}$ is said to be *absolutely continuous* if it has a density and *singular* if the corresponding measure is singular w.r.t. ("with respect to") Lebesgue measure. See Section 8 of the Appendix for more on these notions. An example of a singular distribution is:

Example 1.8   *Uniform distribution on the Cantor set.*   The Cantor set C is defined by removing $(1/3, 2/3)$ from $[0, 1]$ and then removing the middle third of each interval that remains. We define an associated distribution function by setting $F(x) = 0$ for $x \le 0$, $F(x) = 1$ for $x \ge 1$, $F(x) = 1/2$ for $x \in [1/3, 2/3]$, $F(x) = 1/4$ for $x \in [1/9, 2/9]$, $F(x) = 3/4$ for $x \in [7/9, 8/9], \ldots$. The function F that results is called *Lebesgue's singular function* because there is no f for which ($*$) holds. From the definition it is immediate that the corresponding measure has $\mu(C^c) = 0$.

A probability measure P (or its associated distribution function) is said to be *discrete* if there is a countable set S with $P(S^c) = 0$. The simplest example of a discrete distribution is:

Example 1.9   *Pointmass at 0.* $F(x) = 1$ for $x \geq 0$; $F(x) = 0$ for $x < 0$.

The next example shows that the distribution function associated with a discrete probability measure can be quite wild.

Example 1.10   *Dense discontinuities.*   Let $q_1, q_2, \ldots$ be an enumeration of the rationals and let

$$F(x) = \sum_{i=1}^{\infty} 2^{-i} 1_{[q_i, \infty)}$$

where $1_{[\theta, \infty)}(x) = 1$ if $x \in [\theta, \infty)$ and $= 0$ otherwise.

Exercise 1.10   Show that a distribution function has at most countably many discontinuities.

Exercise 1.11   A probability measure P on $(\Omega, \mathscr{F})$ is said to be *nonatomic* if $P(A) > 0$ implies there is a $B \subset A$ with $0 < P(A) < P(B)$. Show that if P is nonatomic, then $\{P(B) : B \subset A\} = [0, P(A)]$.

Exercise 1.12   Let μ be a probability measure on $(\mathbb{R}^d, \mathscr{R}^d)$. x is said to be in the *support* of μ if $\mu(D) > 0$ for all open D containing x. Show that the complement of the support is the largest open set G with $\mu(G) = 0$.

Exercise 1.13   Show that if $F(x) = P(X \leq x)$ is continuous, then $Y = F(X)$ has a uniform distribution on $(0, 1)$. That is, if $y \in [0, 1]$, $P(Y \leq y) = y$.

Exercise 1.14   (i) If X has density f and g is increasing and differentiable, then $g(X)$ has density $f(g^{-1}(x))/g'(g^{-1}(x))$. (ii) Suppose X has a normal distribution and compute the density of $\exp(X)$. (The answer is called the *lognormal distribution*.)

Exercise 1.15   The formula in the last exercise generalizes in a straightforward way to functions h for with $h'(x) = 0$ at only finitely many values. Suppose X has a normal density and compute the density of X^2. (The answer is called the *chi-square distribution*.)

2   Random Variables

In this section we will develop some results that will help us later to prove that quantities we define are random variables; that is, they are measurable. Since most of what we have to say is true for random elements of an arbitrary measurable space $(S, \mathscr{S})$ and the proofs are the same (sometimes easier), we will develop our results in that generality. First we need a definition. A function $X: \Omega \to S$ is said to be a *measurable map* from $(\Omega, \mathscr{F})$ to $(S, \mathscr{S})$ if $X^{-1}(B) \equiv \{\omega : X(\omega) \in B\} \in \mathscr{F}$ for all $B \in \mathscr{S}$. If $(S, \mathscr{S}) = (\mathbb{R}^d, \mathscr{R}^d)$, then X is called a *random vector*.

The next result is useful for proving that maps are measurable.

(2.1) THEOREM   If $\{\omega : X(\omega) \in A\} \in \mathcal{F}$ for all $A \in \mathcal{A}$ and $\mathcal{A}$ generates $\mathcal{S}$ (i.e., $\mathcal{S}$ is the smallest σ-field that contains $\mathcal{A}$), then X is measurable.

Proof   Writing $\{X \in B\}$ as shorthand for $\{\omega : X(\omega) \in B\}$, we have $\{X \in \bigcup_i B_i\} = \bigcup_i \{X \in B_i\}$, and $\{X \in B^c\} = \{X \in B\}^c$. So the class of sets $\mathcal{B} = \{B : \{X \in B\} \in \mathcal{F}\}$ is a σ-field. Since $\mathcal{B} \supset \mathcal{A}$ and $\mathcal{A}$ generates $\mathcal{S}$, $\mathcal{B} \supset \mathcal{S}$. $\qquad\qquad\square$

Exercise 2.1   Show that $\{\{X \in B\} : B \in \mathcal{S}\}$ is a σ-field. It is the smallest σ-field on Ω that makes X a measurable map. It is called the σ-field *generated by* X and denoted $\sigma(X)$.

Example 2.1   If $(S, \mathcal{S}) = (\mathbb{R}, \mathcal{R})$, then possible choices of $\mathcal{A}$ are $\{(-\infty, x] : x \in \mathbb{R}\}$ and $\{(a, b) : a, b \in \mathbb{Q}\}$ where $\mathbb{Q} = $ the rationals.

Example 2.2   If $(S, \mathcal{S}) = (\mathbb{R}^d, \mathcal{R}^d)$, a useful choice of $\mathcal{A}$ is

$$\{(a_1, b_1) \times \cdots \times (a_d, b_d) : -\infty < a_i < b_i < \infty\}$$

or occasionally the larger collection of open sets.

Exercise 2.2   (i) Show that a continuous function from $\mathbb{R}^d \to \mathbb{R}$ is a measurable map from $(\mathbb{R}^d, \mathcal{R}^d)$ to $(\mathbb{R}, \mathcal{R})$. (ii) Show that $\mathcal{R}^d$ is the smallest σ-field that makes all the continuous functions measurable.

Exercise 2.3   A function f is said to be *lower semicontinuous (l.s.c.)* if $\liminf_{y \to x} f(y) \geq f(x)$, and *upper semicontinuous (u.s.c.)* if $-f$ is l.s.c. Prove that semicontinuous functions are measurable.

Exercise 2.4   Let $f : \mathbb{R}^d \to \mathbb{R}$ be an arbitrary function and let $f^\delta(x) = \sup\{f(y) : |y - x| < \delta\}$ and $f_\delta(x) = \inf\{f(y) : |y - x| < \delta\}$ where $|z| = (z_1^2 + \cdots + z_d^2)^{1/2}$. Show that f^δ is l.s.c. and f_δ is u.s.c. Let $f^0 = \lim_{\delta \downarrow 0} f^\delta$, $f_0 = \lim_{\delta \downarrow 0} f_\delta$, and conclude that the set of points at which f is discontinuous $= \{f^0 \neq f_0\}$ is measurable.

(2.2) THEOREM   If $X : (\Omega, \mathcal{F}) \to (S, \mathcal{S})$ and $f : (S, \mathcal{S}) \to (T, \mathcal{T})$ are measurable maps, then $f(X)$ is a measurable map.

Proof   Let $B \in \mathcal{T}$. $\{\omega : f(X(\omega)) \in B\} = \{\omega : X(\omega) \in f^{-1}(B)\} \in \mathcal{F}$, since by assumption $f^{-1}(B) \in \mathcal{S}$. $\qquad\qquad\square$

The next result shows why we wanted to prove the last result for measurable maps.

(2.3) THEOREM   If $X_1, \ldots, X_n$ are random variables and $f : \mathbb{R}^n \to \mathbb{R}$ is measurable, then $f(X_1, \ldots, X_n)$ is a random variable.

Proof   In view of (2.2), it suffices to show that $(X_1, \ldots, X_n)$ is a random vector. To do this we observe that if $A_1, \ldots, A_n$ are Borel sets, then

$$\{(X_1, \ldots, X_n) \in A_1 \times \cdots \times A_n\} = \bigcap_i \{X_i \in A_i\} \in \mathcal{F}.$$

Since sets of the form $A_1 \times \cdots \times A_n$ generate $\mathcal{R}^n$, (2.3) follows from (2.1). $\qquad\square$

(2.4) COROLLARY   If $X_1, \ldots, X_n$ are random variables, then $X_1 + \cdots + X_n$ is a random variable.

Proof   Let $f(x_1, \ldots, x_n) = x_1 + \cdots + x_n$ and use (2.3). $\qquad\square$

To get a feeling for the bare-hands approach to proving measurability, try:

Exercise 2.5   Prove (2.4) in the case $n = 2$ by checking that $\{X_1 + X_2 < x\} \in \mathcal{F}$.

Hint   If $X_1 + X_2 < x$, there are rational numbers r_i with $r_1 + r_2 < x$ and $X_i < r_i$.

(2.5) THEOREM   If $X_1, X_2, \ldots$ are random variables, then so the

$$\inf_n X_n, \quad \sup_n X_n, \quad \liminf_{n\to\infty} X_n, \quad \text{and} \quad \limsup_{n\to\infty} X_n.$$

Proof   Since the infimum of a sequence is $< a$ if and only if some term is $< a$ (if all terms are $\geq a$, then the infimum is), we have

$$\left\{\inf_n X_n < a\right\} = \bigcup_n \left\{X_n < a\right\}.$$

A similar argument shows

$$\left\{\sup_n X_n > a\right\} = \bigcup_n \left\{X_n > a\right\}.$$

For the last two we observe

$$\liminf_{n\to\infty} X_n = \sup_n \left(\inf_{m\geq n} X_m\right),$$

$$\limsup_{n\to\infty} X_n = \inf_n \left(\sup_{m\geq n} X_m\right),$$

and then apply the first two results. $\qquad\square$

Exercise 2.6   A function φ is said to be *simple* if

$$\varphi(x) = \sum_{m=1}^{n} c_m 1_{A_m} \quad \text{where } A_m \in \mathcal{F}.$$

Show that the class of $\mathcal{F}$-measurable functions is the smallest class containing the simple functions and closed under pointwise limits.

Exercise 2.7 The σ-field generated by a random variable was defined in Exercise 2.1. Use the last exercise to conclude that Y is measurable with respect to $\sigma(X)$ if and only if $Y = f(X)$ where $f: \mathbb{R} \to \mathbb{R}$ is measurable.

From (2.5) we see that

$$\Omega_0 \equiv \left\{\omega: \lim_{n\to\infty} X_n \text{ exists}\right\} = \left\{\omega: \limsup_{n\to\infty} X_n - \liminf_{n\to\infty} X_n = 0\right\}$$

is a measurable set. (Here $\equiv$ indicates that the first equality is a definition.) If $P(\Omega_0) = 1$, we say that X_n converges *almost surely* (a.s. for short. This type of convergence is called *almost everywhere* in measure theory). To have a limit defined on the whole space it is convenient to let

$$X_\infty = \limsup_{n\to\infty} X_n,$$

but this random variable may take the value $+\infty$. To accommodate this and future headaches, we will generalize the definition of random variable.

A function whose domain is a set $D \in \mathcal{F}$ and whose range is $\mathbb{R}^* = [-\infty, \infty]$ is said to be a *random variable* if for all $B \in \mathscr{R}^*$ we have $X^{-1}(B) = \{\omega: X(\omega) \in B\} \in \mathcal{F}$. Here $\mathscr{R}^* = $ the Borel subsets of $\mathbb{R}^*$ with $\mathbb{R}^*$ given the usual topology—that is, the one generated by intervals of the form $[-\infty, a)$, (a, b), and $(b, \infty]$ where $a, b \in \mathbb{R}$. The reader should note that the *extended real line* $(\mathbb{R}^*, \mathscr{R}^*)$ is a measurable space, so all the results above generalize immediately.

Exercise 2.8 *Egorov's theorem.* Suppose $X_n \to X$ a.s. If $\delta > 0$ there is a set A with $P(A) < \delta$ so that $X_n \to X$ uniformly on A^c; that is, for any $\varepsilon > 0$ there is an N so that if $n \geq N$, then $|X_n(\omega) - X(\omega)| < \varepsilon$ for $\omega \notin A$.

3 Expected Value

If X is a random variable on $(\Omega, \mathcal{F}, P)$ with $\int |X| \, dP < \infty$, then we define its *expected value* to be $EX = \int X \, dP$. (The integral is defined in Section 4 of the Appendix.) If we only integrate over $A \subset \Omega$, we write

$$E(X; A) = \int_A X \, dP.$$

EX is often called the *mean* of X and denoted by μ. EX is defined by integrating X, so it has all the properties that integrals do.

$$E(X + Y) = EX + EY. \tag{3.1a}$$

$$E(cX) = cE(X) \text{ for any real number } c. \tag{3.1b}$$

If $X \geq Y$, then $EX \geq EY$. $\qquad\qquad$ (3.1c)

$E|X| \geq |EX|$. $\qquad\qquad$ (3.1d)

$E(X^2) \geq (EX)^2$. $\qquad\qquad$ (3.1e)

(3.1a)–(3.1d) are proved in Section 4 of the Appendix. To prove (3.1e), let $\mu = EX$ and observe

$$0 \leq E(X - \mu)^2 = E(X^2) - 2\mu EX + \mu^2 = E(X^2) - (EX)^2.$$

The difference $E(X^2) - \mu^2 = E(X - \mu)^2$ is called the *variance* and will play an important role in what follows. It is often denoted by σ^2. We will use var(X) to avoid confusion with $\sigma(X)$, the σ-field generated by X. It follows easily from the definition that

$$\operatorname{var}(cX) = E(cY - E(cY))^2 = c^2\, E(Y - EY)^2 = c^2\, \operatorname{var}(X). \qquad (3.1f)$$

The next result generalizes and consolidates (3.1d) and (3.1e). The proofs of (3.2)–(3.6) can be found in Section 5 of the Appendix.

(3.2) Jensen's Inequality Suppose φ is convex; that is,

$$\lambda\,\varphi(x) + (1 - \lambda)\,\varphi(y) \geq \varphi(\lambda x + (1 - \lambda)y)$$

for all $\lambda \in (0, 1)$ and $x, y \in \mathbb{R}$. Then $\varphi(EX) \leq E(\varphi(X))$ provided both expectations exist; that is, $E|X|$ and $E|\varphi(X)| < \infty$.

Remark If φ is strictly convex—that is, $>$ holds for $\lambda \in (0, 1)$, then $\varphi(EX) = E\varphi(X)$ implies $X = EX$ a.s.

(3.3) Hölder's Inequality If $p, q \in [1, \infty]$ with $1/p + 1/q = 1$, then

$$E|XY| \leq \|X\|_p \|Y\|_q$$

where $\|X\|_r = (E|X|^r)^{1/r}$ for $r \in [1, \infty)$ and $\|X\|_\infty = \inf\{M: P(|X| > M) = 0\}$.

The special case $p = q = 2$ is called the *Cauchy–Schwarz inequality*.

(3.4) Fatou's Lemma If $X_n \geq 0$, then $\liminf_{n \to \infty} EX_n \geq E(\liminf_{n \to \infty} X_n)$.

(3.5) Monotone Convergence Theorem If $X_n \geq 0$ and $X_n \uparrow X$, then $EX_n \uparrow EX$.

(3.6) Dominated Convergence Theorem If $X_n \to X$ a.s., $|X_n| \leq Y$ for all n, and $EY < \infty$, then $EX_n \to EX$.

The special case of (3.6) in which Y is constant is called the *bounded convergence theorem*.

(3.7) Chebyshev's Inequality   Suppose $\varphi \geq 0$ and let $i_A = \inf\{\varphi(x): x \in A\}$.

$$i_A \, P(X \in A) \leq E(\varphi(X); X \in A) \leq E\varphi(X).$$

Proof   The hypotheses imply that

$$i_A \, 1_{(X \in A)} \leq \varphi(X) \, 1_{(X \in A)} \leq \varphi(X),$$

so taking expected values gives the desired result.   □

Remark   The name Chebyshev's inequality is often reserved for the special case

$$a^2 P(|X| \geq a) \leq EX^2. \tag{$*$}$$

Exercise 3.1   (i) Show that the inequality in ($*$) is sharp by giving an example for which $=$ holds. (ii) Show that the inequality in ($*$) is not sharp by showing that for fixed X,

$$\lim_{a \to \infty} a^2 P(|X| \geq a)/EX^2 = 0.$$

For practice applying (3.2)–(3.7) do:

Exercise 3.2   If $k \geq 1$ is an integer, EX^k is called the *kth moment of X*. Show that

$$E|X|^j \leq (E|X|^k)^{j/k} \quad \text{when } j < k,$$

so if $E|X|^k < \infty$, all moments of lower order are finite.

Exercise 3.3   Apply Jensen's inequality with $\varphi(x) = e^x$ and $P(X = \log y_m) = p(m)$ to conclude that if $\Sigma \, p(m) = 1$ and $p(m)$, $y_m > 0$, then

$$\sum_{m=1}^{n} p(m) y_m \geq \prod_{m=1}^{n} y_m^{P(m)}.$$

When $p(m) = 1/n$, this says the arithmetic mean exceeds the geometric mean.

Exercise 3.4   (i) Let $a > 0$, and suppose $EY = 0$ and $\text{var}(Y) = \sigma^2$. Use (3.7) with $\varphi(x) = (x + \sigma^2/a)^2$ to get $P(Y \geq a) \leq \sigma^2/(\sigma^2 + a^2)$. (ii) Show that this "one-sided" equality is sharp by giving an example for which equality holds.

Exercise 3.5   To get a lower bound complementing Chebyshev's inequality show that if $EX \geq 0$ and $0 \leq \lambda < 1$, then $P(X > \lambda EX) \geq (1 - \lambda)^2 (EX)^2/E(X^2)$. Consequently if $E|Y| = 1$, $P(|Y| > \lambda) \geq (1 - \lambda)^2/E(Y^2)$.

Hint   Observe $(1 - \lambda)EX \leq E(X1_{(X > \lambda EX)})$ and use Cauchy–Schwarz.

Exercise 3.6   If $E|X_1| < \infty$ and $X_n \uparrow X$, then $EX_n \uparrow EX$.

Exercise 3.7   Let $\varphi(\theta) = E \exp(\theta X)$ be the *moment generating function of X*. $\{\theta: \varphi(\theta) < \infty\}$ is an interval and $\varphi(\theta)$ is continuous on the interior of the interval.

Exercise 3.8   Let $X \geq 0$. Show:

$$\lim_{y \to \infty} yE(1/X; X > y) = 0,$$

$$\lim_{y \to 0} yE(1/X; X > y) = 0.$$

Exercise 3.9   If $X_n \geq 0$, then $E(\sum_{n=0}^{\infty} X_n) = \sum_{n=0}^{\infty} EX_n$.

Exercise 3.10   If X is integrable and A_n are disjoint sets with union A, then

$$\sum_{n=0}^{\infty} E(X; A_n) = E(X; A).$$

In the developments below we will need another result on integration to the limit. Perhaps the most important special case of this result is $\varphi(x) = x^2$.

(3.8) THEOREM   Suppose $X_n \to X$ a.s. and there is a continuous $\varphi \geq 0$ with $\varphi(x)/|x| \to \infty$ as $|x| \to \infty$ so that $E \varphi(X_n) \leq K < \infty$ for all n. Then $EX_n \to EX$.

Proof   Let $\psi(x) = x$ if $|x| \leq M$, $= 0$ if $|x| \geq M + 1$, and is linear on $[-M-1, -M]$ and $[M, M + 1]$. Since ψ is continuous, $\psi(X_n) \to \psi(X)$ a.s. Since $|\psi(X_n)| \leq M$, it follows from the bounded convergence theorem that

$$E \psi(X_n) \to E \psi(X). \tag{a}$$

To get from this to the desired result we observe

$$|E \psi(X_n) - EX_n| \leq E|\psi(X_n) - X_n| \leq E(|X_n|; |X_n| \geq M)$$

$$\leq \varepsilon_M E(\varphi(X_n); |X_n| \geq M),$$

where $\varepsilon_M = \sup\{|x|/\varphi(x) : |x| \geq M\}$. So

$$|E \psi(X_n) - EX_n| \leq K\varepsilon_M, \tag{b}$$

where K is the constant in the theorem. To estimate $|E \psi(X) - EX|$, we observe that Fatou's lemma implies

$$E \varphi(X) \leq \liminf_{n \to \infty} E \varphi(X_n) \leq K,$$

so erasing the subscript n in the proof of (b) gives

$$|E \, \psi(X) - EX| \le K\varepsilon_M. \tag{c}$$

Using the triangle inequality with (a)–(c) gives

$$\limsup_{n \to \infty} |EX_n - EX| \le 2K\varepsilon_M,$$

which proves the desired result, since $K < \infty$ and $\varepsilon_M \to 0$ as $M \to \infty$. $\square$

Exercise 3.11   Let $\Omega = (0, 1)$ equipped with the Borel sets and Lebesgue measure. If $\alpha \in (1, 2)$, then $X_n = n^\alpha \, 1_{(1/(n+1), 1/n)} \to 0$ a.s. and has $E(X_n) \to 0$ but the X_n are not dominated by an integrable function. Show that (3.7) can be applied with $\varphi(x) = x^{2/\alpha}$.

The next result is useful in computing expected values.

(3.9) *Change of Variables Formula*   Let X be a random element of $(S, \mathscr{S})$ with distribution μ; that is, $\mu(A) = P(X \in A)$. If f is a measurable function from $(S, \mathscr{S})$ to $(\mathbb{R}, \mathscr{R})$ that is ≥ 0 or has $E|f(X)| < \infty$, then

$$Ef(X) = \int_S f(y) \, \mu(dy). \tag{$*$}$$

Remark   To explain the name, write h for X and $P \circ h^{-1}$ for μ to get

$$\int_\Omega f(h(\omega)) \, dP = \int_S f(y) \, d(P \circ h^{-1}).$$

Proof   We will prove this result by verifying it in four increasingly general special cases. The reader should note the method employed, since it will be used several times below.

   Case 1: If $B \in \mathscr{S}$ and $f = 1_B$, then both sides of $(*)$ are $\mu(B)$.

   Case 2: Since each integral is linear in f, it follows that $(*)$ holds for simple functions—that is, f of the form

$$f(x) = \sum_{m=1}^n c_m 1_{A_m} \qquad c_m \in \mathbb{R}, \quad A_m \in \mathscr{S}.$$

   Case 3: Now if $f \ge 0$ and we let $f_n(x) = ([2^n f(x)]/2^n) \wedge n$, where $[x] =$ the largest integer $\le x$ and $a \wedge b = \min\{a, b\}$, then the f_n are simple and $f_n \uparrow f$, so it follows from the monotone convergence theorem that $(*)$ holds for all $f \ge 0$.

   Case 4: The general case now follows by writing $f(x) = f(x)^+ - f(x)^-$ where $y^+ = \max\{y, 0\}$ and $y^- = \max\{-y, 0\}$. The condition $E|f(X)| < \infty$ guarantees that $Ef(X)^+ - Ef(X)^-$ makes sense—that is, is not $\infty - \infty$. $\square$

A consequence of (3.9) is that we can compute expected values of functions of random variables by performing integrals on the real line. We record now some particular instances that will appear below, leaving the calculus in the first two examples to the reader. (Integrate by parts.)

Example 3.1   If X has an *exponential distribution*, then

$$EX^k = \int_0^\infty x^k e^{-x}\, dx = k!.$$

So the mean and variance of X are 1. If we let $Y = X/\lambda$, then by Exercise 1.14 Y has density $\lambda e^{-\lambda y}$, $y \geq 0$, the exponential density with parameter λ. From (3.1b) and (3.1f) it follows that Y has mean $1/\lambda$ and variance $1/\lambda^2$.

Example 3.2   If X has a *normal distribution*,

$$EX = \int x(2\pi)^{-1/2} \exp(-x^2/2)\, dx = 0 \text{ (by symmetry)},$$

$$EX^2 = \int x^2\, (2\pi)^{-1/2} \exp(-x^2/2)\, dx = 1.$$

If we let $Y = a^{1/2} X + \mu$, then $EY = \mu$ and $\mathrm{var}(Y) = a$. By Exercise 1.14, Y has density

$$(2\pi a)^{-1/2} \exp(-(y-\mu)^2/2a).$$

For obvious reasons this is called the normal distribution with mean μ and variance a, or normal (μ, a) for short.

Example 3.3   We say that X has a *Poisson distribution with parameter λ* if

$$P(X = k) = e^{-\lambda}\, \lambda^k/k! \quad \text{for } k = 0, 1, 2, \ldots.$$

To compute the moments of X we introduce the *generating function* defined for $s \in [0, 1]$ by

$$\varphi(s) = Es^X = \sum_{n=0}^\infty e^{-\lambda} \frac{\lambda^n}{n!} s^n = \exp(-\lambda(1-s)).$$

Since $\varphi(s) = \sum_{n \geq 0} s^n P(X = n)$, it should be easy to believe that

$$\varphi'(s) = \sum_{n \geq 0} n s^{n-1} P(X = n) \uparrow \sum_{n \geq 0} n P(X = n) = EX \quad \text{as } s \uparrow 1.$$

The convergence for $s \uparrow 1$ follows from the monotone convergence theorem. To justify interchanging the differentiation and summation we use the following result.

(3.10) LEMMA   Suppose that for $x \in (y - \delta, y + \delta)$ we have $f(x) = \sum_{n \geq 0} f_n(x)$ with
(i) $\sum_{n \geq 0} |f_n(x)| < \infty$ for each x, (ii) $f_n'(x)$ continuous, and (iii) if $\varepsilon > 0$ and $n \geq N(\varepsilon)$,
$\sum_{n \geq N(\varepsilon)} |f_n'(x)| \leq \varepsilon$ for $x \in (y - \delta, y + \delta)$, then $f'(y) = \sum_{n \geq 0} f_n'(y)$.

Proof   Suppose $h > 0$; then

$$\frac{1}{h}(f(y + h) - f(y)) = \frac{1}{h} \sum_{n \geq 0} (f_n(y + h) - f_n(y)) = \frac{1}{h} \sum_{n \geq 0} \int_y^{y+h} f_n'(x) \, dx.$$

Since f_n' is continuous,

$$\frac{1}{h} \sum_{n=0}^{N-1} \int_y^{y+h} f_n'(x) \, dx \to \sum_{n=0}^{N-1} f_n'(y)$$

for each fixed N. On the other hand, if $h < \delta$, $\varepsilon > 0$, and N is large,

$$\frac{1}{h} \sum_{n=N}^{\infty} \int_y^{y+h} |f_n'(x)| \, dx \leq \varepsilon.$$

Combining the last three results proves (3.10). $\square$

Since $P(X = n) \leq 1$, the last result is more than enough to show that:

(3.11) THEOREM   The kth derivative of the generating function

$$\varphi^{(k)}(s) = \sum_{n \geq k} n(n - 1) \cdots (n - k + 1) s^{n-k} P(X = n) \quad \text{for } s < 1$$

and

$$\varphi^{(k)}(s) \uparrow E(X(X - 1) \cdots (X - k + 1)) \quad \text{as } s \uparrow 1.$$

Exercise 3.12   Use the last exercise to show that the Poisson distribution has
$EX = \lambda$, $E(X(X - 1) \cdots (X - k + 1)) = \lambda^k$, and var $(X) = \lambda$.

4   Independence

I have heard it said that "probability theory is just measure theory plus the notion
of independence." Although I think this statement is about as accurate as saying
that "complex analysis is just real analysis plus $i = \sqrt{-1}$," there is no doubt that
independence is one of the most important concepts in probability, and, as in our
analysis analogy, the game changes considerably when the new concept is introduced.

We begin with what is hopefully a familiar definition and then work our way
up to a definition that is appropriate for our current setting. Two events A and B
are *independent* if $P(A \cap B) = P(A) P(B)$. Two random variables X and Y are

independent if for all Borel sets C and D, $P(X \in C, Y \in D) = P(X \in C) P(Y \in D)$; that is, the events $A = \{X \in C\}$ and $B = \{Y \in D\}$ are independent. Two σ-fields $\mathcal{F}$ and $\mathcal{G}$ are *independent* if for all $A \in \mathcal{F}$ and $B \in \mathcal{G}$, the events A and B are independent.

As the next exercise shows, the third definition is a generalization of the second.

Exercise 4.1   (i) Use the result in Exercise 2.1 to show that if X and Y are independent, then $\sigma(X)$ and $\sigma(Y)$ are. (ii) Conversely if $\mathcal{F}$ and $\mathcal{G}$ are independent, $X \in \mathcal{F}$, and $Y \in \mathcal{G}$, then X and Y are independent.

Exercise 4.2   (i) Use the result of Exercise 4.1 to show that if X and Y are independent and f and g are Borel functions, then $f(X)$ and $g(Y)$ are independent. (ii) Prove (i) directly from the definition.

The second definition above is, in turn, a generalization of the first. Let $X = 1_A$, let $Y = 1_B$, and observe that if A and B are independent, then so are A^c and B, A^c and B^c, A and Ω, A and $\emptyset$, and so on. In view of this, when we say what it means for several things to be independent, we take things in the opposite order. We begin by reducing to the case of finitely many objects. An infinite collection of objects (σ-fields, random variables, or sets) is said to be independent if every finite subcollection is. σ-fields $\mathcal{F}_1, \mathcal{F}_2, \ldots, \mathcal{F}_n$ are *independent* if whenever $A_i \in \mathcal{F}_i$ for $i = 1, \ldots, n$, we have

$$P\left(\bigcap_{i=1}^{n} A_i \right) = \prod_{i=1}^{n} P(A_i).$$

Random variables $X_1, \ldots, X_n$ are *independent* if whenever $B_i \in \mathcal{R}$ for $i = 1, \ldots, n$, we have

$$P\left(\bigcap_{i=1}^{n} \{X_i \in B_i\} \right) = \prod_{i=1}^{n} P(X_i \in B_i).$$

Sets $A_1, \ldots, A_n$ are *independent* if whenever $I \subset \{1, \ldots, n\}$, we have

$$P\left(\bigcap_{i \in I} A_i \right) = \prod_{i \in I} P(A_i).$$

At first glance it might seem that the last definition does not match the other two. But if you think about it for a minute, you will see that it is what we get when the second is specialized to the case $X_i = 1_{A_i}$ and we use $B_i = \{1\}$ for $i \in I$ and $B_i = \{0, 1\}$ for $i \in I^c$.

Exercise 4.3   *Rademacher functions.* Let Ω be the unit interval $(0, 1)$ equipped with the Borel sets $\mathcal{F}$ and Lebesgue measure P. Let $Y_n(\omega) = 1$ if $[2^n \omega]$ is odd and -1 if $[2^n \omega]$ is even. Show that $Y_1, Y_2, \ldots$ are independent.

One of the first things to understand about the definition of independent events is that it is not enough to assume $P(A_i \cap A_j) = P(A_i) P(A_j)$ for all $i \neq j$. A sequence of events $A_1, \ldots, A_n$ with the last property is called *pairwise independent*. It is clear that independent events are pairwise independent. The next example shows that the converse is not true.

Example 4.1   Let X_1, X_2, X_3 be independent random variables with

$$P(X_i = 0) = P(X_i = 1) = 1/2.$$

Let $A_1 = \{X_2 = X_3\}$, $A_2 = \{X_3 = X_1\}$, and $A_3 = \{X_1 = X_2\}$. These events are pairwise independent since if $i \neq j$, then

$$P(A_i \cap A_j) = P(X_1 = X_2 = X_3) = 1/4 = P(A_i) P(A_j),$$

but they are not independent since

$$P(A_1 \cap A_2 \cap A_3) = 1/4 \neq 1/8 = P(A_1) P(A_2) P(A_3).$$

Exercise 4.4   Let $K \geq 3$ be a prime and let X and Y be independent random variables that are uniformly distributed on $\{0, 1, \ldots, K-1\}$. For $0 \leq n < K$ let $Z_n = X + nY$. Show that $Z_0, Z_1, \ldots, Z_{K-1}$ are *pairwise independent*; that is, each pair is independent, but if we know the values of two of the variables, then we know the values of all the variables.

Exercise 4.5   Find four random variables taking values in $\{-1, 1\}$ so that any three are independent but all four are not.

Hint   Consider products of independent random variables.

In order to show that random variables X and Y are independent, we have to check that $P(X \in A, Y \in B) = P(X \in A) P(Y \in B)$ for all Borel sets A and B. Since there are a lot of Borel sets, it is useful to know that we can check the definition for smaller collections of sets. The next result will help us find sufficient conditions for independence. To state it we need a definition that generalizes the notion of independence for σ-fields. $\mathscr{A}_1, \mathscr{A}_2, \ldots, \mathscr{A}_n \subset \mathscr{F}$ are said to be *independent* if whenever $A_i \in \mathscr{A}_i$,

$$P\left(\bigcap_{i=1}^{n} A_i\right) = \prod_{i=1}^{n} P(A_i).$$

(4.1) THEOREM   If $\mathscr{A}_1, \mathscr{A}_2, \ldots, \mathscr{A}_n$ are independent and each $\mathscr{A}_1$ is a π-system (i.e., $\Omega \in \mathscr{A}_i$ and $\mathscr{A}_i$ is closed under intersections), then $\sigma(\mathscr{A}_1), \sigma(\mathscr{A}_2), \ldots, \sigma(\mathscr{A}_n)$ are independent.

The proof of (4.1) is based on Dynkin's π-λ theorem ((2.1) in the Appendix). To state this result we need a definition. $\mathscr{L}$ is a λ-*system* if it satisfies: (i) $\Omega \in \mathscr{L}$. (ii) If $A, B \in \mathscr{L}$ and $A \subset B$, then $B - A \in \mathscr{L}$. (iii) If $A_n \in \mathscr{L}$ and $A_n \uparrow A$, then $A \in \mathscr{L}$.   □

(4.2) π-λ THEOREM   If $\mathscr{P}$ is a π-system and $\mathscr{L}$ is a λ-system that contains $\mathscr{P}$, then $\sigma(\mathscr{P}) \subset \mathscr{L}$.

Proof of (4.1)   Let $A_2, \dots, A_n$ be sets with $A_i \in \mathscr{A}_i$ and let

$$\mathscr{L} = \left\{ A: P(A \cap A_2 \cap \cdots \cap A_n) = P(A) \prod_{i=2}^{n} P(A_i) \right\}.$$

Then (i) $\Omega \in \mathscr{L}$. (ii) If $A, B \in \mathscr{L}$ with $A \subset B$, then $B - A \in \mathscr{L}$. To see this write

$$P(B \cap A_2, \cap \cdots \cap A_n) = P(B) \prod_{i=2}^{n} P(A_i),$$

$$P(A \cap A_2, \cap \cdots \cap A_n) = P(A) \prod_{i=2}^{n} P(A_i),$$

and subtract. (iii) If $B_k \in \mathscr{L}$ and $B_k \uparrow B$, then $B \in \mathscr{L}$. To see this write

$$P(B_k \cap A_2 \cap \cdots \cap A_n) = P(B_k) \prod_{i=2}^{n} P(A_i),$$

let $k \uparrow \infty$, and notice $B_k \cap A_2 \cap \cdots \cap A_n \uparrow B \cap A_2 \cap \cdots \cap A_n$ so

$$P(B \cap A_2 \cap \cdots \cap A_n) = P(B) \prod_{i=2}^{n} P(A_i).$$

Applying the π-λ theorem now gives $\mathscr{L} \supset \sigma(\mathscr{A}_1)$ and since $A_2, \dots, A_n$ are arbitrary members of $\mathscr{A}_2, \dots, \mathscr{A}_n$, it follows that $\sigma(\mathscr{A}_1), \mathscr{A}_2, \dots, \mathscr{A}_n$ are independent.   □

At this point we have shown

(4.1')   If $\mathscr{A}_1, \mathscr{A}_2, \dots, \mathscr{A}_n$ are independent, then $\sigma(\mathscr{A}_1), \mathscr{A}_2, \dots, \mathscr{A}_n$ are independent.   □

Applying (4.1') to $\mathscr{A}_2, \dots, \mathscr{A}_n, \sigma(\mathscr{A}_1)$ shows that $\sigma(\mathscr{A}_2), \mathscr{A}_3, \dots, \mathscr{A}_n, \sigma(\mathscr{A}_1)$ are independent, and after n iterations we have the desired result.

Remark   The reader should note that it is not easy to show that if $A, B \in \mathscr{L}$, then $A \cap B \in \mathscr{L}$, or $A \cup B \in \mathscr{L}$; that is, it is hard to show that $\mathscr{L}$ is a σ-field but it is easy to check that $\mathscr{L}$ is a λ-system.

Exercise 4.6   Give an example of two collections of sets $\mathscr{A}_1$ and $\mathscr{A}_2$ that are independent but the generated σ-fields are not.

Hint   Take $\Omega = \{1, 2, 3, 4\}$ with all points equally likely. Let $\mathscr{A}_1$ have one set and $\mathscr{A}_2$ have two sets.

Having worked to establish (4.1), we get several corollaries.

(4.3) COROLLARY   In order for $X_1, \ldots, X_n$ to be independent, it is sufficient that

$$P(X_1 \le x_1, \ldots, X_n \le x_n) = \prod_{i=1}^{n} P(X_i \le x_i)$$

for all $x_1, \ldots, x_n \in (-\infty, \infty]$.

Proof   Let $\mathscr{A}_i =$ the sets of the form $\{X_i \le x_i\}$. $\mathscr{A}_i$ is a π-system (we need to allow $x_i = \infty$, so $\Omega \in \mathscr{A}_i$) and $\sigma(\mathscr{A}_i) = \sigma(X_i)$, so the result follows from (4.1). $\square$

Exercise 4.7   Suppose $(X_1, X_2, \ldots, X_n)$ has density $f(x_1, x_2, \ldots, x_n)$; that is,

$$P((X_1, X_2, \ldots, X_n) \in A) = \int_A f(x)\, dx \quad \text{for } A \in \mathscr{R}^n.$$

If $f(x)$ can be written as $g_1(x_1) \cdots g_n(x_n)$, then $X_1, X_2, \ldots, X_n$ are independent. Note that the g_m are not assumed to be probability densities.

Exercise 4.8   Let U and V be independent and have uniform distributions on $(0, 1)$. Let $\theta = 2\pi U$, $Z = -\log V$, $R = \sqrt{2Z}$, $X_1 = R \cos \theta$, and $X_2 = R \sin \theta$. Then X_1 and X_2 are independent and have standard normal distributions.

(4.4) COROLLARY   Suppose $\mathscr{F}_{i,j}$, $1 \le i \le n$, $1 \le j \le m(i)$, are independent and let $\mathscr{G}_i = \sigma(\bigcup_j \mathscr{F}_{i,j})$. Then $\mathscr{G}_1, \ldots, \mathscr{G}_n$ are independent.

Proof   Let $\mathscr{A}_i$ be the collection of sets of the form $\bigcap_j A_{i,j}$ where $A_{i,j} \in \mathscr{F}_{i,j}$. $\mathscr{A}_i$ is a π-system and $\sigma(\mathscr{A}_i) = \mathscr{G}_i$, so the result follows from (4.1). $\square$

An immediate consequence of the last result is:

(4.5) COROLLARY   If $X_{i,j}$, $1 \le i \le n$, $1 \le j \le m(i)$, are independent and $f_i: \mathbb{R}^{m(i)} \to \mathbb{R}$ are measurable, then $f_i(X_{i,1}, \ldots, X_{i,m(i)})$ are independent.

The results above were concerned with showing that random variables are independent. The next result, which is an immediate consequence of Fubini's theorem ((6.2) in the Appendix), begins our study of the properties of independent random variables.

(4.6) THEOREM   Suppose X and Y are independent and have distributions μ and v. If $h \ge 0$ or $E|h(X, Y)| < \infty$, then

$$Eh(X, Y) = \iint h(x, y)\, v(dy)\, \mu(dx).$$

(4.7) COROLLARY   If $X_1, \ldots, X_n$ are independent and have $E|X_i| < \infty$, then

$$E\left(\prod_{i=1}^{n} X_i\right) = \prod_{i=1}^{n} E(X_i);$$

that is, the expectation on the left-hand side exists and has the value given on the right.

Proof of (4.7) $X = X_1$ and $Y = X_2 \cdots X_n$ are independent by (4.5). If we suppose that the $X_i \geq 0$, then it follows from (4.6) and induction that the result is true. To get the general case, apply the result for nonnegative r.v.'s to $|X_m|$ to conclude

$$E|X_m \cdots X_n| = E|X_m| \cdots E|X_n| < \infty$$

and then use (4.6) and induction again.   □

As the reader can probably guess, it can happen that $E(XY) = EX\, EY$ without the variables being independent. Suppose the joint distribution of X and Y is given by the following table

$X \backslash Y$	1	0	-1
1	0	a	0
0	b	c	b
-1	0	a	0

where $a, b, c, > 0$ and $2a + 2b + c = 1$. Things are arranged so that $XY \equiv 0$. Symmetry implies $EX = 0, EY = 0$, so $E(XY) = 0 = EX\, EY$. The random variables are not independent since

$$P(X = 1, Y = 1) = 0 \neq ab = P(X = 1)\,P(Y = 1).$$

Two random variables that have $EXY = EX\, EY$ are said to be *uncorrelated*.
Applying (4.6) to $h(x, y) = 1_{(x+y \leq z)}$ gives:

(4.8) COROLLARY   If X and Y are independent, $F(x) = P(X \leq x)$, and $G(y) = P(Y \leq y)$, then

$$P(X + Y \leq z) = \int F(z - y)\, dG(y).$$

The integral on the right-hand side is called the *convolution* of F and G and is denoted $F * G(z)$.

Example 4.2   It should not be surprising that the distribution of $X + Y$ can be $F * G$ without the random variables being independent. A simple example of this is provided by the (X, Y) with a joint density that is 2 on the triangle $(1/2, 0), (1, 0), (1, 1/2)$ and also 2 on the quadrilateral $(0, 0), (0, 1/2), (1/2, 1), (1, 1)$. Figure 1.1 should

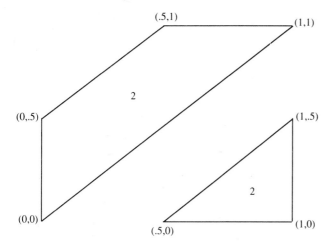

Figure 1.1

make it clear that X and Y have distributions $F = G =$ uniform on $(0, 1)$ and $X + Y$ has distribution $F * G$.

Exercise 4.9 Suppose that X has density f and Y has distribution G. Show that if X and Y are independent, $X + Y$ has density

$$h(z) = \int f(z - y) \, dG(y).$$

When Y has density g, the last formula can be written as

$$h(z) = \int f(z - y) \, g(y) \, dy.$$

Exercise 4.10 The *gamma density* with parameters α and λ is given by

$$f(x) = \lambda^\alpha x^{\alpha-1} e^{-\lambda x}/\Gamma(\alpha) \quad where \ \Gamma(\alpha) = \int_0^\infty x^{\alpha-1} e^{-x} \, dx.$$

Show that if $X =$ gamma(α, λ) and $Y =$ gamma(β, λ) are independent, then $X + Y$ is gamma($\alpha + \beta$, λ). Since gamma(1, λ) is an exponential with parameter λ, it follows from induction that the sum of n independent exponential(λ) r.v.'s has a gamma(n, λ) distribution. As the last observation should suggest, a gamma(α, λ) r.v. has mean α/λ.

Exercise 4.11 The *normal density* with mean μ and variance a is given by

$$(2\pi a)^{-1/2} \exp(-(x - \mu)^2/2a).$$

Show that if $X =$ normal(μ, a) and $Y =$ normal(v, b) are independent, then $X + Y =$ normal($\mu + v$, $a + b$). To simplify this tedious calculation notice that it is enough to prove the result for $\mu = \nu = 0$. In Exercise 3.4 of Chapter 2 we will give a simple proof of this result.

Exercise 4.12   The *Poisson distribution* with parameter λ is given by

$$P(Z = k) = e^{-\lambda} \lambda^k / k! \quad \text{for } k = 0, 1, 2, \dots .$$

Show that if $X = \text{Poisson}(\lambda)$ and $Y = \text{Poisson}(\mu)$ are independent, then $X + Y = \text{Poisson}(\lambda + \mu)$.

Exercise 4.13   Let $X, Y \geq 0$ be independent with distributions F and G. Find the distribution of XY.

The last question that we have to address before we can study independent random variables is: Do they exist? (If they don't exist, then there is no point in studying them!) If we are given a finite number of distribution functions $F_i, 1 \leq i \leq n$, it is easy to construct independent random variables $X_1, \dots, X_n$ with $P(X_i \leq x) = F_i(x)$. Let $\Omega = \mathbb{R}^n$, $\mathscr{F} = \mathscr{R}^n$, $X_i(\omega_1, \dots, \omega_n) = \omega_i$ (the ith coordinate of $\omega \in \mathbb{R}^n$), and let P be the measure on $\mathscr{R}^n$ that has

$$P((a_1, b_1] \times \cdots \times (a_n, b_n]) = (F_1(b_1) - F_1(a_1)) \cdots (F_n(b_n) - F_n(a_n)).$$

To construct an infinite sequence of independent random variables $X_1, X_2, \dots$ with given distribution functions we want to perform the last construction on the infinite product space $\mathbb{R}^{\mathbb{N}} = \{(\omega_1, \omega_2, \dots): \omega_i \in \mathbb{R}\} = \{\text{functions } \omega: \mathbb{N} \to \mathbb{R}\}$, where $\mathbb{N} = \{1, 2, \dots\}$ and $\mathbb{N}$ stands for natural numbers. We define $X_i(\omega) = \omega_i$ and we equip $\mathbb{R}^{\mathbb{N}}$ with the product σ-field $\mathscr{R}^{\mathbb{N}}$, which is generated by the *finite-dimensional sets* = sets of the form $\{\omega: \omega_i \in B_i \text{ for } i = 1, \dots, n\}$ where $B_i \in \mathscr{R}$. It is clear how we want to define P for finite-dimensional sets. To assert the existence of a unique extension to $\mathscr{R}^{\mathbb{N}}$ we use the following result:

(4.9) *Kolmogorov's Extension Theorem*   Suppose we are given probability measures μ_n on $(\mathbb{R}^n, \mathscr{R}^n)$ that are consistent; that is,

$$\mu_{n+1}((a_1, b_1] \times \cdots \times (a_n, b_n] \times \mathbb{R}) = \mu_n((a_1, b_1] \times \cdots \times (a_n, b_n]).$$

Then there is a unique probability measure P on $(\mathbb{R}^{\mathbb{N}}, \mathscr{R}^{\mathbb{N}})$ with

$$P(\omega: \omega_i \in (a_i, b_i], 1 \leq i \leq n) = \mu_n((a_1, b_1] \times \cdots \times (a_n, b_n]).$$

Remark   Consistency is clearly necessary since

$$P(\omega_1 \in (a_1, b_1], \dots, \omega_n \in (a_n, b_n], \omega_{n+1} \in \mathbb{R}) = P(\omega_1 \in (a_1, b_1], \dots, \omega_n \in (a_n, b_n]).$$

Proof   See (7.1) in the Appendix.   □

In what follows we will need to construct sequences of random variables that take values in other measurable spaces $(S, \mathscr{S})$. Unfortunately, Kolmogorov's theorem is not valid for arbitrary measurable spaces. The first example (on an infinite product of different spaces $\Omega_1 \times \Omega_2 \times \cdots$) was due to Andersen and Jessen (1948). (See Halmos 1950, p. 214, or Neveu 1965, p. 84.) For an example in which all the spaces Ω_i are the same, see Wegner (1973). Fortunately, there is a class of spaces that is adequate for all of our results and for which the generalization of Kolmogorov's theorem is trivial.

$(S, \mathscr{S})$ is said to be *nice* if there is a 1–1 map φ from S into $\mathbb{R}$ so that φ and φ^{-1} are both measurable. Such spaces are often called *standard Borel spaces*, but we already have too many things with similar names. The next result shows that most spaces arising in applications are nice.

(4.10) THEOREM   If S is a Borel subset of a complete separable metric space M, and $\mathscr{S}$ is the collection of Borel subsets of S, then $(S, \mathscr{S})$ is nice.

Proof   We begin with the special case $S = [0, 1)^N$ with metric

$$\rho(x, y) = \sum_{n=1}^{\infty} |x_n - y_n|/2^n$$

and $M = \{x \in \mathbb{R}^N : \rho(0, x) < \infty\}$. If $x = (x^1, x^2, x^3, \ldots)$, expand each component in binary

$$x^j = .x_1^j \, x_2^j \, x_3^j \cdots$$

(taking the expansion with an infinite number of 0's). Let

$$\varphi_0(x) = .x_1^1 \, x_2^1 \, x_1^2 \, x_3^1 \, x_2^2 \, x_1^3 \, x_4^1 \, x_3^2 \, x_2^3 \, x_1^4 \ldots .$$

To treat the general case we observe that by letting $d(x, y) = \rho(x, y)/(1 + \rho(x, y))$ we can suppose that the metric has $d(x, y) < 1$ for all x, y. Let $q_1, q_2, \ldots$ be a countable dense set in M. Let

$$\psi(x) = (d(x, q_1), d(x, q_2), \ldots).$$

$\psi : M \to [0, 1)^N$ is continuous and 1–1. $\varphi_0 \circ \psi$ gives the desired mapping. $\square$

5   Weak Laws of Large Numbers

In this section we will prove three versions of the weak law of large numbers. The first order of business is to define the mode of convergence that appears in the conclusion of the theorems and the hypothesis that appears in the first version. We say that Y_n converges to Y *in probability* if for all $\varepsilon > 0$, $P(|Y_n - Y| > \varepsilon) \to 0$ as $n \to \infty$. Given two random variables X and Y with $EX^2, EY^2 < \infty$, we define their *correlation* by

$$\mathrm{corr}(X, Y) = (EXY - EX\ EY)/\{\mathrm{sd}(X)\ \mathrm{sd}(Y)\}$$

where $\mathrm{sd}(X) = \sqrt{\mathrm{var}(X)}$ is the *standard deviation of* X. When $\mathrm{corr}(X, Y) = 0$ (i.e., $EXY - EX\ EY = 0$), we say that X and Y are *uncorrelated*. Random variables X_i, $i \in I$, with $EX_i^2 < \infty$ are said to be *uncorrelated* if each pair is; that is, if $i \neq j$, then $E(X_iX_j) = EX_iEX_j$.

Exercise 5.1   Show that

$$\mathrm{corr}(X, Y) = E\left(\frac{X - EX}{\mathrm{sd}(X)} \cdot \frac{Y - EY}{\mathrm{sd}(Y)}\right)$$

and use this to conclude that (i) $\mathrm{corr}(X, Y) \in [-1, 1]$ and (ii) if $X' = aX + b$ and $Y' = cY + d$ with $a, c > 0$, then $\mathrm{corr}(X, Y) = \mathrm{corr}(X', Y')$. In words, the correlation is independent of the units of measurement.

Example 5.1   Let $\Omega = [0, 1)$, $\mathscr{F} = $ Borel subsets, $P = $ Lebesgue measure. $X_n(\omega) = \sin(2\pi n\omega)$, $n = 1, 2, \ldots$, are uncorrelated but not independent.

Exercise 5.2   Suppose $X \overset{\mathrm{d}}{=} -X$ and $EX^4 < \infty$. Show that X and X^2 are uncorrelated.

The key to the weak law for uncorrelated random variables is:

(5.1) LEMMA   Let $X_1, \ldots, X_n$ be uncorrelated random variables with $E(X_i^2) < \infty$. Then

$$\mathrm{var}(X_1 + \cdots + X_n) = \mathrm{var}(X_1) + \cdots + \mathrm{var}(X_n)$$

where $\mathrm{var}(Y) = $ the variance of Y.

Proof   Let $\mu_i = EX_i$ and $S_n = X_1 + \cdots + X_n$.

$$\mathrm{var}(S_n) = E(S_n - ES_n)^2 = E\left(\sum_{i=1}^{n}(X_i - \mu_i)\right)^2$$

$$= E\left(\sum_{i=1}^{n}\sum_{j=1}^{n}(X_i - \mu_i)(X_j - \mu_j)\right)$$

$$= \sum_{i=1}^{n}E(X_i - \mu_i)^2 + 2\sum_{i=1}^{n}\sum_{j=1}^{i-1}E((X_i - \mu_i)(X_j - \mu_j)).$$

The first term is $\mathrm{var}(X_1) + \cdots + \mathrm{var}(X_n)$, so we want to show that the second is zero. To do this we observe

$$E((X_i - \mu_i)(X_j - \mu_j)) = EX_iX_j - \mu_iEX_j - \mu_jEX_i + \mu_i\mu_j = EX_iX_j - \mu_i\mu_j = 0$$

since X_i and X_j are uncorrelated.   □

In words, (5.1) says that for uncorrelated random variables the variance of the sum is the sum of the variances. A second important property is (3.1f):

$$\text{var}(cY) = c^2 \, \text{var}(Y). \tag{5.2}$$

Combining (5.1) and (5.2), we see that if $S_n = X_1 + \cdots + X_n$ and the X_i are uncorrelated, then

$$\text{var}(S_n/n) = (\text{var}(X_1) + \cdots + \text{var}(X_n))/n^2.$$

This observation leads to:

(5.3) L^2 Weak Law   Let $X_1, X_2, \ldots$ be uncorrelated random variables with $EX_i = \mu$ and $\text{var}(X_i) \le C < \infty$. If $S_n = X_1 + \cdots + X_n$, then as $n \to \infty$, $S_n/n \to \mu$ in L^2 and in probability.

Proof   To prove L^2 convergence observe that $E(S_n/n) = \mu$, so

$$E(S_n/n - \mu)^2 = \text{var}(S_n/n) \le (Cn)/n^2 = C/n \to 0.$$

To conclude there is also convergence in probability, we use Chebyshev's inequality, (3.7), with $\varphi(x) = x^2$ and $X = |S_n/n - \mu|$ to get

$$\varepsilon^2 P(|S_n/n - \mu| \ge \varepsilon) \le E(S_n/n - \mu)^2 \to 0. \qquad \square$$

Exercise 5.3   The L^2 weak law generalizes immediately to certain dependent sequences. Suppose $EX_n = 0$ and $EX_n X_m \le r(n - m)$ for $m \le n$ (no absolute value on the left-hand side!) with $r(k) \to 0$ as $k \to \infty$. Show that $(X_1 + \cdots + X_n)/n \to 0$ in probability.

Exercise 5.4   Show that $d(X, Y) = E(|X - Y|/(1 + |X - Y|))$ defines a metric on the set of random variables; that is, (i) $d(X, Y) = 0$ if and only if $X = Y$ a.s.; (ii) $d(X, Y) = d(Y, X)$; (iii) $d(X, Z) \le d(X, Y) + d(Y, Z)$ and that $d(X_n, X) \to 0$ as $n \to \infty$ if and only if $X_n \to X$ in probability.

Before generalizing the weak law we would like to give two applications to situations that on the surface have nothing to do with randomness. The first one is for comic relief.

Example 5.2   *A high-dimensional cube is almost the boundary of a ball.* Let $X_1, X_2, \ldots$ be independent and uniformly distributed on $(-1, 1)$. $EX_i^2 = 1/3$, so the weak law implies

$$(X_1^2 + \cdots + X_n^2)/n \to 1/3 \text{ in probability as } n \to \infty.$$

Let $A_{n,\varepsilon} = \{x \in \mathbb{R}^n : (1/3 - \varepsilon)\sqrt{n} < |x| < (1/3 + \varepsilon)\sqrt{n}\}$ where $|x| = (x_1^2 + \cdots + x_n^2)^{1/2}$. The last conclusion implies that for any $\varepsilon > 0$, $|A_{n,\varepsilon} \cap (-1, 1)^n|/2^n \to 1$ where

$|S|$ denotes the Lebesgue measure of S. In words, the cube $(-1, 1)^n$ is almost the boundary of the ball of radius $\sqrt{n/3}$.

Example 5.3   *Polynomial approximation.* Let f be a continuous function on $[0, 1]$, and let

$$f_n(x) = \sum_{m=0}^{n} \binom{n}{m} x^m (1 - x)^{n-m} f(m/n) \quad \text{where} \quad \binom{n}{m} = \frac{n!}{m!(n-m)!}$$

be the *Bernstein polynomial of degree n* associated with f. Then as $n \to \infty$,

$$\sup_{x \in [0, 1]} |f_n(x) - f(x)| \to 0.$$

Proof   First observe that if S_n is the sum of n independent random variables with $P(X_i = 1) = p$ and $P(X_i = 0) = 1 - p$, then $EX_i = p$, $\mathrm{var}(X_i) = p(1 - p)$, and

$$P(S_n = m) = \binom{n}{m} p^m (1 - p)^{n-m}, \tag{5.4}$$

so $Ef(S_n/n) = f_n(p)$. (5.3) tells us that as $n \to \infty$, $S_n/n \to p$ in probability. To get from this to the statement given above, let

$$M = \sup_{x \in [0, 1]} |f(x)|,$$

let $\varepsilon > 0$, and pick $\delta > 0$ so that if $|x - y| < \delta$, then $|f(x) - f(y)| < \varepsilon$. (This is possible since a continuous function is uniformly continuous on each bounded interval.) Now

$$|E f(S_n/n) - f(p)| \leq E|f(S_n/n) - f(p)| \leq \varepsilon + M \, P(|S_n/n - p| > \delta),$$

and by Chebyshev's inequality,

$$P(|S_n/n - p| > \delta) \leq \mathrm{var}(S_n/n)/\delta^2 = p(1 - p)/n\delta^2 \leq 1/4n\delta^2,$$

since $p(1 - p) \leq 1/4$ for $p \in [0, 1]$. Combining the last two computations proves the desired result.   □

Exercise 5.5   Generalize the last result to $[0, 1]^d$.

Example 5.4   *Widder's inversion formula for Laplace transform s.* Let μ be a probability measure on $[0, \infty)$ and define its *Laplace transform* by

$$\varphi(\theta) = \int_0^\infty e^{-\theta y} \mu(dy).$$

Using the weak law of large numbers we can recover μ from φ. For $\theta > 0$ we can differentiate the integral k times to get

$$\varphi^{(k)}(\theta) = \int_0^\infty (-y)^k e^{-\theta y} \mu(dy).$$

Setting $\theta = n$ and letting $x > 0$, we have

$$\sum_{k=0}^{[nx]} \frac{(-1)^k}{k!} n^k \varphi^{(k)}(n) = \int_0^\infty \sum_{k=0}^{[nx]} e^{-ny} \frac{(ny)^k}{k!} \mu(dy).$$

The last manipulation was motivated by the fact that $e^{-ny}(ny)^k/k!$ is the probability a Poisson r.v. with mean ny is $= k$, and Exercise 4.12 implies that the sum of n independent Poissons has a Poisson distribution. Using the weak law of large numbers now we get

$$\sum_{k=0}^{[nx]} e^{-ny} \frac{(ny)^k}{k!} \to \begin{cases} 1 & \text{if } x > y \\ 0 & \text{if } x < y \end{cases}$$

as $n \to \infty$. So if $\mu(\{x\}) = 0$, the bounded convergence theorem gives

$$\lim_{n \to \infty} \sum_{k=0}^{[nx]} \frac{(-1)^k}{k!} n^k \varphi^{(k)}(n) = \mu([0, x)).$$

By Exercise 1.10, this gives the distribution function of μ for all but a countable number of values, enough to identify μ.

Exercise 5.6   *Monte Carlo integration.* Let f be a measurable function on $[0, 1]$ with

$$\int_0^1 f^2 \, dx < \infty.$$

Let $U_1, U_2, \ldots$ be independent and uniformly distributed on $[0, 1]$ and let

$$I_n = n^{-1}(f(U_1) + \cdots + f(U_n)).$$

Show that $I_n \to I \equiv \int_0^1 f \, dx$ in probability and use Chebyshev's inequality to estimate

$$P(|I_n - I| > a/n^{1/2}).$$

Let $X_1, X_2, \ldots$ be independent random variables that all have the same distribution. In the jargon, they are *independent and identically distributed*, or *i.i.d.* for short. The L^2 weak law (5.3) tells us that if $EX_i^2 < \infty$, then S_n/n converges to $\mu = EX_i$ in probability as $n \to \infty$. The next result shows that it is sufficient to assume that $E|X_i| < \infty$.

(5.5) *Weak Law of Large Numbers*   Let $X_1, X_2, \ldots$ be i.i.d., and let $S_n = X_1 + \cdots + X_n$. In order that there exist constants μ_n so that $S_n/n - \mu_n \to 0$ in probability, it is necessary and sufficient that

$$xP(|X_1| > x) \to 0 \quad \text{as } x \to \infty.$$

In this case we can take $\mu_n = E(X_1 1_{(|X_1| \le n)})$.

Before proving (5.5), we will consider two examples.

Example 5.5   If $E|X_1| < \infty$, then

$$xP(|X_1| > x) \le E(|X_1| 1_{(|X_1| > x)}) \to 0 \quad \text{as } x \to \infty$$

and

$$\mu_n = E(X_1 1_{(|X_1| \le n)}) \to E(X_1) \quad \text{as } n \to \infty$$

by the dominated convergence theorem, giving us the usual form of the weak law.

(5.6) COROLLARY   Let $X_1, X_2, \ldots$ be i.i.d. with $E|X_1| < \infty$ and $EX_1 = \mu$. As $n \to \infty$,

$$S_n/n \to \mu \text{ in probability.}$$

Example 5.6   For an example where the weak law does not hold, suppose $X_1, X_2,$ ... are independent and have a *Cauchy distribution*:

$$P(X_i \le x) = \int_{-\infty}^{x} \frac{dt}{\pi(1 + t^2)}.$$

As $x \to \infty$,

$$P(|X_1| > x) = 2 \int_{x}^{\infty} \frac{dt}{\pi(1 + t^2)} \sim (2/\pi) \int_{x}^{\infty} t^{-2} \, dt = \frac{2}{\pi} x^{-1},$$

where $f(x) \sim g(x)$ means $f(x)/g(x) \to 1$, so $xP(|X| > x) \to 2/\pi > 0$, and from the necessity of the condition above we can conclude that there is no sequence of constants μ_n so that $S_n/n - \mu_n \to 0$. We will see later that S_n/n has the same distribution as X_1. (See Exercise 3.10 in Chapter 2.)

Exercise 5.7   Let $X_1, X_2, \ldots$ be i.i.d. with $P(X_i = (-1)^{k-1}k) = C/k^2 \log k$ for $k \ge 2$ where C is chosen to make the sum of the probabilities $= 1$. Show that $E|X_i| = \infty$, but there is a finite constant μ so that $S_n/n \to \mu$ in probability.

Proof   We have mentioned the necessity of the condition in (5.5) only to assure the reader that the weak law we are about to prove is the best possible. The techniques involved in proving necessity are not very useful for doing other things, so the reader is referred to Feller, Vol. II (1971), pp. 234–236, for a proof. The first step in proving sufficiency is to truncate. Let

$$X_{n,k} = X_k 1_{(|X_k| \le n)} \quad \text{and} \quad S'_n = X_{n,1} + \cdots + X_{n,n}.$$

Clearly

$$P(|S_n/n - \mu_n| > \varepsilon) \le P(|S'_n/n - \mu_n| > \varepsilon) + P(S_n \ne S'_n).$$

To estimate the second term we observe

$$P(S_n \ne S'_n) \le P(X_k \ne X_{n,k} \text{ for some } k \le n) \le nP(|X_1| > n) \to 0 \qquad \text{(a)}$$

as $n \to \infty$ by hypothesis. To estimate the other term we use Chebyshev's inequality

$$P(|S'_n/n - \mu_n| > \varepsilon) \le \varepsilon^{-2} E(|S'_n/n - \mu_n|^2)$$

$$\le \varepsilon^{-2} n \, \text{var}(X_{n,1})/n^2 \le E(X_{n,1}^2)/(n\varepsilon^2), \qquad \text{(b)}$$

since $\mu_n = E(X_{n,i})$ and $\text{var}(Y) = E(Y^2) - (EY)^2 \le E(Y^2)$. To estimate $E(X_{n,1}^2)$ we need the following result, which will be useful several times below.   □

(5.7) LEMMA   If $Y \ge 0$ and $p > 0$, then $E(Y^p) = \int_0^\infty py^{p-1}P(Y > y)\, dy$.

Proof   Using the definition of expected value, Fubini's theorem (for nonnegative random variables), and then calculating the resulting integrals give:

$$\int_0^\infty py^{p-1}P(Y > y)\, dy = \int_0^\infty \int_\Omega py^{p-1} 1_{(Y>y)}\, dP\, dy$$

$$= \int_\Omega \int_0^\infty py^{p-1} 1_{(Y>y)}\, dy\, dP$$

$$= \int_\Omega \int_0^Y py^{p-1}\, dy\, dP = \int_\Omega Y^p\, dP = EY^p.$$

Returning to the proof of (5.5), we observe

$$E(X_{n,1}^2) = \int_0^\infty 2yP(|X_{n,1}| > y)\, dy \le \int_0^n 2yP(|X_1| > y)\, dy,$$

and our hypothesis implies

$$E(X_{n,1}^2)/n = \frac{1}{n}\int_0^n 2yP(|X_1| > y)\,dy \to 0 \quad \text{as } n \to \infty.$$

To see the last claim, let $g(y) = 2yP(|X_1| > y)$ and $M = \sup g(y)$, which is $< \infty$ since $g(y) \to 0$ as $y \to \infty$. Let $\varepsilon > 0$, pick y_0 so that $g(y) < \varepsilon$ for $y > y_0$, and break the integral into two parts $0 \le y \le y_0$ and $y_0 \le y \le n$ to get

$$E(X_{n,1}^2)/n \le \frac{M}{n} \cdot y_0 + \varepsilon.$$

Using $E(X_{n,1}^2)/n \to 0$ in (b) and combining that result with (a) give (5.5).   □

Remark   When $p = 1$, (5.7) becomes

$$EY = \int_0^\infty P(Y > y)\,dy,$$

a result we will use quite a bit below. Applying (5.7) with $p = 1 - \varepsilon$ and $\varepsilon > 0$, we see that

$$xP(|X_1| > x) \to 0 \text{ implies } E|X_1|^{1-\varepsilon} < \infty,$$

so the assumption in (5.5) is not much weaker than finite mean.

Exercise 5.8   Show that if $X \ge 0$ is integer-valued, $EX = \sum_{n \ge 1} P(X \ge n)$. Find a similar expression for EX^2.

Exercise 5.9   Generalize (5.7) to conclude that if $H(x) = \int_{(-\infty, x]} h(y)\,dy$ with $h(y) \ge 0$, then

$$EH(X) = \int_{-\infty}^\infty h(y)P(X \ge y)\,dy.$$

An important special case in $H(x) = \exp(\theta x)$ with $\theta > 0$.

The proof of (5.5) generalizes easily to give us:

(5.8) *Weak Law for Triangular Arrays*   For each n let $X_{n,k}, 1 \le k \le n$, be independent random variables with distribution functions $F_{n,k}$. Let $b_n > 0$ with $b_n \to \infty$. Suppose that

$$\sum_{k=1}^n P(|X_{n,k}| > b_n) \to 0, \quad \text{and} \tag{i}$$

$$b_n^{-2} \sum_{j=1}^n \int_{|x| \le b_n} x^2\,dF_{n,k}(x) \to 0 \quad \text{as } n \to \infty. \tag{ii}$$

If we let $S_n = X_{n,1} + \cdots + X_{n,n}$ and put $a_n = \sum_{j=1}^{n} \int_{|x| \le b_n} x \, dF_j(x)$, then

$$(S_n - a_n)/b_n \to 0 \quad \text{in probability.}$$

Proof   Let $X'_{n,k} = X_{n,k} 1_{(|X_{n,k}| \le b_n)}$ and let $S'_n = X'_{n,1} + \cdots + X'_{n,n}$. Then

$$P(S_n \ne S'_n) = \sum_{k=1}^{n} P(|X_{n,k}| > b_n) \to 0$$

$$P(|(S'_n - a_n)/b_n| > \varepsilon) \le (b_n \varepsilon)^{-2} \sum_{k=1}^{n} \operatorname{var}(X'_{n,k})$$

$$\operatorname{var}(X'_{n,k}) \le \int_{|x| \le b_n} x^2 \, dF_{n,k}(x)$$

so we have assumed what we need to get the analogues of (a) and (b) from the proof of (5.5) and the desired result follows.   □

(5.8) generalizes (5.5) in two ways. First, normalizing constants $b_n \ne n$ are allowed and second, instead of one sequence $X_1, X_2, \ldots$, we have a triangular array $X_{n,k}$, $1 \le k \le n$. The next two examples will show the usefulness of these generalizations.

Example 5.7   *The "St. Petersburg paradox."* Let $X_1, X_2, \ldots$ be i.i.d. with

$$P(X_1 = 2^n) = 2^{-n} \quad \text{for } n \ge 1.$$

In words, you win 2^n dollars if it takes n tosses to get a heads. The paradox here is that $EX_1 = \infty$, but you clearly wouldn't pay an infinite amount to play this game. An application of (5.8) will tell us how much we should pay to play the game n times. In this example $X_{n,k} = X_k$ and $F_{n,k} = F$, so we will delete the unnecessary subscripts.

To apply (5.8) we have to pick b_n. To do this we observe that if m is an integer,

$$P(X_1 \ge 2^m) = \sum_{n=m}^{\infty} 2^{-n} = 2^{-m+1}.$$

Letting $m(n) = \log_2 n + K(n)$ where $K(n) \to \infty$ and is chosen so that $m(n)$ is an integer, and letting $b_n = 2^{m(n)}$, we have

$$nP(X_1 \ge b_n) = n2^{-m(n)+1} = 2^{-K(n)+1} \to 0,$$

proving (i). To check (ii) we observe that

$$\int_{|x| \le b_n} x^2 \, dF(x) = \sum_{n=1}^{m(n)} 2^{2n} \cdot 2^{-n} = 2^{m(n)} \sum_{k=0}^{m(n)-1} 2^{-k} \le 2b_n,$$

so the expression in (ii) is smaller than $2n/b_n$, which $\to 0$ since

$$b_n = 2^{m(n)} = n2^{K(n)} \quad \text{and} \quad K(n) \to \infty.$$

The last step is to evaluate a_n:

$$\int_{|x|\le b_n} x\, dF_j(x) = \sum_{n=1}^{m(n)} 2^n 2^{-n} = m(n),$$

so $a_n = nm(n)$. $m(n) = \log n + K(n)$ (here and until the end of the proof all logs are base 2), so if we pick $K(n)/\log n \to 0$, then $a_n/n \log n \to 1$ as $n \to \infty$. If we pick

$$K(n) = \sup\{k \le \log\log n \text{ such that } \log n + k \text{ is an integer}\},$$

then $m(n) \le \log n + \log\log n$, so $b_n = 2^{m(n)} \le n \log n$ and (5.8) implies

$$(S_n - n \log n)/n \log n \to 0 \text{ in probability.}$$

That is,

$$S_n/(n \log n) \to 1 \text{ in probability.}$$

Returning to our original question, we see that a fair price for playing n times is $\log_2 n$ dollars per play. When $n = 1024$, this is \$10 per play. Nicolas Bernoulli wrote in 1713, "there ought not to exist any even halfway sensible person who would not sell the right of playing the game for 40 ducates." If the wager were 1 ducat, one would need $2^{40} \approx 10^{12}$ plays to start to break even.

Exercise 5.10   *An unfair "fair game."* Let

$$p_k = 1/2^k k(k+1), \ k = 1, 2, \ldots \quad \text{and} \quad p_0 = 1 - (p_1 + p_2 + \cdots).$$

$$\sum_{k=1}^{\infty} 2^k p_k = \left(1 - \frac{1}{2}\right) + \left(\frac{1}{2} - \frac{1}{3}\right) + \cdots = 1,$$

so if we let $X_1, X_2, \ldots$ be i.i.d. with $P(X_n = -1) = p_0$ and

$$P(X_n = 2^k - 1) = p_k \quad \text{for } k \ge 1,$$

then $EX_n = 0$. Let $S_n = X_1 + \cdots + X_n$. Use (5.8) to conclude that

$$S_n/(n/\log_2 n) \to -1 \text{ in probability.}$$

Exercise 5.11   *Weak law for positive variables.* Suppose $X_1, X_2, \ldots$ are i.i.d. and ≥ 0. Let

$$\mu(s) = \int_0^s x\, dF(x)$$

and

$$v(s) = \mu(s)/s(1 - F(s)).$$

In order that there exist constants a_n so that $S_n/a_n \to 1$ in probability, it is necessary and sufficient that $v(s) \to \infty$ as $s \to \infty$. In this case there are numbers b_n so that $n\mu(b_n) = b_n$ and we can pick $a_n = b_n$. Use (5.8) to prove that the condition is sufficient.

Example 5.8   *Random Permutations.* Let Ω_n consist of the $n!$ permutations (i.e., 1–1 mappings from $\{1, \ldots, n\}$ onto $\{1, \ldots, n\}$) and make this into a probability space by assuming all the permutations are equally likely. This application of the weak law concerns the cycle structure of a random permutation π, so we begin by describing the decomposition of a permutation into cycles. Consider the sequence $1, \pi(1), \pi(\pi(1)), \ldots$. Eventually $\pi^k(1) = 1$. When it does, we say the first cycle is completed and has length k. To start the second cycle, we pick the smallest integer not in the first cycle and look at $i, \pi(i), \pi(\pi(i)), \ldots$ until we come back to i. We repeat the construction until all the elements are accounted for. For example, if the permutation is

i	1	2	3	4	5	6	7	8	9
$\pi(i)$	3	9	6	8	2	1	5	4	7

then the cycle decomposition is (1 3 6) (2 9 7 5) (4 8).

Let $X_{n,k} = 1$ if a right parenthesis occurs after the kth number in the sequence, and let $S_n = X_{n,1} + \cdots + X_{n,n} = $ the number of cycles. (In the example $X_{9,3} = X_{9,7} = X_{9,9} = 1$ and the other $X_{9,m} = 0$.) I claim that $X_{n,1}, \ldots, X_{n,n}$ are independent and

$$P(X_{n,j} = 1) = 1/(n - j + 1)$$

since, independent of what has happened so far, there are $n - j + 1$ values that have not appeared in the range, and only one of them will complete the cycle. To make the last statement precise and prove it, it is useful to generate the permutation in a special way. Let $i_1 = 1$. Pick j_1 at random from $\{1, \ldots, n\}$ and let $\pi(i_1) = j_1$. If $j_1 \neq 1$, let $i_2 = j_1$. If $j_1 = 1$, let $i_2 = 2$. In either case pick j_2 at random from $\{1, \ldots, n\} - \{j_1\}$. In general if $i_1, j_1, \ldots, i_{k-1}, j_{k-1}$ have been selected, then (a) if $j_{k-1} \in \{i_1, \ldots, i_{k-1}\}$ so a cycle has just been completed, we let $i_k = \inf(\{1, \ldots, n\} - \{i_1, \ldots, i_k\})$, and (b) if $j_{k-1} \notin \{i_1, \ldots, i_{k-1}\}$, we let $i_k = j_{k-1}$. In either case we pick j_k at random from $\{1, \ldots, n\} - \{j_1, \ldots, j_{k-1}\}$ and let $\pi(i_k) = j_k$.

To illustrate the definitions, consider the example above:

i_k	1	3	6	2	9	7
j_k	3	6	1	9	7	?

If the ? (i.e., j_6) is 2, a cycle will be completed and we will let $i_7 = 4$. If $j_6 = 5$ as in the example above, we let $i_7 = 5$ and continue. The construction above is tedious to write out or read, but now I can claim with a clear conscience that $X_{n,1}, \ldots, X_{n,n}$ are independent and $P(X_{n,k} = 1) = 1/(n - j + 1)$, since when we pick j_k there are $n - j + 1$ values in $\{1, \ldots, n\} - \{j_1, \ldots, j_{k-1}\}$ and only one of them will complete the cycle.

To check the conditions of (5.8), we reverse the approach used in Example 5.7 and start with

$$a_n = \sum_{j=1}^{n} \int_{|x| \le b_n} x \, dF_{n,k}(x).$$

The truncation level $b_n \to \infty$ by assumption, so when $b_n \ge 1$,

$$a_n = 1/n + 1/(n-1) + \cdots + 1/2 + 1 \sim \log n$$

as $n \to \infty$. Turning to (ii) next, since $X_{n,k} \in \{0, 1\}$,

$$\sum_{j=1}^{n} \int_{|x| \le b_n} x^2 \, dF_{n,k}(x) = \sum_{j=1}^{n} \int_{|x| \le b_n} x \, dF_{n,k}(x) \sim \log n,$$

so if $b_n = (\log n)^{.5+\varepsilon}$, (ii) is satisfied. (i) is satisfied as soon as $b_n \ge 1$, so (5.8) implies that for any $\varepsilon > 0$,

$$(S_n - \log n)/(\log n)^{.5+\varepsilon} \to 0 \text{ in probability.}$$

We will see in Example 4.7 of Chapter 2 that the last conclusion is false if $\varepsilon = 0$.

Our last two examples are situations in which we have a triangular array of dependent variables for which a weak law holds. In each case the result is proved by computing the variance and applying Chebyshev's inequality.

Example 5.9   *An occupancy problem.* Suppose we put r balls at random in n boxes; that is, all n^r assignments of balls to boxes have equal probability. Let A_i be the event that the ith box is empty and $N_n = $ the number of empty boxes. It is easy to see that

$$P(A_i) = (1 - 1/n)^r \quad \text{and} \quad EN_n = n(1 - 1/n)^r.$$

A little calculus (take logarithms) shows that if $r/n \to c$, $EN_n/n \to e^{-c}$. (For a more general fact, see (4.2) in Chapter 2.) To prove that

$$N_n/n \to e^{-c} \text{ in probability}$$

in this case we compute second moments. To do this we observe that

$$EN_n^2 = E\left(\sum_{m=1}^{n} 1_{A_m} \right)^2 = \sum_{1 \le k, m \le n} P(A_k \cap A_m)$$

$$\text{var}(N_n) = EN_n^2 - (EN_n)^2 = \sum_{1 \le k, m \le n} P(A_k \cap A_m) - P(A_k)P(A_m)$$

$$= n(n-1)((1-2/n)^r - (1-1/n)^{2r}) + n((1-1/n)^r - (1-1/n)^{2r}).$$

The first term comes from $k \neq m$ and the second from $k = m$. Since $(1 - 2/n)^r \to e^{-2c}$ and $(1 - 1/n)^r \to e^{-c}$, it follows easily from the last formula that $\text{var}(N_n/n) = \text{var}(N_n)/n^2 \to 0$, and we have proved the desired result.

Exercise 5.12 N is said to have a *geometric distribution* with parameter p if

$$P(N = k) = p(1 - p)^{k-1} \quad \text{for } k = 1, 2, \ldots .$$

N is the number of independent trials needed to observe an event with probability p. Differentiating the identity

$$\sum_{k=0}^{\infty} (1 - p)^k = 1/p$$

gives

$$- \sum_{k=1}^{\infty} k(1 - p)^{k-1} = -1/p^2$$

and

$$\sum_{k=2}^{\infty} k(k - 1)(1 - p)^{k-2} = 2/p^3.$$

Use this to conclude that $EN = 1/p$ and $\text{var}(N) = (1 - p)/p^2$.

Exercise 5.13   Use (3.11) to derive the results in the last exercise.

Example 5.10   *Coupon collector's problem.* Let X_1, X_2, ... be i.i.d. uniform on $\{1, 2, \ldots, n\}$. Let $T_n = \inf\{m: \{X_1, \ldots, X_m\} = \{1, 2, \ldots, n\}\}$ be the first time we have seen all the values. To analyze the asymptotic behavior of T_n, we let $\tau_k^n = \inf\{m: |\{X_1, \ldots, X_m\}| = k\}$. $\tau_1^n = 1$ and for $2 \leq k \leq n$, $\tau_k^n - \tau_{k-1}^n$ are independent and have a geometric distribution with parameter $1 - (k - 1)/n$. Using Exercise 5.12 we see that as $n \to \infty$,

$$ET_n = \sum_{k=1}^{n} 1/(1 - (k - 1)/n) = n \sum_{m=1}^{n} 1/m \sim n \log n$$

$$\text{var}(T_n) = \sum_{k=1}^{n} \frac{k - 1}{n}(1 - (k - 1)/n)^{-2} = n^2 \sum_{m=1}^{n} \frac{n - m}{n} m^{-2} \leq n^2 \sum_{m=1}^{\infty} m^{-2}.$$

So $\text{var}(T_n/ET_n) = \text{var}(T_n)/(ET_n)^2 \to 0$ and it follows that $T_n/n \log n \to 1$ in probability.

6 Borel–Cantelli Lemmas

If A_n is a sequence of subsets of Ω, we let

$$\limsup A_n = \lim_{m \to \infty} \bigcup_{n=m}^{\infty} A_n = \{\omega \text{ that are in infinitely many } A_n\}$$

(the limit exists since the sequence is decreasing in m) and let

$$\liminf_{m \to \infty} A_n = \lim_{m \to \infty} \bigcap_{n=m}^{\infty} A_n = \{\omega \text{ that are in all but finitely many } A_n\}$$

(the limit exists since the sequence is increasing in m). The names limsup and liminf can be explained by noting that

$$\limsup_{n \to \infty} 1_{A_n} = 1_{(\limsup A_n)} \qquad \liminf_{n \to \infty} 1_{A_n} = 1_{(\liminf A_n)}.$$

In probability it is common to write $\limsup A_n = \{\omega : \omega \in A_n \text{ i.o.}\}$ where i.o. stands for infinitely often. An example which illustrates the use of this notation is: "$X_n \to 0$ a.s. if and only if for all $\varepsilon > 0$, $P(|X_n| > \varepsilon \text{ i.o.}) = 0$." The reader will see many other examples below. The next result should be familiar from measure theory even though its name may not be.

(6.1) Borel–Cantelli Lemma   If $\sum_{n=1}^{\infty} P(A_n) < \infty$, then $P(A_n \text{ i.o.}) = 0$.

Proof   Let $N = \sum_k 1_{A_k}$ be the number of events that occur. Fubini's theorem implies $EN = \sum_k P(A_k) < \infty$, so we must have $N < \infty$ a.s.   □

For practice using the Borel–Cantelli lemma, the reader should do:

Exercise 6.1   If X_n is any sequence of random variables, there are constants $c_n \to \infty$ so that $X_n/c_n \to 0$ a.s.

The next result is a typical application of the Borel–Cantelli lemma.

(6.2) THEOREM   If $X_n \to X$ in probability, then there is a sequence of integers $n(k) \uparrow \infty$ so that $X_{n(k)} \to X$ almost surely.

Proof   Let ε_k be a sequence of positive numbers that $\downarrow 0$. For each k there is an $n(k) > n(k-1)$ so that $P(|X_{n(k)} - X| > \varepsilon_k) \leq 2^{-k}$. Since

$$\sum_{k=1}^{\infty} P(|X_{n(k)} - X| > \varepsilon_k) < \infty,$$

the Borel–Cantelli lemma implies $P(|X_{n(k)} - X| > \varepsilon_k \text{ i.o.}) = 0$; that is, $X_{n(k)} \to X$ a.s.   □

Many properties of convergence in probability follow easily from (6.2).

Exercise 6.2   Recall that a sequence of real numbers x_n converges to x if and only if every subsequence has a further subsequence that converges to x. (i) Prove that $X_n \to X$ in probability if and only if every subsequence has a further subsequence which converges a.s. to X. (ii) Use (i) to conclude that if f is continuous and $X_n \to X$

in probability, then $f(X_n) \to f(X)$ in probability. (iii) Since there is a sequence of random variables which converges in probability but not a.s. (for examples, see Exercises 6.9 and 6.10), it follows that a.s. convergence does not come from a metric, or even from a topology.

Exercise 6.3 Suppose $X_n \to X$ in probability. If (i) $|X_n| \le Y$ with $EY < \infty$ or (ii) there is a $\varphi \ge 0$ with $\varphi(x)/|x| \to \infty$ as $|x| \to \infty$ so that $\sup E \varphi(X_n) = A < \infty$, then $EX_n \to EX$.

Exercise 6.4 Show that random variables are a complete space under the metric defined in Exercise 5.4; that is, if $d(X_m, X_n) \to 0$ whenever $m, n \to \infty$, then there is a r.v. X_∞ so that $X_n \to X_\infty$ in probability.

As our second application of the Borel–Cantelli lemma we get our first strong law of large numbers:

(6.3) THEOREM   Let $X_1, X_2, \ldots$ be i.i.d. with $EX_1 = \mu$ and $EX_1^4 < \infty$. If $S_n = X_1 + \cdots + X_n$, then $S_n/n \to \mu$ a.s.

Proof   By letting $X_i' = X_i - \mu$ we can suppose without loss of generality that $\mu = 0$. Now

$$ES_n^4 = E\left(\sum_{i=1}^n X_i\right)^4 = E \sum_{1 \le i,j,k,l \le n} X_i X_j X_k X_l.$$

Terms in the sum of the form $E(X_i^3 X_j)$, $E(X_i^2 X_j X_k)$, and $E(X_i X_j X_k X_l)$ are 0 (if i, j, k, l are distinct), since the expectation of the product is the product of the expectations, and in each case one of the terms has expectation 0. The only terms that do not vanish are those of the form EX_i^4 and $EX_i^2 X_j^2 = (EX_i^2)^2$. There are n and $3n(n-1)$ of these terms, respectively. (In the second case we can pick the two indices in $n(n-1)/2$ ways and with the indices fixed, the term can arise in a total of six ways.) The last observation implies

$$ES_n^4 \le nEX_1^4 + 3(n^2 - n)(EX_1^2)^2 \le Cn^2$$

where $C < \infty$. Chebyshev's inequality gives us

$$P(|S_n| > n\varepsilon) \le E(S_n^4)/(n\varepsilon)^4 \le C/(n^2\varepsilon^4).$$

Summing over n and applying the Borel–Cantelli lemma give $P(|S_n| > n\varepsilon \text{ i.o.}) = 0$. Since ε is arbitrary, the proof is complete. □

The converse of the Borel–Cantelli lemma is trivially false:

Example 6.1   Let $\Omega = (0, 1)$, $\mathscr{F} = $ Borel sets, $P = $ Lebesgue measure. If $A_n = (0, a_n)$ where $a_n \to 0$ as $n \to \infty$, then $\limsup A_n = \varnothing$, but if $a_n \ge 1/n$, we have $\sum a_n = \infty$.

A more general counterexample is provided by:

Exercise 6.5   If $P(A_n) \to 0$ and $\sum_{n=1}^{\infty} P(A_n \cap A_{n+1}^c) < \infty$, then $P(A_n \text{ i.o.}) = 0$.

The examples just given suggest that for general sets we cannot say much more than the following result.

Exercise 6.6 $P(\text{limsup } A_n) \geq \text{limsup}_{n \to \infty} P(A_n)$ and $P(\text{liminf } A_n) \leq \text{liminf}_{n \to \infty} P(A_n)$.

For independent events, however, the necessary condition for $P(\text{limsup } A_n) > 0$ is sufficient for $P(\text{limsup } A_n) = 1$.

(6.4) The Second Borel-Cantelli Lemma   If the events A_n are independent, then $\sum P(A_n) = \infty$ implies $P(A_n \text{ i.o.}) = 1$.

Proof   Let $M < N < \infty$. Independence implies

$$P\left(\bigcap_{n=M}^{N} A_n^c \right) = \prod_{n=M}^{N} (1 - P(A_n)) \leq \prod_{n=M}^{N} \exp(-P(A_n)),$$

since $e^{-x} \geq 1 - x$. The last expression is

$$\exp\left(- \sum_{n=M}^{N} P(A_n) \right) \to 0 \quad \text{as } N \to \infty,$$

so $P(\bigcup_{n=M}^{\infty} A_n) = 1$ for all M, and it follows that $P(\text{limsup } A_n) = 1$.   □

Exercise 6.7   Let A_n be a sequence of independent events with $P(A_n) < 1$ for all n. Show that $P(\bigcup A_n) = 1$ implies $P(A_n \text{ i.o.}) = 1$.

Exercise 6.8   Let $X_1, X_2, \ldots$ be independent. Show that $\sup X_n < \infty$ if and only if

$$\sum_n P(X_n > A) < \infty \quad \text{for some } A.$$

Exercise 6.9   Let $X_1, X_2, \ldots$ be independent with $P(X_n = 1) = p_n$ and $P(X_n = 0) = 1 - p_n$. Show that (i) $X_n \to 0$ in probability if and only if $p_n \to 0$, and (ii) $X_n \to 0$ a.s. if and only if $\sum p_n < \infty$.

Exercise 6.10   Let $Y_1, Y_2, \ldots$ be i.i.d. Show that $Y_n/n \to 0$ a.s. if and only if $E|Y_1| < \infty$, while $Y_n/n \to 0$ in probability if and only if $nP(|Y_1| > n) \to 0$.

The last two exercises give examples to show that we may have $X_n \to X$ in probability without $X_n \to X$ a.s. There is one situation in which the two notions are the same.

Exercise 6.11   Let $X_1, X_2, \ldots$ be a sequence of r.v.'s on $(\Omega, \mathscr{F}, P)$ where Ω is a countable set. Show that $X_n \to X$ in probability implies $X_n \to X$ a.s.

A typical application of the second Borel–Cantelli lemma is:

(6.5) THEOREM   If $X_1, X_2, \ldots$ are i.i.d. with $E|X_i| = \infty$, then $P(|X_n| \geq n \text{ i.o.}) = 1$, so if $S_n = X_1 + \cdots + X_n$, then $P(\lim_{n \to \infty} S_n/n \text{ exists} \in (-\infty, \infty)) = 0$.

Proof   From (5.7) we get

$$E|X_1| = \int_0^\infty P(|X_1| > x)\, dx \leq \sum_{n=0}^\infty P(|X_1| > n).$$

Since $E|X_1| = \infty$ and $X_1, X_2, \ldots$ are i.i.d., it follows from the second Borel–Cantelli lemma that $P(|X_n| \geq n \text{ i.o.}) = 1$. To prove the second claim observe that

$$\frac{S_n}{n} - \frac{S_{n+1}}{n+1} = \frac{S_n}{n(n+1)} - \frac{X_{n+1}}{n+1}$$

and on $C \equiv \{\omega \colon \lim_{n \to \infty} S_n/n \text{ exists} \in (-\infty, \infty)\}$, $S_n/(n(n+1)) \to 0$. So on $C \cap \{\omega \colon |X_n| \geq n \text{ i.o.}\}$ we have

$$\left| \frac{S_n}{n} - \frac{S_{n+1}}{n+1} \right| > 2/3 \text{ i.o.,}$$

contradicting the fact that $\omega \in C$. From the last observation we conclude that

$$\{\omega \colon |X_n| \geq n \text{ i.o.}\} \cap C = \varnothing,$$

and since $P(|X_n| \geq n \text{ i.o.}) = 1$, it follows that $P(C) = 0$.   □

Exercise 6.12   Improve the last argument to show that under the hypotheses of (6.5),

$$\limsup_{n \to \infty} |S_n/n - a_n| = \infty$$

for any sequence of constants a_n.

Exercise 6.13   Show that if X_n is the outcome of the nth play of the St. Petersburg game (Example 5.7), then $\limsup_{n \to \infty} X_n/(n \log_2 n) = \infty$ a.s. and hence the same result holds for S_n. This shows that the convergence $S_n/(n \log_2 n) \to 1$ in probability proved in Section 5 does not occur a.s.

(6.5) shows that $E|X_i| < \infty$ is necessary for the strong law of large numbers. The reader will have to wait until (8.3) to see that condition is also sufficient. The next result extends the second Borel–Cantelli lemma and sharpens its conclusion.

(6.6) THEOREM   If $A_1, A_2, \ldots$ are pairwise independent and $\sum_{n=1}^{\infty} P(A_n) = \infty$, then as $n \to \infty$,

$$\sum_{m=1}^{n} 1_{A(m)} \bigg/ \sum_{m=1}^{n} P(A_m) \to 1 \quad \text{a.s.}$$

Proof   Let $X_m = 1_{A(m)}$ and let $S_n = X_1 + \cdots + X_n$. Since the A_m are pairwise independent, the X_m are uncorrelated, and hence (5.1) implies

$$\text{var}(S_n) = \text{var}(X_1) + \cdots + \text{var}(X_n).$$

If we let $p_m = P(A_m)$, then (5.4) shows $\text{var}(X_m) = p_m(1 - p_m) \leq p_m = E(X_m)$, so $\text{var}(S_n) \leq E(S_n)$. If we let $c_n = E(S_n)$, then Chebyshev's inequality implies

$$P(|S_n - c_n| > \delta c_n) \leq \text{var}(S_n)/(\delta c_n)^2 \leq 1/(\delta^2 c_n) \to 0 \qquad (*)$$

as $n \to \infty$ (since we have assumed $c_n \to \infty$).

The last computation shows that $S_n/c_n \to 1$ in probability. To get almost sure convergence we have to take subsequences. Let $n_k = \inf\{n: ES_n \geq k^2\}$. Let $T_k = S_{n(k)}$. Replacing n by n_k in $(*)$ shows

$$P(|T_k - ET_k| > \delta k^2) \leq 1/(\delta^2 k^2),$$

so

$$\sum_{k=1}^{\infty} P(|T_k - ET_k| > \delta k^2) < \infty,$$

and the Borel–Cantelli lemma implies $P(|T_k - ET_k| > \delta k^2 \text{ i.o.}) = 0$. Since δ is arbitrary and $k^2 \leq ET_k \leq k^2 + 1$, it follows that $T_k/ET_k \to 1$ a.s. To show $S_n/ES_n \to 1$ a.s., we observe that if $n_k \leq n < n_{k+1}$, then

$$\frac{T_k}{ET_{k+1}} \leq \frac{S_n}{ES_n} \leq \frac{T_{k+1}}{ET_k}.$$

To show that the terms at the left and right ends $\to 1$, we recall

$$k^2 \leq ET_k \leq ET_{k+1} \leq (k+1)^2 + 1$$

and observe that

$$\frac{(k+1)^2 + 1}{k^2} = 1 + 2/k + 2/k^2 \to 1. \qquad \square$$

The last observation completes the proof. The moral of the story is that if you want to show that $X_n/c_n \to 1$ a.s. for a sequence $X_n \geq 0$ that is increasing, it is enough to prove the result for a subsequence $n(k)$ that has $c_{n(k+1)}/c_{n(k)} \to 1$. For practice with this technique, try the following.

Exercise 6.14   Let X_n be independent Poisson r.v.'s with $EX_n = \lambda_n$ and let $S_n = X_1 + \cdots + X_n$. Show that if $\sum \lambda_n = \infty$, then $S_n/ES_n \to 1$ a.s.

Exercise 6.15   Let $0 \le X_1 \le X_2 \le \cdots$ be random variables with $EX_n \sim an^\alpha$, $\alpha > 0$, and $\text{var}(X_n) \le Bn^\beta$ where $\beta < 2\alpha$. Show that $X_n/n^\alpha \to a$ a.s.

Example 6.2   *Record Values.* Let $X_1, X_2, \ldots$ be i.i.d. with a distribution $F(x)$ that is continuous. Let $A_k = \{X_k > \sup_{j<k} X_j\}$ be the event that a record occurs at time k. Even though it may seem that the occurrence of a record at time k will make it less likely that one will occur at time $k + 1$, the A_k are independent. The key to proving this is the observation that if π is a permutation of $\{1, 2, \ldots, k\}$, then

$$P\left(X_k > \sup_{j<k} X_j \right) = P\left(X_{\pi(k)} > \sup_{j<k} X_{\pi(j)} \right).$$

If we suppose that $X_1, X_2, \ldots, X_k$ are distinct, then summing over all the permutations gives

$$\sum_\pi 1_{(X_{\pi(k)} > \sup_{j<k} X_{\pi(j)})} = (k-1)!.$$

If F is continuous, then $X_1, X_2, \ldots, X_k$ are distinct with probability 1, so combining the last two observations gives

$$P\left(X_k > \sup_{j<k} X_j \right) = 1/k.$$

Let $j < k$. Summing over all permutations π of $\{1, \ldots, k\}$ and σ of $\{1, \ldots, j\}$ (and regarding the latter as maps on $\{1, \ldots, k\}$ that leave the last $k - j$ points fixed) shows

$$\sum_\pi \sum_\sigma 1_{(X_{\pi\sigma(k)} > \sup_{i<k} X_{\pi\sigma(i)},\, X_{\pi\sigma(j)} > \sup_{i<j} X_{\pi\sigma(i)})} = (k-1)!\,(j-1)!.$$

Reasoning as before, it follows that

$$P(A_j \cap A_k) = \frac{1}{j} \cdot \frac{1}{k} = P(A_j)P(A_k),$$

and the A_i are pairwise independent. This is enough to apply (6.6), so we leave the rest of the proof of independence to the reader. From (6.6) it follows that if

$$R_n = \sum_{m=1}^{n} 1_{A_m} = \text{the number of records at time } n,$$

then as $n \to \infty$,

$$R_n/(\log n) \to 1 \quad \text{a.s.}$$

The reader should note that the last result is independent of the distribution F (as long as it is continuous).

Exercise 6.16   Let $X_1, X_2, \ldots$ be i.i.d. with a distribution that is continuous. Let

$$Y_i = \text{the number of } j \le i \text{ with } X_j > X_i.$$

Prove that the Y_i are independent random variables with $P(Y_i = j) = 1/i$ for $0 \le j < i - 1$.

Comic relief   Let $X_0, X_1, \ldots$ be i.i.d. and imagine they are the offers you get for a car you are going to sell. Let $N = \inf\{n: X_n > X_0\}$. Symmetry implies $P(N > n) \ge 1/n$ (we are not supposing the distribution is continuous), so the expected time you have to wait until you get an offer better than the first one is ∞.

Example 6.3   *Head runs.* Let $X_n, n \in \mathbb{Z}$, be i.i.d. with $P(X_n = 1) = P(X_n = -1) = 1/2$. Let $l_n = \max\{m: X_{n-m+1} = \cdots = X_n = 1\}$ be the length of the run of $+1$'s at time n, and let $L_n = \max_{1 \le m \le n} l_m$ be the longest run at time n. We use a two-sided sequence so that for all n, $P(l_n = k) = (1/2)^{k+1}$ for $k \ge 0$. Since $l_1 < \infty$, all the results below are true for a one-sided sequence.

$$P(l_n \ge (1 + \varepsilon) \log_2 n) \le n^{-(1+\varepsilon)}$$

for any $\varepsilon > 0$, so it follows from the Borel–Cantelli lemma that

$$\limsup_{n \to \infty} l_n/\log_2 n \le 1 \quad \text{a.s.} \tag{i}$$

To get a result in the other direction, let $r_1 = 1, r_2 = 2$, and $r_n = r_{n-1} + [\log_2 n]$ for $n \ge 3$, where $[x] = $ the greatest integer $\le x$. Let $A_m = \{X_m = 1 \text{ for } r_{n-1} < m \le r_n\}$. $P(A_n) \ge 1/n$, so it follows from the second Borel–Cantelli lemma that $P(A_n \text{ i.o.}) = 1$ and

$$\limsup_{n \to \infty} l_n/\log_2 n \ge 1 \quad \text{a.s.} \tag{ii}$$

By breaking the first n trials into disjoint blocks of length $[(1 - \varepsilon)\log_2 n] + 1$, it is easy to see that for large n,

$$P(L_n > (1 - \varepsilon)\log_2 n) \le (1 - n^{-(1-\varepsilon)})^{n/(\log_2 n)} \le \exp(-n^\varepsilon/\log_2 n).$$

So the Borel–Cantelli lemma (and (i)) implies

$$L_n/\log_2 n \to 1 \quad \text{a.s.} \tag{iii}$$

Exercise 6.17   Let $X_1, X_2, \ldots$ be i.i.d. with $P(X_i > x) = e^{-x}$. Show that

$$\limsup_{n \to \infty} X_n / \log n = 1 \quad \text{a.s.}$$

$$\left(\max_{1 \le m \le n} X_m\right) / \log n \to 1 \quad \text{a.s.}$$

Exercise 6.18   Let X_n be i.i.d. with distribution F, let λ_n be an increasing sequence, and let $A_n = \{\max_{1 \le m \le n} X_m > \lambda_n\}$. Show that $P(A_n \text{ i.o.}) = 0$ or 1 according as

$$\sum_{n \ge 1} (1 - F(\lambda_n)) < \infty \text{ or } = \infty.$$

Remark   O. Barndorff–Nielsen (1961) has shown that if in addition $F(\lambda_n)^n$ is decreasing, then $P(A_n^c \text{ i.o.}) = 0$ or 1 according as

$$\sum_{n=3}^{\infty} F(\lambda_n)^n (\log \log n)/n < \infty \text{ or } = \infty.$$

Exercise 6.19   *Kochen–Stone lemma*   (A Borel–Cantelli lemma for dependent events.) Suppose $\sum P(A_k) = \infty$ and

$$\limsup_{n \to \infty} \left(\sum_{k=1}^{n} P(A_k)\right)^2 \Big/ \left(\sum_{1 \le j, k \le n} P(A_j \cap A_k)\right) = \alpha > 0.$$

Then $P(A_n \text{ i.o.}) \ge \alpha$.

Sketch of proof: Let $X_n = \sum_{k \le n} 1_{A_k}$, let $Y_n = X_n/EX_n$, and note that our hypothesis says $\limsup_{n \to \infty} 1/E(Y_n^2) = \alpha$. To take advantage of this use Cauchy–Schwarz to conclude that if $\varepsilon > 0$,

$$EY_n^2 P(Y_n \ge \varepsilon) \ge (EY_n 1_{(Y_n \ge \varepsilon)})^2 \ge (1 - \varepsilon)^2,$$

since $EY_n = 1$ and $Y_n \ge 0$.

7   Convergence of Random Series

The aim of this section is to develop some results that will be useful in proving the strong law of large numbers. Let $\mathcal{F}_n' = \sigma(X_n, X_{n+1}, \ldots) = $ the future after time $n = $ the smallest σ-field with respect to which all the X_m, $m \ge n$, are measurable. Let $\mathcal{T} = \bigcap_n \mathcal{F}_n' = $ the remote future, or *tail σ-field*. Intuitively $A \in \mathcal{T}$ if and only if changing a finite number of values does not affect the occurrence of the event. As usual, we turn to examples to help explain the definition.

Example 7.1   If $B_n \in \mathcal{R}$, then $\{X_n \in B_n \text{ i.o.}\} \in \mathcal{T}$. If we let $X_n = 1_{A(n)}$ and $B_n = \{1\}$, this example becomes $\{A_n \text{ i.o.}\}$.

Example 7.2   Let $S_n = X_1 + \cdots + X_n$. $\{\lim_{n \to \infty} S_n \text{ exists}\} \in \mathcal{T}$.

Example 7.3   If $c_n \to \infty$, then $\{\limsup_{n \to \infty} S_n/c_n > x\} \in \mathcal{T}$.

The next result shows that all examples are trivial.

(7.1) Kolmogorov's 0–1 Law   If $X_1, X_2, \ldots$ are independent and $A \in \mathcal{T}$, then $P(A) = 0$ or 1.

Proof   We will show that A is independent of itself; that is, $P(A \cap A) = P(A)P(A)$, so $P(A) = P(A)^2$, and hence $P(A) = 0$ or 1. To do this we will identify A and B for which

$$P(A \cap B) = P(A)P(B). \tag{$*$}$$

(a) If $A \in \sigma(X_1, \ldots, X_k)$ and $B \in \sigma(X_{k+1}, X_{k+2}, \ldots)$, then $(*)$ holds.

Proof of (a)   If $B \in \sigma(X_{k+1}, \ldots, X_{k+j})$ for some j, this follows from (4.4). Since $\sigma(X_1, \ldots, X_k)$ and $\bigcup_j \sigma(X_{k+1}, \ldots, X_{k+j})$ are π-systems, (a) follows from (4.1).

(b) If $A \in \sigma(X_1, X_2, \ldots)$ (which contains $\mathcal{T}$) and $B \in \mathcal{T}$, then $(*)$ holds.

Proof of (b)   If $A \in \sigma(X_1, \ldots, X_k)$ for some k, this follows from (a). $\bigcup_k \sigma(X_1, \ldots, X_k)$ and $\mathcal{T}$ are π-systems, so (b) (and hence (7.1)) follows from (4.1).   □

If $A_1, A_2, \ldots$ are independent, then (7.1) implies $P(A_n \text{ i.o.}) = 0$ or 1. Applying (7.1) to Example 7.2 gives $P(\lim_{n \to \infty} S_n \text{ exists}) = 0$ or 1. The next result will help us prove the probability is 1 in certain situations.

(7.2) Kolmogorov's Inequality   If $X_1, \ldots, X_n$ are independent with mean 0 and finite variance, then

$$P\left(\max_{1 \le k \le n} |S_k| \ge x \right) \le x^{-2} \, \mathrm{var}(S_n).$$

Remark   Under the same hypotheses, Chebyshev's inequality (3.7) gives only

$$P(|S_n| \ge x) \le x^{-2} \, \mathrm{var}(S_n).$$

Proof   Let $A_k = \{|S_k| \ge x \text{ but } |S_j| < x \text{ for } j < k\}$; that is, we break things down according to the time that $|S_k|$ first exceeds x. Since the A_k are disjoint,

$$ES_n^2 \ge \sum_{k=1}^{n} \int_{A_k} S_n^2 \, dP$$

$$= \sum_{k=1}^{n} \int_{A_k} S_k^2 + 2S_k(S_n - S_k) + (S_n - S_k)^2 \, dP$$

$$\ge \sum_{k=1}^{n} \int_{A_k} S_k^2 \, dP + \sum_{k=1}^{n} \int 2S_k 1_{A_k} \cdot (S_n - S_k) \, dP.$$

$S_k 1_{A_k} \in \sigma(X_1, \ldots, X_k)$ and $S_n - S_k \in \sigma(X_{k+1}, \ldots, X_n)$ are independent by (4.5), so using (4.7) and $E(S_n - S_k) = 0$ shows

$$\int 2 S_k 1_{A_k} \cdot (S_n - S_k) \, dP = 0.$$

Using now the fact that $|S_k| \geq x$ on A_k shows

$$ES_n^2 \geq \sum_{k=1}^{n} \int_{A_k} S_k^2 \, dP \geq \sum_{k=1}^{n} x^2 P(A_k) = x^2 P\left(\max_{1 \leq k \leq n} |S_k| \geq x \right),$$

and the proof is complete. $\qquad\qquad\qquad\qquad\qquad\qquad\qquad\qquad\qquad\qquad$ □

The next result is also due to Kolmogorov:

(7.3) THEOREM   Suppose $X_1, X_2, \ldots$ are independent with $EX_n = 0$. If $\sum \mathrm{var}(X_n) < \infty$, then $\sum X_n$ converges a.s.; that is, $\lim_{N \to \infty} \sum_{n \leq N} X_n$ exists a.s.

Proof   Let $S_N = \sum_{n \leq N} X_n$. From (7.2) we get

$$P\left(\max_{M \leq m \leq N} |S_m - S_M| > \varepsilon \right) \leq \varepsilon^{-2} \mathrm{var}(S_N - S_M) = \varepsilon^{-2} \sum_{n=M+1}^{N} \mathrm{var}(X_n).$$

Letting $N \to \infty$ in the last result, we get

$$P\left(\max_{M \leq m < \infty} |S_m - S_M| > \varepsilon \right) \leq \varepsilon^{-2} \sum_{n=M+1}^{\infty} \mathrm{var}(X_n) \to 0 \quad \text{as } M \to \infty.$$

If we let $w_M = \max_{m, n \geq M} |S_m - S_n|$, then $w_M \downarrow$ as $M \uparrow$ and

$$P(w_M > 2\varepsilon) \leq P\left(\max_{M \leq m < \infty} |S_m - S_M| > \varepsilon \right) \to 0$$

as $M \to \infty$, so $w_M \downarrow 0$ almost surely. But $w_M(\omega) \downarrow 0$ implies $\lim_{n \to \infty} S_n(\omega)$ exists, so we have proved (7.3). $\qquad\qquad\qquad\qquad\qquad\qquad\qquad\qquad\qquad\qquad\qquad\qquad\qquad$ □

Example 7.4   Let $X_1, X_2, \ldots$ be independent with

$$P(X_n = n^{-\alpha}) = P(X_n = -n^{-\alpha}) = 1/2.$$

$EX_n = 0$ and $\mathrm{var}(X_n) = n^{-2\alpha}$, so if $\alpha > 1/2$ it follows from (7.3) that $\sum X_n$ converges. The next result shows that $\alpha > 1/2$ is also necessary for this conclusion. Notice that there is absolute convergence; that is, $\sum |X_n| < \infty$ if and only if $\alpha > 1$.

Exercise 7.1   Let $X_1, X_2, \ldots$ be i.i.d. standard normals. Show that for any t,

$$\sum_{n=1}^{\infty} X_n(\sin n\pi t)/n \text{ converges a.s.}$$

We will see this series again at the end of Section 1 of Chapter 7.

Exercise 7.2   Let $X_n \geq 0$. Show that $\sum X_n < \infty$ a.s. if and only if $\sum_n E(X_n/(1 + X_n)) < \infty$.

(7.4) Kolmogorov's Three Series Theorem   Let $X_1, X_2, \ldots$ be independent. Let $A > 0$ and let $Y_i = X_i 1_{(|X_i| \leq A)}$. In order that $\sum X_n$ converges (i.e., $\lim_{N \to \infty} \sum_{n \leq N} X_n$) a.s., it is necessary and sufficient that

(i) $\displaystyle\sum_{n=1}^{\infty} P(|X_n| > A) < \infty$,   (ii) $\displaystyle\sum_{n=1}^{\infty} EY_n$ converges,   and   (iii) $\displaystyle\sum_{n=1}^{\infty} \text{var}(Y_n) < \infty$.

Proof   We will prove the necessity in Example 4.8 of Chapter 2 as an application of the central limit theorem. To prove the sufficiency, let $\mu_n = EY_n$. (iii) and (7.3) imply that

$$\sum_{n=1}^{\infty} (Y_n - \mu_n) \text{ converges a.s.}$$

Using (ii) now gives that $\sum Y_n$ converges a.s. (i) and the Borel–Cantelli lemma imply that

$$P(X_n \neq Y_n \text{ i.o.}) = 0 \quad \text{so} \quad \sum_{n=1}^{\infty} X_n \text{ converges a.s.} \qquad \square$$

Exercise 7.3   Show that if $X_1, X_2, \ldots$ are independent with $EX_n = 0$ and

$$\sum_{n=1}^{\infty} E(X_n^2 1_{(|X_n| \leq 1)} + |X_n| 1_{(|X_n| > 1)}) < \infty,$$

then $\sum X_n$ converges a.s.

Exercise 7.4   (Loève) Suppose $\sum E|X_n|^{p(n)} < \infty$ where $0 < p(n) \leq 2$ for all n and $EX_n = 0$ when $p(n) > 1$. Show that $\sum X_n$ converges a.s.

Exercise 7.5   (i) Let X_n be i.i.d. with a symmetric distribution: $P(X > x) = P(X < -x)$ for all x. Show that $\sum X_n/n$ converges a.s. if and only if $E|X_1| < \infty$. (ii) Give an example of an i.i.d. sequence with $EX_n = 0$ so that $\sum_{n \leq N} X_n/n \to \infty$ a.s.

Exercise 7.6   Show that for an i.i.d. sequence the radius of convergence of the power series $\sum_{n \geq 1} X_n(\omega)z^n$ (i.e., $r(\omega) = \sup\{c: \sum |X_n(\omega)|c^n < \infty\}$) is 1 a.s. or 0 a.s. according as $E \log^+ |X_1| < \infty$ or $= \infty$ where $\log^+ x = \max(\log x, 0)$.

Exercise 7.7   Let $X_1, X_2, \ldots$ be independent and let $S_{m,n} = X_{m+1} + \cdots + X_n$. (i) Show that

$$P\left(\max_{m < j \le n} |S_{m,j}| > 2a\right) \min_{m < k \le n} P(|S_{k,n}| \le a) \le P(|S_{m,n}| > a).$$

(ii) Use this to prove a theorem of P. Lévy: If $\lim_{N \to \infty} \sum_{n \le N} X_n$ exists in probability, then it also exists a.s.

Exercise 7.8   Use the inequality in (i) of the last exercise to conclude that if $S_n/n \to 0$ in probability, then $(\max_{1 \le m \le n} S_m)/n \to 0$ in probability.

Exercise 7.9   Let $X_1, X_2, \ldots$ be i.i.d. and $S_n = X_1 + \cdots + X_n$. (i) Show that if $S_n/n \to 0$ in probability and $S_{2^n}/2^n \to 0$ a.s., then $S_n/n \to 0$ a.s. (ii) Use Chebyshev's inequality to conclude that if we assume in addition that $EX_1 = 0$ and $EX_1^2 < \infty$, then $S_n/n \to 0$ a.s.

8 The Strong Law of Large Numbers

The link between the strong law of large numbers and convergence of series is provided by:

(8.1) _Kronecker's Lemma_   If $a_n \uparrow \infty$ and $\sum_{n \ge 1} x_n/a_n$ converges (i.e., $\lim_{N \to \infty} \sum_{n \le N} x_n/a_n$ exists), then $a_n^{-1} \sum_{m \le n} x_m \to 0$.

Proof   Let $a_0 = 0$, $b_0 = 0$, and for $n \ge 1$, let

$$b_n = \sum_{j=1}^{n} x_j/a_j.$$

Then $x_n = a_n(b_n - b_{n-1})$ and so

$$a_n^{-1} \sum_{m=1}^{n} x_m = a_n^{-1}\left\{\sum_{m=1}^{n} a_m b_m - \sum_{m=1}^{n} a_m b_{m-1}\right\} = b_n - \sum_{m=1}^{n} \frac{(a_m - a_{m-1})}{a_n} b_{m-1}.$$

(Recall $a_0 = 0$.) By hypothesis $b_n \to b_\infty$ as $n \to \infty$. Since $a_m - a_{m-1} \ge 0$, the last sum is an average of $b_0, \ldots, b_n$. If $\varepsilon > 0$ and $M < \infty$ are fixed and n is large, the average assigns mass $\ge 1 - \varepsilon$ to the b_m with $m \ge M$, so

$$\sum_{m=1}^{n} \frac{(a_m - a_{m-1})}{a_n} b_{m-1} \to b_\infty.$$

More formally, let $B = \sup |b_n|$, pick M so that $|b_m - b_\infty| < \varepsilon/2$ for $m \ge M$, and then pick N so that $a_M/a_n < \varepsilon/4B$ for $n \ge N$. With this done, if $n \ge N$, we have

$$\left| \sum_{m=1}^{n} \frac{(a_m - a_{m-1})}{a_n} b_{m-1} - b_\infty \right| \leq \sum_{m=1}^{n} \frac{(a_m - a_{m-1})}{a_n} |b_{m-1} - b_\infty| \leq \frac{a_M}{a_n} 2B + \varepsilon/2 < \varepsilon,$$

proving the desired result since ε is arbitrary.   □

Combining (8.1) with (7.3) gives:

(8.2) THEOREM   Let $Z_1, Z_2, \ldots$ be independent with $EZ_i = 0$, and let $T_n = Z_1 + \cdots + Z_n$. If a_n increases to ∞ and $\sum EZ_n^2/a_n^2 < \infty$, then $T_n/a_n \to 0$ a.s.

Exercise 8.1   Let $\sigma_n^2 \geq 0$ have $\sum \sigma_n^2/n^2 = \infty$, and without loss of generality $\sigma_n^2 \leq n^2$ for all n. Show that there are independent random variables X_n with $EX_n = 0$ and $\text{var}(X_n) \leq \sigma_n^2$ so that X_n/n and hence $n^{-1} \sum_{m \leq n} X_m$ does not converge to 0 a.s.

If $X_1, X_2, \ldots$ are i.i.d. with $EX_i = \mu$ and we apply (8.2) with $Z_i = X_i - \mu$ and $a_n = n$, we get the strong law of large numbers for variables with $E(X_i^2) < \infty$. If we replace $a_n = n$ by $a_n = n^{1/2}(\log n)^{1/2+\varepsilon}$, we get

$$(S_n - n\mu)/n^{1/2}(\log n)^{1/2+\varepsilon} \to 0 \quad \text{a.s.}$$

where $S_n = X_1 + \cdots + X_n$. (9.6) in Chapter 7 will show that

$$\limsup_{n \to \infty} (S_n - n\mu)/n^{1/2}(\log \log n)^{1/2} = c > 0 \quad \text{a.s.,}$$

so the last result is not far from the best possible.

We come now to:

(8.3) *The Strong Law of Large Numbers*   Let $X_1, X_2, \ldots$ be i.i.d. with $E|X_i| < \infty$. Let $EX_i = \mu$ and $S_n = X_1 + \cdots + X_n$. Then $S_n/n \to \mu$ a.s. as $n \to \infty$.

Proof   As in the proof of (5.5) we truncate and then use the proof for variables with finite variance. Let $Y_k = X_k 1_{(|X_k| \leq k)}$.

$$\sum_{k=1}^{\infty} P(|X_k| > k) \leq \int_0^\infty P(|X_1| > t)\, dt = E|X_1| < \infty,$$

so $P(X_k \neq Y_k \text{ i.o.}) = 0$. Let $Z_k = Y_k - EY_k$, so that we can apply (8.2) if we can estimate

$$\sum_{k=1}^{\infty} \text{var}(Z_k)/k^2 \leq \sum_{k=1}^{\infty} E(Y_k^2)/k^2.$$

To bound the right-hand side, we observe

$$E(Y_k^2) = \int_0^\infty 2y P(|Y_k| \geq y)\, dy \leq \int_0^k 2y P(|X_1| \geq y)\, dy,$$

so

$$\sum_{k=1}^{\infty} E(Y_k^2)/k^2 \leq \sum_{k=1}^{\infty} k^{-2} \int_0^k 2yP(|X_1| \geq y)\, dy$$

$$= \int_0^{\infty} \left\{ \sum_{k=1}^{\infty} k^{-2} 1_{(y \leq k)} \right\} 2yP(|X_1| \geq y)\, dy$$

by Fubini's theorem (since everything is ≥ 0). To estimate this we observe that

$$\sum_{k \geq m} k^{-2} \leq \int_{m-1}^{\infty} x^{-2}\, dx = (m-1)^{-1}$$

for $m \geq 2$, and when $m = 1$ the sum is bounded by

$$1 + \sum_{k \geq 2} k^{-2} \leq 1 + 1 = 2.$$

Considering $y > 1$ and $0 < y \leq 1$ separately, we see that $2y \sum_{k \leq y} k^{-2} \leq 4$, so

$$\sum_{k=1}^{\infty} E(Y_k^2)/k^2 \leq 4 \int_0^{\infty} P(|X_1| \geq y)\, dy < \infty.$$

Applying (8.2) with $Z_k = Y_k - EY_k$ and $a_k = k$ gives

$$n^{-1} \sum_{k=1}^{n} (Y_k - EY_k) \to 0 \quad \text{a.s.}$$

The dominated convergence theorem implies $EY_k \to \mu$ as $k \to \infty$, so

$$\left| \left\{ n^{-1} \sum_{k=1}^{n} Y_k \right\} - \mu \right| \leq \left| n^{-1} \sum_{k=1}^{n} (Y_k - EY_k) \right| + n^{-1} \sum_{k=1}^{n} |EY_k - \mu| \to 0.$$

Since $P(X_k \neq Y_k \text{ i.o.}) = 0$, it follows that

$$n^{-1} \sum_{k=1}^{n} X_k \to \mu \quad \text{a.s.},$$

completing the proof. $\qquad\qquad\qquad\qquad\qquad\qquad\qquad\qquad\qquad\qquad\qquad$ □

In the discussion after (8.2), we considered the rate at which $|S_n/n| \to 0$ when $EX_i = 0$ and $EX_i^2 < \infty$. The next two exercises, due to Marcinkiewicz and Zygmund, investigate the situation when $E|X_i|^p < \infty$ with $1 < p < 2$. Note that the result from Chapter 7 quoted above shows that the first conclusion is false when $p = 2$.

Exercise 8.2   Let X_1, X_2, ... be i.i.d. with $EX_1 = 0$ and $E|X_1|^p < \infty$ where $1 < p < 2$. If $S_n = X_1 + \cdots + X_n$, then $S_n/n^{1/p} \to 0$ a.s.
Hint   Truncate at $n^{1/p}$. To simplify the problem you may suppose that the distribution of X_i is symmetric about 0; that is, $X_i \stackrel{d}{=} -X_i$.

Exercise 8.3   The converse of the last exercise is much easier. Let $p \in (0, 2)$. If there are constants c_n so that $(S_n - c_n)/n^{1/p} \to 0$ a.s., then $E|X_1|^p < \infty$.

Hint   In Exercise 6.12, we considered the special case $p = 1$.

The strong law generalizes easily to cover one situation in which $E|X_i| = \infty$.

Exercise 8.4   Let $X_1, X_2, \ldots$ be i.i.d. with $EX_i^+ = \infty$ and $EX_i^- < \infty$. Then $S_n/n \to \infty$ a.s.

The next result, due to Feller (1946), shows that when $E|X_1| = \infty$, S_n/a_n cannot converge to a nonzero limit.

Exercise 8.5   Let $X_1, X_2, \ldots$ be i.i.d. with $E|X_1| = \infty$ and let $S_n = X_1 + \cdots + X_n$. Let a_n be a sequence of positive numbers with a_n/n increasing. Then $\limsup_{n\to\infty} |S_n|/a_n = 0$ or ∞ according as $\sum_n P(|X_1| \geq a_n) < \infty$ or $= \infty$.

Sketch   Since $a_n/n \uparrow$, $a_{kn} \geq ka_n$ for any integer k and hence

$$\sum_{n=1}^{\infty} P(|X_1| \geq ka_n) \geq \sum_{n=1}^{\infty} P(|X_1| \geq a_{kn}).$$

The last observation shows that if the sum is infinite, $\limsup_{n\to\infty} |X_n|/a_n = \infty$. To prove the other half, observe that

$$\sum_{n=1}^{\infty} P(|X_1| \geq a_n) = \sum_{n=1}^{\infty} nP(a_n \leq |X_1| < a_{n+1}).$$

Let $Y_n = X_n 1_{(|X_n| \leq a_n)}$, $T_n = Y_1 + \cdots + Y_n$, and estimate the second moment of Y_n/a_n as in (8.3), to conclude $(T_n - ET_n)/a_n \to 0$ a.s. The last step is to show $ET_n/a_n \to 0$. To do this, it is useful to notice that the sum being $< \infty$ and $E|X_1| = \infty$ imply $a_n/n \to \infty$.

Etemadi's proof of the strong law of large numbers   Our next topic is a proof of the strong law of large numbers that not only is more direct since it does not use the results on convergence of random series, but is also more applicable because we only require that the random variables are pairwise independent. We have not thrown away the old proof because some of the ingredients in the old proof are quite important (e.g., (7.1) and (7.2)) and the new proof does not give rates of convergence of $(S_n/n - \mu) \to 0$.

(8.4) THEOREM   Let X_1, X_2, ... be pairwise independent and identically distributed random variables (r.v.'s) with $E|X_i| < \infty$, and let $S_n = X_1 + \cdots + X_n$. As $n \to \infty$, $S_n/n \to EX_1$ a.s.

Proof   Since $X_n^+, n \geq 1$, and $X_n^-, n \geq 1$, satisfy the assumptions of the theorem and $X_n = X_n^+ - X_n^-$, we can without loss of generality suppose $X_n \geq 0$. As in the proof of (8.3), we begin by truncating $Y_n = X_n 1_{\{X_n \leq n\}}$ and observing $P(Y_n \neq X_n \text{ i.o.}) = 0$, so it suffices to show that $T_n = Y_1 + \cdots + Y_n$ has $T_n/n \to EX_1$ a.s. The next step is completely different. Let $\varepsilon > 0$, $\alpha > 1$, and $k(n) = [\alpha^n]$. Chebyshev's inequality implies

$$\sum_{n=1}^{\infty} P(|T_{k(n)} - ET_{k(n)}| > \varepsilon k(n)) \leq \varepsilon^{-2} \sum_{n=1}^{\infty} \text{var}(T_{k(n)})/k(n)^2$$

$$= \varepsilon^{-2} \sum_{n=1}^{\infty} k(n)^{-2} \sum_{m=1}^{k(n)} \text{var}(Y_m)$$

$$= \varepsilon^{-2} \sum_{m=1}^{\infty} \text{var}(Y_m) \sum_{n\,:\,k(n) \geq m} k(n)^{-2}$$

where $n : k(n) \geq m$ indicates that the sum is over n with $k(n) \geq m$, and we have used Fubini's theorem to interchange the two summations (everything is ≥ 0). Now $k(n) = [\alpha^n]$ and $[\alpha^n] \geq \alpha^n/2$ for $n \geq 1$, so

$$\sum_{n\,:\,\alpha^n \geq m} [\alpha^n]^{-2} \leq 4 \sum_{n\,:\,\alpha^n \geq m} \alpha^{-2n} \leq 4(1 - \alpha^{-2})^{-1} m^{-2},$$

the last inequality coming from summing the geometric series. Now

$$\sum_{n=1}^{\infty} P(|T_{k(n)} - ET_{k(n)}| > \varepsilon k(n)) \leq 4(1 - \alpha^{-2})^{-1} \varepsilon^{-2} \sum_{m=1}^{\infty} E(Y_m^2) m^{-2} < \infty$$

by the computation in the proof of (8.3), and $EY_k \to EX_1$ as $k \to \infty$, so $ET_{k(n)}/k(n) \to EX_1$ and we have shown $T_{k(n)}/k(n) \to EX_1$ a.s. To handle the intermediate values, we observe that if $k(n) \leq m < k(n + 1)$,

$$T_{k(n)}/k(n + 1) \leq T_m/m \leq T_{k(n+1)}/k(n)$$

(here we use $Y_i \geq 0$), so recalling the definition of $k(n)$, we have

$$\frac{1}{\alpha} EX_1 \leq \liminf_{m \to \infty} T_m/m < \limsup_{m \to \infty} T_m/m \leq \alpha EX_1.$$

Since $\alpha > 1$ is arbitrary, the proof is complete. $\square$

The rest of this section is devoted to two applications of the strong law of large numbers. The first is renewal theory, a subject that we will consider in detail in Section 4 of Chapter 3. Let $X_1, X_2, \ldots$ be i.i.d. with $X_i > 0$. Let $T_n = X_1 + \cdots + X_n$ and think of T_n as the time of the nth occurrence of some event. For a concrete situation consider a diligent janitor who replaces a light bulb the instant it burns out. Suppose the first bulb is put in at time 0 and let X_i be the lifetime of the ith lightbulb. In this interpretation T_n is the time the nth light bulb burns out, and $N_t = \sup\{n: T_n \le t\}$ is the number of light bulbs that have burned out by time t.

(8.5) THEOREM   If $EX_1 = \mu \le \infty$, then as $t \to \infty$, $N_t/t \to 1/\mu$ a.s. $(1/\infty = 0)$.

Proof   By (8.3) (and Exercise 8.4 for the case $\mu = \infty$), $T_n/n \to \mu$ a.s. From the definition of N_t it follows that $T(N_t) \le t < T(N_t + 1)$, so dividing through by N_t gives

$$\frac{T(N_t)}{N_t} \le \frac{t}{N_t} \le \frac{T(N_t + 1)}{N_t + 1} \cdot \frac{N_t + 1}{N_t}.$$

Since $N_t \uparrow \infty$ as $t \uparrow \infty$, the left- and right-hand sides converge to μ a.s. as $t \to \infty$, and it follows that $t/N_t \to \mu$ a.s.   □

Exercise 8.6   (i) Show that if $X_n \to X$ a.s. and $N(n) \to \infty$ a.s., then $X_{N(n)} \to X_\infty$ a.s. (ii) Give an example with $X_n \to 0$ in probability, $N(n) \uparrow \infty$ a.s., and $X_{N(n)} \to 1$ a.s.

Our next topic is the convergence of the *empirical distribution function*. Let $X_1, X_2, \ldots$ be i.i.d. with distribution F and let

$$F_n(x) = n^{-1} \sum_{m=1}^{n} 1_{(X_m \le x)}.$$

$F_n(x) = $ the observed frequency of values that are $\le x$—hence the name given above. The next result shows that F_n converges uniformly to F as $n \to \infty$.

(8.6) *The Glivenko-Cantelli Theorem*   As $n \to \infty$, $\sup_x |F_n(x) - F(x)| \to 0$ a.s.

Proof   Suppose first that F is continuous. For $1 \le j \le k - 1$, let $x_{j,k} = \inf\{y: F(y) \ge j/k\}$. The strong law of large numbers implies that for each j, $F_n(x_{j,k}) \to F(x_{j,k})$, so if $n \ge N_k(\omega)$, then $|F_n(x_{j,k}) - F(x_{j,k})| < k^{-1}$ for $1 \le j \le k - 1$. Let $x_{0,k} = -\infty$ and $x_{k,k} = \infty$ and observe the last inequality is trivial if $j = 0$ or k. If $x \in (x_{j-1,k}, x_{j,k})$ with $1 \le j \le k$ and $n \ge N_k(\omega)$, then

$$F_n(x) \le F_n(x_{j,k}) \le F(x_{j,k}) + k^{-1} = F(x_{j-1,k}) + 2k^{-1} \le F(x) + 2k^{-1}$$

$$F_n(x) \ge F_n(x_{j-1,k}) \ge F(x_{j-1,k}) - k^{-1} = F(x_{j,k}) - 2k^{-1} \ge F(x) - 2k^{-1}$$

so $\sup_x |F_n(x) - F(x)| \le 2k^{-1}$, proving the result in the case F is continuous. To prove the result when there are jumps, we consider

$$F_n(x_{j,k}-) = \lim_{y \uparrow x_{j,k}} F_n(y) = n^{-1} |\{m \le n : X_m < x_{j,k}\}|.$$

The strong law of large numbers implies that if $1 \le j \le k - 1$, then $F_n(x_{j,k}-) \to$

$F(x_{j,k}-)$ (and the result is trivial if $j = k$ since $x_{j,k}- = \infty$), so by picking $N_k(\omega)$ larger, we can assume that $|F_n(x_{j,k}-) \to F(x_{j,k}-)| < k^{-1}$ for $1 \leq j \leq k$. Repeating the argument above, we see that if $x \in (x_{j-1,k}, x_{j,k})$ with $1 \leq j \leq k$ and $n \geq N_k(\omega)$, then

$$F_n(x) \leq F_n(x_{j,k}-) \leq F(x_{j,k}-) + k^{-1} \leq F(x_{j-1,k}) + 2k^{-1} \leq F(x) + 2k^{-1}$$

$$F_n(x) \geq F_n(x_{j-1,k}) \geq F(x_{j-1,k}) - k^{-1} \geq F(x_{j,k}-) - 2k^{-1} \geq F(x) - 2k^{-1}$$

so $\sup_x |F_n(x) - F(x)| \leq 2k^{-1}$, and we have proved the result in general.   □

The fact that $\log(\prod_{m \leq n} x_m) = \sum_{m \leq n} \log x_m$ can be used to reduce problems about random products to problems about random sums.

Exercise 8.7   Let $X_0 = 1$ and define X_n inductively by declaring that X_{n+1} is uniformly distributed over $(0, X_n)$. Prove that $n^{-1} \log X_n \to c$ a.s. and compute c.

Exercise 8.8   *Shannon's theorem.* Let $X_1, X_2, \ldots \in \{1, \ldots, r\}$ be independent with $P(X_i = k) = p(k) > 0$ for $1 \leq k \leq r$. Let $\pi_n(i_1, \ldots, i_n) = p(i_1) \cdots p(i_n)$, and $\pi_n(\omega) = \pi_n(X_1(\omega), \ldots, X_n(\omega))$ is the probability that someone else would see what we observed in the first n trials. Show that

$$-n^{-1} \log \pi_n(\omega) \to H \equiv -\sum_{k=1}^{r} p(k) \log p(k)   \text{a.s.}$$

Remark   In information theory $1, \ldots, r$ are interpreted as the letters of an alphabet; $X_1, X_2, \ldots$ are the successive letters produced by an information source; and H is the *entropy* of the source. The last result is the *asymptotic equipartition property*: If $\varepsilon > 0$, then for large n there is probability $> 1 - \varepsilon$ that $\pi_n(\omega)$ lies in the range $\exp(-n(H \pm \varepsilon))$. We will give a more general version of this result in Chapter 6. See (4.1).

Exercise 8.9   *An Investment Problem.* We assume that at the beginning of each year you can buy bonds for \$1 that are worth \$$a$ at the end of the year or stocks that are worth a random amount $V \geq 0$. If you always invest a fixed proportion p of your wealth in bonds, then your wealth at the end of year $n + 1$ is $W_{n+1} = (ap + (1 - p)V_n)W_n$. Suppose $V_1, V_2, \ldots$ are i.i.d. with $EV_n < \infty$ and $E(V_n^{-2}) < \infty$. (i) Show that $n^{-1} \log W_n \to c(p)$ a.s. (ii) $c(p)$ is concave. By investigating $c'(0)$ and $c'(1)$, give conditions on V that guarantee that the optimal choice of p is in $(0, 1)$. (iii) Suppose $P(V = 1) = P(V = 4) = 1/2$. Find the optimal p as a function of a.

*9   Large Deviations

Let $X_1, X_2, \ldots$ be i.i.d. and let $S_n = X_1 + \cdots + X_n$. In this section we will investigate the rate at which $P(S_n > na) \to 0$ for $a > \mu = EX_i$. We will ultimately conclude that if the *moment generating function* $\varphi(\theta) = E \exp(\theta X_i) < \infty$ for some $\theta > 0$, $P(S_n \geq na) \to 0$ exponentially rapidly, and we will identify

$$\gamma(a) = \lim_{n \to \infty} \frac{1}{n} \log P(S_n \geq na).$$

Our first step is to prove that the limit exists. This is based on an observation that will be useful several times below. Let $\pi_n = P(S_n \geq na)$.

$$\pi_{m+n} \geq P(S_m \geq ma, S_{n+m} - S_m \geq na) = \pi_m \pi_n,$$

since S_m and $S_{n+m} - S_m$ are independent. Letting $\gamma_n = (1/n) \log \pi_n$ transforms multiplication into addition.

(9.1) LEMMA   If $\gamma_{m+n} \geq \gamma_m + \gamma_n$, then as $n \to \infty$ $\gamma_n/n \to \sup_{m \geq 1} \gamma_m/m$.

Proof   Clearly the $\limsup \gamma_n/n \leq \sup \gamma_m/m$. To complete the proof, it suffices to prove that for any m, $\liminf \gamma_n/n \geq \gamma_m/m$. Writing $n = km + l$ with $0 \leq l < m$ and making repeated use of the hypothesis give $\gamma_n \geq k\gamma_m + \gamma_l$. Dividing by $n = km + l$ gives

$$\frac{\gamma(n)}{n} \geq \left(\frac{km}{km + l} \right) \frac{\gamma(m)}{m} + \frac{\gamma(l)}{n}.$$

Letting $n \to \infty$ and recalling $n = km + l$ with $0 \leq l < m$ give the desired result.   □

(9.1) implies that

$$\lim_{n \to \infty} \frac{1}{n} \log P(S_n \geq na) = \gamma(a) \text{ exists } \leq 0.$$

It follows from the formula for the limit that

$$P(S_n \geq na) \leq e^{n\gamma(a)}. \tag{9.2}$$

The last two observations give us some useful information about γ.

Exercise 9.1   The following are equivalent: (i) $\gamma(a) = -\infty$, (ii) $P(X_1 \geq a) = 0$, and (iii) $P(S_n \geq na) = 0$ for all n.

Exercise 9.2 $\gamma((a + b)/2) \geq (\gamma(a) + \gamma(b))/2$ and hence γ is concave.

The conclusions above are valid for any distribution. For the rest of this section we will suppose:

$$\varphi(\theta) = E \exp(\theta X_i) < \infty \quad \text{for some } \theta > 0. \tag{H1}$$

Clearly (H1) implies that $EX_i^+ < \infty$, so $\mu = EX^+ - EX^- \in [-\infty, \infty)$. If $\theta > 0$, Chebyshev's inequality implies

$$e^{\theta na} P(S_n \geq na) \leq E \exp(\theta S_n) = \varphi(\theta)^n,$$

or letting $\kappa(\theta) = \log \varphi(\theta)$,

$$P(S_n \geq na) \leq \exp(-n(a\theta - \kappa(\theta))). \tag{9.3}$$

Our first goal is to show:

(9.4) LEMMA   If $a > \mu$ and θ is small, $a\theta - \kappa(\theta) > 0$.

Proof $\kappa(0) = \log \varphi(0) = 0$, so it suffices to show that $\kappa'(\theta) = \varphi'(\theta)/\varphi(\theta) \to \mu$ as $\theta \to 0$. The first step is to show that the derivatives exist. Let $F(x) = P(X_i \leq x)$. Since

$$|e^{hx} - 1| = \left| \int_0^{hx} e^y \, dy \right| \leq |hx|(e^{hx} + 1),$$

an application of the dominated convergence theorem shows that

$$\varphi'(\theta) = \int x e^{\theta x} \, dF(x) \quad \text{for } \theta \in (0, \theta_0).$$

Letting $\theta \downarrow 0$ using the monotone convergence theorem for $x < 0$ and the dominated convergence theorem for $x > 0$ shows that $\varphi'(\theta) \to \mu$ as $\theta \to 0$. Since $\varphi(\theta) \to 1$ as $\theta \to 0$, we have shown $\kappa'(\theta) \to \mu$ and proved (9.4). $\square$

Having found an upper bound on $P(S_n \geq na)$, it is natural to optimize it by finding the minimum of $-\theta a + \kappa(\theta)$.

$$\frac{d}{d\theta} \{\theta a - \log \varphi(\theta)\} = a - \varphi'(\theta)/\varphi(\theta),$$

so (assuming things are nice) the maximum occurs when $a = \varphi'(\theta)/\varphi(\theta)$. To turn the parenthetical clause into a mathematical hypothesis, we begin by defining

$$F_\theta(x) = \frac{1}{\varphi(\theta)} \int_{-\infty}^x e^{\theta y} \, dF(y).$$

F_θ is a distribution function with mean

$$\int x \, dF_\theta(x) = \frac{1}{\varphi(\theta)} \int_{-\infty}^\infty x e^{\theta x} \, dF(x) = \varphi'(\theta)/\varphi(\theta).$$

Differentiating again,

$$\frac{d}{d\theta} \frac{\varphi'(\theta)}{\varphi(\theta)} = \frac{\varphi''(\theta)}{\varphi(\theta)} - \left(\frac{\varphi'(\theta)}{\varphi(\theta)}\right)^2 = \int x^2 \, dF_\theta(x) - \left(\int x \, dF_\theta(x)\right)^2 \geq 0,$$

so if we assume

the distribution F is not a point mass at μ,   (H2)

$\varphi'(\theta)/\varphi(\theta)$ is strictly increasing. Since $\varphi'(0)/\varphi(0) = \mu$, this shows that for each $a > \mu$ there is at most one $\theta_a \geq 0$ that solves $a = \varphi'(\theta_a)/\varphi(\theta_a)$ and this value of θ maximizes the concave function $a\theta - \log \varphi(\theta)$. Before discussing the existence of θ_a, we will consider some examples:

Example 9.1   *The normal distribution.*

$$\int e^{\theta x}(2\pi)^{-1/2} \exp(-x^2/2) \, dx = \exp(\theta^2/2) \int (2\pi)^{-1/2} \exp(-(x-\theta)^2/2) \, dx.$$

The integrand in the last integral is the density of a normal distribution with mean θ and variance 1, so $\varphi(\theta) = \exp(\theta^2/2)\theta \in (-\infty, \infty)$. In this case $\varphi'(\theta)/\varphi(\theta) = \theta$ and F_θ is a normal distribution with mean θ and variance 1.

Example 9.2   *Exponential distribution with parameter λ. If $\theta < \lambda$,*

$$\int_0^\infty e^{\theta x} \lambda e^{-\lambda x} \, dx = \lambda/(\lambda - \theta).$$

$\varphi'(\theta)/\varphi(\theta) = 1/(\lambda - \theta)$ and F_θ is an exponential distribution with parameter $\lambda - \theta$.

Example 9.3   *Coin flips.* $P(X_i = 1) = P(X_i = -1) = 1/2$,

$$\varphi(\theta) = (e^\theta + e^{-\theta})/2$$

$$\varphi'(\theta)/\varphi(\theta) = (e^\theta - e^{-\theta})/(e^\theta + e^{-\theta}) \to 1.$$

$F_\theta(\{1\}) = e^\theta/(e^\theta + e^{-\theta})$ and $F_\theta(\{-1\}) = e^{-\theta}/(e^\theta + e^{-\theta})$. As $\theta \to \infty$, all the mass moves to 1.

Example 9.4   *Perverted exponential.* Let $g(x) = Cx^{-3}e^{-x}$ for $x \geq 1$, $g(x) = 0$ otherwise, and choose C so that g is a probability density. In this case

$$\varphi(\theta) = \int e^{\theta x} g(x) \, dx < \infty \quad \text{if and only if } \theta \leq 1.$$

$$\varphi'(\theta)/\varphi(\theta) \leq \varphi'(1)/\varphi(1) = \int_1^\infty Cx^{-2} \, dx \bigg/ \int_1^\infty Cx^{-3} \, dx = 2.$$

Let $x_0 = \sup\{x: F(x) < 1\}$ and $\theta_1 = \sup\{\theta: \varphi(\theta) < \infty\}$. Examples 9.1–9.3 are nice in the sense that as θ increases to θ_1, $\varphi'(\theta)/\varphi(\theta)$ increases to x_0. Example 9.4 does not have this property and we cannot solve $a = \varphi'(\theta)/\varphi(\theta)$ when $a > 2$.

(9.5) THEOREM Let $\mu < a < x_0$. Suppose in addition to (H1) and (H2) that we have:

there is a $\theta_a \in (0, \theta_1)$ so that $a = \varphi'(\theta_a)/\varphi(\theta_a)$.   (H3)

Then as $n \to \infty$,

$$n^{-1} \log P(S_n \geq na) \to -a\theta_a + \log \varphi(\theta_a).$$

Remark We do not have to worry about the case $a > x_0$ because then $P(S_n \geq na) = 0$. The case $a = x_0$ will be dealt with in Exercise 9.6. (9.6) will deal with the values

$$a \geq \lim_{\theta \uparrow \theta_1} \varphi'(\theta)/\varphi(\theta).$$

Proof The fact that limsup left-hand side $\leq$ right-hand side follows from (9.3). To prove the other inequality, pick $\lambda \in (\theta_a, \theta_1)$, let $X_1^\lambda, X_2^\lambda \ldots$ be i.i.d. with distribution F_λ, and let $S_n^\lambda = X_1^\lambda + \cdots + X_n^\lambda$. Writing dF/dF_λ for the Radon–Nikodym derivative of the associated measures, it is immediate from the definition that $dF/dF_\lambda = e^{-\lambda x}\varphi(\lambda)$. If we let F_λ^n and F^n denote the distributions of S_n^λ and S_n, a little thought reveals that $dF^n/dF_\lambda^n = e^{-\lambda x}\varphi(\lambda)^n$. To check this for $n = 2$ we observe

$$F_\lambda * F_\lambda(z) = \int_{-\infty}^\infty dF_\lambda(x) \int_{-\infty}^{z-x} dF_\lambda(x) = \int_{-\infty}^\infty dF(x) \int_{-\infty}^{z-x} dF(x) e^{-\lambda(x+y)}\varphi(\lambda)^2$$

$$= \int_{-\infty}^z d(F * F)(u)e^{-\lambda u}\varphi(\lambda)^2.$$

One way to justify the last equality is to break the integral over $x + y \leq z$ into strips $a < x + y \leq b$ and use the fact that in the strip,

$$e^{-\lambda a} < e^{-\lambda(x+y)} \leq e^{-\lambda b}.$$

Let $v > \varphi'(\lambda)/\varphi(\lambda)$, which is $> a$. Now

$$P(S_n \geq na) = \int_{na}^\infty e^{-\lambda x}\varphi(\lambda)^n \, dF_\lambda^n(x) \geq \varphi(\lambda)^n e^{-\lambda nv}(F_n^\lambda(nv) - F_n^\lambda(na)).$$

F_n^λ has mean $\lambda \in (a, v)$, so the weak law of large numbers implies

$$F_n^\lambda(nv) - F_n^\lambda(na) \to 1 \quad \text{as } n \to \infty.$$

From the last conclusion it follows that

$$\liminf_{n\to\infty} n^{-1} \log P(S_n > na) \geq -\lambda v + \log \varphi(\lambda).$$

Since $\lambda > \theta_a$ and $v > \varphi'(\lambda)/\varphi(\lambda)$ are arbitrary, the proof is complete.   □

To get a feel for what the answers look like, we consider our examples. First, the normal distribution (Example 9.1):

$$\kappa(\theta) = \theta^2/2, \quad \varphi'(\theta)/\varphi(\theta) = \theta, \quad \theta_a = a$$

$$\gamma(a) = -a\theta_a + \kappa(\theta_a) = -a^2/2.$$

Exercise 9.3   Check the last result by observing that S_n has a normal distribution with mean 0 and variance n, and using (1.3).

Turning to the exponential distribution with $\lambda = 1$ (Example 9.2),

$$\kappa(\theta) = \log(1 - \theta), \quad \varphi'(\theta)/\varphi(\theta) = 1/(1 - \theta), \quad \theta_a = 1 - 1/a$$

$$\gamma(a) = -a\theta_a + \kappa(\theta_a) = -a + 1 - \log a.$$

Finally we look at coin flips (Example 9.3). In Exercise 1.3 of Chapter 2, we will show by direct computation that if $c \in [0, 1]$, then

$$\frac{1}{n} \log P(S_n \geq cn) \to -\gamma(c)$$

where

$$\gamma(c) = \frac{1}{2}\{(1 + c) \log (1 + c) + (1 - c) \log (1 - c)\}.$$

Exercise 9.4   Derive the last result by solving $(1 + c)/2 = e^\theta/(e^\theta + e^{-\theta})$.

Hint   Let $x = e^\theta$ to get a quadratic equation.

Exercise 9.5   Show that $P(S_n \geq an) \leq \exp(-na^2/4)$ for all $a \in [0, 1]$.

Sketch $P(S_n \geq na) \leq \inf_{\theta > 0} e^{-\theta na} \varphi(\theta)^n$ and $\varphi(\theta) \leq \exp(\varphi(\theta) - 1) \leq \exp(\theta^2)$ for $\theta \leq 1$.

Exercise 9.6   Suppose $x_0 = \sup\{x: F(x) < 1\} < \infty$. As $a \uparrow x_0, \gamma(a) \downarrow \log P(X_i = x_0)$.

Exercise 9.7   Let $X_1, X_2, \ldots$ be i.i.d. Poisson with mean 1, and let $S_n = X_1 + \cdots + X_n$. Find $\lim_{n\to\infty} \frac{1}{n} \log P(S_n \geq na)$ for $a > 1$.

To finish off Example 9.4, we will use:

(9.6) THEOREM   Let $\theta_1 = \sup\{\theta: \varphi(\theta) < \infty\}$ and suppose $\varphi'(\theta)/\varphi(\theta)$ increases to a finite limit x_1 as $\theta \uparrow \theta_1$. If $x_1 \le a < \infty$,

$$n^{-1} \log P(S_n \ge na) \to -a\theta_1 + \log \varphi(\theta_1).$$

That is, $\gamma(a)$ is linear for $a \ge x_1$.

Proof   Since $(\log \varphi(\theta))' = \varphi'(\theta)/\varphi(\theta)$, $\varphi(\theta_1) < \infty$. Letting $\theta = \theta_1$ in (9.3) shows that the limsup left-hand side $\le$ right-hand side. To get the other direction we will use the transformed distribution F_λ for $\lambda = \theta_1$. The monotone and dominated convergence theorems imply that G has mean x_1. By the reasoning that led to (9.5), if $x_1 \le a < v = a + 3\varepsilon$,

$$P(S_n \ge na) \ge \varphi(\lambda)^n e^{-n\lambda v}(F_\lambda^n(nv) - F_\lambda^n(na)).$$

Letting $X_1^\lambda, X_2^\lambda, \ldots$ be i.i.d with distribution F_λ and $S_n^\lambda = X_1^\lambda + \cdots + X_n^\lambda$, we have

$$P(S_n^\lambda \in (na, nv]) \ge P\{S_{n-1}^\lambda \in ((x_1 - \varepsilon)n, (x_1 + \varepsilon)n]\}$$

$$\cdot P\{X_n^\lambda \in ((a - x_1 + \varepsilon)n, (a - x_1 + 2\varepsilon)n]\}$$

$$\ge \frac{2}{3} P\{X_n^\lambda \in ((a - x_1 + \varepsilon)n, (a - x_1 + \varepsilon)n + 1]\}$$

for large n by the weak law of large numbers. To finish the argument now we observe that

$$\limsup_{t \to \infty} \frac{1}{t} \log P(X_1^\lambda \in (t, t + 1]) = 0,$$

for if the limsup was <0, we would have $E \exp(\varepsilon X_1^\lambda) < \infty$ for some $\lambda > 0$, contradicting the definition of θ_1. $\square$

By adapting the proof of the last result, you can show that (H1) is necessary for exponential convergence:

Exercise 9.8   Suppose $EX_i = 0$ and $E \exp(\varepsilon X_i) = \infty$ for all $\varepsilon > 0$. Then

$$\frac{1}{n} \log P(S_n \ge na) \to 0 \quad \text{for all } a > 0.$$

Exercise 9.9   Suppose $EX_i = 0$. Show that

$$\liminf_{n \to \infty} P(S_n \ge na)/nP(X_1 \ge na) \ge 1.$$

2 Central Limit Theorems

The first four sections of this chapter develop the central limit theorem. The last five treat various extensions and complements. We begin this chapter by going back to our gambling roots.

1 The DeMoivre Laplace Theorem

Let $X_1, X_2, \ldots$ be i.i.d. with $P(X_1 = 1) = P(X_1 = -1) = 1/2$ and let $S_n = X_1 + \cdots + X_n$. In words, we are betting \$1 on the flipping of a fair coin and S_n is our winnings at time n. If n and k are integers,

$$P(S_{2n} = 2k) = \binom{2n}{n+k} 2^{-2n},$$

since $S_{2n} = 2k$ if and only if there are $n + k$ +1's and $n - k$ −1's in the first $2n$ outcomes. The first factor gives the number of such outcomes and the second the probability of each one. *Stirling's formula* (see Feller, Vol. I (1968), p. 50) tells us

$$n! \sim n^n e^{-n} \sqrt{2\pi n} \quad \text{as } n \to \infty \tag{1.1}$$

where $a_n \sim b_n$ means $a_n/b_n \to 1$ as $n \to \infty$, so

$$\binom{2n}{n+k} = \frac{(2n)!}{(n+k)!(n-k)!}$$

$$\sim \frac{(2n)^{2n}}{(n+k)^{n+k}(n-k)^{n-k}} \cdot \frac{(2\pi(2n))^{1/2}}{(2\pi(n+k))^{1/2}(2\pi(n-k))^{1/2}}$$

and

$$\binom{2n}{n+k} 2^{-2n} \sim \left(1 + \frac{k}{n}\right)^{-n-k} \cdot \left(1 - \frac{k}{n}\right)^{-n+k} \cdot (\pi n)^{-1/2} \cdot \left(1 + \frac{k}{n}\right)^{-1/2} \cdot \left(1 - \frac{k}{n}\right)^{-1/2}.$$

The first two terms on the right are

$$= \left(1 - \frac{k^2}{n^2}\right)^{-n} \cdot \left(1 + \frac{k}{n}\right)^{-k} \cdot \left(1 - \frac{k}{n}\right)^{k}.$$

A little calculus shows (take logarithms or see (4.2) below) that:

(1.2)   If $j \to \infty$ and $jc_j \to \lambda$, then $(1 + c_j)^j \to e^\lambda$.   □

If $2k = x\sqrt{2n}$—that is, $k = x\sqrt{n/2}$, then

$$\left(1 - \frac{k^2}{n^2}\right)^{-n} = (1 + x^2/2n)^{-n} \to e^{x^2/2}$$

$$\left(1 + \frac{k}{n}\right)^{-k} = (1 + x/\sqrt{2n})^{-x\sqrt{n/2}} \to e^{-x^2/2}$$

$$\left(1 - \frac{k}{n}\right)^{k} = (1 - x/\sqrt{2n})^{x\sqrt{n/2}} \to e^{-x^2/2}.$$

For this choice of k, $k/n \to 0$, so putting things together gives:

(1.3)   If $2k/\sqrt{2n} \to x$, then $P(S_{2n} = 2k) \sim (\pi n)^{-1/2} e^{-x^2/2}$.

Our next step is to compute

$$P(a\sqrt{2n} \leq S_{2n} \leq b\sqrt{2n}) = \sum_{m \in [a\sqrt{2n}, b\sqrt{2n}] \cap 2\mathbb{Z}} P(S_{2n} = m).$$

Changing variables $m = x\sqrt{2n}$, we have that the above is

$$\approx \sum_{x \in [a,b] \cap (2\mathbb{Z}/\sqrt{2n})} (2\pi)^{-1/2} e^{-x^2/2} \cdot (2/n)^{1/2},$$

where $2\mathbb{Z}/\sqrt{2n} = \{2z/\sqrt{2n}; z \in \mathbb{Z}\}$. We have multiplied and divided by $\sqrt{2}$, since the space between points in the sum is $(2/n)^{1/2}$, so if n is large, the sum above is

$$\approx \int_a^b (2\pi)^{-1/2} e^{-x^2/2} dx.$$

The integrand is the density of the (standard) normal distribution, so changing notation, we can write the last quantity as $P(a \leq \chi \leq b)$ where χ is a random variable with that distribution.

It is not hard to get from the heuristic argument above to:

(1.4)   *The DeMoivre Laplace Theorem*   If $a < b$, then as $m \to \infty$,

$$P(a \le S_m/\sqrt{m} \le b) \to \int_a^b (2\pi)^{-1/2} e^{-x^2/2} dx.$$

(To remove the restriction to even integers observe $S_{2n+1} = S_{2n} \pm 1$.) The last result is a special case of the central limit theorem given in Section 4, so further details are left to the reader.

To motivate some of the developments in Section 3 that will lead to the proof of the central limit theorem, we will now present the proof that will be given in Section 4 for the special case under consideration. We begin by defining the *characteristic function* $\varphi(t) = E \exp(itX_1)$. When $P(X_1 = 1) = P(X_1 = -1) = 1/2$,

$$\varphi(t) = \tfrac{1}{2}(e^{it} + e^{-it}) = \cos t,$$

since $e^{it} = \cos t + i \sin t$. If $S_n = X_1 + \cdots + X_n$, its characteristic function is

$$E \exp(itS_n) = E \prod_{m=1}^n \exp(itX_m) = \prod_{m=1}^n E \exp(itX_m) = \varphi(t)^n. \tag{1.5}$$

When t is small,

$$\cos t = 1 - t^2/2 + O(t^4)$$

where $O(t^4)$ indicates a quantity $g(t)$ with $\limsup_{t \to 0} g(t)/t^4 < \infty$, so

$$E \exp(itS_n/\sqrt{n}) = \varphi(t/\sqrt{n})^n = \left(1 - \frac{t^2}{2n} + O(n^{-2})\right)^n \to e^{-t^2/2}.$$

At this point the reader can probably guess that

$$e^{-t^2/2} = \int e^{itx} (2\pi)^{-1/2} e^{-x^2/2} dx. \tag{1.6}$$

So we have shown that the characteristic function of $S_n/\sqrt{n}$ converges to that of χ. In Section 3 we will show that this implies the conclusion in (1.4).

To prepare for the proof of the central limit theorem in Section 4, we will now prove (1.6). Let

$$\varphi(t) = \int e^{itx} (2\pi)^{-1/2} e^{-x^2/2} dx = \int \cos tx (2\pi)^{-1/2} e^{-x^2/2} dx.$$

Differentiating with respect to t and then integrating by parts give

$$\varphi'(t) = \int -x \sin tx (2\pi)^{-1/2} e^{-x^2/2} dx$$

$$= -\int t \cos tx (2\pi)^{-1/2} e^{-x^2/2} dx = -t \, \varphi(t).$$

The last equality implies

$$\frac{d}{dt} \{\varphi(t) \exp(t^2/2)\} = 0.$$

Since $\varphi(0) = 1$, it follows that $\log \varphi(t) = -t^2/2$.

Computations with Stirling's formula can be used to prove special cases of other results that we will prove later.

Exercise 1.1   *Poisson convergence, Section 6.* Let $X_{n,1}, \ldots, X_{n,n}$ be independent with $P(X_{n,i} = 1) = \lambda/n$, $P(X_{n,i} = 0) = 1 - \lambda/n$, and let $S_n = X_{n,1} + \cdots + X_{n,n}$. Use Stirling's formula to show

$$P(S_n = k) \to e^{-\lambda} \lambda^k/k!.$$

Exercise 1.2   *Local central limit theorem, Section 5.* Let S_n have a Poisson distribution with mean n; that is, $P(S_n = k) = e^{-n} n^k/k!$ Use Stirling's formula to show that if $(k - n)/\sqrt{n} \to x$, then $\sqrt{2\pi n} \, P(S_n = k) \to \exp(-x^2/2)$.

Stirling's formula can also be used to compute some large deviations probabilities considered in Section 9 in Chapter 1. In Exercises 1.3–1.5, $X_1, X_2, \ldots$ are i.i.d. and $S_n = X_1 + \ldots + X_n$. In each case you should begin by considering $P(S_n = [na] + j)$ for $j = 1$ and then relate the value for j to the value for $j - 1$.

Exercise 1.3   Suppose $P(X_i = 1) = P(X_i = -1) = 1/2$. Show that if $a \in (-1, 1)$,

$$\frac{1}{n} \log P(S_n \geq na) \to -\gamma(a)$$

where $\gamma(a) = \frac{1}{2}\{(1 + a) \log(1 + a) + (1 - a) \log(1 - a)\}$.

Exercise 1.4   Suppose $P(X_i = k) = e^{-1}/k!$ for $k = 0, 1, \ldots$. Show that if $a > 1$,

$$\frac{1}{n} \log P(S_n \geq na) \to a - 1 - a \log a.$$

Exercise 1.5   A more refined analysis of the last exercise shows that

$$P(S_n > na) \sim e^{-n} (e/a)^{na} (2\pi na)^{-1/2} w_n/(1 - 1/a)$$

where $w_n = (1/a)^{1-\{na\}}$ and $\{x\} = x - [x]$ is the fractional part of x. Here w is for wobbles. If, for example, $a = \pi$, $1 - \{n\pi\}$ is dense in $(0, 1)$.

2   Weak Convergence

In this section we will define the type of convergence that appears in the central limit theorem and explore some of its properties. A sequence of distribution functions is said to *converge weakly* to a limit F (written $F_n \Rightarrow F$) if $F_n(y) \to F(y)$ for all y that are continuity points of F. A sequence of random variables X_n is said to *converge in distribution* to a limit X_∞ (written $X_n \Rightarrow X_\infty$) if their distribution functions $F_n(x) = P(X_n \le x)$ converge weakly. To see that convergence at continuity points is enough to identify the limit, observe that the discontinuities of F are at most a countable set, since at each discontinuity F jumps over a different rational number $\in (0, 1)$.

Two examples of weak convergence that we have seen are:

Example 2.1   Let $X_1, X_2, \ldots$ be i.i.d. with $P(X_i = 1) = P(X_i = -1) = 1/2$ and let $S_n = X_1 + \cdots + X_n$. Then (1.4) implies

$$F_n(y) = P(S_n/\sqrt{n} \le y) \to \int_{-\infty}^{y} (2\pi)^{-1/2} e^{-x^2/2} dx.$$

Example 2.2   Let $X_1, X_2, \ldots$ be i.i.d. with distribution F. The Glivenko–Cantelli theorem ((8.6) in Chapter 1) implies that for almost every ω,

$$F_n(y) = n^{-1} \sum_{m=1}^{n} 1_{(X_m(\omega) \le y)} \to F(y) \quad \text{for all } y.$$

In the last two examples convergence occurred for all y, even though in the second case the distribution function could have discontinuities. The next example shows why we restrict our attention to continuity points.

Example 2.3   Let X have distribution F. Then $X + 1/n$ has distribution

$$F_n(x) = P(X + 1/n \le x) = F(x - 1/n).$$

As $n \to \infty$, $F_n(x) \to F(x-) = \lim_{y \uparrow x} F(y)$, so convergence occurs only at continuity points.

Example 2.4   *Waiting for rare events.* Let X_p be the number of trials needed to get a success in a sequence of independent trials with success probability p. Then

$$P(X_p \ge n) = (1 - p)^{n-1} \quad \text{for } n = 1, 2, 3, \ldots$$

and it follows from (1.2) that

as $p \to 0$, $P(pX_p > x) \to e^{-x}$ for all $x \geq 0$.

In words, (the distribution of) pX_p converges weakly to an exponential distribution. In what follows we will often drop the parenthetical phrase.

Example 2.5   *Birthday problem.* Let X_1, X_2, ... be independent and uniformly distributed on $\{1, ..., N\}$, and let $T_N = \min\{n : X_n = X_m \text{ for some } m < n\}$.

$$P(T_N > n) = \prod_{m=2}^{n} \left(1 - \frac{m-1}{N}\right).$$

When $N = 365$ this is the probability that two people in a group of size n do not have the same birthday (assuming all birthdays are equally likely). Taking logarithms it is easy to see that

$$P(T_N/N^{1/2} > x) \to \exp(-x^2/2)   \text{ for all } x \geq 0.$$

For $N = 365$ this says, for instance, that

$$P(T_{365} > 22) \approx e^{-.6630} \approx .515$$

This answer is 2% smaller than the true probability .524.

Before giving our sixth example we need a simple observation called *Scheffé's theorem.* Suppose $X_n 1 \leq n \leq \infty$ have densities f_n and $f_n \to f_\infty$ pointwise as $n \to \infty$. Then for all Borel sets B,

$$\left| \int_B f_n(x)\, dx - \int_B f_\infty(x)\, dx \right| \leq \int |f_n(x) - f_\infty(x)|\, dx$$

$$= 2 \int (f_\infty(x) - f_n(x))^+ \, dx \to 0$$

by the dominated convergence theorem, the equality following from the fact that f_n and f_∞ are ≥ 0 and have integral $= 1$. Writing μ_n for the corresponding measures, we have shown that the *total variation norm*

$$\|\mu_n - \mu_\infty\| \equiv \sup_B |\mu_n(B) - \mu_\infty(B)| \to 0,$$

a conclusion much stronger than weak convergence. (Take $B = (-\infty, x]$.) The example $\mu_n = $ a point mass at $1/n$ (with $1/\infty = 0$) shows that we may have $\mu_n \Rightarrow \mu_\infty$ with $\|\mu_n - \mu_\infty\| = 1$ for all n.

Example 2.6   *Central order statistic.* Put $(2n + 1)$ points at random in $(0, 1)$—that is, with locations that are independent and uniformly distributed. Let V_{n+1} be the

$(n + 1)$th largest point. It is easy to see that V_{n+1} has density

$$(2n + 1)\binom{2n}{n} x^n(1 - x)^n.$$

(There are $2n + 1$ ways to pick the observation that falls at x; then we have to pick n indices for observations $< x$ ....) Changing variables $x = 1/2 + y/2\sqrt{2n}$, we see $2(V_{n+1} - 1/2)\sqrt{2n}$ has density

$$(2n + 1)\binom{2n}{n}\left(\frac{1}{2} + \frac{y}{2\sqrt{2n}}\right)^n \left(\frac{1}{2} - \frac{y}{2\sqrt{2n}}\right)^n \frac{1}{2\sqrt{2n}}$$

$$= \sqrt{n/2} \cdot 2^{-2n} \binom{2n}{n} \cdot (1 - y^2/2n)^n \cdot \frac{2n + 1}{2n}.$$

The second factor is $P(S_{2n} = 0)$ for a simple random walk, so (1.3) and (1.2) imply that

$$P(2(V_{n+1} - 1/2)\sqrt{2n} = y) \to (2\pi)^{-1/2}\exp(-y^2/2) \quad \text{as } n \to \infty.$$

Exercise 2.1   If $F_n \Rightarrow F$ and F is continuous, then $\sup_x |F_n(x) - F(x)| \to 0$.

Exercise 2.2   Let X_n, $1 \le n \le \infty$, be integer-valued. Show that $X_n \Rightarrow X_\infty$ (i.e., the associated distributions converge weakly) if and only if $P(X_n = j) \to P(X_\infty = j)$ for all j.

Exercise 2.3   Give an example of random variables X_n with densities f_n so that $X_n \Rightarrow$ a uniform distribution on $(0, 1)$ but $f_n(x)$ does not converge to 1 for any $x \in [0, 1]$.

Exercise 2.4   Suppose that $X_n = (X_n^1, \ldots, X_n^n)$ is uniformly distributed over the surface of the sphere of radius $\sqrt{n}$ in $\mathbb{R}^n$. Show that $X_n^1 \Rightarrow$ a standard normal. *Hint*   Let $Y_1, Y_2, \ldots$ be i.i.d. standard normals and let $X_n^i = Y_i(n/\sum_{m=1}^n Y_m^2)^{1/2}$.

The next result is useful for proving things about weak convergence.

(2.1) THEOREM   If $F_n \Rightarrow F_\infty$, then there are random variables Y_n, $1 \le n \le \infty$, with distribution F_n so that $Y_n \to Y_\infty$ a.s.

Proof   Let $\Omega = (0, 1)$, $\mathscr{F} = $ Borel sets, $P = $ Lebesgue measure, and let $Y_n(x) = \sup\{y : F_n(y) < x\}$. By (1.1) in Chapter 1, Y_n has distribution F_n. We will now show that $Y_n(x) \to Y_\infty(x)$ for all but a countable number of x. To do this it is convenient to write $Y_n(x)$ as $F_n^{-1}(x)$ and drop the subscript ∞ from the limit. We begin by identifying the exceptional set. Let $a_x = \sup\{y : F_n(y) < x\}$, $b_x = \inf\{y : F_n(y) > x\}$, and $\Omega_0 = \{x : (a_x, b_x) = \varnothing\}$ where (a_x, b_x) is the open interval with the indicated endpoints. $\Omega - \Omega_0$ is countable since the (a_x, b_x) are disjoint and each nonempty interval contains a different rational number. If $x \in \Omega_0$, then $F(y) < x$ for $y < F^{-1}(x)$ and $F(z) > x$ for $z > F^{-1}(x)$. To prove that $F_n^{-1}(x) \to F^{-1}(x)$ for $x \in \Omega_0$, there are two things to show:

$$\liminf_{n \to \infty} F_n^{-1}(x) \geq F^{-1}(x). \tag{a}$$

Proof of (a)   Let $y < F^{-1}(x)$ be such that F is continuous at y. Since $x \in \Omega_0$, $F(y) < x$, and if n is sufficiently large, $F_n(y) < x$; that is, $F_n^{-1}(x) \geq y$. Since this holds for all y satisfying the indicated restrictions, the result follows.

$$\limsup_{n \to \infty} F_n^{-1}(x) \leq F^{-1}(x). \tag{b}$$

Proof of (b)   Let $y > F^{-1}(x)$ be such that F is continuous at y. Since $x \in \Omega_0$, $F(y) > x$, and if n is sufficiently large, $F_n(y) > x$; that is, $F_n^{-1}(x) \leq y$. Since this holds for all y satisfying the indicated restrictions, the result follows and we have completed the proof of (2.1).   □

The next result illustrates the usefulness of (2.1) and gives an equivalent definition of weak convergence that makes sense in any topological space.

(2.2) THEOREM $X_n \Rightarrow X_\infty$ if and only if for every bounded continuous function g we have $Eg(X_n) \to Eg(X_\infty)$.

Proof   Let Y_n have the same distribution as X_n and converge a.s. Since g is continuous, $g(X_n) \to g(X_\infty)$ a.s. and the bounded convergence theorem implies

$$Eg(X_n) = Eg(Y_n) \to Eg(Y_\infty) = Eg(X_\infty).$$

To prove the converse let

$$g_{x,\varepsilon}(y) = \begin{cases} 1 & y \leq x \\ 0 & y \geq x + \varepsilon \\ \text{linear} & x \leq y \leq x + \varepsilon. \end{cases}$$

Since $g_{x,\varepsilon}(y) = 1$ for $y \leq x$,

$$\limsup_{n \to \infty} P(X_n \leq x) \leq \limsup_{n \to \infty} Eg_{x,\varepsilon}(X_n) = Eg_{x,\varepsilon}(X_\infty) \leq P(X_\infty \leq x + \varepsilon).$$

Letting $\varepsilon \to 0$ gives

$$\limsup_{n \to \infty} P(X_n \leq x) \leq P(X_\infty \leq x).$$

The last conclusion is valid for any x. To get the other direction we observe

$$\liminf_{n\to\infty} P(X_n \le x) \ge \liminf_{n\to\infty} Eg_{x-\varepsilon,\varepsilon}(X_n) = Eg_{x-\varepsilon,\varepsilon}(X_\infty) \ge P(X_\infty \le x + \varepsilon).$$

Letting $\varepsilon \to 0$ gives

$$\liminf_{n\to\infty} P(X_n \le x) \ge P(X_\infty < x) = P(X_\infty \le x)$$

if x is a continuity point.   □

The next result is a trivial but useful generalization of (2.2).

Exercise 2.5   Let g be a function and $D_g = \{x : g \text{ is discontinuous at } x\}$. ($D_g$ is a Borel set. See Exercise 2.4 in Chapter 1.) If $X_n \Rightarrow X$ and $P(X \in D_g) = 0$, then $Eg(X_n) \to Eg(X)$ and $g(X_n) \Rightarrow g(X)$.

Exercise 2.6   *Fatou's lemma revisited.* Let $g \ge 0$ be continuous. If $X_n \Rightarrow X_\infty$, then

$$\liminf_{n\to\infty} Eg(X_n) \ge Eg(X_\infty).$$

The next three exercises show that the criterion in (2.2) is useful for proving things.

Exercise 2.7   If $X_n \to X$ in probability, then $X_n \Rightarrow X$. The converse is false with one notable exception: X is constant.

Exercise 2.8   Suppose $X_n \Rightarrow X$ and $Y_n \Rightarrow Y$. It does not follow that $X_n + Y_n \Rightarrow X + Y$. (Give a counterexample.) The conclusion is true if we assume Y is a constant. A useful consequence of this result (called *Slutsky's theorem* or *the converging together lemma*) is that if $X_n \Rightarrow X$ and $Z_n - X_n \Rightarrow 0$, then $Z_n \Rightarrow X$.

Exercise 2.9   Suppose $X_n \Rightarrow X$ and $Y_n \to c$ in probability, then $X_n Y_n \Rightarrow cX$.

The next result makes explicit some observations implicit in the proof of (2.2).

(2.3) THEOREM   The following statements are equivalent:

(i) $X_n \Rightarrow X_\infty$.
(ii) For all open sets G, $\liminf_{n\to\infty} P(X_n \in G) \ge P(X_\infty \in G)$.
(iii) For all closed sets K, $\limsup_{n\to\infty} P(X_n \in K) \le P(X_\infty \in K)$.
(iv) For all sets A with $P(X_\infty \in \partial A) = 0$, $\lim_{n\to\infty} P(X_n \in A) = P(X_\infty \in A)$.

Proof   We will prove four things and leave it to the reader to check that we have proved the result given above.
        (i) implies (ii): Let Y_n have the same distribution as X_n and $Y_n \to Y_\infty$ a.s. Since G is open,

$$\liminf_{n\to\infty} 1_G(Y_n) \ge 1_G(Y_\infty),$$

so Fatou's lemma implies

$$\liminf_{n\to\infty} P(Y_n \in G) \geq P(Y_\infty \in G).$$

(ii) is equivalent to (iii): A is open if and only if A^c is closed and $P(A) + P(A^c) = 1$.

(ii) and (iii) imply (iv): Let $K = \bar{A}$ and $G = A^0$ be the closure and interior of A, respectively. The boundary of A, $\partial A = \bar{A} - A^0$ and $P(X_\infty \in \partial A) = 0$, so

$$P(X_\infty \in K) = P(X_\infty \in A) = P(X_\infty \in G).$$

Using (ii) and (iii) now,

$$\limsup_{n\to\infty} P(X_n \in A) \leq \limsup_{n\to\infty} P(X_n \in K) \leq P(X_\infty \in K) = P(X_\infty \in A)$$

$$\liminf_{n\to\infty} P(X_n \in A) \geq \liminf_{n\to\infty} P(X_n \in G) \geq P(X_\infty \in G) = P(X_\infty \in A).$$

(iv) implies (i): Let x be such that $P(X_\infty = x) = 0$ and let $A = (-\infty, x]$. $\quad\square$

The next result is useful in studying limits of sequences of distributions.

(2.4) Helly's Selection Theorem For every sequence F_n of distribution functions there is a subsequence $F_{n(k)}$ and a right continuous nondecreasing function F so that

$$\lim_{k\to\infty} F_{n(k)}(y) = F(y)$$

at all continuity points of y.

Remark The limit may not be a distribution function. For example, if $a + b + c = 1$ and

$$F_n(x) = a1_{(x \geq n)} + b1_{(x \geq -n)} + cG(x)$$

where G is a distribution function, then

$$F_n(x) \to F(x) = b + cG(x)$$

$$\lim_{x\downarrow-\infty} F(x) = b \quad \text{and} \quad \lim_{x\uparrow\infty} F(x) = b + c = 1 - a.$$

In words, an amount of mass a escapes to $+\infty$, and mass b escapes to $-\infty$. The type of convergence that occurs in (2.4) is sometimes called *vague convergence* and denoted $\xrightarrow{v}$. It is a weaker form of weak convergence that allows mass to "escape."

Exercise 2.10 Show that $F_n \xrightarrow{v} F$ if and only if for all continuous functions g with compact support, $\int g(x)\, dF_n(x) \to \int g(x)\, dF(x)$.

Proof of (2.4)   Let $q_1, q_2, \ldots$ be an enumeration of the rationals. For each k there is a sequence $m_k(i) \to \infty$ that is a subsequence of $m_{k-1}(j)$ (let $m_0(j) \equiv j$) so that

$$F_{m_k(i)}(q_k) \text{ converges to } G(q_k) \text{ as } i \to \infty.$$

Let $F_{n(k)} = F_{m_k(k)}$. By construction $F_{n(k)}(q) \to G(q)$ for all rational q. The function G may not be right continuous, but $F(x) = \inf\{G(q) : q \in \mathbb{Q}, q > x\}$ is, since

$$\lim_{x_n \downarrow x} F(x_n) = \inf\{G(q) : q \in \mathbb{Q}, q > x_n \text{ for some } n\}$$

$$= \inf\{G(q) : q \in \mathbb{Q}, q > x\} = F(x).$$

To complete the proof, let x be a continuity point of F. Pick rationals $r_1 < r_2 < x < s$ so that

$$F(x) - \varepsilon < F(r_1) \le F(r_2) \le F(x) \le F(s) < F(x) + \varepsilon.$$

Since $F_n(r_2) \to G(r_2) \ge F(r_1)$ and $F_n(s) \to G(s) \le F(s)$, it follows that if n is large,

$$F(x) - \varepsilon < F_n(r_2) \le F_n(x) \le F_n(s) < F(x) + \varepsilon,$$

which is the desired conclusion.   □

The last result raises a question: When can we conclude that no mass is lost?

(2.5) THEOREM   Every subsequential limit is a probability measure if and only if the sequence F_n is *tight*; that is, for all $\varepsilon > 0$ there is an M so that

$$\limsup_{n \to \infty} 1 - F_n(M) + F_n(-M) \le \varepsilon.$$

Proof   Exercise 2.11.   □

The following sufficient condition for tightness is often useful.

(2.6) THEOREM   If there is a $\varphi \ge 0$ so that $\varphi(x) \to \infty$ as $|x| \to \infty$ and

$$C = \sup_n \int \varphi(x) \, dF_n(x) < \infty,$$

then F_n is tight.

Proof $1 - F_n(M) + F_n(-M) \le C / \inf_{|x| \ge M} \varphi(x)$.   □

Exercise 2.12   *Integration to the limit.* Suppose $h(x)$ is continuous, $g(x) > 0$, and $|h(x)|/g(x) \to 0$ as $|x| \to \infty$. If $F_n \Rightarrow F$ and

$$\limsup_{n \to \infty} \int g(x)\, dF_n(x) < \infty,$$

then

$$\int h(x)\, dF_n(x) \to \int h(x)\, dF(x).$$

An important special case is $g(x) = x^2$, $h(x) = x$, or more generally $h(x) = x^k$ with $k \in \mathbb{Z}$ and $g(x) = |x|^p$ with $p > k$.

Exercise 2.13 Suppose $Y_n \geq 0$, $EY_n^\alpha \to 1$, and $EY_n^\beta \to 1$ for some $\alpha < \beta$. Show that $Y_n \to 1$ in probability.

Exercise 2.14 For each $K < \infty$ there is a $c_K > 0$ so that $EX = 0$, $EX^2 = 1$, $EX^4 \leq K$ implies $P(X > 0) \geq c_K$.
Hint Suppose there is a sequence X_n with the indicated properties and $P(X_n > 0) \to 0$.

Exercise 2.15 *The Lévy metric.* Show that

$$\rho(F, G) = \inf\{\varepsilon : F(x - \varepsilon) - \varepsilon \leq G(x) \leq F(x + \varepsilon) + \varepsilon \text{ for all } x\}$$

defines a metric on the space of distributions and $\rho(F_n, F) \to 0$ if and only if $F_n \Rightarrow F$.

Exercise 2.16 *The Ky Fan metric* on random variables is defined by

$$\alpha(X, Y) = \inf\{\varepsilon \geq 0 : P(|X - Y| > \varepsilon) \leq \varepsilon\}.$$

Show that if $\alpha(X, Y) = \alpha$, then the corresponding distributions have Levy distance $\rho(F, G) \leq \alpha$. Conversely, if $\rho(F, G) = \rho$, we can construct r.v.'s X and Y with these distributions so that $\alpha(X, Y) \leq \rho$.

Exercise 2.17 Let $\alpha(X, Y)$ be the metric in the previous exercise and let $\beta(X, Y) = E(|X - Y|/(1 + |X - Y|))$ be the metric of Exercise 5.4 in Chapter 1. If $\alpha(X, Y) = a$, then

$$a^2/(1 + a) \leq \beta(X, Y) \leq a + (1 - a)\, a/(1 + a).$$

Exercise 2.18 *Convergence of Maxima.* Let $X_1, X_2, \ldots$ be independent with distribution F, and let $M_n = \max_{m \leq n} X_m$. Then $P(M_n \leq x) = F(x)^n$. Prove the following limit laws for M_n:

(i) If $F(x) = 1 - x^{-\alpha}$ for $x \geq 1$ where $\alpha > 0$, then for $y > 0$,

$$P(M_n/n^{1/\alpha} \leq y) \to \exp(-y^{-\alpha}).$$

(ii) If $F(x) = 1 - (-x)^\beta$ for $-1 \leq x \leq 0$ where $\beta > 0$, then for $y < 0$,

$$P(n^{1/\beta} M_n \leq y) \to \exp(-|y|^\beta).$$

(iii) If $F(x) = 1 - e^{-x}$ for $x \geq 0$, then for all $y \in (-\infty, \infty)$,

$$P(M_n - \log n \leq y) \to \exp(-e^{-y}).$$

The limits that appear above are called the *extreme value distributions*. The last one is called the *double exponential* or *Gumbel distribution*. Necessary and sufficient conditions for $(M_n - b_n)/a_n$ to converge to these limits were obtained by Gnedenko (1943). For a recent treatment, see Resnick (1987).

Exercise 2.19   (i) From (1.3) in Chapter 1 we know

$$\int_x^\infty e^{-u^2/2}\, du \sim \frac{1}{x} e^{-x^2/2}.$$

Use this to conclude that if $X_1, X_2, \ldots$ are i.i.d. and have a normal distribution, then

$$M_n/(2 \log n)^{1/2} \to 1 \text{ in probability.}$$

(ii) The asymptotic formula in (i) implies that if $x \to \infty$, then

$$P\left(X_i > x + \frac{\theta}{x}\right)\Big/ P(X_i > x) \to e^{-\theta} \quad \text{for all } \theta \in \mathbb{R}.$$

Show that if we define b_n by $P(X_i > b_n) = 1/n$, then $b_n \sim (2 \log n)^{1/2}$ and

$$P((2 \log n)^{1/2}(M_n - b_n) > x) \to \exp(-e^{-x}).$$

3 Characteristic Functions

This long section is divided into three parts. In part a we show that the characteristic function $\varphi(t) = E \exp(itX)$ determines $F(x) = P(X \leq x)$, and we give a recipe for computing F from φ. In part b we relate weak convergence of distributions to the behavior of the corresponding characteristic functions. Finally, in part c we relate the behavior of $\varphi(t)$ at 0 to the moments of X.

a   Definition, Inversion Formula

If X is a random variable, we define its *characteristic function* (ch.f.) by

$$\varphi(t) = Ee^{itX} = E(\cos tX + i \sin tX).$$

Some properties are immediate from the definition:

$$\varphi(0) = 1 \tag{3.1a}$$

$$\varphi(-t) = E(\cos(-tX) + i \sin(-tX)) = \overline{\varphi(t)} \tag{3.1b}$$

where $\bar{z}$ denotes the *complex conjugate* of z, $\overline{a + bi} = a - bi$.

$$|\varphi(t)| = |Ee^{itX}| \leq E|e^{itX}| = 1. \tag{3.1c}$$

Here $|z|$ denotes the *modulus* of the complex number z, $|a + bi| = (a^2 + b^2)^{1/2}$. The inequality above is obvious if you recall that the numbers e^{itx} lie on the unit circle in the complex plane and any average of these numbers—for example, Ee^{itX}—must lie inside the circle.

$$|\varphi(t + h) - \varphi(t)| = |Ee^{i(t+h)X} - e^{itX}|$$

$$\leq E|e^{i(t+h)X} - e^{itX}| = E|e^{ihX} - 1|, \tag{3.1d}$$

since $|zw| = |z| \cdot |w|$. The last quantity $\to 0$ as $h \to 0$ by the bounded convergence theorem, so $\varphi(t)$ is uniformly continuous on $(-\infty, \infty)$.

$$Ee^{it(aX+b)} = e^{itb}\varphi(at). \tag{3.1e}$$

Finally recall from (1.5) that:

(3.1f)   If X_1 and X_2 are independent and have ch.f.'s φ_1 and φ_2, then $X + Y$ has ch.f. $\varphi_1(t)\varphi_2(t)$.

The next order of business is to give some examples. The first was computed at the end of Section 1.

Example 3.1   *The normal distribution*

Density $(2\pi)^{-1/2} \exp(-x^2/2)$
Ch.f. $\exp(-t^2/2)$

Combining this result with (3.1e), we see that a normal distribution with mean μ and variance σ^2 has ch.f. $\exp(i\mu t - \sigma^2 t^2/2)$. Similar scalings can be applied to other examples, so we will often give the ch.f. for only one member of the family.

In the next three examples the density is 0 outside the indicated range.

Example 3.2   *Uniform distribution on (a, b)*

Density $1/(b - a)$ $x \in (a, b)$
Ch.f. $(e^{itb} - e^{ita})/it(b - a)$

Example 3.3   *Triangular distribution*

Density $1 - |x|$ $x \in (-1, 1)$
Ch.f. $2(1 - \cos t)/t^2$

To see this, notice that if X and Y are independent and uniform on $(-1/2, 1/2)$, then $X + Y$ has a triangular distribution and

$$[(e^{it/2} - e^{-it/2})/it]^2 = [2\sin(t/2)/t]^2.$$

Example 3.4   *Exponential distribution*

> Density e^{-x} $x \in (0, \infty)$
> Ch.f. $1/(1 - it)$

since expanding e^{itx} in power series gives

$$\int_0^\infty e^{itx} e^{-x}\, dx = \sum_{n=0}^\infty \int_0^\infty \frac{(itx)^n}{n!}\, e^{-x}\, dx = \sum_{n=0}^\infty (it)^n.$$

Example 3.5   *Bilateral exponential* or *Laplace distribution*

> Density $\frac{1}{2} e^{-|x|}$ $x \in (-\infty, \infty)$
> Ch.f. $1/(1 + t^2)$

This follows from a more general fact:

(3.1g)   If $F_1, \ldots, F_n$ have ch.f. $\varphi_1, \ldots, \varphi_n$ and $\lambda_i \geq 0$ have $\lambda_1 + \cdots + \lambda_n = 1$, then

$$\sum_{i=1}^n \lambda_i F_i \text{ has ch.f. } \sum_{i=1}^n \lambda_i \varphi_i.$$

The formula in Example 3.5 comes from applying (3.1g) with F_1 the distribution of an exponential random variable X_1, F_2 the distribution of $X_2 = -X_1$, and $\lambda_1 = \lambda_2 = 1/2$.

Exercise 3.1   Show that if φ is a ch.f., then Re φ is also. Here $\text{Re}(a + bi) = a$ is the real part.

The first issue to be settled is that the characteristic function uniquely determines the distribution. This (and more) is provided by:

(3.2) *The Inversion Formula*   Let μ be a probability measure and let

$$\varphi(t) = \int e^{itx} \mu\,(dx).$$

If $a < b$, then

$$\lim_{T \to \infty} (2\pi)^{-1} \int_{-T}^T \frac{e^{-ita} - e^{-itb}}{it}\, \varphi(t)\, dt = \mu(a, b) + \frac{1}{2} \mu(\{a, b\}).$$

Remark   The existence of the limit is part of the conclusion. If $\mu = \delta_0$, a point mass at 0, $\varphi(t) \equiv 1$. In this case if $a = -1$ and $b = 1$, the integrand is $(2 \sin t)/t$ and the integral does not converge absolutely.

Proof   Let

$$I_T = \int_{-T}^{T} \frac{e^{-ita} - e^{-itb}}{it} \varphi(t) \, dt = \int_{-T}^{T} \int \frac{e^{-ita} - e^{-itb}}{it} e^{itx} \mu(dx) \, dt.$$

If we observe that

$$\frac{e^{-ita} - e^{-itb}}{it} = \int_a^b e^{-ity} \, dy,$$

we see that the modulus of the integrand is bounded by $b - a$. Since μ is a probability measure and $[-T, T]$ is a finite interval, it follows from Fubini's theorem that

$$I_T = \int \int_{-T}^{T} \frac{e^{-ita} - e^{-itb}}{it} e^{itx} \, dt \, \mu(dx)$$

$$= \int \left[\int_{-T}^{T} \frac{\sin(t(x - a))}{t} \, dt - \int_{-T}^{T} \frac{\sin(t(x - b))}{t} \, dt \right] \mu(dx), \qquad \text{(a)}$$

since $\cos(-x) = \cos x$ and $\sin(-x) = -\sin x$. If we let

$$S(T) = \int_0^T \frac{\sin x}{x} \, dx,$$

then for $\theta > 0$,

$$\int_{-T}^{T} \frac{\sin t\theta}{t} \, dt = 2 \int_0^{T\theta} \frac{\sin x}{x} \, dx = 2S(T\theta)$$

and for $\theta < 0$,

$$\int_{-T}^{T} \frac{\sin t\theta}{t} \, dt = -\int_{-T}^{T} \frac{\sin t|\theta|}{t} \, dt = -2S(T|\theta|).$$

Introducing the function sgn x, which is 1 if $x > 0$, -1 if $x < 0$, and 0 if $x = 0$, we can write the last two formulas together as

$$\int_{-T}^{T} \frac{\sin t\theta}{t} \, dt = 2(\text{sgn } \theta) S(T|\theta|). \qquad \text{(b)}$$

Let $R(\theta, T) = 2(\text{sgn } \theta) S(T|\theta|)$. Using (b) in (a) gives

$$I_T = \int R(x - a, T) - R(x - b, T) \, \mu(dx).$$

As $T \to \infty$, $S(T) \to \pi/2$ (see Exercise 6.7 in the Appendix) and $R(\theta, T) \to \pi \operatorname{sgn} \theta$, so

$$
R(x - a, T) - R(x - b, T) \to \begin{cases} 0 & x < a \\ \pi & x = a \\ 2\pi & a < x < b \\ \pi & x = b \\ 0 & x > b. \end{cases}
$$

$R(\theta, T) \leq 2 \sup S(y) < \infty$, so the bounded convergence theorem implies

$$
(2\pi)^{-1} I_T \to \mu(a, b) + \tfrac{1}{2} \mu(\{a, b\}),
$$

proving (3.2). $\square$

Exercise 3.2   (i) Imitate the proof of (3.2) to show that

$$
\mu(\{a\}) = \lim_{T \to \infty} \frac{1}{2\pi} \int_{-T}^{T} e^{-ita} \varphi(t) \, dt.
$$

(ii) If $P(X \in h\mathbb{Z}) = 1$ where $h > 0$, then its ch.f. has $\varphi(2\pi/h + t) = \varphi(t)$, so

$$
P(X = x) = \frac{h}{2\pi} \int_{-\pi/h}^{\pi/h} e^{-itx} \varphi(t) \, dt \quad \text{for } x \in h\mathbb{Z}.
$$

(iii) If $Y = Z + b$, then $E \exp(itY) = e^{itb} E \exp(itZ)$. So if $P(X \in b + h\mathbb{Z}) = 1$, the inversion formula in (ii) is valid for $x \in b + h\mathbb{Z}$.

Two trivial consequences of the inversion formula are:

Exercise 3.3   If φ is real, then X and $-X$ have the same distribution.

Exercise 3.4   If X_i, $i = 1, 2$, are independent and have normal distributions with mean 0 and variance σ_i^2, then $X_1 + X_2$ has a normal distribution with mean 0 and variance $\sigma_1^2 + \sigma_2^2$.

Our next topic is to investigate what happens when $\int |\varphi(t)| \, dt < \infty$. Since

$$
\left| \frac{e^{-ita} - e^{-itb}}{it} \right| = \left| \int_a^b e^{-ity} \, dy \right| \leq |b - a|,
$$

the integral in (3.2) converges absolutely in this case and

$$
\mu(a, b) + \frac{1}{2} \mu(\{a, b\}) \leq \frac{(b - a)}{2\pi} \int_{-\infty}^{\infty} |\varphi(t)| \, dt.
$$

The last result implies μ has no point masses and

$$\mu(x, x + h) = \frac{1}{2\pi} \int \frac{e^{-itx} - e^{-it(x+h)}}{it} \varphi(t) \, dt$$

$$= \frac{1}{2\pi} \int \left[\int_x^{x+h} e^{-ity} \, dy \right] \varphi(t) \, dt$$

$$= \int_x^{x+h} \left[\frac{1}{2\pi} \int e^{-ity} \varphi(t) \, dt \right] dy$$

by Fubini's theorem, so the distribution μ has density function

$$f(y) = \frac{1}{2\pi} \int e^{-ity} \varphi(t) \, dt.$$

The proof of the continuity of φ given at the beginning of this section can be applied to conclude that f is continuous. Putting things together, we have:

(3.3) THEOREM   If $\int |\varphi(t)| \, dt < \infty$, then μ has bounded continuous density

$$f(y) = \frac{1}{2\pi} \int e^{-ity} \varphi(t) \, dt.$$

Exercise 3.5   Give an example of a measure μ with a density but for which $\int |\varphi(t)| \, dt = \infty$.

Hint   Two of our first five examples have this property.

Exercise 3.6   Show that if $X_1, X_2, \ldots$ are independent and uniformly distributed on $(-1, 1)$, then for $n \geq 2$, $X_1 + \cdots + X_n$ has density

$$f(x) = \frac{1}{\pi} \int_0^\infty \left[\frac{\sin t}{t} \right]^n \cos tx \, dt.$$

Although it is not obvious from the formula, f is a polynomial in each interval $(k, k + 1)$, $k \in \mathbb{Z}$, and vanishes on $[-n, n]^c$.

(3.3) and the next result show that the behavior of φ at infinity is related to the smoothness of the underlying measure.

Exercise 3.7   Apply Exercise 3.2 to $X - Y$ to get

$$\lim_{T \to \infty} \frac{1}{2\pi} \int_{-T}^T |\varphi(t)|^2 \, dt = P(X - Y = 0) = \sum_x \mu(\{x\})^2$$

where the sum is over all real numbers x. (All but countably many of the terms are 0.)

Exercise 3.8   The last result implies that if $\varphi(t) \to 0$ as $t \to \infty$, μ has no point masses. Give an example to show that the converse is false. Use the Riemann–Lebesgue lemma (Exercise 4.5 in the Appendix) to show that if μ has a density, $\varphi(t) \to 0$ as $t \to \infty$.

Exercise 3.9   *Plancherel identity.* Show that if $|\varphi(t)|^2$ is integrable, then so is f^2 and

$$\frac{1}{2\pi} \int |\varphi(t)|^2 \, dt = \int f(x)^2 \, dx.$$

Applying the inversion formula (3.3) to the ch.f. in Examples 3.3 and 3.5 gives us two more examples. The first one does not have an official name, so we gave it one to honor its role in the proof of Polya's criterion (see Exercise 3.25).

Example 3.6   *Polya's distribution*

> Density $\dfrac{1}{\pi} \cdot \dfrac{1 - \cos x}{x^2}$

> Ch.f. $1 - |t|$ $|t| \le 1$, 0 otherwise

Example 3.7   *The Cauchy distribution*

> Density $\dfrac{1}{\pi} \cdot \dfrac{1}{1 + x^2}$

> Ch.f. $\exp(-|t|)$

Exercise 3.10   Use the last result to conclude that if $X_1, X_2, \ldots$ are independent and have the Cauchy distribution, then $(X_1 + \cdots + X_n)/n$ has the same distribution as X_1.

b   Weak Convergence

Our next step toward the central limit theorem is to relate convergence of characteristic functions to weak convergence.

(3.4) Continuity Theorem   Let μ_n, $1 \le n \le \infty$, be probability measures with ch.f. φ_n. (i) If $\mu_n \Rightarrow \mu_\infty$, then $\varphi_n(t) \to \varphi_\infty(t)$ for all t. (ii) If $\varphi_n(t)$ converges pointwise to a limit $\varphi(t)$ that is continuous at 0, then the associated sequence of distributions μ_n is tight and converges weakly to a limit μ with characteristic function φ.

To see why continuity of the limit is needed in (ii), let μ_n have a normal distribution with variance n. In this case $\varphi_n(t) = \exp(-nt^2/2) \to 0$ for $t \neq 0$, and $\varphi_n(0) = 1$ for all n, but the measures do not converge weakly, since $\mu_n((-\infty, x]) \to 1/2$ for all x.

Proof   (i) is easy. e^{itx} is bounded and continuous, so if $\mu_n \Rightarrow \mu_\infty$, then $\varphi_n(t) \to \varphi_\infty(t)$. To prove (ii) we begin by observing

$$\int_{-u}^{u} 1 - e^{itx}\, dt = 2u - \int_{-u}^{u} (\cos tx + i \sin tx)\, dt = 2u - \frac{2 \sin ux}{x}.$$

Dividing both sides by u and integrating $\mu_n(dx)$ give

$$u^{-1}\int_{-u}^{u}(1 - \varphi_n(t))\, dt = 2\int 1 - \frac{\sin ux}{ux}\,\mu_n(dx)$$

$$\geq 2\int_{|x|\geq 2/u} 1 - \frac{1}{|ux|}\,\mu_n(dx) \geq \mu_n(\{x: |x| > 2/u\}).$$

(Recall $|\sin x| \leq 1$.) Let φ_∞ be the ch.f. of μ_∞. Since $\varphi_\infty(t) \to 1$ as $t \to 0$,

$$u^{-1}\int_{-u}^{u}(1 - \varphi_\infty(t))\, dt \to 0 \quad \text{as } u \to 0.$$

Pick u so that the integral is $< \varepsilon$. Since $\varphi_n(t) \to \varphi_\infty(t)$ for each t, it follows from the dominated convergence theorem that for $n \geq N$,

$$2\varepsilon \geq u^{-1}\int_{-u}^{u}(1 - \varphi_n(t))\, dt \geq \mu_n\{x: |x| > 2/u\}.$$

Since ε is arbitrary, the sequence μ_n is tight.

To complete the proof we observe that if $\mu_{n(k)} \Rightarrow \mu$, then it follows from the first sentence of the proof that μ has ch.f. φ. The last observation and tightness imply that every subsequence has a further subsequence that converges to μ. I claim that this implies the whole sequence converges to μ. To see this, observe that we have shown that if f is bounded and continuous, then every subsequence of $\int f\, d\mu_n$ has a further subsequence that converges to $\int f\, d\mu$, so the whole sequence converges to that limit. $\qquad\square$

Exercise 3.11 *Continuity theorem for Laplace transforms.* Recall that Example 5.4 in Chapter 1 shows that the distribution of $X_n \geq 0$ is determined by its Laplace transform $\psi_n(\lambda) = E \exp(-\lambda X_n)$. Imitate the proof of (3.4) to show: (i) If $X_n \Rightarrow X_\infty$, then $\psi_n(\lambda) \to \psi_\infty(\lambda)$ for all $\lambda \geq 0$. (ii) If $\psi_n(t)$ converges pointwise to a limit $\psi(t)$ that is continuous at 0, then the associated sequence of distributions μ_n is tight and converges weakly to a limit μ with Laplace transform ψ.

Exercise 3.12 Suppose $\mu_n \Rightarrow \mu_\infty$. Use (3.1d) to conclude that the corresponding ch.f.'s are equicontinuous and hence by the Arzela–Ascoli theorem converge uniformly on compact sets. Give an example to show that the convergence need not be uniform on the whole real line.

Exercise 3.13 Suppose that $X_n \Rightarrow X$ and X_n has a normal distribution with mean 0 and variance σ_n^2. Prove that $\sigma_n^2 \to \sigma^2 \in [0, \infty)$.

Exercise 3.14   Show that if X_n and Y_n are independent for $1 \le n \le \infty$, $X_n \Rightarrow X_\infty$, and $Y_n \Rightarrow Y_\infty$, then $X_n + Y_n \Rightarrow X_\infty + Y_\infty$.

Exercise 3.15   Using the identity $\sin \theta = 2 \sin(\theta/2) \cos(\theta/2)$ repeatedly leads to

$$\frac{\sin \theta}{\theta} = \prod_{n=1}^{\infty} \cos(\theta/2^n).$$

Prove the last identity by interpreting each side as a characteristic function.

Exercise 3.16   Let $X_1, X_2, \ldots$ be independent, taking values 0 and 1 with probability $1/2$ each. $X = 2\sum_{j \ge 1} X_j/3^j$ has the Cantor distribution. Compute the ch.f. φ of X and notice that φ has the same value at all points $t = 3^k \pi$.

c   Moments and Derivatives

The inequality

$$\mu\{x: |x| > 2/u\} \le u^{-1} \int_{-u}^{u} (1 - \varphi(t))\, dt \tag{3.5}$$

relates the smoothness of the characteristic function at 0 to the decay of the measure at ∞. The next result continues this theme. We leave the proof to the reader, since we will give a more precise result in (3.6) below.

Exercise 3.17   If $\int |x|^n \mu(dx) < \infty$, then its characteristic function φ has a continuous derivative of order n given by

$$\varphi^n(t) = \int (ix)^n e^{itx}\, \mu(dx).$$

Exercise 3.18   Use the last exercise to show that the standard normal distribution has

$$EX^{2n} = 2n!/2^n n! = (2n - 1)(2n - 3) \cdots 3 \cdot 1 \equiv (2n - 1)!!$$

The result in Exercise 3.17 shows that if $E|X|^n < \infty$, then its characteristic function is n times differentiable at 0, and $\varphi^n(0) = E(iX)^n$. Expanding φ in a Taylor series about 0 leads to

$$\varphi(t) = \sum_{m=0}^{n} \frac{E(itX)^m}{m!} + o(t^n)$$

where $o(t^n)$ indicates a quantity $g(t)$ that has $g(t)/t^n \to 0$ as $t \to 0$. For our purposes below it will be important to have a good estimate on the error term, so we will now derive the last result. The starting point is a little calculus. Integrating by parts gives

$$\int_0^x (x - s)^n e^{is}\, ds = \frac{x^{n+1}}{n+1} + \frac{i}{n+1}\int_0^x (x - s)^{n+1} e^{is}\, ds.$$

When $n = 0$, this says

$$\int_0^x e^{is}\, ds = x + i\int_0^x (x - s)e^{is}\, ds.$$

The left-hand side is $(e^{ix} - 1)/i$, so rearranging gives

$$e^{ix} = 1 + ix + i^2 \int_0^x (x - s)e^{is}\, ds.$$

Using the result for $n = 1$ now gives

$$e^{ix} = 1 + ix + \frac{i^2 x^2}{2} + \frac{i^3}{2}\int_0^x (x - s)^2 e^{is}\, ds,$$

and iterating we arrive at

$$e^{ix} = \sum_{m=0}^n \frac{(ix)^m}{m!} + \frac{i^{n+1}}{n!}\int_0^x (x - s)^n e^{is}\, ds.$$

Since $|x - s| \le |x|$ for $s \in [0, x]$, the second term is smaller than $|x|^{n+1}/(n + 1)!$ The last estimate is good when x is small. The next is designed for large x. Integrating by parts,

$$\frac{i}{n}\int_0^x (x - s)^n e^{is}\, ds = -\frac{x^n}{n} + \int_0^x (x - s)^{n-1} e^{is}\, ds.$$

Noticing

$$-x^n/n = \int_0^x (x - s)^{n-1}\, ds$$

now gives

$$\frac{i^{n+1}}{n!}\int_0^x (x - s)^n e^{is}\, ds = \frac{i^n}{(n-1)!}\int_0^x (x - s)^{n-1}(e^{is} - 1)\, ds,$$

and since $|e^{ix} - 1| \le 2$, the last expression is $\le 2|x|^n/n!$ Putting things together we have

$$\left| e^{ix} - \sum_{m=0}^n \frac{(ix)^m}{m!} \right| \le \min\left[\frac{|x|^{n+1}}{(n+1)!}, \frac{2|x|^n}{n!} \right]. \tag{3.6}$$

Applying the last result to $x = tX$ and taking expected values,

$$\left| E e^{itX} - \sum_{m=0}^{n} E \frac{(itX)^m}{m!} \right| \le E \left| e^{itX} - \sum_{m=0}^{n} \frac{(itX)^m}{m!} \right| \tag{3.7}$$

$$\le E \min \left[\frac{|tX|^{n+1}}{(n+1)!}, \frac{2|tX|^n}{n!} \right]$$

$$= \frac{|t|^n}{(n+1)!} E \min \left(|t| \cdot |X|^{n+1}, 2(n+1)|X|^n \right).$$

In the next section the following special case will be useful:

(3.8) THEOREM   If $EX = 0$ and $E|X|^2 = \sigma^2 < \infty$, then $\varphi(t) = 1 - t^2\sigma^2/2 + o(t^2)$.

Proof   The error term is $\le t^2 E(|t| \cdot |X|^3 \wedge 6|X|^2)/3!$. The variable in parentheses is smaller than $6|X|^2$ and converges to 0 as $t \to 0$, so the desired conclusion follows from the dominated convergence theorem.   □

Remark   The point of the estimate in (3.7) is that we get the conclusion in (3.8) under the assumption $E|X|^2 < \infty$; that is, we do not have to assume $E|X|^3 < \infty$.

Exercise 3.19   Let $X_1, X_2, \ldots$ be i.i.d. with characteristic function φ. (i) If $\varphi'(0) = ia$ and $S_n = X_1 + \cdots + X_n$, then $S_n/n \to a$ in probability. (ii) If $S_n/n \to a$ in probability, then $\varphi(t/n)^n \to e^{iat}$. Use this to conclude that $\varphi'(0) = ia$, so the weak law holds if and only if $\varphi'(0)$ exists. This is due to E. J. G. Pitman (1956).

The last exercise in combination with (5.6) from Chapter 1 shows that $\varphi'(0)$ may exist when $E|X| = \infty$. For an explicit example consider

$$\varphi(t) = C \sum_{n=2}^{\infty} \frac{\cos nt}{n^2 \log n},$$

the ch.f. of the distribution with mass $C/2n^2 \log n$ at $-n$ and n for $n = 2, 3, 4, \ldots$. Differentiating the series (or consulting Zygmund (1947), p. 272) shows $\varphi'(t)$ exists and is continuous but

$$E|X| = C \sum_{n=2}^{\infty} \frac{n}{n^2 \log n} = \infty.$$

Exercise 3.20 $\dfrac{2}{\pi} \displaystyle\int \frac{1 - \operatorname{Re} \varphi(t)}{t^2} \, dt = \int |x| \, dF(x).$

The next two results show that the existence of second derivatives implies the existence of second moments. The second is easier than the first and suffices for most purposes.

Exercise 3.21   Show that if $\lim_{h \downarrow 0} (\varphi(h) - 2\varphi(0) + \varphi(-h))/h^2$ exists, then $E|X|^2 < \infty$.

Exercise 3.22 Show that if $\lim_{t \downarrow 0} (\varphi(t) - 1)/t^2 = -\sigma^2/2 > -\infty$, then $EX = 0$ and $E|X|^2 = \sigma^2 < \infty$. In particular, if $\varphi(t) = 1 + o(t^2)$, then $\varphi(t) \equiv 1$.

Exercise 3.23 If Y_n are r.v.'s with ch.f.'s φ_n, then $Y_n \Rightarrow 0$ if and only if there is a $\delta > 0$ so that $\varphi_n(t) \to 1$ for $|t| \leq \delta$.

Exercise 3.24 Let $X_1, X_2, \ldots$ be independent. If $S_n = \sum_{m \leq n} X_m$ converges in distribution, then it converges in probability (and hence a.s. by Exercise 7.7 in Chapter 1).

Sketch The last exercise implies that if $m, n \to \infty$, then $S_m - S_n \to 0$ in probability. Now use Exercise 6.4 in Chapter 1.

The next result is useful for constructing examples of ch.f.'s.

Exercise 3.25 *Polya's criterion.* Let $\varphi(t)$ be real nonnegative and have $\varphi(0) = 1$, $\varphi(t) = \varphi(-t)$, and φ convex on $(0, \infty)$. Then there is a probability measure v on $(0, \infty]$, so that

$$\varphi(t) = \int_0^\infty \left(1 - \left|\frac{t}{s}\right|\right)^+ v(ds) \qquad (*)$$

and hence φ is a characteristic function.

Sketch If $(*)$ holds, then φ is differentiable at points with $v(\{t\}) = 0$ and

$$\varphi'(t) = \int_t^\infty -s^{-1} v(ds);$$

that is,

$$v[0, t] = \int_0^t s \, d\varphi'(s).$$

Taking the last equation as the definition of v and reversing our steps, $(*)$ holds. v is a measure since φ' is decreasing. Setting $t = 0$ in $(*)$ shows v has total mass 1. If φ is piecewise linear, v has a finite number of atoms and the result follows from Example 3.6 and (3.1g). The general result follows by taking limits and using part (ii) of (3.4).

A classic application of Polya's criterion is that $\exp(-|t|^\alpha)$ is a characteristic function for $0 < \alpha \leq 1$. (The case $\alpha = 1$ corresponds to the Cauchy distribution.) The next argument, which we learned from Frank Spitzer, proves that this is true for $0 < \alpha \leq 2$. The case $\alpha = 2$ corresponds to a normal distribution, so that case can be safely ignored in the proof.

Exercise 3.26 For any β and $|x| < 1$,

$$(1 - x)^\beta = \sum_{n=0}^\infty \binom{\beta}{n}(-x)^n$$

where

$$\binom{\beta}{n} = \frac{\beta(\beta - 1)\cdots(\beta - n + 1)}{1 \cdot 2 \cdot \cdots \cdot n}.$$

Let

$$\psi(t) = 1 - (1 - \cos t)^{\alpha/2} = \sum_{n=1}^{\infty} c_n (\cos t)^n$$

where

$$c_n = \binom{\alpha/2}{n}(-1)^{n+1}.$$

Since $c_n \geq 0$ (here we use $\alpha < 2$), ψ is a ch.f. and it follows that

$$\exp(-|t|^\alpha) = \lim_{n \to \infty} \{\psi(t \cdot 2^{-1/2} \cdot n^{-1/\alpha})\}^n \text{ is a ch.f.}$$

Exercise 3.22 shows that $\exp(-|t|^\alpha)$ is not a ch.f. when $\alpha > 2$. A reason for interest in these characteristic functions is explained by the following generalization of Exercise 3.10.

Exercise 3.27 If X_1, X_2, ... are independent and have characteristic function $\exp(-|t|^\alpha)$, then $(X_1 + \cdots + X_n)/n^{1/\alpha}$ has the same distribution as X_1.

We will return to this topic in Section 3.7. Polya's criterion can also be used to construct some "pathological examples."

Exercise 3.28 Let φ_1 and φ_2 be ch.f.'s. Show that $A = \{t: \varphi_1(t) = \varphi_2(t)\}$ is closed, contains 0, and is symmetric about 0. Show that if A is a set with these properties and $\varphi_1(t) = e^{-|t|}$, there is a φ_2 so that $\{t: \varphi_1(t) = \varphi_2(t)\} = A$.

For some purposes it is nice to have an explicit example of two ch.f.'s that agree on $[-1, 1]$. From Example 3.6 we know that $1 - |t|$ is the ch.f. of the density $(1 - \cos x)/\pi x^2$. Define $\psi(t)$ to be equal to φ on $[-1, 1]$ and periodic with period 2; that is, $\psi(t) = \psi(t + 2)$. The Fourier series for ψ is

$$\psi(u) = \frac{1}{2} + \sum_{n=-\infty}^{\infty} \frac{2}{\pi^2(2n - 1)^2} \exp(i(2n - 1)\pi u).$$

The right-hand side is the ch.f. of a discrete distribution with

$$P(X = 0) = 1/2 \quad \text{and} \quad P(X = (2n - 1)\pi) = 2\pi^{-2}(2n - 1)^{-2} \; n \in \mathbb{Z}.$$

Exercise 3.29 Find independent r.v.'s X, Y, and Z so that Y and Z do not have the same distribution but $X + Y$ and $X + Z$ do.

Exercise 3.30 Show that if X and Y are independent and $X + Y$ and X have the same distribution, then $Y \equiv 0$.

For more curiosities see Feller, Vol. II (1971), Section XV.2a.

Method of moments Suppose $\int x^k \, dF_n(x)$ has a limit μ_k for each k. Then the sequence of distributions is tight by (2.6) and every subsequential limit has the moments μ_k by Exercise 2.12, so we can conclude the sequence converges weakly if there is only one distribution with these moments. It is easy to see that this is true if F is concentrated on a finite interval $[-M, M]$, since every continuous function can be approximated uniformly on $[-M, M]$ by polynomials. The result is false in general.

Counterexample 1   Heyde (1963). Consider the *lognormal density*

$$f_0(x) = (2\pi)^{-1/2} x^{-1} \exp(-(\log x)^2/2), \qquad x \geq 0,$$

and for $-1 \leq a \leq 1$, let

$$f_a(x) = f_0(x)\{1 + a \sin(2\pi \log x)\}.$$

To see that f_a is a density and has the same moments as f_0, it suffices to show that

$$\int_0^\infty x^r f_0(x) \sin(2\pi \log x) \, dx = 0 \quad \text{for } r = 0, 1, 2, \ldots .$$

Changing variables $x = \exp(s + r), s = \log x - r, ds = dx/x$, the integral becomes

$$(2\pi)^{-1/2} \int_0^\infty \exp(rs + r^2) \exp(-(s+r)^2/2) \sin(2\pi(s+r)) \, ds$$

$$= (2\pi)^{-1/2} \exp(r^2/2) \int_0^\infty \exp(-s^2/2) \sin(2\pi s) \, ds = 0.$$

The two equalities hold because r is an integer and the integrand is odd. From the proof it should be clear that we could let

$$g(x) = f_0(x)\left\{1 + \sum_{k=1}^\infty a_k \sin(k\pi \log x)\right\} \quad \text{if } \sum_{k=1}^\infty |a_k| \leq 1$$

to get a large family of densities having the same moments as the lognormal.

The moments of the lognormal are easy to compute, for if $X = \exp(\chi)$,

$$EX^n = E \exp(n\chi) = \exp(n^2/2).$$

Somewhat remarkably there is a family of discrete random variables with these moments. Let $a > 0$ and

$$P(Y_a = ae^k) = a^{-k} \exp(-k^2/2)/c_a \quad \text{for } k \in \mathbb{Z}$$

where c_a is chosen to make the total mass 1. Computing the ch.f. and differentiating give

$$\left[\frac{d}{dt}\right]^n E \exp(it\,Y_a) = \sum_k (iae^k)^n a^{-k} \exp(-k^2/2) \exp(itae^k)/c_a.$$

So from Exercise 3.19,

$$\exp(-n^2/2)\,EY_a^n = \sum_k a^{-(k-n)} \exp(-(k-n)^2/2)/c_a = 1$$

by the definition of c_a. This example is due to Leipnik (1981).

The lognormal density decays fairly slowly as $|x| \to \infty$. The next counter-example has more rapid decay.

Counterexample 2   Let $\lambda \in (0, 1)$ and for $-1 \le a \le 1$, let

$$f_{a,\lambda}(x) = c_\lambda \exp(-|x|^\lambda)\{1 + a\,\sin(\beta|x|^\lambda \mathrm{sgn}(x))\}$$

where

$$\beta = \tan(\lambda\pi/2) \quad \text{and} \quad c_\lambda = 1\Big/\int \exp(-|x|^\lambda)\,dx.$$

To prove that these are density functions and that for a fixed value of λ they have the same moments, it suffices to show

$$\int x^n \exp(-|x|^\lambda)\,\sin(\beta|x|^\lambda \mathrm{sgn}(x))\,dx = 0 \quad \text{for } n = 0, 1, 2, \dots .$$

This is clear for even n since the integrand is odd. To prove the result for odd n, it suffices to integrate over $[0, \infty)$. Using the identity

$$\int_0^\infty t^{p-1} e^{-qt}\,dt = \Gamma(p)/q^p \quad \text{when Re } q > 0$$

with $p = (n+1)/\lambda$, $q = 1 + \beta i$, and changing variables $t = x^\lambda$, we get

$$\Gamma((n+1)/\lambda)/(1 + \beta i)^{(n+1)/\lambda}$$

$$= \int_0^\infty x^{\lambda\{(n+1)/\lambda - 1\}} \exp(-(1 + \beta i)x^\lambda)\lambda x^{\lambda - 1}\,dx$$

$$= \lambda \int_0^\infty x^n \exp(-x^\lambda)\cos(\beta x^\lambda)\,dx - i\lambda \int_0^\infty x^n \exp(-x^\lambda)\sin(\beta x^\lambda)\,dx.$$

Since $\beta = \tan(\lambda\pi/2)$,

$$(1 + \beta i)^{(n+1)/\lambda} = (\cos \lambda\pi/2)^{-(n+1)/\lambda} (\exp(i\lambda\pi/2))^{(n+1)/\lambda}.$$

The right-hand side is real since $\lambda < 1$ and $(n + 1)$ is even so

$$\int_0^\infty x^n \exp(-x^\lambda) \sin(\beta x^\lambda)\, dx = 0.$$

A useful sufficient condition for a distribution to be determined by its moments is:

(3.9) THEOREM   If $\limsup_{k\to\infty} \mu_{2k}^{1/2k}/2k = r < \infty$, then there is at most one distribution function (d.f.) F with

$$\mu_k = \int x^k\, dF(x).$$

Remark   This is slightly stronger than *Carleman's condition*,

$$\sum_{k=1}^\infty 1/\mu_{2k}^{1/2k} = \infty,$$

which is also sufficient for the conclusion of (3.9). If Y and Z are independent exponentials with mean 1, and $X = Y \log (1 + Z)$, then Carleman's condition holds but (3.9) does not. Examples like this are rare so we will be content to prove (3.9).

Proof   Let F be any d.f. with the moments μ_k and let $v_k = \int |x|^k\, dF(x)$. The Cauchy–Schwarz inequality implies $v_{2k+1}^2 \le \mu_{2k}\mu_{2k+2}$ so

$$\limsup_{k\to\infty} v_k^{1/k} = r < \infty.$$

Multiplying (3.6) by $e^{i\theta x}$, we have

$$\left| e^{i\theta x}\left[e^{itx} - \sum_{m=0}^{n-1} \frac{(itx)^m}{m!} \right] \right| \le \frac{|tx|^n}{n!}.$$

Taking expected values gives

$$\left| \varphi(\theta + t) - \varphi(\theta) - t\varphi'(\theta) - \cdots - \frac{t^{n-1}}{(n-1)!} \varphi^{(n-1)}(\theta) \right| \le \frac{|t|^n}{n!} v_n.$$

Using the last result and the trivial bound $e^k \ge k^k/k!$ (expand the left-hand side in its power series), we see that for any θ,

$$\varphi(\theta + t) = \varphi(\theta) + \sum_{m=1}^{\infty} \frac{t^m}{m!} \varphi^{(m)}(t) \quad \text{for } |t| < 1/er. \tag{*}$$

Let G be another distribution with the given moments and ψ its ch.f. Since $\varphi(0) = \psi(0) = 1$, it follows from (*) and induction that $\varphi(t) = \psi(t)$ for $|t| \le k/3r$ for all k, so the two ch.f's coincide and the distributions must be equal. $\square$

Combining (3.9) with the discussion that began our consideration of the method of moments gives:

(3.10) THEOREM   Suppose $\int x^k \, dF_n(x)$ has a limit μ_k for each k and

$$\limsup_{k \to \infty} \mu_{2k}^{1/2k}/2k < \infty;$$

then F_n converges weakly to the unique distribution with these moments.

Exercise 3.31   Let $G(x) = P(|X| < x)$, $\lambda = \sup\{x: G(x) < 1\}$, and $v_k = E|X|^k$. Show that $v_k^{1/k} \to \lambda$, so (3.9) holds if $\lambda < \infty$.

Exercise 3.32   Suppose $|X|$ has density $Cx^\alpha \exp(-x^\lambda)$ on $(0, \infty)$. Changing variables $y = x^\lambda$,

$$E|X|^n = \int_0^\infty y^{(n+\alpha)/\lambda} \exp(-y)y^{1/\lambda-1} \, dy = \Gamma((n + \alpha + 1)/\lambda).$$

Use the identity $\Gamma(x + 1) = x\Gamma(x)$ for $x \ge 0$ to conclude that (3.9) is satisfied for $\lambda \ge 1$ but not for $\lambda < 1$. This shows the normal ($\lambda = 2$) and gamma ($\lambda = 1$) distributions are determined by their moments. Recalling Counterexample 2 (and generalizing the last computation slightly), we see that (3.9) is not too far from the best possible result.

Remark   Our results so far have been for the so-called *Hamburger moment problem*. If we assume a priori that the distribution is concentrated on $[0, \infty)$, we have the *Stieltjes moment problem*. There is a 1–1 correspondence between $X \ge 0$ and symmetric distributions on $\mathbb{R}$ given by $X \to \xi X^2$ where $\xi \in \{-1, 1\}$ is independent of X and takes its two values with equal probability. From this we see that

$$\limsup_{k \to \infty} v_k^{1/2k}/2k < \infty$$

is sufficient for there to be a unique distribution on $[0, \infty)$ with the given moments.

The next example shows that for nonnegative random variables, the last result is close to the best possible.

Counterexample 3   Let $\lambda \in (0, 1/2)$, $\beta = \tan(\lambda\pi)$, $-1 \le a \le 1$, and

$$f_a(x) = c_\lambda \exp(-x^\lambda)(1 + a \sin(\beta x^\lambda)) \quad \text{for } x \ge 0$$

where

$$c_\lambda = 1 / \int_0^\infty \exp(-x^\lambda) \, dx.$$

By imitating the calculations in Counterexample 2, it is easy to see that the f_a are probability densities that have the same moments. This example seems to be due to Stoyanov (1987), pp. 92–93. The special case $\lambda = 1/4$ is widely known.

4 The Lindeberg–Feller Theorem

Given the developments in the last section, it is now easy to generalize the proof given at the end of Section 1 to establish:

(4.1) *The Central Limit Theorem*   Let X_1, X_2, ... be i.i.d. with $EX_i = \mu$ and $\text{var}(X_i) = \sigma^2 \in (0, \infty)$. If $S_n = X_1 + \cdots + X_n$, then

$$(S_n - n\mu)/\sigma n^{1/2} \Rightarrow \chi,$$

where χ has the standard normal distribution.

Mnemonic   To explain our notation, recall that the square of a standard normal has a chi-square distribution.

Proof   By considering $X_i' = X_i - \mu$, it suffices to prove the result when $\mu = 0$. From (3.8),

$$\varphi(t) = E \exp(itX_1) = 1 - \frac{\sigma^2 t^2}{2} = o(t^2),$$

so

$$E \exp(itS_n/\sigma n^{1/2}) = \left(1 - \frac{t^2}{2n} + o(n^{-1})\right)^n.$$

From (1.2) it should be clear that the last quantity $\to \exp(-t^2/2)$ as $n \to \infty$, which with (3.4) completes the proof. However (1.2) is a fact about real numbers, so we need to extend it to the complex case to complete the proof. $\square$

(4.2) THEOREM   If $c_n \to c \in \mathbb{C}$, then $(1 + c_n/n)^n \to e^c$.

The proof is based on two simple facts:

(4.3) LEMMA   Let $z_1, \ldots, z_n$ and $w_1, \ldots, w_n$ be complex numbers of modulus ≤ 1. Then

$$\left| \prod_{m=1}^{n} z_m - \prod_{m=1}^{n} w_m \right| \leq \sum_{m=1}^{n} |z_m - w_m|.$$

Proof   The result is true for $n = 1$. To prove it for $n > 1$ observe that

$$\left| \prod_{m=1}^{n} z_m - \prod_{m=1}^{n} w_m \right| \leq \left| z_1 \sum_{m=2}^{n} z_m - z_1 \prod_{m=2}^{n} w_m \right| + \left| z_1 \prod_{m=2}^{n} w_m - w_1 \prod_{m=2}^{n} w_m \right|$$

$$\leq \left| \prod_{m=2}^{n} z_m - \prod_{m=2}^{n} w_m \right| + |z_1 - w_1|$$

and use induction.   □

(4.4) LEMMA   If b is a complex number with $|b| < 1$, then $|e^{-b} - (1 - b)| \leq |b|^2$.

Proof

$$e^{-b} - (1 - b) = b^2/2 - b^3/3! + b^4/4! - \ldots,$$

so

$$|e^{-b} - (1 - b)| \leq |b|^2/2(1 + 1/2 + 1/2^2 + \ldots) = |b|^2.$$ □

The proof of (4.2) is easy now: Let $z_m = (1 - c_n/n)$ and $w_m = \exp(-c_n/n)$. It follows from (4.3) and (4.4) that

$$|(1 - c_n/n)^n - \exp(-c_n)| \leq n|c_n/n|^2 \to 0 \quad \text{as } n \to \infty.$$

From the last argument it should be clear that:

(4.5) THEOREM   If $\sum_{m=1}^{n} a_{n,m} \to a$ and $\sum_{m=1}^{n} |a_{n,m}|^2 \to 0$, then $\prod_{m=1}^{n} (1 - a_{n,m}) \to e^{-a}$.

In the first two exercises $X_1, X_2, \ldots$ and S_n are as in (4.1).

Exercise 4.1   Suppose $\mu = 1$. Show that $\dfrac{2}{\sigma} \left[\sqrt{S_n} - \sqrt{n} \right] \Rightarrow \chi$.

Exercise 4.2 Suppose $\mu = 0$. (i) Use the central limit theorem and Kolmogorov's 0–1 law to conclude that $\limsup S_n/\sqrt{n} = \infty$ a.s. (ii) Show that $S_n/\sqrt{n}$ does not converge in probability.

Exercise 4.3 Let $X_1, X_2, \ldots$ be i.i.d. and let $S_n = X_1 + \cdots + X_n$. Assume that $S_n/\sqrt{n} \Rightarrow$ a limit and conclude that $EX_i^2 < \infty$ and $EX_i = 0$.

Sketch Suppose $EX_i^2 = \infty$. Let $X_1', X_2', \ldots$ be an independent copy of the original sequence. Let $Y_i = X_i - X_i'$, $U_i = Y_i 1_{(|Y_i| \le A)}$, $V_i = Y_i 1_{(|Y_i| > A)}$, and observe

$$P\left(\sum_{m=1}^{n} Y_m \ge K\sqrt{n}\right) \ge P\left(\sum_{m=1}^{n} U_m \ge K\sqrt{n}, \sum_{m=1}^{n} V_m \ge 0\right)$$

$$\ge \frac{1}{2} P\left(\sum_{m=1}^{n} U_m \ge K\sqrt{n}\right) \ge \frac{1}{5}$$

for large n if A is large enough.

To get a feel for what the central limit theorem says, we will look at some concrete cases.

Example 4.1 *Coin flips.* Let $X_1, X_2, \ldots$ be i.i.d. with $P(X_i = 0) = P(X_i = 1) = 1/2$. If $X_i = 1$ indicates that a heads occurred on the ith toss, then $S_n = X_1 + \cdots + X_n$ is the total number of heads at time n.

$$EX_i = 1/2 \quad \text{and} \quad \text{var}(X) = EX^2 - (EX)^2 = 1/2 - 1/4 = 1/4,$$

so the central limit theorem tells us $(S_n - n/2)/\sqrt{n/4} \Rightarrow \chi$. From our little table of the normal distribution (see Table 2.1),

$$P(\chi > 2) = 1 - .9773 = .0227,$$

Table 2.1. Table of the normal distribution function

x	$P(\chi \le x)$	x	$P(\chi \le x)$	x	$P(\chi \le x)$
0.1	.5398	1.1	.8643	2.1	.9821
0.2	.5793	1.2	.8849	2.2	.9861
0.3	.6179	1.3	.9032	2.3	.9893
0.4	.6554	1.4	.9192	2.4	.9918
0.5	.6915	1.5	.9332	2.5	.9938
0.6	.7257	1.6	.9452	2.6	.9953
0.7	.7580	1.7	.9554	2.7	.9965
0.8	.7881	1.8	.9641	2.8	.9974
0.9	.8159	1.9	.9713	2.9	.9981
1.0	.8413	2.0	.9773	3.0	.9987

so

$$P(|\chi| \le 2) = 1 - 2(.0227) = .9546,$$

or plugging into the central limit theorem,

$$.95 \approx P((S_n - n/2)/\sqrt{n/4} \in [-2, 2]) = P(S_n - n/2 \in [-\sqrt{n}, \sqrt{n}]).$$

Taking $n = 10{,}000$ this says that 95% of the time the number of heads will between 4900 and 5100.

Example 4.2   *Roulette.* A roulette wheel has slots numbered 1–36 (18 red and 18 black) and two slots numbered 0 and 00 that are painted green. Players can bet $1 that the ball will land in a red (or black) slot and win $1 if it does. If we let X_i be the winnings on the ith play, then $X_1, X_2, \dots$ are i.i.d. with $P(X_i = 1) = 18/38$ and $P(X_i = -1) = 20/38$.

$$EX_i = -1/19 \quad \text{and} \quad \text{var}(X) = EX^2 - (EX)^2 = 1 - (1/19)^2 = .9972.$$

We are interested in

$$P(S_n \ge 0) = P(S_n - n\mu)/\sigma\sqrt{n} \ge -n\mu/\sigma\sqrt{n}).$$

Taking $n = 361 = 19^2$ and replacing σ by 1 to keep computations simple, the central limit theorem tells us that

$$P(S_n \ge 0) \approx P(\chi \ge 1) = 1 - .8413 = .1587.$$

In words, after 361 spins of the roulette wheel, the casino will have won $19 of your money on the average, but there is a probability of about .16 that you will be ahead.

Example 4.3   *Normal approximation to the binomial.* Let $X_1, X_2, \dots$ and S_n be as in Example 4.1. To estimate $P(S_{16} = 8)$ using the central limit theorem, we regard 8 as the interval [7.5, 8.5]. Since $\mu = 1/2$ and $\sigma\sqrt{n} = 2$ for $n = 16$,

$$P(|S_{16} - 8| \le .5) = P(|S_n - n\mu|/\sigma\sqrt{n} \le .25) \approx P(|\chi| \le .25)$$

$$= 2(.5987 - .5) = .1974.$$

Even though n is small, this agrees well with the exact probability

$$\begin{bmatrix} 16 \\ 8 \end{bmatrix} 2^{-16} = \frac{13 \cdot 11 \cdot 10 \cdot 9}{65{,}536} = .1964.$$

The computations above motivate the *histogram correction*, which is important in using the normal approximation for small n. If we are going to approximate

$P(S_{16} \le 11)$, we should regard this probability as $P(S_{16} \le 11.5)$. One obvious reason for doing this is to get the same answer if we regard $P(S_{16} \le 11) = 1 - P(S_{16} \ge 12)$.

Exercise 4.4   Suppose you roll a die 180 times. Use the normal approximation (with the histogram correction) to estimate the probability you will get fewer than 25 sixes.

Example 4.4   1,359,670 boys and 1,285,086 girls were born in Switzerland from 1871 to 1900. Are these data consistent with the hypothesis that both types of birth are equally likely?

$$n = 2,644,756, \quad n\mu = 1,322,378, \quad \sigma\sqrt{n} \approx 813,$$

$$(1,359,670 - 1,322,378)/813 = 45.86.$$

Since $P(\chi \ge 45) \approx \frac{1}{45}e^{-(45)^2/2}$ by (1.3) in Chapter 1, we can safely say NO.

Example 4.5   *Normal approximation to the Poisson.* Let Z_λ have a Poisson distribution with mean λ. If $X_1, X_2, \ldots$ are independent and have Poisson distributions with mean 1, then $S_n = X_1 + \cdots + X_n$ has a Poisson distribution with mean n. Since $\text{var}(X_i) = 1$, the central limit theorem implies:

$$(S_n - n)/n^{1/2} \Rightarrow \chi \quad \text{as } n \to \infty.$$

Comparing Z_λ with $S_{[\lambda]}$ and $S_{[\lambda]+1}$, it follows that

$$(Z_\lambda - \lambda)/\lambda^{1/2} \Rightarrow \chi \quad \text{as } \lambda \to \infty.$$

Using the last result with $\lambda = n$(or the one above it), it follows that

$$e^{-n}(1 + n + n^2/2! + \cdots + n^n/n!) \to 1/2.$$

Exercise 4.5   Sharpen the conclusion of Example 5.4 in Chapter 1 by showing

$$\lim_{\theta \to \infty} \sum_{k=0}^{[\theta x]} \frac{(-1)^k}{k!} \theta^k \varphi^{(k)}(\theta) = \frac{1}{2}\{\mu([0, x]) + \mu([0, x))\}.$$

Example 4.6   *Pairwise independence* is good enough for the strong law of large numbers (see(8.4) in Chapter 1). It is not good enough for the central limit theorem. Let $\xi_1, \xi_2, \ldots$ be i.i.d. with $P(\xi_i = 1) = P(\xi_i = -1) = 1/2$. We will arrange things so that

$$S_{2^n} = \xi_1(1 + \xi_2)\cdots(1 + \xi_{n+1}) = \begin{cases} \pm 2n & 2^{-n} \\ & \text{with prob} \\ 0 & 1 - 2^{-n+1} \end{cases}$$

To do this we let $X_1 = \xi_1$, $X_2 = \xi_1\xi_2$, and for $m = 2^{n-1} + j, 0 < j \le 2^{n-1}, n \ge 2$,

$$X_m = X_j \xi_{n+1}.$$

Each X_m is a product of a different set of ξ_j's, so they are pairwise independent.

Our next step is to generalize the central limit theorem to "triangular arrays":

(4.6) The Lindeberg–Feller Theorem   For each n, let $X_{n,m}, 1 \le m \le n$, be independent random variables with $EX_{n,m} = 0$. Suppose

$$\sum_{m=1}^{n} EX_{n,m}^2 \to \sigma^2 > 0, \text{ and} \tag{i}$$

$$\text{for all } \varepsilon > 0, \lim_{n \to \infty} \sum_{m=1}^{n} E(|X_{n,m}|^2; |X_{n,m}| > \varepsilon) = 0. \tag{ii}$$

Then $S_n = X_{n,1} + \cdots + X_{n,n} \Rightarrow \sigma \chi$ as $n \to \infty$.

Remarks   In words, a sum of a large number of small independent errors has approximately a normal distribution. (ii) implies

$$\lim_{n \to \infty} \sum_{m=1}^{n} P(|X_{n,m}| > \varepsilon) = 0. \tag{iii}$$

If we assume (iii), then (ii) is necessary for the conclusion. This is Feller's contribution. As usual we will not prove the necessity. See Billingsley (1979), pp. 314–315. To see that (4.6) contains our first central limit theorem, let $Y_1, Y_2, \ldots$ be i.i.d. with $EY_i = 0$ and $EY_i^2 = \sigma^2 \in (0, \infty)$, and let $X_{n,m} = Y_m / n^{1/2}$. Then $\sum_{m \le n} EX_{n,m}^2 = \sigma^2$ and if $\varepsilon > 0$,

$$\sum_{m=1}^{n} E(|X_{n,m}|^2; |X_{n,m}| > \varepsilon) = nE(|Y_1/n^{1/2}|^2; |Y_1/n^{1/2}| > \varepsilon)$$

$$= E(|Y_1|^2; |Y_1| > \varepsilon n^{1/2}) \to 0$$

by the dominated convergence theorem, since $EY_1^2 < \infty$.

(4.6) is so important that we will give two proofs. The first is based on characteristic functions. The second uses truncation and the method of moments.

First proof   Let $\varphi_{n,m}(t) = E \exp(itX_{n,m})$, $\sigma_{n,m}^2 = EX_{n,m}^2$. We want to show that

$$\prod_{m=1}^{n} \varphi_{n,m}(t) \to \exp(-t^2/2).$$

Let $z_m = \varphi_{n,m}(t)$ and $w_m = (1 - t^2 \sigma_{n,m}^2/2)$. By (3.6),

$$|z_m - w_m| \le E(|tX_{n,m}|^3/3! \wedge 2|tX_{n,m}|^2/2!)$$

$$\le E(|tX_{n,m}|^3/6; |X_{n,m}| \le \varepsilon) + E(|tX_{n,m}|^2; |X_{n,m}| > \varepsilon)$$

$$\le \frac{\varepsilon t^3}{6} E(|X_{n,m}|^2; |X_{n,m}| \le \varepsilon) + t^2 E(|X_{n,m}|^2; |X_{n,m}| > \varepsilon).$$

Summing $m = 1$ to n, letting $n \to \infty$, and using (i) and (ii) give

$$\limsup_{n \to \infty} \sum_{m=1}^{n} |z_m - w_m| \le \varepsilon t^3 \sigma^2/6.$$

Since $\varepsilon > 0$ is arbitrary, it follows that the sequence converges to 0 and using (4.3) gives

$$\left| \prod_{m=1}^{n} \varphi_{n,m}(t) - \prod_{m=1}^{n} (1 - t^2 \sigma_{n,m}^2/2) \right| \to 0.$$

To complete the proof now we apply (4.5) with $a_{n,m} = t^2 \sigma_{n,m}^2/2$. (i) implies

$$\sum_{m=1}^{n} a_{n,m} \to \sigma^2 t^2/2.$$

To check the other condition we observe that

$$\sigma_{n,m}^2 \le \varepsilon^2 + E(|X_{n,m}|^2; |X_{n,m}| > \varepsilon),$$

so (ii) implies $\sup_m \sigma_{n,m}^2 \to 0$, and

$$\sum_{m=1}^{n} t^4 \sigma_{n,m}^4/4 \le (t^4/4) \left(\sup_m \sigma_{n,m}^2 \right) \sum_{m=1}^{n} \sigma_{n,m}^2 \to 0.$$

Second proof   The first step is to truncate to get r.v.'s that satisfy:

$$|Z_{n,m}| \le \delta_n \quad \text{with } \delta_n \to 0. \tag{ii'}$$

The first step is to let

$$h_n(\varepsilon) = \sum_{m=1}^{n} E(X_{n,m}^2; |X_{n,m}| > \varepsilon)$$

and observe:

(4.7) LEMMA $h_n(\varepsilon) \to 0$ for each fixed $\varepsilon > 0$, so we can pick $\varepsilon_n \to 0$ so that $h_n(\varepsilon_n) \to 0$.

Proof of (4.7)   Let N_m be chosen so that $h_n(1/m) \leq 1/m$ for $n \geq N_m$ and $m \to N_m$ is increasing. Let $\varepsilon_n = 1/m$ for $N_m \leq n < N_{m+1}$ and $= 1$ for $n < N_1$. When $N_m \leq n < N_{m+1}$, $\varepsilon_n = 1/m$, so $|h_n(\varepsilon_n)| \leq |h_n(1/m)| \leq 1/m$ and the desired result follows.

Let $X'_{n,m} = X_{n,m} 1_{(|X_{n,m}| > \varepsilon_n)}$

$\qquad Y_{n,m} = X_{n,m} 1_{(|X_{n,m}| \leq \varepsilon_n)}$

$\qquad Z_{n,m} = Y_{n,m} - E Y_{n,m}.$

Clearly $|Z_{n,m}| \leq 2\varepsilon_n$, so (ii') holds with $\delta_n = 2\varepsilon_n$. Since $E Y_{n,m} = -E X'_{n,m}$,

$$E\left(\sum_{m=1}^{n} X_{n,m} - \sum_{m=1}^{n} Z_{n,m}\right)^2 = E\left(\sum_{m=1}^{n} X'_{n,m} - E X'_{n,m}\right)^2$$

$$\leq E \sum_{m=1}^{n} (X'_{n,m} - E X'_{n,m})^2 \leq E \sum_{m=1}^{n} (X'_{n,m})^2 \to 0$$

as $n \to \infty$, so by Exercise 2.8 it suffices to show $T_n \equiv \sum_{m \leq n} Z_{n,m} \Rightarrow \sigma\chi$. Notice that the last computation and (i) imply

$$E T_n^2 = \sum_{m=1}^{n} E Z_{n,m}^2 \to \sigma^2.$$

To compute higher moments we observe

$$T_n^r = \sum_{k=1}^{r} \sum{}' \frac{r!}{r_1! \cdots r_k!} \frac{1}{k!} \sum{}'' Z_{n,i_1}^{r_1} \cdots Z_{n,i_k}^{r_k}$$

where $\sum'$ extends over all k-tuples of positive integers with $r_1 + \cdots + r_k = r$ and $\sum''$ extends over all k-tuples of distinct integers $1 \leq i \leq n$. If we let

$$A_n(r_1, \ldots, r_k) = \sum{}'' E Z_{n,i_1}^{r_1} \cdots E Z_{n,i_k}^{r_k},$$

then

$$E T_n^r = \sum_{k=1}^{r} \sum{}' \frac{r!}{r_1! \cdots r_k!} \frac{1}{k!} A_n(r_1, \ldots, r_k).$$

To evaluate the limit of $E T_n^r$ we observe:

(a) If some $r_j = 1$, then $A_n(r_1, \ldots, r_k) = 0$, since $E Z_{n,i_j} = 0$.

(b) If all $r_j = 2$, then

$$\sum{}'' E Z_{n,i_1}^2 \cdots E Z_{n,i_k}^2 \leq \left[\sum_{m=1}^{n} E Z_{n,m}^2\right]^k \to \sigma^{2k},$$

and so

$$\left[\sum_{m=1}^{n} EZ_{n,m}^2\right]^k - \sum'' EZ_{n,i_1}^2 \cdots EZ_{n,i_k}^2 \le \binom{k}{2}\delta_n^2\left[\sum_{m=1}^{n} EZ_{n,m}^2\right]^{k-1} \to 0.$$

(c) If some $r_j > 2$, then

$$A_n(r_1, \ldots, r_k) \le \sigma_n^{r-2k} A_n(2, \ldots, 2) \to 0.$$

When r is odd, some r_j must be $= 1$ or ≥ 3, so $ET_n^r \to 0$ by (a) and (c). If $r = 2k$ is even, (a)–(c) imply

$$ET_n^r \to (2k!/2^k k!)\sigma^{2k} = E(\sigma\chi)^r,$$

and the result follows from (3.10). $\qquad\square$

Exercise 4.6   Use (4.6) to prove *Lyapunov's theorem*. Let $X_1, X_2, \ldots$ be independent and let $\alpha_n = \{\text{var}(S_n)\}^{1/2}$. If there is a $\delta > 0$ so that

$$\lim_{n\to\infty} \alpha_n^{-(2+\delta)} \sum_{m=1}^{n} E(|X_m - EX_m|^{2+\delta}) = 0,$$

then $(S_n - ES_n)/\alpha_n \Rightarrow \chi$.

Exercise 4.7   Let $X_1, X_2, \ldots$ be independent and have $|X_i| \le M$. If $\sum \text{var}(X_n) = \infty$, there are constants a_n, b_n so that $(S_n - b_n)/a_n \Rightarrow \chi$.

Exercise 4.8   Let $X_1, X_2, \ldots$ be independent with

$$P(X_m = m) = P(X_m = -m) = m^{-2}/2$$

$$P(X_m = 1) = P(X_m = -1) = \tfrac{1}{2}(1 - m^{-2}).$$

Let $S_n = X_1 + \cdots + X_n$. Show that $\text{var}(S_n)/n \Rightarrow 2$ but $S_n/\sqrt{n} \Rightarrow \chi$. The trouble here is that $X_{n,m} = X_m/\sqrt{n}$ does not satisfy (ii) in (4.6).

Exercise 4.9   Let $Y_1, Y_2, \ldots$ be independent with $P(Y_j = j) = P(Y_j = -j) = 1/2j^\beta$ and $P(Y_j = 0) = 1 - j^{-\beta}$, and let $S_n = Y_1 + \cdots + Y_n$. Show: (i) If $\beta > 1$, then $S_n \to S_\infty$ a.s. (ii) If $\beta < 1$, then $S_n/n^{(3-\beta)/2} \Rightarrow c\chi$. (iii) If $\beta = 1$, then $S_n/n \Rightarrow \aleph$ where

$$E \exp(it\aleph) = \exp\left(-\int_0^1 x^{-1}(1 - \cos xt)\, dx\right).$$

How are the conclusions in (i)–(iii) affected if we let $P(Y_j = j^\alpha) = P(Y_j = -j^\alpha) = 1/2j^\beta$ with $\alpha > 0$?

Exercise 4.10   Let $X_1^n, \ldots, X_n^n$ be independent and uniformly distributed over $[-n, n]$. Let

$$F_n = \sum_{m=1}^{n} \text{sgn}(X_m^n)/|X_m^n|^p$$

where $p < 1/2$. Show that $F_n/n^{1/2-p} \Rightarrow c\chi$. See Exercise 4.11 for the case $p = 1/2$, Exercise 7.7 for the case $p > 1/2$.

Example 4.7   *Cycles in a random permutation and record values.* Continuing the analysis of Examples 5.8 and 6.2 in Chapter 1, let $Y_1, Y_2, \ldots$ be independent with $P(Y_m = 1) = 1/m$, and $P(Y_m = 0) = 1 - 1/m$. $EY_m = 1/m$ and $\text{var}(Y_m) = 1/m - 1/m^2$. So if $S_n = Y_1 + \cdots + Y_n$, then $ES_n \sim \log n$ and $\text{var}(S_n) \sim \log n$. Let

$$X_{n,m} = (Y_m - 1/m)/(\log n)^{1/2}.$$

$EX_{n,m} = 0$, $\sum_{m=1}^{n} EX_{n,m}^2 \to 1$, and for any $\varepsilon > 0$,

$$\sum_{m=1}^{n} E(|X_{n,m}|^2; |X_{n,m}| > \varepsilon) \to 0$$

since the sum is 0 as soon as $(\log n)^{-1/2} < \varepsilon$. Applying (4.6) now gives

$$\left(S_n - \sum_{m=1}^{n} \frac{1}{m}\right)/(\log n)^{1/2} \Rightarrow \chi.$$

Observing that

$$\sum_{m=1}^{n-1} \frac{1}{m} \geq \int_1^n x^{-1}\, dx = \log n \geq \sum_{m=2}^{n} \frac{1}{m}$$

shows

$$\left|\log n - \sum_{m=1}^{n} \frac{1}{m}\right| \leq 1,$$

so the conclusion can be written as

$$(S_n - \log n)/(\log n)^{1/2} \Rightarrow \chi.$$

The last result gives a central limit theorem for the number of records by time n or the number of cycles in a random permutation of length n.

Example 4.8   *The converse of the three series theorem.* Let $X_1, X_2, \ldots$ be independent, let $A > 0$, and let $Y_m = X_m 1_{(|X_m| \leq A)}$. In order that $\sum X_m$ converges (i.e., $\lim_{n\to\infty} \sum_{m \leq n} X_m$ exists), it is necessary that:

(i) $\sum_{m=1}^{\infty} P(|X_m| > A) < \infty$,   (ii) $\sum_{m=1}^{\infty} EY_m$ converges,

and   (iii) $\sum_{m=1}^{\infty} \text{var}(Y_m) < \infty$.

Proof   The necessity of the first condition is clear, for if that sum is infinite, $P(|X_n| > A$ i.o.) > 0 and $\lim_{n\to\infty} \sum_{m\leq n} X_m$ cannot exist. Suppose next that the sum in (i) is finite but the sum in (iii) is infinite. Let

$$c_n = \sum_{m=1}^{n} \text{var}(Y_m) \quad \text{and} \quad X_{n,m} = (Y_m - EY_m)/c_n^{1/2}.$$

$EX_{n,m} = 0$, $\sum_{m\leq n} EX_{n,m}^2 = 1$, and for any $\varepsilon > 0$,

$$\sum_{m=1}^{n} E(|X_{n,m}|^2; |X_{n,m}| > \varepsilon) \to 0,$$

since the sum is 0 as soon as $2A/c_n^{1/2} < \varepsilon$. Applying (4.6) now gives that if $S_n = X_{n,1} + \cdots + X_{n,n}$, then $S_n \Rightarrow \chi$. Now if $\lim_{n\to\infty} \sum_{m\leq n} X_m$ exists, $\lim_{n\to\infty} \sum_{m\leq n} Y_m$ does, and if we let $T_n = (\sum_{m\leq n} Y_m)/c_n^{1/2}$, then $T_n \Rightarrow 0$. The last two results and Exercise 2.8 imply $(S_n - T_n) \Rightarrow \chi$. Since $S_n - T_n = (\sum_{m\leq n} EY_m)/c_n^{1/2}$ is nonrandom, this is absurd. Finally assume the series in (i) and (iii) are finite. (7.3) in Chapter 1 implies that $\lim_{n\to\infty} \sum_{m\leq n} (Y_m - EY_m)$ exists, so if $\lim_{n\to\infty} \sum_{m\leq n} X_m$ and hence $\lim_{n\to\infty} \sum_{m\leq n} Y_m$ do, taking differences shows that (ii) holds.   □

Example 4.9   *Infinite variance.* Suppose X_1, X_2, ... are i.i.d. with $P(X_1 > x) = P(X_1 < -x)$ and $P(|X_1| > x) = x^{-2}$ for $x \geq 1$.

$$E|X_1|^2 = \int_0^{\infty} 2xP(|X_1| > x)\, dx = \infty,$$

but it turns out that when $S_n = X_1 + \cdots + X_n$ is suitably normalized, it converges to a normal distribution. Let

$$Y_{n,m} = X_m 1_{(|X_m| \leq n^{1/2} \log \log n)}.$$

The truncation level $c_n = n^{1/2} \log \log n$ is chosen large enough to make

$$\sum_{m=1}^{n} P(Y_{n,m} \neq X_m) \leq nP(|X_1| > c_n) \to 0.$$

However, we want the variance of $Y_{n,m}$ to be as small as possible, so we keep the truncation close to the lowest possible level.

$$EY^2_{m,n} = \int_0^{c_n} 2y[P(|X_1| > y) - P(|X_1| > c_n)] \, dy$$

$$\leq 1 + \int_1^{c_n} 2/y \, dy \sim \log n$$

and

$$EY^2_{m,n} \geq (1 - (\log \log n)^{-2}) \int_1^{\sqrt{n}} 2/y \, dy \sim \log n,$$

so $EY^2_{n,m} \sim \log n$. If $S'_n = Y_{n,1} + \cdots + Y_{n,n}$, then $\mathrm{var}(S'_n) \sim n \log n$, so we apply (4.6) to $X_{n,m} = Y_{n,m}/(n \log n)^{1/2}$. Things have been arranged so that (i) is satisfied. Since $|Y_{n,m}| \leq n^{1/2} \log \log n$, the sum in (ii) is 0 for large n, and it follows that $S'_n/(n \log n)^{1/2} \Rightarrow \chi$. Since the choice of c_n guarantees $P(S_n \neq S'_n) \to 0$, the same result holds for S_n.

Exercise 4.11   Show that if $p = 1/2$ in Exercise 4.10, $F_n/(\log n)^{1/2} \Rightarrow c\chi$.

Remark   In Section 6 we will see that if we replace the assumption $P(|X_1| > x) = x^{-2}$ in Example 4.9 by $P(|X_1| > x) = x^{-\alpha}$ where $0 < \alpha < 2$, then $S_n/n^{1/\alpha} \Rightarrow$ a limit which is not χ. The last word on convergence to the normal distribution is the next result due to Lévy.

(4.8) THEOREM   Let $X_1, X_2, \ldots$ be i.i.d. and $S_n = X_1 + \cdots + X_n$. In order that there exist constants a_n and $b_n > 0$ so that $(S_n - a_n)/b_n \Rightarrow \chi$, it is necessary and sufficient that

$$y^2 P(|X_1| > y)/E(|X_1|; |X_1| \leq y) \to 0.$$

A proof can be found in Gnedenko and Kolmogorov (1954), a reference that contains the last word on many results about sums of independent random variables.

Example 4.10   *Prime divisors.* Our aim here is to prove a result of Erdös and Kac that an integer picked at random from $\{1, 2, \ldots, n\}$ has about

$$\log \log n + \chi \, (\log \log n)^{1/2}$$

prime divisors. Since $\exp(e^4) = 5.15 \times 10^{23}$, this result does not apply to most numbers we encounter in "everyday life." Let P_n denote the uniform distribution on $\{1, \ldots, n\}$. If $P_\infty(A) \equiv \lim P_n(A)$ exists, the limit is called the *density* of $A \subset \mathbb{Z}$. Let A_p be the set of integers divisible by p. Clearly if p is a prime, $P_\infty(A_p) = 1/p$ and $q \neq p$ is another prime

$$P_\infty(A_p \cap A_q) = 1/pq = P_\infty(A_p)P_\infty(A_q).$$

We can interpret this as saying that the events of being divisible by p and q are independent. Let $\delta_p(n) = 1$ if n is divisible by p and $= 0$ otherwise, and

$$g(n) = \sum_{p \leq n} \delta_p(n) \text{ be the number of prime divisors of } n,$$

this and future sums on p being over the primes. Intuitively the $\delta_p(n)$ behave like X_p that are i.i.d. with

$$P(X_p = 1) = 1/p \quad \text{and} \quad P(X_p = 0) = 1 - 1/p.$$

The mean and variance of $\sum_{p \leq n} X_p$ are

$$\sum_{p \leq n} 1/p \quad \text{and} \quad \sum_{p \leq n} 1/p(1 - 1/p),$$

respectively, and it is known that

$$\sum_{p \leq n} 1/p = \log \log n + O(1) \tag{*}$$

(see Hardy and Wright (1959), Chapter XXII). Anyone can see that $\sum_p 1/p^2 < \infty$, so applying (4.6) to X_p and making a small leap of faith give us:

(4.9) Erdös–Kac Central Limit Theorem

$$P_n[m: g(m) - \log \log n \leq x(\log \log n)^{1/2}] \to P(\chi \leq x).$$

Proof   We begin by showing that we can ignore the primes "near" n. Let

$$\alpha_n = n^{1/\log \log n}$$

$$\log \alpha_n = \log n / \log \log n$$

$$\log \log \alpha_n = \log \log n - \log \log \log n.$$

The sequence α_n has two nice properties:

$$\left(\sum_{\alpha_n < p \leq n} 1/p \right) / (\log \log n)^{1/2} \to 0 \quad \text{by}(*). \tag{a}$$

If $\varepsilon > 0$, then $\alpha_n \leq n^\varepsilon$ for large n and hence $\alpha_n^r/n \to 0$ for all $r < \infty$. \tag{b}

Let $g_n(m) = \sum_{p \leq \alpha_n} \delta_p(m)$ and let E_n denote expected value with respect to P_n.

$$E_n\left[\sum_{p > \alpha_n} \delta_p \right] = \sum_{\alpha_n < p \leq n} P_n(m: \delta_p(m) = 1) \leq \sum_{\alpha_n < p \leq n} 1/p,$$

so by (a) it is enough to prove the result for g_n. Let

$$S_n = \sum_{p \le a_n} X_p$$

where the X_p are the independent random variables introduced above. Let $b_n = ES_n$ and $a_n^2 = \mathrm{var}(S_n)$. (a) tells us that b_n and a_n^2 are both

$$\log \log n + o((\log \log n)^{1/2}),$$

so it suffices to show

$$P_n[m: g_n(m) - b_n \le xa_n] \to P(\chi \le x).$$

An application of (4.6) shows $(S_n - b_n)/a_n \Rightarrow \chi$, and since $|X_p| \le 1$, it follows from the second proof of (4.6) that

$$E[(S_n - b_n)/a_n]^r \to E\chi^r \quad \text{for all } r.$$

Using notation from that proof

$$ES_n^r = \sum_{k=1}^{r} \sum' \frac{r!}{r_1! \cdots r_k! \, k!} \sum'' E(X_{p_1}^{r_1} \cdots X_{p_k}^{r_k}).$$

Since $X_p \in \{0, 1\}$, the summand is

$$E(X_{p_1} \cdots X_{p_k}) = 1/(p_1 \cdots p_k).$$

A little thought reveals that

$$E_n(\delta_{p_1} \cdots \delta_{p_k}) = \frac{1}{n}[n/(p_1 \cdots p_k)].$$

The two moments differ by $\le 1/n$, so

$$|E(S_n^r) - E_n(g_n^r)| \le \frac{1}{n}\left[\sum_{p \le \alpha(n)} 1\right]^r \le \alpha_n^r/n \to 0$$

by (b). Now

$$E(S_n - b_n)^r = \sum_{m=0}^{r} \binom{r}{m} ES_n^m (-b_n)^{r-m},$$

so

$$|E(S_n - b_n)^r - E(g_n - b_n)^r| \le \sum_{m=0}^{r} \binom{r}{m} \frac{1}{n} \alpha_n^m b_n^{r-m} = (\alpha_n + b_n)^r/n \to 0,$$

since $b_n \leq \alpha_n$. This is more than enough to conclude that

$$E[(g_n - b_n)/a_n]^r \to E\chi^r$$

and the desired result follows from (3.10).   □

The next four exercises treat various extensions of the central limit theorem.

Exercise 4.12   *Random index central limit theorem.* Let $X_1, X_2, \ldots$ be i.i.d. with $EX_i = 0$ and $EX_i^2 = \sigma^2 \in (0, \infty)$, and let $S_n = X_1 + \cdots + X_n$. Let N_n be a sequence of random variables and a_n a sequence of constants with $N_n/a_n \to 1$ in probability. Show that

$$S_{N_n}/\sigma\sqrt{a_n} \Rightarrow \chi.$$

Sketch   Kolmogorov's inequality ((7.2) in Chapter 1) implies

$$P\left(\sup_{(1-\varepsilon)a_n \leq m \leq (1+\varepsilon)a_n} |S_m - S_{(1-\varepsilon)a_n}| > \delta\sqrt{a_n} \right) \leq 2\varepsilon\sigma^2/\delta^2.$$

Use this to conclude that if

$$X_n = S_{N_n}/\sigma\sqrt{a_n} \quad \text{and} \quad Y_n = S_{a_n}/\sigma\sqrt{a_n},$$

then $X_n - Y_n \Rightarrow 0$, so the desired conclusion follows from Exercise 2.8.

Exercise 4.13   *A central limit theorem (CLT) in renewal theory.* Let $X_1, X_2, \ldots$ be i.i.d. positive random variables with $EX_i = \mu$ and $\text{var}(X_i) = \sigma^2$. Let $S_n = X_1 + \cdots + X_n$ and $N_t = \sup\{m: S_m \leq t\}$. Prove that as $n \to \infty$,

$$(\mu N_n - n)/(\sigma^2 n/\mu)^{1/2} \Rightarrow \chi.$$

Sketch of proof $N_n/(n/\mu) \to 1$ by (8.4) in Chapter 1, so by the last exercise

$$(S_{N_n} - \mu N_n)/(\sigma^2 n/\mu)^{1/2} \Rightarrow \chi.$$

To complete the proof now, we observe that $EX_i^2 < \infty$ implies $nP(X_i > \varepsilon n^{1/2}) \to 0$ for all $\varepsilon > 0$, which implies $(S_{N_n} - \mu n)/n^{1/2} \to 0$.

Exercise 4.14   *A second proof of the renewal CLT.* Let $X_1, X_2, \ldots, S_n$, and N_t be as in the last exercise. Let $D_t = S_{t/\mu} - t$. The central limit theorem implies $D_t/\sigma\sqrt{t/\mu} \Rightarrow \chi$, and by the weak law we expect $N_t \approx (t - D_t)/\mu$. To translate this idea into a proof, notice that Kolmogorov's inequality implies

$$P(|X_m - m\mu| > t^{2/5} \text{ for some } m \in [0, t^{3/5}]) \to 0 \quad \text{as } t \to \infty,$$

and apply this to the X_m near t/μ to conclude

$$|N_t - (t - D_t)/\mu|/t^{1/2} \to 0 \text{ in probability.}$$

Exercise 4.15   *Self-normalized sums.* Let $X_1, X_2, \ldots$ be i.i.d. with $EX_i = 0$ and $EX_i^2 = \sigma^2 \in (0, \infty)$. Then

$$\sum_{m=1}^{n} X_m \Big/ \left[\sum_{m=1}^{n} X_m^2 \right]^{1/2} \Rightarrow \chi.$$

Our next result gives a rate of convergence in the central limit theorem.

(4.10) Berry–Esseen Theorem   Let $X_1, X_2, \ldots$ be i.i.d. with $EX_i = 0$, $EX_i^2 = \sigma^2$, and $E|X_i|^3 = \rho < \infty$. If $F_n(x)$ is the distribution of $(X_1 + \cdots + X_n)/\sigma\sqrt{n}$ and $\mathcal{N}(x)$ is the standard normal distribution, then

$$|F_n(x) - \mathcal{N}(x)| \le 3\rho/\sigma^3 \sqrt{n}.$$

Remarks   The reader should note that the inequality holds for all n and x, but since $\rho \ge \sigma^3$, it has only nontrivial content for $n \ge 10$. It is easy to see that the rate cannot be faster than $n^{-1/2}$. When $P(X_i = 1) = P(X_i = -1) = 1/2$,

$$F_{2n}(0) = \frac{1}{2}\{1 + P(S_{2n} = 0)\} = \frac{1}{2} + (2\pi(2n))^{-1/2} + o(n^{-1/2}).$$

The constant 3 is not the best known (now about 0.8, see van Beek (1972)), but as Feller (1971) brags on page 542, "our streamlined method yields a remarkably good bound even though it avoids the usual messy numerical calculations." The hypothesis $E|X|^3$ is needed to get the rate $n^{-1/2}$. Heyde (1967) has shown that for $0 < \delta < 1$,

$$\sum_{n=1}^{\infty} n^{-1+\delta/2} \sup_x |F_n(x) - \mathcal{N}(x)| < \infty$$

if and only if $E|X|^{2+\delta} < \infty$. For this and more on rates of convergence, see Hall (1982).

Proof   Since neither side of the inequality is affected by scaling, we can suppose without loss of generality that $\sigma^2 = 1$. The first phase of the argument is to derive an inequality, (4.13), that relates the difference between the two distributions to the distance between their ch.f.'s. Polya's density (Example 3.6)

$$h_L(x) = \frac{1}{\pi} \frac{1 - \cos Lx}{Lx^2}$$

has ch.f.

$$\omega_L(\theta) = 1 - |\theta|/L \quad \text{for } |\theta| \le L.$$

We will use H_L for its distribution function. We will convolve the distributions under consideration with H_L to get ch.f.'s that have compact support. The first step is to show that convolution with H_L does not reduce the difference between the distributions too much.

(4.11) LEMMA   Let F and G be distribution functions with $G'(x) \le \lambda < \infty$. Let $\Delta(x) = F(x) - G(x)$, $\eta = \sup|\Delta(x)|$, $\Delta_L = \Delta * H_L$, and $\eta_L = \sup|\Delta_L(x)|$. Then

$$\eta_L \geq \frac{\eta}{2} - \frac{12\lambda}{\pi L} \quad \text{or} \quad \eta \leq 2\eta_L + \frac{24\lambda}{\pi L}.$$

Proof Δ goes to 0 at $\pm\infty$, G is continuous, and F is a d.f., so there is an x_0 with $\Delta(x_0) = \eta$ or $\Delta(x_0-) = -\eta$. By looking at the d.f.'s of (-1) times the r.v.'s in the second case, we can suppose without loss of generality that $\Delta(x_0) = \eta$. Since $G'(x) \leq \lambda$ and F is nondecreasing,

$$\Delta(x_0 + s) \geq \eta - \lambda s.$$

Letting $\delta = \eta/2\lambda$ and $t = x_0 + \delta$, we have

$$\Delta(t - x) \geq \frac{\eta}{2} + \lambda x \text{ for } |x| \leq \delta, \quad \geq -\eta \text{ otherwise.}$$

To estimate the convolution Δ_L we observe

$$2 \int_\delta^\infty h_L(x)\, dx \leq 2 \int_\delta^\infty 2/\pi L x^2 \, dx = 4/\pi L \delta,$$

so

$$\eta_L \geq \Delta_L(x_0) \geq \frac{\eta}{2}\left[1 - \frac{4}{\pi L \delta}\right] - \eta\frac{4}{\pi L \delta} = \frac{\eta}{2} - \frac{6}{\pi L \delta} = \frac{\eta}{2} - \frac{12\lambda}{\pi L},$$

proving (4.11). ∎

(4.12) LEMMA   If K is a d.f. whose ch.f. κ is integrable,

$$K(x) = (2\pi)^{-1} \int \{-e^{-itx}/it\}\, \kappa(t)\, dt.$$

Proof If κ is integrable, the inversion formula (3.3) implies

$$K(x) - K(a) = (2\pi)^{-1} \int \{(e^{-ita} - e^{-itx})/it\}\, \kappa(t)\, dt.$$

Letting $a \to -\infty$ and using the Riemann–Lebesgue lemma (Exercise 4.5 in the Appendix) give (4.12). ∎

Applying (4.12) to $F * H_L$ and $G * H_L$ and using (4.11), we have

$$|F(x) - G(x)| \leq \frac{1}{\pi} \int_{-L}^{L} \left|\frac{\varphi(\theta) - \psi(\theta)}{\theta}\right|\, d\theta + \frac{24\lambda}{\pi L} \tag{4.13}$$

where φ and ψ are the ch.f.'s of F and G.

To get (4.10) we will apply (4.13) with $F = F_n$ and $G = \mathcal{N}$ to get

$$|F_n(x) - \mathcal{N}(x)| \leq \frac{1}{\pi} \int_{-L}^{L} \left| \frac{\varphi''(\theta) - \psi(\theta)}{\theta} \right| d\theta + \frac{24\lambda}{\pi L} \tag{4.14}$$

and then estimate the right-hand side. This phase of the argument is routine but there is a fair amount of algebra. To save the reader from trying to improve the inequalities along the way in hopes of getting a better bound, we would like to observe that we have used the fact that $C = 3$ to get rid of the case $n \leq 9$, and we use $n \geq 10$ in part (e).

To estimate the second term in (4.14) we observe that

$$\max G'(x) = G'(0) = (2\pi)^{-1/2} < 2/5. \tag{a}$$

For the first we observe that if $|\alpha|, |\beta| \leq \gamma$,

$$|\alpha^n - \beta^n| \leq \sum_{m=0}^{n-1} |\alpha^{n-m}\beta^m - \alpha^{n-m-1}\beta^{m+1}| \leq n|\alpha - \beta|\gamma^{n-1}. \tag{b}$$

Using (3.7) now gives (recall we are supposing $\sigma^2 = 1$),

$$|\varphi(t) - 1 + t^2/2| \leq \rho|t|^3/6 \tag{c}$$

$$|\varphi(t)| \leq 1 - t^2/2 + \rho|t|^3/6 \quad \text{if } t^2 \leq 2. \tag{d}$$

Let $L = 4\sqrt{n}/3\rho$. If $|\theta| \leq L$, then by (d) and the fact that $\rho|\theta|/\sqrt{n} \leq 4/3$,

$$|\varphi(\theta/\sqrt{n})| \leq 1 - \theta^2/2n + \rho|\theta|^3/6n^{3/2}$$

$$\leq 1 - 5\theta^2/18n \leq \exp(-5\theta^2/18n).$$

We will now apply (b) with

$$\alpha = \varphi(\theta/\sqrt{n}), \quad \beta = \exp(-\theta^2/2n), \quad \gamma = \exp(-5\theta^2/18n).$$

Since we are supposing $n \geq 10$,

$$\gamma^{n-1} \leq \exp(-\theta^2/4). \tag{e}$$

For the other part of (b) we write

$$n|\alpha - \beta| \leq n|\varphi(\theta/\sqrt{n}) - 1 + \theta^2/2n| + n|1 - \theta^2/2n - \exp(-\theta^2/2n)|.$$

To bound the right-hand side, observe (c) implies

$$n|\varphi(\theta/\sqrt{n}) - 1 + \theta^2/2n| \leq \rho|\theta|^3/6n^{1/2},$$

while a standard fact about alternating series gives

$$n|1 - \theta^2/2n - \exp(-\theta^2/2n)| \le \theta^4/8n \quad \text{for } |\theta| \le L \le \sqrt{2n},$$

so

$$n|\alpha - \beta| \le \rho|\theta|^3/6n^{1/2} + \theta^4/8n. \tag{f}$$

Using (f) and (e) in (b) gives

$$|\varphi''(\theta/\sqrt{n}) - \exp(-\theta^2/2)|/|\theta| \le \exp(-\theta^2/4)\{\rho\theta^2/6n^{1/2} + |\theta|^3/8n\} \tag{g}$$

$$\le \frac{1}{L}\exp(-\theta^2/4)\{2\theta^2/9 + |\theta|^3/18\},$$

since $\rho/\sqrt{n} = 4/3L$, and $1/n = 1/\sqrt{n} \cdot 1/\sqrt{n} \le 4/3L \cdot 1/3$, since $\rho \ge 1$ and $n \ge 10$. Using (g) and (b) in (4.13) gives

$$\pi L|F(x) - G(x)| \le \int \exp(-\theta^2/4)\{2\theta^2/9 + |\theta|^3/18\}\, d\theta + 9.6.$$

Recalling $L = 4\sqrt{n}/3\rho$, we see that the last result is of the form $|F(x) - G(x)| \le C\rho/\sqrt{n}$. To evaluate the constant we observe

$$\int (2\pi a)^{-1/2}x^2 \exp(-x^2/2a)\, dx = a,$$

and writing $x^3 = 2x^2 \cdot x/2$ and integrating by parts give

$$2\int_0^\infty x^3 \exp(-x^2/4)\, dx = 2\int_0^\infty 4x \exp(-x^2/4)\, dx = 16.$$

This gives us

$$|F(x) - G(x)| \le \frac{1}{\pi}\cdot\frac{3}{4}\left[\frac{2}{9}2\sqrt{4\pi} + \frac{16}{18} + 9.6\right]\rho/\sqrt{n} < 3\rho/\sqrt{n}.$$

*5 Local Limit Theorems

In Section 1 we saw that if $X_1, X_2, \ldots$ are i.i.d. with $P(X_1 = 1) = P(X_1 = -1) = 1/2$ and k_n is a sequence of integers with $2k_n/(2n)^{1/2} \to x$, then

$$P(S_{2n} = 2k_n) \sim (\pi n)^{1/2} \exp(-x^2/2).$$

In this section we will prove two theorems that generalize the last result. We begin with two definitions. A random variable X has a *lattice distribution* if there are constants b and $h > 0$ so that $P(X \in b + h\mathbb{Z}) = 1$, where $b + h\mathbb{Z} = \{b + hz : z \in \mathbb{Z}\}$. The largest h for which the last statement holds is called the *span* of the distribution.

Example 5.1   If $P(X = 1) = P(X = -1) = 1/2$, then X has a lattice distribution with span 2. When h is 2, one possible choice is $b = -1$.

The next result relates the last definition to the characteristic function.

(5.1) THEOREM   There are only three possibilities.
   (i) $|\varphi(t)| < 1$ for all $t \neq 0$.
   (ii) There is a $\lambda > 0$ so that $|\varphi(\lambda)| = 1$ and $|\varphi(t)| < 1$ for $0 < t < \lambda$. In this case X has a lattice distribution with span $2\pi/\lambda$.
   (iii) $|\varphi(t)| = 1$ for all t. In this case $X = b$ a.s. for some b.

Proof   We begin with (ii). First if X has a lattice distribution with span h,

$$\varphi(t) = Ee^{itX} = e^{itb} \sum_n e^{ithn} P(X = b + hn) = e^{itb}$$

when $t = 2\pi/h$. Conversely, if $|\varphi(t)| = 1$, then there is equality in the inequality $|Ee^{itX}| \leq E|e^{itX}|$, so the distribution of e^{itX} must be concentrated at some point e^{itb}, and $P(X \in b + (2\pi/t)\mathbb{Z}) = 1$.

To complete the proof now we suppose that (i) and (ii) do not hold; that is, there is a sequence $t_n \downarrow 0$ so that $|\varphi(t_n)| = 1$. The last result shows that there is a b_n so that $P(X \in b_n + (2\pi/t_n)\mathbb{Z}) = 1$. Without loss of generality we can pick $b_n \in (-\pi/t_n, \pi/t_n]$. As $n \to \infty$, $P(X \notin (-\pi/t_n, \pi/t_n]) \to 0$, so it follows that $P(X = b_n) \to 1$. This is only possible if $b_n = b$ for $n \geq N$ and $P(X = b) = 1$.   □

We call the three cases in (5.1): (i) *nonlattice*, (ii) *lattice*, and (iii) *degenerate*. The reader should notice that this means that lattice random variables are by definition nondegenerate. Before we turn to the main business of this section, we would like to introduce one more special case. If X is a lattice distribution with span h and we can take $b = 0$—that is, $P(X \in h\mathbb{Z}) = 1$, then X is said to be *arithmetic*. In this case if $\lambda = 2\pi/h$, then $\varphi(\lambda) = 1$ and φ is periodic: $\varphi(t + \lambda) = \varphi(t)$.

Our first local limit theorem is for the lattice case. Let $X_1, X_2, \dots$ be i.i.d. with $EX_i = 0$, $EX_i^2 = \sigma^2 \in (0, \infty)$, and having a common lattice distribution with span h. If $S_n = X_1 + \cdots + X_n$ and $P(X \in b + h\mathbb{Z}) = 1$, then $P(S_n \in nb + h\mathbb{Z}) = 1$. We put

$$p_n(x) = P(S_n/\sqrt{n} = x) \quad \text{for } x \in \mathscr{L}_n = \{(nb + hz)/\sqrt{n}: z \in \mathbb{Z}\}$$

and

$$n(x) = (2\pi\sigma^2)^{-1/2} \exp(-x^2/2\sigma^2) \quad \text{for } x \in (-\infty, \infty).$$

(5.2) THEOREM   Under the hypotheses above, as $n \to \infty$,

$$\sup_{x \in \mathscr{L}_n} \left| \frac{n^{1/2}}{h} p_n(x) - \varkappa(x) \right| \to 0.$$

Proof   Let Y be a random variable with $P(Y \in a + \theta\mathbb{Z}) = 1$ and $\psi(t) = E \exp(itY)$. It follows from part (iii) of Exercise 3.2 that

$$P(Y = y) = \frac{1}{2\pi/\theta} \int_{-\pi/\theta}^{\pi/\theta} e^{-ity} \psi(t) \, dt.$$

Using this formula with $\theta = h/\sqrt{n}$, $\psi(t) = E \exp(itS_n/\sqrt{n}) = \varphi^n(t/\sqrt{n})$ gives

$$\frac{n^{1/2}}{h} p_n(x) = \frac{1}{2\pi} \int_{-\pi\sqrt{n}/h}^{\pi\sqrt{n}/h} e^{-itx} \varphi^n(t/\sqrt{n}) \, dt.$$

Using the inversion formula (3.3) for $\varkappa(x)$, which has ch.f. $\exp(-\sigma^2 t^2/2)$, gives

$$\varkappa(x) = \frac{1}{2\pi} \int e^{-itx} \exp(-\sigma^2 t^2/2) \, dt.$$

Subtracting the last two equations gives (recall $\pi > 1$)

$$\left| \frac{n^{1/2}}{h} p_n(x) - \varkappa(x) \right| \leq \int_{-\pi\sqrt{n}/h}^{\pi\sqrt{n}/h} |\varphi^n(t/\sqrt{n}) - \exp(-\sigma^2 t^2/2)| \, dt$$

$$+ \int_{\pi\sqrt{n}/h}^{\infty} \exp(-\sigma^2 t^2/2) \, dt.$$

The right-hand side is independent of x, so to prove (2) it suffices to show that it approaches 0. The second integral clearly $\to 0$. To estimate the first integral, we observe that

$$\varphi^n(t/\sqrt{n}) \to \exp(-\sigma^2 t^2/2),$$

so the bounded convergence theorem implies that for any $A < \infty$, the integral over $(-A, A)$ approaches 0.

To estimate the integral over $(-A, A)^c$, we observe that since $EX_i = 0$ and $EX_i^2 = \sigma^2$, formula (3.7) implies that

$$|\varphi(u)| \leq |1 - \sigma^2 u^2/2| + \frac{u^2}{2} E(\min(|u| \cdot |X|^3, 6|X|^2)).$$

The last expected value $\to 0$ as $u \to 0$. This means we can pick $\delta > 0$ so that if $|u| < \delta$,

$$|\varphi(u)| \leq 1 - \sigma^2 u^2/2 + \sigma^2 u^2/4 = 1 - \sigma^2 u^2/4 \leq \exp(-\sigma^2 u^2/4),$$

since $1 - x \leq e^{-x}$. Applying the last result to $u = t/\sqrt{n}$, we see that for $t \leq \delta\sqrt{n}$,

$$|\varphi(t/\sqrt{n})^n| \leq \exp(-\sigma^2 t^2/4). \tag{5.3}$$

So the integral over $(-\delta\sqrt{n}, \delta\sqrt{n}) - (-A, A)$ is smaller than

$$2 \int_A^{\delta\sqrt{n}} 2 \exp(-\sigma^2 t^2/4) \, dt,$$

which is small if A is large.

To estimate the rest of the integral we observe that since X has span h, $|\varphi(u)| \neq 1$ for $u \in [\delta, \pi/h]$. φ is continuous so there is an $\eta < 1$ so that $|\varphi(u)| \leq \eta < 1$ for $|u| \in [\delta, \pi/h]$. Letting $u = t/\sqrt{n}$ again, we see that the integral over $[-\pi\sqrt{n}/h, \pi\sqrt{n}/h] - [-\delta\sqrt{n}, \delta\sqrt{n}]$ is smaller than

$$2 \int_{\delta\sqrt{n}}^{\pi\sqrt{n}/h} \eta^n + \exp(-\sigma^2 t^2/2) \, dt,$$

which $\to 0$ as $n \to \infty$. This completes the proof of (5.2). $\square$

We turn now to the nonlattice case. Let $X_1, X_2, \ldots$ be i.i.d. with $EX_i = 0$, $EX_i^2 = \sigma^2 \in (0, \infty)$, and having a common characteristic function $\varphi(t)$ that has $|\varphi(t)| < 1$ for all $t \neq 0$. Let $S_n = X_1 + \cdots + X_n$ and $n(x) = (2\pi\sigma^2)^{-1/2} \exp(-x^2/2\sigma^2)$.

(5.4) THEOREM   Under the hypotheses above, if $x_n/\sqrt{n} \to x$ and $a < b$,

$$\sqrt{n} \, P(S_n \in (x_n + a, x_n + b)) \to (b - a) \, n(x).$$

The proof of (5.4) has to be a little devious because the assumption above does not give us much control over the behavior of φ. For a bad example, let $q_1, q_2, \ldots$ be an enumeration of the positive rationals which has $q_n \leq n$. Suppose

$$P(X = q_n) = P(X = -q_n) = 1/2^{n+1}.$$

In this case $EX = 0, EX^2 < \infty$, and the distribution is nonlattice. However the characteristic function has $\limsup_{n\to\infty} |\varphi(t)| = 1$.

To tame bad examples like this we use a trick. Let

$$h_0(y) = \frac{1}{\pi} \cdot \frac{1 - \cos y}{y^2}$$

be the density of the Polya's distribution and let

$$h_\theta(x) = e^{i\theta y} h_0(y).$$

If we introduce the Fourier transform

$$\hat{g}(u) = \int e^{iuy} g(y) \, dy,$$

then it follows from Example 3.6 that

$$\hat{h}_0(u) = \begin{cases} 1 - |u| & \text{if } |u| \le 1 \\ 0 & \text{otherwise} \end{cases}$$

and $\hat{h}_\theta(u) = h_0(u + \theta)$. We will show that for any θ,

$$\sqrt{n} \, E h_\theta(S_n - x_n) \to \varkappa(x) \int h_\theta(y) \, dy. \tag{*}$$

Before proving $(*)$ we will explain why this is enough to prove (5.4). Let

$$\mu_n(A) = \sqrt{n} \, P(S_n - x_n \in A)$$

and

$$\mu(A) = \varkappa(x)|A|,$$

where $|A|$ = Lebesgue measure of A. Let

$$\alpha_n = \sqrt{n} \, E h_0(S_n - x_n)$$

and

$$\alpha = \varkappa(x) \int h_0(y) \, dy = \varkappa(x).$$

Finally, define probability measures by

$$v_n(B) = \frac{1}{\alpha_n} \int_B h_0(y) \, \mu_n(dy)$$

and

$$v(B) = \frac{1}{\alpha} \int_B h_0(y) \, \mu(dy).$$

Taking $\theta = 0$ in $(*)$, we see $\alpha_n \to \alpha$ and so $(*)$ implies

$$\int e^{i\theta y} v_n(dy) \to \int e^{i\theta y} v(dy). \tag{**}$$

Since this holds for all θ, it follows from (3.4) that $v_n \Rightarrow v$. Now the function

$$k(y) = \frac{1}{h_0(y)} \cdot 1_{(a,b)}(y)$$

is bounded and continuous a.s. with respect to v, so it follows from Exercise 2.5 that

$$\int k(y) \, v_n(dy) \to \int k(y) \, v(dy).$$

Since $\alpha_n \to \alpha$, this implies

$$\sqrt{n} \, P(S_n \in (x_n + a, x_n + b)) \to (b - a) \, \varkappa(x),$$

the desired conclusion.

Turning now to the proof of (*), the inversion formula (3.3) implies

$$h_0(x) = \int e^{-iux} \hat{h}_0(u) \, du.$$

Since $\hat{h}_\theta(u) = h_0(u + \theta)$ and $h_\theta(x) = e^{i\theta x} h_0(x)$, the last formula generalizes trivially to $\theta \neq 0$. Letting F_n be the distribution of $S_n - x_n$ and integrating give

$$Eh_\theta(S_n - x_n) = \int\int e^{-iux} \hat{h}_\theta(u) \, du \, dF_n(x)$$

$$= \frac{1}{2\pi} \int\int e^{-iux} \, dF_n(x) \, \hat{h}_\theta(u) \, du$$

by Fubini's theorem. (Recall $\hat{h}_\theta(u)$ has compact support.) Using (3.1e) we see that the last expression

$$= \frac{1}{2\pi} \int \varphi(-u)^n e^{iux_n} \hat{h}_\theta(u) \, du.$$

To take the limit as $n \to \infty$ of this integral, let $[-M, M]$ be an interval with $\hat{h}_\theta(u) = 0$ for $u \notin [-M, M]$. By (5.3) above we can pick δ so that for $|u| < \delta$,

$$|\varphi(u)| \leq \exp(-\sigma^2 u^2 / 4). \tag{5.5}$$

Let $I = [-\delta, \delta]$ and $J = [-M, M] - I$. Since $|\varphi(u)| < 1$ for $u \neq 0$ and φ is continuous, there is a constant $\eta < 1$ so that $|\varphi(u)| \leq \eta < 1$ for $u \in J$. Since $|\hat{h}_\theta(u)| \leq 1$, this implies that

$$\left| \sqrt{n} \, \frac{1}{2\pi} \int_J \varphi(-u)^n e^{iux_n} \hat{h}_\theta(u) \, du \right| \leq \sqrt{n} \, \frac{2M}{2\pi} \eta^n \to 0$$

as $n \to \infty$. For the integral over I, change variables $u = t/\sqrt{n}$ to get

$$\frac{1}{2\pi} \int_{-\delta\sqrt{n}}^{\delta\sqrt{n}} \varphi(-t/\sqrt{n})^n e^{iux_n/\sqrt{n}} \hat{h}_\theta(t/\sqrt{n}) \, dt/\sqrt{n}.$$

The central limit theorem implies $\varphi(-t/\sqrt{n})^n \to \exp(-\sigma^2 t^2/2)$. Using (5.5) now and the dominated convergence theorem gives (recall $x_n/\sqrt{n} \to x$)

$$\sqrt{n}\frac{1}{2\pi} \int_I \varphi(-u)^n e^{iux_n} \hat{h}_\theta(u) \, du \to \frac{1}{2\pi} \int \exp(-\sigma^2 t^2/2) e^{itx} \hat{h}_\theta(0) \, dt$$

$$= \varkappa(x)\hat{h}_\theta(0) = \varkappa(x) \int h_\theta(y) \, dy$$

by the inversion formula (3.3) and the definition of $\hat{h}_\theta(0)$. This proves (*) and completes the proof.

6 Poisson Convergence

Much of this section is devoted to giving two proofs of the following result, which is sometimes facetiously called the "weak law of small numbers" or "law of rare events." These names derive from the fact that the Poisson appears as the limit of a sum of random variables that are positive with small probabilities.

(6.1) THEOREM For each n let $X_{n,m}$, $1 \le m \le n$, be independent random variables with

$$P(X_{n,m} = 1) = p_{n,m}, \quad P(X_{n,m} = 0) = 1 - p_{n,m}, \tag{i}$$

$$\sum_{m=1}^{n} p_{n,m} \to \lambda \in (0, \infty), \quad \text{and} \tag{ii}$$

$$\max_{1 \le m \le n} p_{n,m} \to 0. \tag{iii}$$

If $S_n = X_{n,1} + \cdots + X_{n,n}$, then $S_n \Rightarrow Z$ where Z is Poisson(λ). Here Poisson(λ) is shorthand for Poisson distribution with mean λ; that is,

$$P(Z = k) = e^{-\lambda}\lambda^k/k!.$$

Note that in the spirit of the Lindeberg–Feller theorem, no single term contributes very much to the sum. In contrast to that theorem, the contributions when positive are not small.

First Proof Let $\varphi_{n,m}(t) = E(\exp(itX_{n,m})) = (1 - p_{n,m}) + p_{n,m}e^{it}$. If $S_n = X_{n,1} + \cdots + X_{n,n}$, then

$$E \exp(itS_n) = \prod_{m=1}^{n} (1 + p_{n,m}(e^{it} - 1)).$$

Using (4.3) and (4.4),

$$\left| \exp\left(\sum_{m=1}^{n} p_{n,m}(e^{it} - 1) \right) - \prod_{m=1}^{n} (1 + p_{n,m}(e^{it} - 1)) \right|$$

$$\leq \sum_{m=1}^{n} \left| \exp(p_{n,m}(e^{it} - 1)) - (1 + p_{n,m}(e^{it} - 1)) \right|$$

$$\leq \sum_{m=1}^{n} p_{n,m}^2 |e^{it} - 1|^2.$$

Since $|e^{it} - 1| \leq 2$, it follows that the last expression

$$\leq 4 \left(\sum_{m=1}^{n} p_{n,m} \right) \left(\max_{1 \leq m \leq n} p_{n,m} \right) \to 0$$

by assumptions (ii) and (iii). The last conclusion implies

$$E \exp(itS_n) \to \exp(\lambda(e^{it} - 1)).$$

To complete the proof now, we observe

$$E e^{itZ} = \sum_{k=0}^{\infty} e^{-\lambda} \frac{\lambda^k}{k!} e^{itk} = \sum_{k=0}^{\infty} e^{-\lambda}(\lambda e^{it})^k/k! = \exp(\lambda(e^{it} - 1))$$

and apply (3.4).   □

Some concrete situations in which (6.1) can be applied are:

Example 6.1   In a calculus class with 400 students, the number of students who have their birthday on the day of the final exam has approximately a Poisson distribution with mean $400/365 = 1.096$. This means that the probability no one was born on that date is about .334. Similar reasoning shows that the number of babies born on a given day or the number of people who arrive at a bank between $1:15$ and $1:30$ should have a Poisson distribution.

Example 6.2   Suppose we roll two dice 36 times. The probability of "double ones" (one on each die) is $1/36$, so the number of times this occurs should have approximately a Poisson distribution with mean 1. Comparing the Poisson approximation with exact probabilities shows that the agreement is good even though the number of trials is small.

k	0	1	2	3
Poisson	.3678	.3678	.1839	.0613
Exact	.3627	.3730	.1865	.0604

After we give the second proof of (6.1) (see (6.5)), we will discuss rates of convergence. Those results show that for large n the largest discrepancy occurs for $k = 1$ and is about $1/2en$ ($= .0051$ in this case).

Example 6.3   Let $\xi_{n,1}, \ldots, \xi_{n,n}$ be independent and uniformly distributed over $[-n, n]$. Let $X_{n,1} = 1$ if $\xi_n \in (a, b)$, $= 0$ otherwise. S_n is the number of points that land in (a, b). $p_{n,m} = (b - a)/2n$ so $\sum_m p_{n,m} = (b - a)$, (iii) holds, and we conclude that $S_n \Rightarrow Z$, a Poisson r.v. with mean $(b - a)/2$. A two-dimensional version of the last theorem might explain why the statistics of flying bomb hits in the south of London during World War II fit a Poisson distribution. As Feller (Vol. I (1968), pp. 160–161) reports, the area was divided into 576 areas of 1/4 square kilometers each. The total number of hits was 537, for an average of .9323 per cell. The table below compares N_k, the number of cells with k hits, with the predictions of the Poisson approximation.

k	0	1	2	3	4	≥ 5
N_k	229	211	93	35	7	1
Poisson	226.74	211.39	98.54	30.62	7.14	1.57

For other observations fitting the Poisson distribution, see Feller, Vol. I (1968), Section VI.7.

Our second proof of (6.1) requires a little more work but provides information about the rate of convergence. (See (6.5) below.) We begin by defining the *total variation distance* between two measures on a countable set S.

$$\|\mu - v\| = \sum_z |\mu(z) - v(z)| = 2 \sup_{A \subset S} |\mu(A) - v(A)|.$$

To see that the two expressions are equal, observe that

$$\sum_z |\mu(z) - v(z)| \geq |\mu(A) - v(A)| + |\mu(A^c) - v(A^c)| = 2|\mu(A) - v(A)|,$$

and there is equality when $A = \{z: \mu(z) \geq v(z)\}$.

Exercise 6.1   Show that $\| \ \|$ defines a metric on probability measures on $\mathbb{Z}$ and $\|\mu_n - \mu\| \to 0$ if and only if $\mu_n \Rightarrow \mu$.

Exercise 6.2   Show that $\|\mu - v\| \leq \delta$ if and only if there are random variables X and Y with distributions μ and v so that $P(X \neq Y) \leq \delta$.

The next three lemmas are the keys to our second proof.

(6.2) LEMMA   If $\mu_1 \times \mu_2$ denotes the product measure on $\mathbb{Z} \times \mathbb{Z}$ with $\mu_1 \times \mu_2(x, y) = \mu_1(x)\mu_2(y)$, then

$$\|\mu_1 \times \mu_2 - \nu_1 \times \nu_2\| \leq \|\mu_1 - \nu_1\| + \|\mu_2 - \nu_2\|.$$

Proof $\|\mu_1 \times \mu_2 - \nu_1 \times \nu_2\| = \sum_{x,y} |\mu_1(x)\mu_2(y) - \nu_1(x)\nu_2(y)|$

$$\leq \sum_{x,y} |\mu_1(x)\mu_2(y) - \nu_1(x)\mu_2(y)| + \sum_{x,y} |\nu_1(x)\mu_2(y) - \nu_1(x)\nu_2(y)|$$

$$\leq \sum_y \mu_2(y) \sum_x |\mu_1(x) - \nu_1(x)| + \sum_x \nu_1(x) \sum_y |\mu_2(y) - \nu_2(y)|$$

$$= \|\mu_1 - \nu_1\| + \|\mu_2 - \nu_2\|. \qquad \square$$

(6.3) LEMMA   If $\mu_1 * \mu_2$ denotes the convolution of μ_1 and μ_2—that is,

$$\mu_1 * \mu_2(x) = \sum_y \mu_1(x - y)\mu_2(y),$$

then $\|\mu_1 * \mu_2 - \nu_1 * \nu_2\| \leq \|\mu_1 \times \mu_2 - \nu_1 \times \nu_2\|$.

Proof $\|\mu_1 * \mu_2 - \nu_1 * \nu_2\| = \sum_x \left| \sum_y \mu_1(x - y)\mu_2(y) - \sum_y \nu_1(x - y)\nu_2(y) \right|$

$$\leq \sum_x \sum_y \left| \mu_1(x - y)\mu_2(y) - \nu_1(x - y)\nu_2(y) \right|$$

$$= \|\mu_1 \times \mu_2 - \nu_1 \times \nu_2\|. \qquad \square$$

(6.4) LEMMA   Let μ be the measure with $\mu(1) = p$ and $\mu(0) = 1 - p$. Let ν be a Poisson distribution with mean p. Then $\|\mu - \nu\| \leq 2p^2$.

Proof $\|\mu - \nu\| = |\mu(0) - \nu(0)| + |\mu(1) - \nu(1)| + \sum_{n \geq 2} \nu(n)$

$$= |1 - p - e^{-p}| + |p - p e^{-p}| + 1 - e^{-p}(1 + p).$$

Since $1 - x \leq e^{-x} \leq 1$ for $x \geq 0$, the above

$$= e^{-p} - 1 + p + p(1 - e^{-p}) + 1 - e^{-p} - pe^{-p}$$

$$= 2p(1 - e^{-p}) \leq 2p^2. \qquad \square$$

Second Proof of (6.1)   Let $\mu_{n,m}$ be the distribution of $X_{n,m}$. Let μ_n be the distribution of S_n. Let $\nu_{n,m}$, ν_n, and ν be Poisson distributions with means $p_{n,m}$, $\lambda_n = \sum_{m \leq n} p_{n,m}$, and λ, respectively. Since $\mu_n = \mu_{n,1} * \cdots * \mu_{n,n}$ and $\nu_n = \nu_{n,1} * \cdots * \nu_{n,n}$, (6.2), (6.3), and (6.4) imply

$$\|\mu_n - \nu_n\| \le 2 \sum_{m=1}^{n} p_{n,m}^2.$$

Using the definition of total variation distance now gives

$$\sup_{A \subset \mathbb{Z}} |\mu_n(A) - \nu_n(A)| \le \sum_{m=1}^{n} p_{n,m}^2. \tag{6.5}$$

(ii) and (iii) in (6.1) imply that the right-hand side $\to 0$. Since $\nu_n \Rightarrow \nu$ as $n \to \infty$, the result follows. □

Remark   The proof above is due to Hodges and LeCam (1960). By different methods C. Stein ((1987, see (43) on p. 89) has proved

$$\sup_{A \subset \mathbb{Z}} |\mu_n(A) - \nu_n(A)| \le (\lambda \vee 1)^{-1} \sum_{m=1}^{n} p_{n,m}^2.$$

Rates of convergence   When $p_{n,m} = 1/n$, (6.5) becomes

$$\sup_{A \subset \mathbb{Z}} |\mu_n(A) - \nu_n(A)| \le 1/n.$$

To assess the quality of this bound, we will compare the Poisson and binomial probabilities for k successes.

k	Poisson	Binomial
0	e^{-1}	$\left(1 - \dfrac{1}{n}\right)^n$
1	e^{-1}	$n n^{-1}\left(1 - \dfrac{1}{n}\right)^{n-1} = \left(1 - \dfrac{1}{n}\right)^{n-1}$
2	$e^{-1}/2!$	$\binom{n}{2} n^{-2}\left(1 - \dfrac{1}{n}\right)^{n-2} = \left(1 - \dfrac{1}{n}\right)^{n-1} \Big/ 2!$
3	$e^{-1}/3!$	$\binom{n}{3} n^{-3}\left(1 - \dfrac{1}{n}\right)^{n-3} = \left(1 - \dfrac{2}{n}\right)\left(1 - \dfrac{1}{n}\right)^{n-2} \Big/ 3!$

Since $(1 - x) \le e^{-x}$, we have $\mu_n(0) - \nu_n(0) \le 0$. Expanding

$$\log(1 + x) = x - \frac{x^2}{2} + \frac{x^3}{3} - \cdots$$

gives

$$(n - 1) \log\left(1 - \frac{1}{n}\right) = -\frac{n-1}{n} - \frac{n-1}{2n^2} - \cdots = -1 + \frac{1}{2n} + O(n^{-2}).$$

So

$$n\left(\left(1 - \frac{1}{n}\right)^{n-1} - e^{-1}\right) \approx ne^{-1}(e^{1/2n} - 1) \to e^{-1}/2$$

and it follows that

$$n(\mu_n(1) - \nu_n(1)) \to e^{-1}/2$$

$$n(\mu_n(2) - \nu_n(2)) \to e^{-1}/4.$$

For $k \geq 3$, using $(1 - x) \leq e^{-x}$ shows $\mu_n(k) - \nu_n(k) \leq 0$, so

$$\sup_{A \subset Z} |\mu_n(A) - \nu_n(A)| \approx 3/4en.$$

There is a large literature on Poisson approximations for dependent events. Here we consider two examples that can be treated by exact calculations.

Example 6.4   *Matching.* Let π be a random permutation of $\{1, 2, \ldots, n\}$, let $X_{n,m} = 1$ if m is a fixed point (0 otherwise), and let $S_n = X_{n,1} + \cdots + X_{n,n}$ be the number of fixed points. We want to compute $P(S_n = 0)$. (For a more exciting story, consider men checking hats or wives swapping husbands.) Let $A_{n,m} = \{X_{n,m} = 1\}$. The inclusion–exclusion formula implies

$$P\left(\bigcup_{m=1}^{n} A_m\right) = \sum_m P(A_m) - \sum_{l < m} P(A_l \cap A_m) + \sum_{k < l < m} P(A_k \cap A_l \cap A_m) - \cdots$$

$$= n\frac{1}{n} - \binom{n}{2}\frac{(n-2)!}{n!} + \binom{n}{3}\frac{(n-3)!}{n!} - \cdots,$$

since the number of permutations with k specified fixed points is $(n - k)!$. Cancelling some factorials gives

$$P(S_n > 0) = \sum_{m=1}^{n} \frac{(-1)^{m-1}}{m!}, \qquad P(S_n = 0) = \sum_{m=0}^{n} \frac{(-1)^m}{m!}.$$

Recognizing the second sum as the first $n + 1$ terms in the expansion of e^{-1} gives

$$|P(S_n = 0) - e^{-1}| = \left|\sum_{m=n+1}^{\infty} \frac{(-1)^m}{m!}\right|$$

$$\leq \frac{1}{(n+1)!}\left|\sum_{k=0}^{\infty} (n+2)^{-k}\right| = \frac{1}{(n+1)!} \cdot \left(1 - \frac{1}{n+2}\right)^{-1},$$

a much better rate of convergence than $1/n$. To compute the other probabilities, we observe that by considering the locations of the fixed points,

$$P(S_n = k) = \binom{n}{k} n^{-k} P(S_{n-k} = 0) \to e^{-1}/k!.$$

Example 6.5 *Occupancy problem.* Suppose that r balls are placed at random into n boxes. It follows from the Poisson approximation to the binomial that if $n \to \infty$ and $r/n \to c$, then the number of balls in a given box will approach a Poisson distribution with mean c. The last observation should explain why the fraction of occupied boxes approached e^{-c} in Example 5.9 of Chapter 1. Here we will show:

(6.6) THEOREM If $ne^{-r/n} \to \lambda \in [0, \infty)$, the number of empty boxes approaches a Poisson distribution with mean λ. To see where the answer comes from, notice that in the Poisson approximation the probability that a given box is empty is $e^{-r/n} \approx \lambda/n$, so if the occupancy of the various boxes were independent, the result would follow from (6.1). To prove the result we begin by observing

$$P(\text{boxes } i_1, i_2, \ldots, i_k \text{ are empty}) = \left(1 - \frac{k}{n}\right)^r,$$

so if we let $p_m(r, n) = $ the probability exactly m boxes are empty when r balls are put in n boxes, then $P(\text{no empty box}) = 1 - P(\text{at least one empty box})$, so by inclusion–exclusion,

$$p_0(r, n) = \sum_{k=0}^{n} (-1)^k \binom{n}{k}\left(1 - \frac{k}{n}\right)^r, \tag{a}$$

and by considering the locations of the empty boxes,

$$p_m(r, n) = \binom{n}{m}\left(1 - \frac{m}{n}\right)^r p_0(r, n - m). \tag{b}$$

To evaluate the limit of $p_0(r, n)$, we begin by showing that if $ne^{-r/n} \to \lambda$, then

$$\binom{n}{m}\left(1 - \frac{m}{n}\right)^r \to \lambda^m/m!. \tag{c}$$

One half of this is easy:

$$\binom{n}{m}\left(1 - \frac{m}{n}\right)^r \le \frac{n^m}{m!}e^{-mr/n} \to \lambda^m/m! \tag{d}$$

by assumption. For the other direction, observe

$$\binom{n}{m} \ge (n - m)^m/m! = \left(1 - \frac{m}{n}\right)^m n^m/m!$$

and

$$\log(1 - t) = -t - t^2/2 - t^3/3 - \cdots \ge -t - t^2$$

for $0 \le t \le 1/2$. $(1 - m/n)^m \to 1$ as $n \to \infty$ and $1/m!$ is a constant. For the rest we observe

$$\log\left(\left(1 - \frac{m}{n}\right)^r\right) \ge m \log n - rm/n - r(m/n)^2.$$

Our assumption $ne^{-r/n} \to \lambda$ means

$$r = n \log n - (\log \lambda)n + o(n),$$

so $r(m/n)^2 \to 0$, $\log n - r/n \to \log \lambda$, and (c) follows. From (a), (c), (d), and the dominated convergence theorem, we get

$$\text{if } e^{-r/n} \to \lambda, \text{ then } p_0(r, n) \to \sum_{k=0}^{\infty} (-1)^k \frac{\lambda^k}{k!} = e^{-\lambda}. \tag{e}$$

For fixed m, $(n - m)e^{-r/(n-m)} \to \lambda$, so it follows from (e) that $p_0(r, n - m) \to e^{-\lambda}$. Combining this with (b) and (c) completes the proof of (6.6).

Exercise 6.3   What is the probability that in a village of 2190 ($= 6\cdot365$) people all birthdays are represented? Do you think the answer is much different for 1825 ($= 5\cdot365$) people?

Our next topic is an application of Example 6.5.

Example 6.6   *Coupon collector's problem.* Let X_1, X_2, ... be i.i.d. uniform on $\{1, 2, ..., n\}$ and $T_n = \inf\{m: \{X_1, ..., X_m\} = \{1, 2, ..., n\}\}$. Since $T_n \le m$ if and only if m balls fill up all n boxes, it follows from (6.6) that

$$P(T_n - n \log n \le x) \to \exp(-e^{-x}).$$

Exercise 6.4   As we observed in Example 5.10 of Chapter 1, if we let $\tau_k^n = \inf\{m: |\{X_1, ..., X_m\}| = k\}$, then $\tau_1^n = 1$ and for $2 \le k \le n$, $\tau_k^n - \tau_{k-1}^n$ are independent and have a geometric distribution with parameter $1 - (k - 1)/n$. Use ch.f.'s to prove the last result.

Exercise 6.5   Let τ_k^n be as in the last exercise. Suppose $k/n^{1/2} \to \lambda \in [0, \infty)$ and show that $\tau_k^n - k \Rightarrow \text{Poisson}(\lambda^2/2)$.

Exercise 6.6   Let τ_k^n be as in the last two exercises. Let $\mu_{n,k} = E\tau_k^n$ and $\sigma_{n,k}^2 = \text{var}(\tau_k^n)$. Suppose $k/n \to a \in (0, 1)$, and show $(\tau_k^n - \mu_{n,k})/\sigma_{n,k} \Rightarrow \chi$.

The last result is true when $k/n^{1/2} \to \infty$ and $n - k \to \infty$; see Baum and Billingsley (1966). Results for $k = n - j$ can be obtained from (6.6), so we have examined all the possibilities.

(6.1) generalizes trivially to give the following result.

(6.7) THEOREM   For each n let $X_{n,m}$, $1 \le m \le n$, be independent nonnegative integer-valued random variables with

$$P(X_{n,m} = 1) = p_{n,m}, \quad P(X_{n,m} \ge 2) = \varepsilon_{n,m}, \tag{i}$$

$$\sum_{m=1}^{n} p_{n,m} \to \lambda \in (0, \infty), \tag{ii}$$

$$\max_{1 \le m \le n} p_{n,m} \to 0, \quad \text{and} \tag{iii}$$

$$\sum_{m=1}^{n} \varepsilon_{n,m} \to 0. \tag{iv}$$

If $S_n = X_{n,1} + \cdots + X_{n,n}$, then $S_n \Rightarrow Z$ where Z is Poisson(λ).

Proof   Let $X'_{n,m} = 1$ if $X_{n,m} = 1$, and $= 0$ otherwise. Let $S'_n = X'_{n,1} + \cdots + X'_{n,n}$. (i)–(iii) and (6.1) imply $S_n \Rightarrow Z$, (iv) tells us $P(S_n \neq S'_n) \to 0$, and the result follows from the converging together lemma, Exercise 2.8.   □

The next result, which uses (6.7), explains why the Poisson distribution comes up so frequently in applications. Let $N(s, t)$ be the number of arrivals at a bank or an ice cream parlor in the time interval $(s, t]$. Suppose

(i) the number of arrivals in disjoint intervals are independent,
(ii) the distribution of $N(s, t)$ depends only on $t - s$,
(iii) $P(N(0, h) = 1) = \lambda h + o(h)$, and
(iv) $P(N(0, h) \ge 2) = o(h)$.

Here, the two $o(h)$'s stand for functions $g_1(h)$ and $g_2(h)$ with $g_i(h)/h \to 0$ as $h \to 0$.

(6.8) THEOREM   If (i)–(iv) hold, then $N(0, t)$ has a Poisson distribution with mean λt.

Proof   Let $X_{n,m} = N((m - 1)t/n, mt/n)$ for $1 \le m \le n$ and apply (6.7).   □

A family of random variables N_t, $t \ge 0$, satisfying:

(i) if $0 = t_0 < t_1 < \cdots < t_n$, $N(t_k) - N(t_{k-1}) 1 \le k \le n$ are independent, and
(ii) $N(t) - N(s)$ is Poisson($\lambda(t - s)$),

is called a *Poisson process with rate* λ. To understand how N_t behaves it is useful to have another method to construct it. Let $\xi_1, \xi_2, \ldots$ be independent random variables with $P(\xi_i > t) = e^{-\lambda t}$ for $t \ge 0$. Let $T_n = \xi_1 + \cdots + \xi_n$ and $N_t = \sup\{n: T_n \le t\}$ where $T_0 = 0$. In the language of renewal theory (see (8.5) in Chapter 1), T_n is the time of the nth arrival and N_t is the number of arrivals by time t. To check that N_t is a Poisson process we begin by recalling (see Exercise 4.8 in Chapter 1)

$$P(T_n = s) = \frac{\lambda^n s^{n-1}}{(n - 1)!} e^{-\lambda s} \quad \text{for } s \ge 0;$$

that is, the distribution of T_n has a density given by the right-hand side. Now

$$P(N_t = 0) = P(T_1 > t) = e^{-\lambda t},$$

and for $n \geq 1$,

$$P(N_t = n) = P(T_n \leq t < T_{n+1}) = \int_0^t P(T_n = s) P(\xi_{n+1} > t - s) \, ds$$

$$= \int_0^t \frac{\lambda^n s^{n-1}}{(n-1)!} e^{-\lambda s} e^{-\lambda(t-s)} \, ds = e^{-\lambda t} \frac{(\lambda t)^n}{n!}.$$

The last two formulas show that N_t has a Poisson distribution with mean λt. To check that the number of arrivals in disjoint intervals is independent, we observe

$$P(T_{n+1} \geq u | N_t = n) = P(T_{n+1} \geq u, T_n \leq t)/P(N_t = n).$$

To compute the numerator we observe

$$P(T_{n+1} \geq u, T_n \leq t) = \int_0^t P(T_n = s) P(\xi_{n+1} \geq u - s) \, ds$$

$$= \int_0^t \frac{\lambda^n s^{n-1}}{(n-1)!} e^{-\lambda s} e^{-\lambda(u-s)} ds = e^{-\lambda u} \frac{(\lambda t)^n}{n!}.$$

The denominator is

$$P(N_t = n) = e^{-\lambda t} \frac{(\lambda t)^n}{n!},$$

so

$$P(T_{n+1} \geq u | N_t = n) = e^{-\lambda u} / e^{-\lambda t} = e^{-\lambda(u-t)},$$

or rewriting things, $P(T_{n+1} - t \geq s | N_t = n) = e^{-\lambda s}$. Let $T' < T_{N(t)+1} - t$, and $T'_k = T_{N(t)+k} - T_{N(t)+k-1}$ for $k \geq 2$. The last computation shows that T'_1 is independent of N_t. If we observe that

$$P(T_n \leq t, T_{n+1} \geq u, T_{n+k} - T_{n+k-1} \geq v_k, k = 2, \dots, K)$$

$$= P(T_n \leq t, T_{n+1} \geq u) \prod_{k=2}^K P(\xi_{n+k} \geq v_k),$$

then it follows that:

(6.9) $T'_1, T'_2, \dots$ are i.i.d. and independent of N_t.

The last observation shows that the arrivals after time t are independent of N_t and have the same distribution as the original sequence. From this it follows easily that:

(6.10)   If $0 = t_0 < t_1 \ldots < t_n$, then $N(t_i) - N(t_{i-1})$, $i = 1, \ldots, n$, are independent.

To see this, observe that the vector $(N(t_2) - N(t_1), \ldots, N(t_n) - N(t_{n-1})) \in \sigma(T_k')$, $k \geq 1$) is independent of $N(t_1)$, and use induction to conclude

$$P(N(t_i) - N(t_{i-1}) = k_i, i = 1, \ldots, n) = \prod_{i=1}^{n} \exp(-\lambda(t_i - t_{i-1}))(\lambda(t_i - t_{i-1}))^k / k!.$$

Remark   The key to the proof of (6.9) is the lack of memory property of the exponential distribution:

$$P(T > t + s \mid T > t) = P(T > s), \tag{$*$}$$

which implies that the location of the first arrival after t is independent of what occurred before time t and has an exponential distribution.

Exercise 6.7   Show that if $P(T > 0) = 1$ and $(*)$ holds, then there is a $\lambda > 0$ so that $P(T > t) = e^{-\lambda t}$ for $t \geq 0$.

Hint   First show that this holds for $t = m2^{-n}$.

Exercise 6.8   Assumptions (iii) and (iv) in (6.7) can be replaced by

$$\text{If } N_{s-} = \lim_{r \uparrow s} N_r, \text{ then } P(N_s - N_{s-} \geq 2 \text{ for some } s) = 0. \tag{v}$$

That is, if (i), (ii), and (v) hold, then there is a $\lambda \geq 0$ so that $N(0, t)$ has a Poisson distribution with mean λt. Prove this by showing: (a) If $u(s) = P(N_s = 0)$, then (i) and (ii) imply $u(r)u(s) = u(r + s)$. It follows that $u(s) = e^{-\lambda s}$ for some $\lambda \geq 0$, so (iii) holds. (b) if $v(s) = P(N_s \geq 2)$ and $A_n = \{N_{k/n} - N_{(k-1)/n} \geq 2 \text{ for some } k \leq n\}$, then (v) implies $P(A_n \text{ i.o.}) = 0$. From (i) and (ii) we get

$$P(A_n) = 1 - (1 - v(1/n))^n,$$

so (iv) holds.

Exercise 6.9   Let T_n be the time of the nth arrival in a rate λ Poisson process. Let $U_1, U_2, \ldots, U_n$ be uniform on $(0, 1)$ and let V_k be the kth smallest number in $\{U_1, \ldots, U_n\}$. Show that $(V_1, \ldots, V_n)$ and $(T_1/T_{n+1}, \ldots, T_n/T_{n+1})$ have the same distribution.

Exercise 6.10   The last result can be used to study the spacings between the order statistics $V_1, \ldots, V_n$ defined in the last exercise. Suppose $\lambda = 1$ and let $V_0 = 0$ and $V_{n+1} = 1$. Using the fact that $T_{n+1}/n \to 1$ a.s., it is easy to prove results of Smirnov (1949): $nV_k \Rightarrow T_k$, and Weiss (1955):

$$n^{-1} \sum_{m=1}^{n} 1_{(n(V_i - V_{i-1}) > x)} \to e^{-x} \text{ in probability.}$$

Using results from Exercise 2.18, one can also show that

$$\frac{n}{\log n} \max_{1 \le m \le n+1} V_m - V_{m-1} \to 1 \text{ in probability}$$

$$P\left(n^2 \min_{1 \le m \le n+1} V_m - V_{m-1} > x\right) \to e^{-x}.$$

Exercise 6.11   Let $V_1, \ldots, V_{n+1}$ be as in the last two exercises. Show that $(V_m/V_{m+1})^m$, $1 \le m \le n$, are i.i.d. uniform on $(0, 1)$.

Exercise 6.12   *Thinning.* Let N have a Poisson distribution with mean λ and let X_1, $X_2, \ldots$ be an independent i.i.d. sequence with $P(X_i = k) = p_k$ for $k = 0, 1, 2, \ldots$. Let $N_k = |\{m \le N : X_m = k\}|$. Show that $N_0, N_1, N_2, \ldots$ are independent and N_k has a Poisson distribution with mean λp_k.

Hint   Write down finite-dimensional distributions starting with the important special case $X_i \in \{0, 1\}$. In this case the result says that if we thin a Poisson process by flipping a coin with probability p of heads to see if we keep the arrival, then the result is a Poisson process with rate λp.

Exercise 6.13   *Poissonization and the occupancy problem.* If we put a Poisson number of balls with mean r in n boxes and let N_i be the number of balls in box i, then the last exercise implies $N_1, \ldots, N_n$ are independent and have a Poisson distribution with mean r/n. Use this observation to prove (6.6).

Hint   If $r = n \log n + (\log \lambda) n + o(n)$ and $s = n \log n + (\log \mu) n$ with $\mu < \lambda$, then the normal approximation to the Poisson tells us $P(\text{Poisson}(s) < r) \to 1$ as $n \to \infty$.

Exercise 6.14   *Compound Poisson process.* At the arrival times $T_1, T_2, \ldots$ of a Poisson process with rate λ, groups of customers of size $\xi_1, \xi_2, \ldots$ arrive at an ice cream parlor. Suppose the ξ_i are i.i.d. and independent of the T_j's. This is a *compound.* Use the result of Exercise 6.12 to show that $N_t^k =$ the number of groups of size k to arrive in $[0, t]$ is a Poisson process with rate $p_k \lambda$.

A *Poisson process on a measure space* $(S, \mathcal{S}, \mu)$ is a random map $m \colon \mathcal{S} \to \{0, 1, \ldots\}$ with the following property: If $A_1, \ldots, A_n$ are disjoint sets with $\mu(A_i) < \infty$, then $m(A_1), \ldots, m(A_n)$ are independent and have Poisson distributions with means $\mu(A_i)$. μ is called the *mean measure* of the process. If $\mu(S) < \infty$, we can construct m by the following recipe: Let $X_1, X_2, \ldots$ be i.i.d. elements of S with distribution $\nu(\cdot) = \mu(\cdot)/\mu(S)$, let N be an independent Poisson random variable with mean $\mu(S)$, and let $m(A) = |A \cap \{X_1, \ldots, X_N\}|$ is the desired Poisson process. To extend the construction to infinite measure spaces—for example, $S = \mathbb{R}^d$, $\mathcal{S} =$ Borel sets, $\mu =$ Lebesgue measure, divide the space up into disjoint sets of finite measure and put independent Poisson processes on each set.

Exercise 6.15   Use the results in Exercise 6.12 to show that the processes just constructed have the desired properties.

*7 Stable Laws

Let $X_1, X_2, \ldots$ be i.i.d. and $S_n = X_1 + \cdots + X_n$. In Section 4 we saw that if $EX_1 = \mu$ and var $(X_1) = \sigma^2 \in (0, \infty)$, then

$$(S_n - n\mu)/\sigma n^{1/2} \Rightarrow \chi.$$

In this section we will investigate the case $EX_1^2 = \infty$ and give necessary and sufficient conditions for the existence of constants a_n and b_n so that

$$(S_n - b_n)/a_n \Rightarrow Y$$

where Y is nondegenerate. We begin with an example. Suppose the distribution of X_i has

$$P(X_1 > x) = P(X_1 < -x) \tag{7.1a}$$

$$P(|X_1| > x) = x^{-\alpha} \quad \text{for } x \geq 1 \tag{7.1b}$$

where $0 < \alpha < 2$. If $\varphi(t) = E \exp(itX_1)$, then

$$1 - \varphi(t) = \int_1^\infty (1 - e^{itx}) \frac{\alpha}{2|x|^{\alpha+1}} dx + \int_{-\infty}^{-1} (1 - e^{itx}) \frac{\alpha}{2|x|^{\alpha+1}} dx$$

$$= \alpha \int_1^\infty \frac{1 - \cos(tx)}{x^{\alpha+1}} dx.$$

Changing variables $tx = u$, the last integral becomes

$$= \alpha \int_t^\infty \frac{1 - \cos u}{(u/t)^{\alpha+1}} \frac{du}{t} = t^\alpha \alpha \int_t^\infty \frac{1 - \cos u}{u^{\alpha+1}} du.$$

As $u \to 0$, $1 - \cos u \sim u^2/2$. So

$$\frac{1 - \cos u}{u^{\alpha+1}} \sim u^{-\alpha+1},$$

which is integrable, since $\alpha < 2$ implies $-\alpha + 1 > -1$. If we let

$$C = \alpha \int_0^\infty \frac{1 - \cos u}{u^{\alpha+1}} du < \infty$$

and observe (7.1a) implies $\varphi(t) = \varphi(-t)$, then the results above show

$$1 - \varphi(t) \sim C|t|^\alpha \quad \text{as } t \to 0. \tag{7.2}$$

Let $X_1, X_2, \ldots$ be i.i.d. with the distribution given in (7.1), and let $S_n = X_1 + \cdots + X_n$.

$$E \exp(itS_n/n^{1/\alpha}) = \varphi(t/n^{1/\alpha})^n = (1 - (1 - \varphi(t/n^{1/\alpha})))^n.$$

As $n \to \infty$, $n(1 - \varphi(t/n^{1/\alpha})) \to C|t|^\alpha$, so it follows from (4.2) that

$$E \exp(itS_n/n^{1/\alpha}) \to \exp(-C|t|^\alpha).$$

From part (ii) of (3.4) it follows that the expression on the right is the characteristic function of some Y and

$$S_n/n^{1/\alpha} \Rightarrow Y. \tag{7.3}$$

To prepare for our general result, we will now give another proof of (7.3). If $0 < a < b$ and $an^{1/\alpha} > 1$, then

$$P(an^{1/\alpha} < X_1 < bn^{1/\alpha}) = \frac{1}{2}(a^{-\alpha} - b^{-\alpha}) n^{-1},$$

so it follows from (6.1) that

$$N_n(a, b) \equiv |\{m \le n : X_m/n^{1/\alpha} \in (a, b)\}| \Rightarrow N(a, b)$$

where $N(a, b)$ has a Poisson distribution with mean $(a^{-\alpha} - b^{-\alpha})/2$. An easy extension of the last result shows that if $A \subset \mathbb{R} - (-\delta, \delta)$ and n is large, then

$$P(X_1/n^{1/\alpha} \in A) = n^{-1} \int_A \frac{\alpha}{2|x|^{\alpha+1}} dx,$$

so

$$N_n(A) \equiv |\{m \le n : X_m/n^{1/\alpha} \in A\}| \Rightarrow N(A)$$

where $N(A)$ has a Poisson distribution with mean

$$\mu(A) = \int_A \frac{\alpha}{2|x|^{\alpha+1}} dx < \infty.$$

The limiting family of random variables $N(A)$ is called a *Poisson process on* $(-\infty, \infty)$ *with mean measure* μ. (See the end of Section 6 for more on this process.) Notice that $\mu(\varepsilon, \infty) < \infty$, so $N(\varepsilon, \infty) < \infty$ for any $\varepsilon > 0$.

The last paragraph describes the limiting behavior of the random set

$$\mathcal{X}_n = \{X_m/n^{1/\alpha} : 1 \le m \le n\}.$$

To describe the limit of $S_n/n^{1/\alpha}$, we will "sum up the points." Let $\varepsilon > 0$ and

$$I_n(\varepsilon) = \{m \leq n : |X_m| > \varepsilon n^{1/\alpha}\}$$

$$\hat{S}_n(\varepsilon) = \sum_{m \in I_n(\varepsilon)} X_m$$

$$\bar{S}_n(\varepsilon) = S_n - \hat{S}_n(\varepsilon).$$

$I_n(\varepsilon) =$ the indices of the big terms—that is, those $> \varepsilon n^{1/\alpha}$ in magnitude. $\hat{S}_n(\varepsilon)$ is the sum of the big terms, and $\bar{S}_n(\varepsilon)$ is the rest of the sum. The first thing we will do is show that the contribution of $\bar{S}_n(\varepsilon)$ is small if ε is. Let

$$\bar{X}_m(\varepsilon) = X_m 1_{(|X_m| \leq \varepsilon n^{1/\alpha})}.$$

Symmetry implies $E\bar{X}_m(\varepsilon) = 0$, so $E(\bar{S}_n(\varepsilon)^2) = nE\bar{X}_1(\varepsilon)^2$.

$$E\bar{X}_1(\varepsilon)^2 = \int_0^\infty 2yP(\bar{X}_1(\varepsilon) > y)\,dy \leq \int_0^1 2y\,dy + \int_1^{\varepsilon n^{1/\alpha}} 2yy^{-\alpha}dy$$

$$= 1 + \frac{2}{2-\alpha}\varepsilon^{2-\alpha}n^{2/\alpha-1} - \frac{2}{2-\alpha} \leq \frac{2}{2-\alpha}\varepsilon^{2-\alpha}n^{2/\alpha-1},$$

since $\alpha > 0$. From this it follows that

$$E(\bar{S}_n(\varepsilon)/n^{1/\alpha})^2 \leq \frac{2}{2-\alpha}\varepsilon^{2-\alpha}. \tag{7.4}$$

To compute the limit of $\hat{S}_n(\varepsilon)/n^{1/\alpha}$, we observe that $|I_n(\varepsilon)| \Rightarrow$ a Poisson distribution with mean $\varepsilon^{-\alpha}$. Given $|I_n(\varepsilon)| = m$, $\hat{S}_n(\varepsilon)/n^{1/\alpha}$ is the sum of m independent random variables with a distribution F_n^ε that is symmetric and has

$$1 - F_n^\varepsilon(x) = P(X_1/n^{1/\alpha} > x \mid |X_1|/n^{1/\alpha} > \varepsilon) = x^{-\alpha}/2\varepsilon^{-\alpha} \quad \text{for } x \geq \varepsilon.$$

The last distribution is the same as that of εX_1, so if $\varphi(t) = E\exp(itX_1)$, the distribution F_n^ε has characteristic function $\varphi(\varepsilon t)$. Combining the observations above and using (6.1) give

$$E\exp(it\hat{S}_n(\varepsilon)/n^{1/\alpha}) = \sum_{m=0}^n \binom{n}{m}(\varepsilon^{-\alpha}/n)^m(1 - \varepsilon^{-\alpha}/n)^{n-m}\varphi(\varepsilon t)^m$$

$$\to \sum_{m=0}^\infty \exp(-\varepsilon^{-\alpha})(\varepsilon^{-\alpha})^m\varphi(\varepsilon t)^m/m!$$

$$= \exp(-\varepsilon^{-\alpha}(1 - \varphi(\varepsilon t))). \tag{7.5}$$

To get (7.3) now, we use the following generalization of (4.7):

(7.6) LEMMA   If $h_n(\varepsilon) \to g(\varepsilon)$ for each $\varepsilon > 0$ and $g(\varepsilon) \to g(0)$ as $\varepsilon \to 0$, then we can pick $\varepsilon_n \to 0$ so that $h_n(\varepsilon_n) \to g(0)$.

Proof   Let N_m be chosen so that $|h_n(1/m) - g(1/m)| \leq 1/m$ for $n \geq N_m$ and $m \to N_m$ is increasing. Let $\varepsilon_n = 1/m$ for $N_m \leq n < N_{m+1}$ and $= 1$ for $n < N_1$. When $N_m \leq n < N_{m+1}$, $\varepsilon_n = 1/m$, so it follows from the triangle inequality and the definition of ε_n that

$$|h_n(\varepsilon_n) - g(0)| \leq |h_n(1/m) - g(1/m)| + |g(1/m) - g(0)|$$

$$\leq 1/m + |g(1/m) - g(0)|.$$

Let $h_n(\varepsilon) = E \exp(it\hat{S}_n(\varepsilon)/n^{1/\alpha})$ and $g(\varepsilon) = \exp(-\varepsilon^{-\alpha}(1 - \varphi(\varepsilon t)))$. (7.2) implies

$$g(\varepsilon) \to \exp(-C|t|^\alpha) \quad \text{as } \varepsilon \to 0,$$

so (7.6) implies we can pick $\varepsilon_n \to 0$ with $h_n(\varepsilon_n) \to \exp(-C|t|^\alpha)$. Introducing Y with $E \exp(itY) = \exp(-C|t|^\alpha)$, it follows that

$$\hat{S}_n(\varepsilon_n)/n^{1/\alpha} \Rightarrow Y.$$

If $\varepsilon_n \to 0$, then (7.4) implies

$$\bar{S}_n(\varepsilon_n)/n^{1/\alpha} \Rightarrow 0,$$

and (7.3) follows from the converging together lemma, Exercise 2.8.   □

Once we give one final definition, we will state and prove the general result alluded to above. L is said to be *slowly varying* if

$$\lim_{x \to \infty} L(tx)/L(x) = 1 \quad \text{for all } t > 0.$$

Exercise 7.1   Show that $L(t) = 3 + \sin(\sqrt{t})$ and $L(t) = \log t$ are slowly varying but $L(t) = t^\varepsilon$ is not if $\varepsilon \neq 0$.

(7.7) THEOREM   Suppose $X_1, X_2, \ldots$ are i.i.d. with a distribution satisfies

$$\lim_{x \to \infty} P(X_1 > x)/P(|X_1| > x) = \theta \in [0, 1] \tag{i}$$

$$P(|X_1| > x) = x^{-\alpha}L(x) \tag{ii}$$

where $\alpha < 2$ and L is slowly varying. Let $S_n = X_1 + \cdots + X_n$,

$$a_n = \inf\{x : P(|X_1| > x) \leq n^{-1}\},$$

and

$$b_n = nE(X_1 1_{(|X_1| \le a_n)}).$$

As $n \to \infty$, $(S_n - b_n)/a_n \Rightarrow Y$ where Y has a nondegenerate distribution.

Remark   This is not much of a generalization of the example, but the conditions are necessary for the existence of constants a_n and b_n so that

$$(S_n - b_n)/a_n \Rightarrow Y$$

where Y is nondegenerate. Proofs of necessity can be found in Chapter 9 of Breiman (1968) or in Gnedenko and Kolmogorov (1954). (7.13) gives the ch.f. of Y. The reader can skip to that point now without much loss.

Proof   It is not hard to see that (ii) implies

$$nP(|X_1| > a_n) \to 1. \tag{7.8}$$

To prove this note that $nP(|X_1| > a_n) \le 1$ and let $\varepsilon > 0$. Taking $x = a_n/(1 + \varepsilon)$ and $t = 1 + 2\varepsilon$, (ii) implies

$$(1 + 2\varepsilon)^{-\alpha} = \lim_{n \to \infty} \frac{P(|X_1| > (1 + 2\varepsilon) a_n/(1 + \varepsilon))}{P(|X_1| > a_n/(1 + \varepsilon))} \le \liminf_{n \to \infty} \frac{P(|X_1| > a_n)}{1/n},$$

proving (7.8) since ε is arbitrary. Combining (7.8) with (i) and (ii) gives

$$nP(X_1 > xa_n) \to \theta x^{-\alpha} \quad \text{for } x > 0, \tag{7.9}$$

so $|\{m \le n : X_m > xa_n\}| \Rightarrow \text{Poisson}(\theta x^{-\alpha})$. The last result leads, as before, to the conclusion that

$$\mathscr{X}_n = \{X_m/n^{1/\alpha} : 1 \le m \le n\}$$

converges to a Poisson process on $(-\infty, \infty)$ with mean measure

$$\mu(A) = \int_{A \cap (0,\infty)} \theta \alpha |x|^{-(\alpha+1)} dx + \int_{A \cap (-\infty,0)} (1 - \theta) \alpha |x|^{-(\alpha+1)} dx.$$

To sum up the points, let

$$I_n(\varepsilon) = \{m \le n : |X_m| > \varepsilon n^{1/\alpha}\}$$

$$\hat{\mu}(\varepsilon) = EX_m 1_{(\varepsilon a_n < |X_m| < a_n)}$$

$$\hat{S}_n(\varepsilon) = \sum_{m \in I_n(\varepsilon)} X_m$$

$$\bar{\mu}(\varepsilon) = EX_m 1_{(|X_m| < \varepsilon a_n)}$$

$$\bar{S}_n(\varepsilon) = (S_n - b_n) - (\hat{S}_n(\varepsilon) - n\hat{\mu}(\varepsilon)) = \sum_{m=1}^{n} (X_m 1_{(|X_m| \le \varepsilon a_n)} - \bar{\mu}(\varepsilon)).$$

If we let $\bar{X}_m(\varepsilon) = X_m 1_{(|X_m| \le \varepsilon a_n)}$, then

$$E(\bar{S}_n(\varepsilon)/a_n)^2 = n \, \text{var}(\bar{X}_1(\varepsilon)/a_n) \le nE(\bar{X}_1(\varepsilon)/a_n)^2$$

$$E(\bar{X}_1(\varepsilon)/a_n)^2 = \int_0^\infty 2yP(\bar{X}_1(\varepsilon) > ya_n) \, dy$$

$$\le P(\bar{X}_1(\varepsilon) > a_n) \int_0^\varepsilon 2y \frac{P(\bar{X}_1(\varepsilon) > ya_n)}{P(\bar{X}_1(\varepsilon) > a_n)} dy.$$

So (ii) and (7.9) imply

$$nE(\bar{X}_1(\varepsilon)/a_n)^2 \to \int_0^\varepsilon 2yy^{-\alpha} \, dy = \frac{2}{2-\alpha} \varepsilon^{2-\alpha},$$

and we have

$$\limsup_{n\to\infty} E(\bar{S}_n(\varepsilon)/a_n)^2 \le \frac{2}{2-\alpha} \varepsilon^{2-\alpha}. \tag{7.10}$$

To compute the limit of $\hat{S}_n(\varepsilon)$, we observe that $|I_n(\varepsilon)| \Rightarrow \text{Poisson}(\varepsilon^{-\alpha})$. Given $|I_n(\varepsilon)| = m$, $\hat{S}_n(\varepsilon)/n^{1/\alpha}$ is the sum of m independent random variables with distribution F_n^ε with

$$1 - F_n^\varepsilon(x) = P(X_1/n^{1/\alpha} > x || X_1|/n^{1/\alpha} > \varepsilon) \to \theta x^{-\alpha}/\varepsilon^{-\alpha}$$

$$F_n^\varepsilon(-x) = P(X_1/n^{1/\alpha} < -x || X_1|/n^{1/\alpha} > \varepsilon) \to (1 - \theta) |x|^{-\alpha}/\varepsilon^{-\alpha}$$

for $x \ge \varepsilon$. If we let $\psi_n^\varepsilon(t)$ denote the characteristic function of F_n^ε, then

$$\psi_n^\varepsilon(t) \to \psi^\varepsilon(t) = \int_\varepsilon^\infty e^{itx}\theta\varepsilon^\alpha x^{-(\alpha+1)}dx + \int_{-\infty}^{-\varepsilon} e^{itx}(1-\theta)\, \varepsilon^\alpha|x|^{-(\alpha+1)}dx$$

as $n \to \infty$, so repeating the proof of (7.5) gives

$$E \exp(it\hat{S}_n(\varepsilon)/a_n) \to \exp(-\varepsilon^{-\alpha}(1 - \psi^\varepsilon(t)))$$

$$= \exp\left[\int_\varepsilon^\infty e^{itx}\theta x^{-(\alpha+1)}dx + \int_\varepsilon^\infty e^{itx}\theta x^{-(\alpha+1)}dx\right].$$

To bring in

$$\hat{\mu}(\varepsilon) = EX_m 1_{(\varepsilon a_n < |X_m| \le a_n)},$$

we observe that (7.9) implies $nP(xa_n < X_m \le ya_n) \to \theta(x^{-\alpha} - y^{-\alpha})$. So

$$n\hat{\mu}(\varepsilon)/a_n \to \int_\varepsilon^1 x\theta\alpha x^{-(\alpha+1)}dx + \int_{-1}^{-\varepsilon} x(1-\theta)\,\alpha|x|^{-(\alpha+1)}\,dx.$$

From this it follows that

$$E\exp(it(\hat{S}_n(\varepsilon) - n\hat{\mu}(\varepsilon))) \to \exp\left[\int_1^\infty (e^{itx}-1)\,\theta\alpha x^{-(\alpha+1)}dx\right.$$

$$+\int_\varepsilon^1 (e^{itx}-1-itx)\,\theta\alpha x^{-(\alpha+1)}dx$$

$$+\int_{-1}^{-\varepsilon} (e^{itx}-1-itx)(1-\theta)\,\alpha|x|^{-(\alpha+1)}dx$$

$$\left.+\int_{-\infty}^{-1} (e^{itx}-1)(1-\theta)\,\alpha|x|^{-(\alpha+1)}dx\right]. \qquad (7.11)$$

The last expression is messy, but $e^{itx}-1-itx \sim -t^2x^2/2$ as $t \to 0$, so we need to subtract the itx to make

$$\int_0^1 (e^{itx}-1-itx)x^{-(\alpha+1)}\,dx$$

converge when $\alpha \ge 1$. To reduce the number of integrals from four to two, we can write the limit as $\varepsilon \to 0$ of the left-hand side of (7.11) as

$$\exp\left[itc + \int_0^\infty \left(e^{itx}-1-\frac{itx}{1+x^2}\right)\theta\alpha x^{-(\alpha+1)}\,dx\right.$$

$$\left.+\int_{-\infty}^0 \left(e^{itx}-1-\frac{itx}{1+x^2}\right)(1-\theta)\alpha|x|^{-(\alpha+1)}\,dx\right] \qquad (7.12)$$

where c is a constant. Combining (7.10) and (7.11) using (7.6) it follows easily that

$$(S_n - b_n)/a_n \Rightarrow Y$$

where Ee^{itY} is given in (7.12).   □

Exercise 7.2 Show that when $\alpha < 1$, centering is unnecessary; that is, we can let $b_n = 0$.

By doing some calculus (see Breiman (1968), pp. 204–206) one can rewrite (7.12) as

$$\exp(itc - b|t|^{\alpha}(1 + i\kappa\ \mathrm{sgn}(t)w_{\alpha}(t)))\tag{7.13}$$

where $-1 \le \kappa \le 1$ $(\kappa = 2\theta - 1)$ and

$$w_{\alpha}(t) = \begin{cases} \tan(\pi\alpha/2) & \text{if}\quad \alpha \ne 1 \\ \dfrac{2}{\pi}\log|t| & \text{if}\quad \alpha = 1. \end{cases}$$

The reader should note that while we have assumed $0 < \alpha < 2$ throughout the developments above, if we set $\alpha = 2$, then the term with κ vanishes and (7.13) reduces to the characteristic function of the normal distribution with mean c and variance $2b$.

The distributions whose characteristic functions are given in (7.13) are called *stable laws*. α is commonly called the *index*. When $\alpha = 1$ and $\kappa = 0$, we have the Cauchy distribution. Apart from the Cauchy and the normal, there is only one other case in which the density is known: $\alpha = 1/2$, $\kappa = 1$. A quick and dirty way of generating a r.v. with this distribution is to let $Y = 1/\chi^2$ where χ has the standard normal density. Changing variables, we see that Y has density

$$(2\pi y^3)^{-1/2}\exp(-1/2y).\tag{7.14}$$

One can calculate the ch.f. and verify our claim. However, later (see Section 4 of Chapter 7) we will be able to check the claim without effort, so we leave the somewhat tedious calculation to the reader. The next exercise allows us to build a few more examples.

Exercise 7.3   Show that if X has a stable law with index $\alpha > 0$ and $Y \ge 0$ is independent and has a stable law with index β, then $XY^{1/\alpha}$ has a stable law with index $\alpha\beta$. In particular if W_1 and W_2 are independent standard normals, then $W_1/|W_2|$ has a Cauchy distribution.

Exercise 7.4   Let Y have a stable law with $\kappa = 1$. Use the limit theorem (7.7) to conclude that $Y \ge 0$ if $\alpha < 1$ but takes both positive and negative values when $\alpha \ge 1$. Can you extract this information from the inversion formula?

Exercise 7.5   Let X have a symmetric stable law with index α. (i) Use (3.5) to show that $E|X|^p < \infty$ for $p < \alpha$. (ii) Use the construction at the beginning of this section to show $P(|X| \ge x) \ge Cx^{-\alpha}$ so $E|X|^{\alpha} = \infty$.

Turning now to applications of (7.7):

Exercise 7.6 Let $X_1, X_2, \ldots$ be i.i.d. with a density that is continuous at 0. Show

$$\frac{1}{n}\left(\frac{1}{X_1} + \cdots + \frac{1}{X_n}\right) \Rightarrow \text{a Cauchy distribution } (\alpha = 1, \kappa = 0).$$

Exercise 7.7 (i) Assume n objects are placed independently and at random in $[-n, n]$. Let

$$F_n = \sum_{m=1}^{n} \text{sgn}(X_m)/|X_m|^p$$

be the net force exerted on 0. Show that if $p > 1/2$,

$$\lim_{n \to \infty} E \exp(i\theta F_n) = \exp(-c|\theta|^{1/p}).$$

(ii) Generalize the last result to $\mathbb{R}^d$, $d > 1$. We will discuss the case $d = 3$, $p = 2$ at the end of this section.

Exercise 7.8 Let $X_1, X_2, \ldots$ be i.i.d. with $P(X_i = 1) = P(X_i = -1) = 1/2$, let $S_n = X_1 + \cdots + X_n$, and let $\tau = \inf\{n \geq 1 : S_n = 1\}$. In Chapter 3 (see (3.4)) we will show

$$P(\tau > 2n) \sim 2\pi^{-1/2}n^{-1/2} \quad \text{as } n \to \infty.$$

Let $\tau_1, \tau_2, \ldots$ be independent with the same distribution as τ, and let $T_n = \tau_1 + \cdots + \tau_n$. Results in Section 3.1 imply that T_k has the same distribution as the kth time S_n first hits n. Use (7.7) to conclude that $T_k/k^2 \Rightarrow$ the stable law with $\alpha = 1/2$, $\kappa = 1$.

Our next result explains the name *stable laws*. A random variable Y is said to have a *stable law* if for every integer $k > 0$ there are constants a_k and b_k so that if $Y_1, \ldots, Y_k$ are i.i.d. and have the same distribution as Y, then $(Y_1 + \cdots + Y_k - b_k)/a_k \overset{d}{=} Y$. The last definition makes half of the next result obvious.

(7.15) THEOREM Y is the limit of $(X_1 + \cdots + X_k - b_k)/a_k$ for some i.i.d. sequence if and only if Y has a stable law.

Proof If Y has a stable law, we can take $X_1, X_2, \ldots$ i.i.d. with distribution Y. To go the other way, let

$$Z_n = (X_1 + \cdots + X_n - b_n)/a_n$$

and

$$S_n^j = X_{(j-1)n+1} + \cdots + X_{jn}.$$

A little arithmetic shows

$$Z_{nk} = (S_1^k + \cdots + S_n^k - b_{nk})/a_{nk}$$

$$a_{nk}Z_{nk} = (S_n^1 - b_n) + \cdots + (S_n^k - b_n) + (kb_n - b_{nk})$$

$$a_{nk}Z_{nk}/a_n = (S_n^1 - b_n)/a_n + \cdots + (S_n^k - b_n)/a_n + (kb_n - b_{nk})/a_n.$$

The first k terms on the right-hand side $\Rightarrow Y_1 + \cdots + Y_k$ as $n \to \infty$ where $Y_1, \ldots, Y_k$ are independent and have the same distribution as Y, and $Z_{nk} \Rightarrow Y$. So the desired result follows from:

(7.16) *The Convergence of Types Theorem*   If $Y_n \Rightarrow Y$ and there are constants $\alpha_n > 0$, β_n so that $\alpha_n Y_n + \beta_n \Rightarrow Y'$ where Y and Y' are nondegenerate, then there are constants α and β so that $\alpha_n \to \alpha$ and $\beta_n \to \beta$.   □

Remark   Before we get lost in the details of the proof, the reader should note that (7.16) implies that the constants in (7.15) satisfy: $a_{nk}/a_n \to$ a limit.

Exercise 7.9   Use the remark to conclude that if Y is stable, then $a_k = k^{1/\alpha}$ for some $\alpha \in (0, 2]$.

Hint   Let Y' be an independent copy and look at $Y - Y'$ to reduce to the case $b_k = 0$.

Proof of (7.16)   Let $\varphi_n(t) = E \exp(it Y_n)$.

$$\psi_n(t) = E \exp(it(\alpha_n Y_n + \beta_n)) = \exp(it\beta_n)\varphi_n(\alpha_n t).$$

If φ and ψ are the characteristic functions of Y and Y', then

$$\varphi_n(t) \to \varphi(t), \qquad \psi_n(t) = \exp(it\beta_n)\varphi_n(\alpha_n t) \to \psi(t). \tag{a}$$

Take a subsequence $\alpha_{n(m)}$ that converges to a limit $\alpha \in [0, \infty]$. Our first step is observe $\alpha = 0$ is impossible. If this happens,

$$|\psi_n(t)| = |\varphi_n(\alpha_n t)| \to 1, \tag{b}$$

$|\psi(u)| \equiv 1$, and the limit is degenerate by (5.1). Letting $t = u/\alpha_n$ and interchanging the roles of φ and ψ show $\alpha = \infty$ is impossible. If α is a subsequential limit, arguing as in (b) gives $|\psi(t)| = |\varphi(\alpha t)|$. If there are two subsequential limits $\alpha' < \alpha$, using the last equation for both limits implies $|\varphi(u)| = |\varphi(u\alpha'/\alpha)|$. Iterating gives $|\varphi(u)| = |\varphi(u(\alpha'/\alpha)^k)| \to 1$ as $k \to \infty$, contradicting our assumption that Y' is nondegenerate, as $\alpha_n \to \alpha \in [0, \infty)$.

To conclude that $\beta_n \to \beta$ now, we observe that Exercise 3.12 implies $\varphi_n(\alpha_n t) \to \varphi(\alpha t)$. If δ is small enough so that $|\varphi(\alpha t)| > 0$ for $|t| \le \delta$, it follows from (a) and another use of Exercise 3.12 that $\exp(it\beta_n) \to \psi(t)/\varphi(\alpha t)$ uniformly on $[-\delta, \delta]$. $\exp(it\beta_n)$ is the ch.f. of a point mass at β_n. Using (3.5) now as in the proof of (3.4), it follows that β_n is tight; that is, β_n is bounded. If $\beta_{n_m} \to \beta$, then $\exp(it\beta) = \psi(t)/\varphi(\alpha t)$ for $|t| \le \delta$, so there can only be one subsequential limit.   □

(7.15) justifies calling the distributions with characteristic functions given by (7.13) or (7.12) stable laws. To complete the story we should mention that these are the only stable laws. Again, see Chapter 9 of Breiman (1968) or Gnedenko and Kolmogorov (1954). The next example shows that it is sometimes useful to know what all the possible limits are.

Example 7.1   *The Holtsmark distribution* ($\alpha = 3/2$, $\kappa = 0$). Suppose stars are distributed in space according to a Poisson process with density t and their masses are i.i.d. Let X_t be the x-component of the gravitational force at 0 when the density is t. A change of density $1 \to t$ corresponds to a change of length $1 \to t^{-1/3}$, and gravitational attraction follows an inverse square law, so

$$X_t \overset{\mathrm{d}}{=} t^{2/3} X_1. \tag{7.17}$$

If we imagine thinning the Poisson process by rolling an n-sided die, then Exercise 6.12 implies

$$X_t \overset{\mathrm{d}}{=} X_{t/n}^1 + \cdots + X_{t/n}^n$$

where the random variables on the right-hand side are independent and have the same distribution as $X_{t/n}$. It follows from (7.15) that X_t has a stable law. The scaling property (7.17) implies $\alpha = 3/2$. Since $X_t \overset{\mathrm{d}}{=} -X_t$, $\kappa = 0$.

*8   Infinitely Divisible Distributions

In the last section we described the distributions that can appear as the limit of normalized sums of i.i.d. r.v.'s. In this section we will describe those that are limits of sums

$$S_n = X_{n,1} + \cdots + X_{n,n} \tag{$*$}$$

where the $X_{n,m}$ are i.i.d. Note the verb "describe." We will prove almost nothing in this section, just state some of the most important facts to bring the reader up to cocktail party literacy.

A sufficient condition for Z to be a limit of sums of the form (*) is that Z has an *infinitely divisible distribution*; that is, for each n there is an i.i.d. sequence $Y_{n,1}$, ..., $Y_{n,n}$ so that

$$Z \overset{\mathrm{d}}{=} Y_{n,1} + \cdots + Y_{n,n}.$$

Our first result shows that this condition is also necessary.

(8.1) THEOREM Z is a limit of sums of type (*) if and only if Z has an infinitely divisible distribution.

Proof   As remarked above we only have to prove necessity. Write

$$S_{2n} = (X_{2n,1} + \cdots + X_{2n,n}) + (X_{2n,n+1} + \cdots + X_{2n,2n}) \equiv Y_n + Y_n'.$$

The random variables Y_n and Y_n' are independent and have the same distribution. If $S_n \Rightarrow Z$, then the distributions of Y_n are a tight sequence, since

$$P(Y_n > y)^2 = P(Y_n > y)P(Y_n' > y) \le P(S_{2n} > 2y),$$

and similarly,

$$P(Y_n < -y)^2 \le P(S_{2n} < -2y).$$

If we take a subsequence n_k so that $Y_{n_k} \Rightarrow Y$ (and hence $Y_{n_k}' \Rightarrow Y'$), then $Z \overset{d}{=} Y + Y'$. A similar argument shows that Z can be divided into $n > 2$ pieces, and the proof is complete. $\square$

With (8.1) established, we turn now to examples. In the first three cases the distribution is infinitely divisible because it is a limit of sums of the form (*). The number gives the relevant limit theorem.

Example 8.1   *Normal distribution.* (4.1)

Example 8.2   *Stable Laws.* (7.7)

Example 8.3   *Poisson distribution.* (6.1)

Example 8.4   *Compound Poisson distribution.* Let $X_1, X_2, \ldots$ be i.i.d. and $N(\lambda)$ be an independent Poisson r.v. with mean λ. Then $Z = X_1 + \cdots + X_{N(\lambda)}$ has an infinitely divisible distribution. (Let $Y_{n,j} \overset{d}{=} X_1 + \cdots + X_{N(\lambda/n)}$.) For developments below we would like to observe that if $\varphi(t) = E \exp(itX)$, then

$$E \exp(itZ) = \sum_{n=0}^{\infty} e^{-\lambda} \frac{\lambda^n}{n!} \varphi(t)^n = \exp(-\lambda(1 - \varphi(t))). \tag{8.2}$$

Exercise 8.1   *Gamma distribution.* Generalize the computation in Example 3.4 to show that the

Density	$e^{-x} x^{\alpha-1}/\Gamma(\alpha)$
Ch.f.	$(1 - it)^{-\alpha}$

and hence is infinitely divisible.

The next two exercises give examples of distributions that are not infinitely divisible.

Exercise 8.2   The distribution of a bounded r.v. X is infinitely divisible if and only if X is constant.

Exercise 8.3   If μ is infinitely divisible, its ch.f. φ never vanishes.

Hint   Look at $\psi = |\varphi|^2$, which is also infinitely divisible, to avoid taking nth roots of complex numbers.

Example 8.4 is a son of Example 8.3 but a father of Examples 8.1 and 8.2. To explain this remark we observe that if $X = \varepsilon$ and $-\varepsilon$ with probability 1/2 each, then $\varphi(t) = (e^{i\varepsilon t} + e^{-i\varepsilon t})/2 = \cos(\varepsilon t)$. So if $\lambda = \varepsilon^{-2}$, then (8.2) implies

$$E \exp(itZ) = \exp(-\varepsilon^2(1 - \cos(\varepsilon t))) \to \exp(-t^2/2)$$

as $\varepsilon \to 0$. In words, the normal distribution is a limit of compound Poisson distributions. To see that Example 8.2 is also a special case (using the notation from the proof of (7.7)), let

$$I_n(\varepsilon) = \{m \le n : |X_m| > a_n\}$$

$$\hat{S}_n(\varepsilon) = \sum_{m \in I_n(\varepsilon)} X_m$$

$$\bar{S}_n(\varepsilon) = S_n - \hat{S}_n(\varepsilon).$$

If $\varepsilon_n \to 0$, then $\bar{S}_n(\varepsilon_n)/a_n \Rightarrow 0$, and if ε is fixed, then as $n \to \infty$, $|I_n(\varepsilon)| \Rightarrow$ Poisson $(\varepsilon^{-\alpha})$ and $\hat{S}_n(\varepsilon) \Rightarrow$ a compound Poisson distribution:

$$E \exp(it\hat{S}_n(\varepsilon)/a_n) \to \exp(-\varepsilon^{-\alpha}\{1 - \psi^\varepsilon(t)\}).$$

Combining the last two observations and using (7.6) show that stable laws are limits of compound Poisson distributions. The formula (7.13) for the limiting ch.f.

$$\exp\left(itc + \int_0^\infty \left(e^{itx} - 1 - \frac{itx}{1 + x^2} \right) \theta \alpha x^{-(\alpha+1)} \, dx \right.$$
$$\left. + \int_{-\infty}^0 \left(e^{itx} - 1 - \frac{itx}{1 + x^2} \right)(1 - \theta)\alpha|x|^{-(\alpha+1)} \, dx \right) \tag{8.3}$$

helps explain:

(8.4) Lévy–Khinchin Theorem Z has an infinitely divisible distribution if and only if its characteristic function has

$$\log \varphi(t) = ict - \frac{\sigma^2 t^2}{2} + \int \left(e^{itx} - 1 - \frac{itx}{1 + x^2} \right) \mu(dx)$$

where μ is a measure with $\mu(\{0\})$ and $\int \frac{x^2}{1 + x^2} \mu(dx) < \infty$.

For a proof see Breiman (1968), Section 9.5, or Feller, vol. II (1971), Section XVII.2 μ is called the *Lévy measure* of the distribution. Comparing with (8.3) suggests the following interpretation of μ: If $\sigma^2 = 0$, then Z can be built up by making a Poisson process on $\mathbb{R}$ with mean measure μ and then summing up the points. As in the case of stable laws we have to sum the points in $[-\varepsilon, \varepsilon]^c$, subtract an appropriate constant, and let $\varepsilon \to 0$.

Exercise 8.4   What is the Lévy measure for the limit $\aleph$ in part (iii) of Exercise 4.9?

The theory of infinitely divisible distributions is simpler in the case of finite variance. In this case we have:

(8.5) Kolmogorov's Theorem Z has an infinitely divisible distribution with mean 0 and finite variance if and only if its ch.f. has

$$\log \varphi(t) = \int (e^{itx} - 1 - itx)x^{-2}v(dx).$$

Here the integrand is $-t^2/2$ at 0, v is called the *canonical measure*, and $\mathrm{var}(Z) = v(\mathbb{R})$.

To explain the formula, notice that if Z_λ has a Poisson distribution with mean λ,

$$E \exp(itx(Z_\lambda - \lambda)) = \exp(\lambda(e^{itx} - 1 - itx)),$$

so the measure for $Z = x(Z_\lambda - \lambda)$ has $v(\{x\}) = \lambda x^2$.

Exercise 8.5   Let Y have an exponential distribution with mean 1. The ch.f. of Y is

$$\varphi(t) = \int_0^\infty e^{itx}e^{-x}dx = (1 - it)^{-1}.$$

Use the fact that $(d/dt) \log \varphi(t) = i \varphi(t)$ to show that the canonical measure of $Z = Y - 1$ has density xe^{-x} on $(0, \infty)$.

Exercise 8.6   Generalize the last result to the gamma distribution.

Hint   Think; don't compute!

9   Limit Theorems in $\mathbb{R}^d$, $d > 1$

Let $X = (X^1, \ldots, X^d)$ be a random vector. We define its distribution function by $F(x) = P(X \le x)$. Here $x \in \mathbb{R}^d$, and $X \le x$ means $X^i \le x^i$ for $i = 1, \ldots, d$. As in one dimension, F has three obvious properties:

(i) It is nondecreasing; that is, if $x \le y$, then $F(x) \le F(y)$.
(ii) $\lim_{x \to \infty} F(x) = 1$, $\lim_{x^i \to -\infty} F(x) = 0$.
(iii) F is right continuous; that is, $\lim_{y \downarrow x} F(y) = F(x)$.

Here $x \to \infty$ means each coordinate $x^i \to \infty$, $y \downarrow x$ means each coordinate $y^i \downarrow x^i$, and $x^i \to -\infty$ means we let $x^i \to -\infty$, keeping the other coordinates fixed.

In one dimension any function with properties (i)–(iii) was the distribution of some random variable. In $d \geq 2$ this is not the case. Suppose $d = 2$ and let $a_1 < b_1$, $a_2 < b_2$.

$$P(X \in (a_1, b_1] \times (a_2, b_2]) = F(b_1, b_2) - F(a_1, b_2) - F(b_1, a_2) + F(a_1, a_2),$$

so if F is going to be a distribution function, the last quantity has to be ≥ 0. The next example shows that this is not guaranteed by (i)–(iii).

$$F(x_1, x_2) = \begin{cases} 1 & \text{if } x^1, x^2 \geq 1 \\ 2/3 & \text{if } x^1 \geq 1, 0 \leq x^2 < 1, \text{or } x^2 \geq 1, 0 \leq x^1 < 1 \\ 0 & \text{otherwise.} \end{cases}$$

If $0 < a_1, a_2 < 1 \leq b_1, b_2 < \infty$, then

$$F(b_1, b_2) - F(a_1, b_2) - F(b_1, a_2) + F(a_1, a_2) = 1 - 2/3 - 2/3 + 0 = -1/3.$$

A little thought reveals that F is the distribution function of the measure with

$$\mu(\{(0, 1)\}) = \mu(\{(1, 0)\}) = 2/3, \qquad \mu(\{(1, 1)\}) = -1/3.$$

To formulate the additional condition we need to guarantee that F is the distribution function of a probability measure, let

$$A = (a_1, b_1] \times \cdots \times (a_d, b_d]$$

$$V = \{a_1, b_1\} \times \cdots \times \{a_d, b_d\}$$

where $V =$ the vertices of the rectangle A. If $v \in V$, let

$$\text{sgn}(v) = (-1)^{\text{\# of } a's \text{ in } v}.$$

The inclusion–exclusion formula implies

$$P(X \in A) = \sum_{v \in V} \text{sgn}(v) F(v).$$

So if we use $\Delta_A F$ to denote the right-hand side, we need

(iv) $\Delta_A F \geq 0$ for all rectangles A.

The last condition guarantees that the measure assigned to each rectangle is ≥ 0. A standard result from measure theory now implies there is a unique probability measure with distribution F.

Exercise 9.1   Prove the "standard result from measure theory" by combining the proofs of (1.3) and Exercise 1.7 from the Appendix.

Exercise 9.2   If F is the distribution of $(X_1, \ldots, X_d)$, then $F_i(x) = P(X_i \leq x)$ are its *marginal distributions*. How can they be obtained from F?

Exercise 9.3   Let $F_1, \ldots, F_d$ be distributions on $\mathbb{R}$. Show that for any $\alpha \in [-1, 1]$,

$$F(x_1, \ldots, x_d) = \left\{ 1 + \alpha \prod_{i=1}^{d} (1 - F_i(x_i)) \right\} \prod_{j=1}^{d} F_j(x_j)$$

is a d.f. with the given marginals. The case $\alpha = 0$ corresponds to independent r.v.'s.

Exercise 9.4   A distribution F is said to have a density f if

$$F(x_1, \ldots, x_k) = \int_{-\infty}^{x_1} \cdots \int_{-\infty}^{x_k} f(y) \, dy_k \ldots dy_1.$$

Show that if f is continuous, $\partial^k F / \partial x_1 \cdots \partial x_k = f$.

If F_n and F are distribution functions on $\mathbb{R}^d$, we say that F_n *converges weakly* to F and write $F_n \Rightarrow F$ if $F_n(x) \to F(x)$ at all continuity points of F. Our first task is to show that there are enough continuity points for this to be a sensible definition. For a concrete example, consider

$$F(x, y) = \begin{cases} 1 & \text{if } x \geq 0, y \geq 1 \\ y & \text{if } x \geq 0, 0 \leq y < 1 \\ 0 & \text{otherwise.} \end{cases}$$

F is the distribution function of $(0, Y)$ where Y is uniform on $(0, 1)$. Notice that this distribution has no atoms, but F is discontinuous at $(0, y)$ when $y > 0$.

Keeping the last example in mind, observe that if $x_n < x$—that is, $x_n^i < x^i$ for all i, and $x^n \uparrow x$ as $n \to \infty$, then

$$F(x) - F(x_n) = P(X \leq x) - P(X \leq x_n) \downarrow P(X \leq x) - P(X < x).$$

In $d = 2$ the last expression is the probability X lies in

$$\{(a, x^2) : a \leq x^1\} \cup \{(x^1, b) : b \leq x^2\}.$$

Let $H_c^i = \{x : x^i = c\}$. For each i, $D^i = \{c : P(X \in H_c^i) > 0\}$ is at most countable. It is easy to see that if x has $x^i \notin D^i$ for all i, then F is continuous at x. This gives us more than enough points to reconstruct F.

As in Section 2, it will be useful to have several equivalent definitions of weak convergence.

(9.1) THEOREM   The following statements are equivalent to $X_n \Rightarrow X_\infty$.

(i) $Ef(X_n) \to Ef(X_\infty)$ for all bounded continuous f.
(ii) For all closed sets K, $\limsup_{n \to \infty} P(X_n \in K) \leq P(X_\infty \in K)$.
(iii) For all open sets G, $\liminf_{n \to \infty} P(X_n \in G) \geq P(X_\infty \in G)$.

(iv) For all sets A with $P(X_\infty \in \partial A) = 0$, $\lim_{n\to\infty} P(X_n \in A) = P(X_\infty \in A)$.
(v) Let D_f = the set of discontinuities of f. For all bounded functions f with $P(X_\infty \in D_f) = 0$, we have $Ef(X_n) \to Ef(X_\infty)$.

Proof We will begin by showing that (i)–(v) are equivalent. In Chapter 7 we will need to know that this is valid for an arbitrary metric space (S, ρ), so we will prove the result in that generality. The metric appears only in the first step.
(i) implies (ii): Let $\rho(x, K) = \inf\{\rho(x, y) : y \in K\}$, $\varphi_j(r) = (1 - jr)^+$, and $f_j(x) = \varphi_j(\rho(x, K))$. f_j is continuous, has values in $[0, 1]$, and $\downarrow 1_K(x)$ as $j \uparrow \infty$. So

$$\limsup_{n\to\infty} P(X_n \in K) \le \lim_{n\to\infty} Ef_j(X_n) = Ef_j(X_\infty) \downarrow P(X_\infty \in K) \quad \text{as } j \uparrow \infty.$$

(ii) is equivalent to (iii): A is open only if A^c is closed. $P(A) + P(A^c) = 1$.
(ii) and (iii) imply (iv): Let $K = \bar{A}$ and $G = A^0$, and reason as in the proof of (2.3).
(iv) implies (v): Suppose $|f(x)| \le K$ and pick $\alpha_0 < \alpha_1 < \cdots < \alpha_l$ so that $P(f(X_\infty) = \alpha_i) = 0$ for $0 \le i \le l$, $\alpha_0 < -K < K < \alpha_l$, and $\alpha_i - \alpha_{i-1} < \varepsilon$. This is always possible, since $\{\alpha : P(f(X_\infty) = \alpha) > 0\}$ is a countable set. Let $A_i = \{x : \alpha_{i-1} < f(x) \le \alpha_i\}$. $\partial A_i \subset \{x : f(x) \in \{\alpha_{i-1}, \alpha_i\}\} \cup D_f$, so $P(X_\infty \in \partial A_i) = 0$, and it follows from (iv) that

$$\sum_{i=1}^l \alpha_i P(X_n \in A_i) \to \sum_{i=1}^l \alpha_i P(X_\infty \in A_i).$$

The definition of the α_i implies

$$0 \ge \sum_{i=1}^l \alpha_i P(X_n \in A_i) - Ef(X_n) \ge -\varepsilon \quad \text{for } 1 \le n \le \infty.$$

Since ε is arbitrary, it follows that $Ef(X_n) \to Ef(X_\infty)$.
   The final implication, (v) implies (i), is trivial, so it remains to show that the five conditions are equivalent to weak convergence ($\Rightarrow$).
(iv) implies ($\Rightarrow$): If F is continuous at x, then $A = (-\infty, x^1] \times \cdots \times (-\infty, x^d]$ has $\mu(\partial A) = 0$, so $F_n(x) = P(X_n \in A) \to P(X_\infty \in A) = F(x)$.
($\Rightarrow$) implies (iii): Let $D^i = \{c : P(X_\infty \in H_c^i) > 0\}$ where $H_c^i = \{x : x^i = c\}$. We say a rectangle $A = (a_1, b_1] \times \cdots \times (a_d, b_d]$ is good if $a_i, b_i \notin D^i$ for all i. ($\Rightarrow$) implies that for all good rectangles $P(X_n \in A) \to P(X_\infty \in A)$. This is also true for B that are a finite disjoint union of good rectangles. Now any open set G is an increasing limit of B_k's that are a finite disjoint union of good rectangles, so

$$\liminf_{n\to\infty} P(X_n \in G) \ge \liminf_{n\to\infty} P(X_n \in B_k) = P(X_\infty \in B_k) \uparrow P(X_\infty \in G) \quad \text{as } k \to \infty.$$

The proof of (9.1) is complete. $\square$

Remark In Section 2 we proved that (i)–(v) are consequences of weak convergence by constructing r.v.'s with the given distributions so that $X_n \to X_\infty$ a.s. This can be done in $\mathbb{R}^d$ (or any complete separable metric space) but the construction is rather messy. See Billingsley (1979), pp. 337–340, for a proof in $\mathbb{R}^d$.

Exercise 9.5   Show that if $F_n \Rightarrow F$, then the marginal distributions $F_{n,i} \Rightarrow F_i$.

A sequence of probability measures μ_n is said to be *tight* if for any $\varepsilon > 0$ there is an M so that $\liminf_{n \to \infty} \mu_n([-M, M]^d) \geq 1 - \varepsilon$.

(9.2) THEOREM   If μ_n is tight, then there is a weakly convergent subsequence.

Proof   Let F_n be the associated distribution functions, and let $q_1, q_2, \ldots$ be an enumeration of $\mathbb{Q}^d =$ the points in $\mathbb{R}^d$ with rational coordinates. By a diagonal argument like the one in the proof of (2.4), we can pick a subsequence so that $F_{n(k)} \to G(q)$ for all $q \in \mathbb{Q}^d$. Let

$$F(x) = \inf\{G(q) : q \in \mathbb{Q}^d, q > x\}$$

where $q > x$ means $q^i > x^i$ for all i. It is easy to see that F is right continuous. To check that it is a distribution function, we observe that if A is a rectangle with vertices in $\mathbb{Q}^d$, then $\Delta_A F_n \geq 0$ for all n, so $\Delta_A G \geq 0$, and taking limits we see that the last conclusion holds for F for all rectangles A. Tightness implies that F has properties (i) and (ii) of a distribution F. We leave it to the reader to check that $F_n \Rightarrow F$. The proof of (2.4) works if you read inequalities such as $r_1 < r_2 < x < s$ as the corresponding relations between vectors.   □

The *characteristic function* of $(X^1, \ldots, X^d)$ is $\varphi(t) = E \exp(it \cdot X)$ where

$$t \cdot X = t^1 X^1 + \cdots + t^d X^d.$$

(9.3) *Inversion Formula*   If $A = [a_1, b_1] \times \cdots \times [a_d, b_d]$ with $\mu(\partial A) = 0$, then

$$\mu(A) = \lim_{T \to \infty} (2\pi)^{-d} \int_{[-T, T]^d} \prod_{j=1}^{d} \psi_j(t^j) \, \varphi(t) \, dt$$

where

$$\psi_j(s) = (\exp(-isa_j) - \exp(-isb_j))/is.$$

Proof   Fubini's theorem implies

$$\int_{[-T, T]^d} \prod_{j=1}^{d} \psi_j(t^j) \int \exp(it \cdot x) \, \mu(dx) \, dt = \int \prod_{j=1}^{d} \int_{-T}^{T} \psi_j(t^j) \exp(it \cdot x) \, dt^j \mu(dx).$$

It follows from the proof of (3.2) that

$$\int_{-T}^{T} \psi_j(t^j) \exp(it \cdot x) \, dt^j \to \pi[1_{(a_j, b_j)}(x) + 1_{[a_j, b_j]}(x)],$$

so the desired conclusion follows from the bounded convergence theorem.   □

Exercise 9.6 Let φ be the ch.f. of a distribution F on $\mathbb{R}$. What is the distribution on $\mathbb{R}^k$ that corresponds to the ch.f. $\psi(t_1, \ldots, t_k) = \varphi(t_1 + \cdots + t_k)$?

Exercise 9.7 Show that random variables $X_1, \ldots, X_k$ are independent if and only if

$$\varphi X_1, \ldots, x_k^{(t)} = \prod_{j=1}^{k} \varphi x_j(t_j).$$

(9.4) *Convergence Theorem* Let X_n, $1 \le n \le \infty$, be random vectors with characteristic functions φ_n. A necessary and sufficient condition for $X_n \Rightarrow X_\infty$ is that $\varphi_n(t) \to \varphi_\infty(t)$.

Proof $\exp(it \cdot x)$ is bounded and continuous, so if $X_n \Rightarrow X_\infty$, then $\varphi_n(t) \to \varphi_\infty(t)$. To prove the other direction, it suffices, as in the proof of (3.4), to prove that the sequence is tight. To do this, we observe that if we fix $\theta \in \mathbb{R}^d$, then for all $s \in \mathbb{R}$, $\varphi_n(s\theta) \to \varphi_\infty(s\theta)$, so it follows from (3.4) that the distributions of $\theta \cdot X_n$ are tight. Applying the last observation to the d unit vectors $e_1, \ldots, e_d$ shows that the distributions of X_n are tight and completes the proof. □

Remark As before, if $\varphi_n(t) \to \varphi_\infty(t)$ with $\varphi_\infty(t)$ continuous at 0, then $\varphi_\infty(t)$ is the ch.f. of some X_∞ and $X_n \Rightarrow X_\infty$.

(9.4) has an important corollary.

(9.5) *Cramér–Wold Device* A sufficient condition for $X_n \Rightarrow X_\infty$ is that $\theta \cdot X_n \Rightarrow \theta \cdot X_\infty$ for all $\theta \in \mathbb{R}^d$.

Proof The indicated condition implies $E \exp(i\theta \cdot X_n) \to E \exp(i\theta \cdot X_\infty)$ for all $\theta \in \mathbb{R}^d$. □

(9.5) leads immediately to:

(9.6) *The Central Limit Theorem in* $\mathbb{R}^d$ Let $X_1, X_2, \ldots$ be i.i.d. random vectors with $EX_n = \mu$ and finite covariances $\Gamma_{ij} = E((X_n^i - \mu^i)(X_n^j - \mu^j))$. If $S_n = X_1 + \cdots + X_n$, then

$$(S_n - n\mu)/n^{1/2} \Rightarrow \chi$$

where χ has a multivariate normal distribution with mean 0 and covariance Γ; that is,

$$E \exp(it \cdot \chi) = \exp\left(-\sum_i \sum_j \theta^i \theta^j \Gamma_{ij}/2\right).$$

Proof By considering $X_n' = X_n - \mu$ we can suppose without loss of generality that $\mu = 0$. Let $\theta \in \mathbb{R}^d$. $\theta \cdot X_n$ is a random variable with mean 0 and variance

$$E\left(\sum_i \theta^i X_n^i\right)^2 = \sum_i \sum_j E(\theta^i \theta^j X_n^i X_n^j) = \sum_i \sum_j \theta^i \theta^j \Gamma_{ij},$$

so it follows from the one-dimensional central limit theorem and (9.5) that $S_n/n^{1/2} \Rightarrow \chi$ where

$$E \exp(i\theta \cdot \chi) = \exp\left(-\sum_i \sum_j \theta^i \theta^j \Gamma_{ij}/2\right).$$

$\square$

To illustrate the use of (9.6), we consider two examples.

Example 9.1   *Simple random walk on $\mathbb{Z}^d$.* Let $X_1, X_2, \ldots$ be i.i.d. with

$$P(X_n = +e_i) = P(X_n = -e_i) = 1/2d \quad \text{for } i = 1, \ldots, d.$$

$EX_n^i = 0$ and if $i \neq j$, $EX_n^i X_n^j = 0$, since both components cannot be nonzero simultaneously. So the covariance matrix is $\Gamma_{ij} = (1/2d)I$.

Example 9.2   Let $X_1, X_2, \ldots$ be i.i.d. with $P(X_n = e_i) = 1/6$ for $i = 1, 2, \ldots, 6$. In words, we are rolling a die and keeping track of the numbers which come up. $EX_n^i = 1/6$ and $EX_n^i X_n^j = 0$ for $i \neq j$, so $\Gamma_{ij} = (1/6)(5/6)$ when $i = j$ and $= -(1/6)^2$ when $i \neq j$. In this case the limiting distribution is concentrated on $\{x : \sum_i x_i = 0\}$.

Exercise 9.8   Let $X_1, X_2, \ldots$ be i.i.d. with $P(X_n = e_i) = p_i$ for $1 \leq i \leq k$. Find the limit distribution of

$$\sum_{i=1}^k (S_n^i - np_i)^2/np_i.$$

Our treatment of the central limit theorem would not be complete without some discussion of the multivariate normal distribution. We begin by observing that $\Gamma_{ij} = \Gamma_{ji}$ and

$$\sum_i \sum_j \theta^i \theta^j \Gamma_{ij} = E\left(\sum_i \theta^i X_n^i\right)^2 \geq 0,$$

so Γ is symmetric and nonnegative definite. A well-known result implies that there is an orthogonal matrix U (i.e., one with $U^t U = I$, the identity) so that $\Gamma = U^t V U$, where $V \geq 0$ is a diagonal matrix. Let W be the nonnegative diagonal matrix with $W^2 = V$. If we let $A = WU$, then $\Gamma = A^t A$. Let Y be a d-dimensional vector whose components are independent and have normal distributions with mean 0 and variance 1. If we view vectors as $1 \times d$ matrices and let $\chi = YA$, then χ has the desired normal distribution. To check this observe that

$$\theta \cdot YA = \sum_i \theta^i \sum_j Y_j A_{ji}$$

has a normal distribution with mean 0 and variance

$$\sum_j \left(\sum_i A_{ji} \theta^i \right)^2 = \sum_j \left(\sum_i \theta^i A_{ij}^t \right) \left(\sum_k A_{jk} \theta^k \right) = \theta A^t A \theta^t = \theta \Gamma \theta^t,$$

so

$$E(\exp(i\theta \cdot \chi)) = \exp(-(\theta \Gamma \theta^t)/2).$$

If the covariance matrix has rank d we say that the normal distribution is *nondegenerate*. In this case its density function is given by

$$(2\pi)^{-d/2} (\det \Gamma)^{-1/2} \exp\left(-\sum_{i,j} y^i \Gamma_{ij}^{-1} y^j \right) dy.$$

The joint distribution in degenerate cases can be computed by using a linear transformation to reduce to the nondegenerate case. For instance, in Example 9.2 we can look at the distribution of $(X_1, \ldots, X_5)$. Finally we would like to observe that $X_1, \ldots, X_d$ are independent if and only if $\Gamma_{ij} = 0$ for $i \neq j$; that is, uncorrelated random variables with a joint normal distribution are independent.

Exercise 9.9 Show that $(X_1, \ldots, X_k)$ has a multivariate normal distribution with mean vector θ and covariance Γ if and only if every linear combination $c_1 X_1 + \cdots + c_k X_k$ has a normal distribution with mean $c\,\theta^t$ and variance $c\Gamma c^t$.

Exercise 9.10 Let U and V be independent standard normals. Let

$$Y = U \quad \text{and} \quad Z = \rho U + (1 - \rho^2) V.$$

Show that the correlation of Y and Z is ρ. To get normals X_i with mean μ_i and variance σ_i^2 that have correlation ρ, let $X_1 = \mu_1 + \sigma_1 Y$ and $X_2 = \mu_2 + \sigma_2 Z$.

3 Random Walks

Let $X_1, X_2, \ldots$ be i.i.d. taking values in $\mathbb{R}^d$ and let $S_n = X_1 + \cdots + X_n$. S_n is a *random walk*. In the last chapter we were primarily concerned with the distribution of S_n. In this one we will look at properties of the sequence $S_1(\omega), S_2(\omega), \ldots$. For example, does the last sequence return to (or near) 0 infinitely often? The first section introduces stopping times, a concept that will be very important in this and the next two chapters. After that section is completed, the remaining three can be read in any order or skipped without much loss.

1 Stopping Times

Most of the results in this section are valid for i.i.d. X's taking values in some nice measurable space $(S, \mathscr{S})$ and will be proved in that generality. Before taking up our main topic, we will prove a 0–1 law that, in the i.i.d. case, generalizes Kolmogorov's. To state the new 0–1 law we need two definitions. A *finite permutation* of $\mathbb{N} = \{1, 2, \ldots\}$ is a map π from $\mathbb{N}$ onto $\mathbb{N}$ so that $\pi(i) \neq i$ for only finitely many i. If π is a finite permutation of $\mathbb{N}$ and $\omega \in S^{\mathbb{N}}$, we define $(\pi\omega)_i = \omega_{\pi(i)}$. In words, the coordinates of ω are rearranged according to π. An event A is *permutable* if $\pi^{-1}A \equiv \{\omega : \pi\omega \in A\}$ is equal to A for any finite permutation π. The collection of permutable events is a σ-field. It is called the *exchangeable σ-field* and denoted by $\mathscr{E}$.

To see the reason we are interested in permutable events, let $X_1, X_2, \ldots \in \mathbb{R}$ be i.i.d. and let $S_n = X_1 + \cdots + X_n$. Two examples of permutable events are

(i) $\{\omega : S_n(\omega) \in A \text{ i.o.}\}$ and

(ii) $\left\{\omega : \limsup_{n \to \infty} S_n(\omega)/c_n \geq 1\right\}$.

In each case the event is permutable because $S_n(\omega) = S_n(\pi\omega)$ for large n. The list of examples can be enlarged considerably by observing:

(iii) All events in the tail σ-field $\mathscr{T}$ are permutable.

To see this observe that if $B \in \sigma(X_{n+1}, X_{n+2}, \ldots)$, then the occurrence of B is unaffected by a permutation of $\{1, \ldots, n\}$. (i) shows that the converse of (iii) is false.

The next result shows that for an i.i.d. sequence there is no difference between $\mathscr{E}$ and $\mathscr{T}$. They are both trivial.

(1.1) Hewitt–Savage 0–1 Law   If $X_1, X_2, \ldots$ are i.i.d. and $A \in \mathscr{E}$, then $P(A) \in \{0, 1\}$.

Proof   Let $A \in \mathscr{E}$. As in the proof of Kolmogorov's 0–1 law, we will show A is independent of itself; that is, $P(A) = P(A \cap A) = P(A)P(A)$, so $P(A) \in \{0, 1\}$. Let $A_n \in \sigma(X_1, \ldots, X_n)$ so that $P(A_n \triangle A) \to 0$. Here $A \triangle B = (A - B) \cup (B - A)$ is the symmetric difference. The existence of the A_n's is proved in Exercise 3.1 in the Appendix. A_n can be written as $\{\omega : (\omega_1, \ldots, \omega_n) \in B_n\}$ with $B_n \in \mathscr{S}^n$. Let

$$\pi(j) = \begin{cases} j + n & \text{if } 1 \le j \le n \\ j - n & \text{if } n + 1 \le j \le 2n \\ j & \text{if } j \ge 2n + 1 \end{cases}$$

Observing that π^2 is the identity (so we don't have to worry about whether to write π or π^{-1}) and the coordinates are i.i.d. (so the permuted coordinates are) gives

$$P(\omega : \omega \in A_n \triangle A) = P(\omega : \pi\omega \in A_n \triangle A).$$

Now $\{\omega : \pi\omega \in A\} = \{\omega : \omega \in A\}$, since A is permutable, and

$$\{\omega : \pi\omega \in A_n\} = \{\omega : (\omega_{n+1}, \ldots, \omega_{2n}) \in B_n\}.$$

If we use A_n' to denote the last event, then we have

$$\{\omega : \pi\omega \in A_n \triangle A\} = \{\omega : \omega \in A_n' \triangle A\}.$$

Combining the last three equalities gives $P(A_n \triangle A) = P(A_n' \triangle A)$. The last equality implies

$$P(A_n \triangle A_n') \le P(A_n \triangle A) + P(A \triangle A_n') \to 0.$$

(To see this recall $A \triangle B = (A - B) \cup (B - A)$ and observe $A - C \subset (A - B) \cup (B - C)$.) The last result implies

$$P(A_n \cup A_n') - P(A_n \cap A_n') = P(A_n \triangle A_n') \to 0,$$

so

$$P(A_n) - P(A_n \cap A_n') \to 0$$

and

$$P(A_n \cap A_n') \to P(A).$$

But A_n and A'_n are independent, so

$$P(A_n \cap A'_n) = P(A_n)P(A'_n) \to P(A)^2.$$

(Recall $P(A'_n) = P(A_n)$.) This shows $P(A) = P(A)^2$ and completes the proof.   □

Exercise 1.1   Show that the conclusion of (1.1) holds if $X_1, X_2, \ldots$ are independent and for each j there is a $k > j$ with $X_k \overset{d}{=} X_j$.

A typical application of (1.1) is:

(1.2) THEOREM   For a random walk on $\mathbb{R}$, there are only four possibilities, one of which has probability one.

(i) $S_n = 0$ for all n.
(ii) $S_n \to \infty$.
(iii) $S_n \to -\infty$.
(iv) $-\infty = \liminf S_n < \limsup S_n = \infty$.

Proof   The Hewitt–Savage 0–1 law implies $\limsup S_n$ is a constant $c \in [-\infty, \infty]$. Let $S'_n = S_{n+1} - X_1$. Since S'_n has the same distribution as S_n, it follows that $c = X_1 + c$. If c is finite, subtracting c from both sides, we conclude $X_1 \equiv 0$ and (i) occurs. Turning the last statement around, we see that if X_1 is not $\equiv 0$, then $c = -\infty$ or ∞. The same analysis applies to the liminf. Discarding the impossible combination $\limsup S_n = -\infty$ and $\liminf S_n = +\infty$, we have proved the result.   □

Exercise 1.2   *Symmetric random walk.* Let $X_1, X_2, \ldots \in \mathbb{R}$ be i.i.d. symmetric and nondegenerate (i.e., $X_i \overset{d}{=} -X_i$ and $P(X_i = 0) < 1$). Show that

$$\limsup_{n \to \infty} S_n = \infty, \qquad \liminf_{n \to \infty} S_n = -\infty \text{ a.s.}$$

A corollary of this is that the *simple random walk* $(P(X_i = 1) = P(X_i = -1) = 1/2)$ visits every integer infinitely many times.

Exercise 1.3   Let $X_1, X_2, \ldots$ be i.i.d. and have a stable law with $\alpha < 2$ and $|\kappa| < 1$. Use (1.1) to conclude that

$$\limsup_{n \to \infty} S_n = \infty, \qquad \liminf_{n \to \infty} S_n = -\infty \text{ a.s.}$$

Hint   Use the following observations from Chapter 2: Exercise 7.2 implies that we can take $b_k = 0$ in (7.15). For $\alpha > 1$, see Exercise 2.6 below.

Let $\mathscr{F}_n = \sigma(X_1, \ldots, X_n) =$ the information known at time n. A random variable N taking values in $\{1, 2, \ldots\} \cup \{\infty\}$ is said to be a *stopping time* or *optional random variable* if for every $n < \infty$, $\{N = n\} \in \mathscr{F}_n$. If we think of S_n as giving the price of a stock at time n, and N as the time we sell it, then the last definition says that the decision to sell at time n must be based on the information known at that time. The last interpretation gives one explanation for the second name. N is a time at which we can exercise an option to buy a stock. Chung (1974) prefers the second name because N is "usually rather a momentary pause after which the process proceeds again: time marches on!"

The canonical example of a stopping time is $N = \inf\{n: S_n \in A\}$, *the hitting time of A*. To check that this is a stopping time, we observe that

$$\{N = n\} = \{S_1 \in A^c, \ldots, S_{n-1} \in A^c, S_n \in A\} \in \mathscr{F}_n.$$

Two concrete examples of hitting times that have appeared above are:

Example 1.1 $N = \min\{n: S_n > \lambda\}$ from the proof of (7.2) in Chapter 1.

Example 1.2   If the $X_i \geq 0$ and $N_t = \sup\{n: S_n \leq t\}$ is the random variable that first appeared in (8.5) in Chapter 1, then $N_t + 1 = \inf\{n: S_n > t\}$ is a stopping time.

The next result allows us to construct new examples from the old ones.

Exercise 1.4   If S and T are stopping times, then $S \wedge T$ and $S \vee T$ are stopping times. Since constant times are stopping times, it follows that $S \wedge n$ and $S \vee n$ are stopping times.

Exercise 1.5   Suppose S and T are stopping times. Is $S + T$ a stopping time? Give a proof or a counterexample.

Associated with each stopping time N is a σ-field $\mathscr{F}_N$ = the information known at time N. Formally $\mathscr{F}_N$ is the collection of sets A that have $A \cap \{N = n\} \in \mathscr{F}_n$ for all $n < \infty$; that is, when $N = n$, A must be measurable with respect to the information known at time n. Trivial but important examples of sets in $\mathscr{F}_N$ are $\{N \leq n\}$; that is, N is measurable with respect to $\mathscr{F}_N$.

Exercise 1.6   Show that if $Y_n \in \mathscr{F}_n$ and N is a stopping time, $Y_N \in \mathscr{F}_N$. As a corollary of this result we see that if $f: S \to \mathbb{R}$ is measurable, $T_n = \sum_{m \leq n} f(X_m)$, and $M_n = \max_{m \leq n} T_m$, then T_N and $M_N \in \mathscr{F}_N$. An important special case is $S = \mathbb{R}^d$, $f(x) = x$.

Exercise 1.7   Show that if $M \leq N$ are stopping times, then $\mathscr{F}_M \subset \mathscr{F}_N$.

Our first result about $\mathscr{F}_N$ is:

(1.3) THEOREM   On $\{N < \infty\}$, $\{X_{N+n}, n \geq 1\}$ is independent of $\mathscr{F}_N$ and has the same distribution as the original sequence.

Proof   What we need to show is that if $A \in \mathscr{F}_N$ and $B_j \in \mathscr{S}$ for $1 \leq j \leq k$, then

$$P(A, X_{N+j} \in B_j, 1 \leq j \leq k) = P(A \cap \{N < \infty\}) \prod_{j=1}^{k} \mu(B_j)$$

where $\mu(B) = P(X_i \in B)$. The method ("divide and conquer") is one that we will see many times below. We break things down according to the value of N.

$$P(A, N = n, X_{N+j} \in B_j, 1 \le j \le k) = P(A, N = n, X_{n+j} \in B_j, 1 \le j \le k)$$

$$= P(A \cap \{N = n\}) \prod_{j=1}^{k} \mu(B_j),$$

since $A \cap \{N = n\} \in \mathcal{F}_n$ and that σ-field is independent of $X_{n+1}, \ldots, X_{n+k}$. Summing over n now gives the desired result. $\square$

To delve further into properties of stopping times, it is convenient to use the special probability space from the proof of Kolmogorov's extension theorem:

$$\Omega = \{(\omega_1, \omega_2, \ldots) : \omega_i \in S\}$$

$$\mathcal{F} = \mathcal{S} \times \mathcal{S} \times \cdots$$

$$P = \mu \times \mu \times \cdots \qquad \mu \text{ is the distribution of } X_i.$$

$$X_n(\omega) = \omega_n$$

The chief advantage of this concrete Ω, which we call *sequence space*, is that it enables us to define the *shift* $\theta: \Omega \to \Omega$ by

$$(\theta\omega)(n) = \omega(n + 1), \quad n = 1, 2, \ldots.$$

In words, we drop the first coordinate and shift the others one place to the left. The iterates of θ are defined by composition

$$\theta^k = \theta \circ \theta^{k-1} \quad \text{for } k \ge 2, \text{ where } \theta^1 = \theta.$$

Clearly $(\theta^k\omega)(n) = \omega(n + k)$, $n = 1, 2, \ldots$. To extend the last definition to stopping times, we let

$$\theta^N\omega = \theta^n\omega \quad \text{on } \{N = n\}.$$

It is tempting to leave θ^N undefined on $\{N = \infty\}$, but things work better if we let

$$\theta^N\omega = \Delta \quad \text{on } \{N = \infty\}$$

where Δ is an extra point which we add to Ω. According to the only joke in Blumenthal and Getoor (1968), Δ is a "cemetery or heaven depending upon your point of view." Seriously, Δ is a convenience in making definitions.

Example 1.3   *Returns to 0.* For a concrete example of the use of θ, suppose $S = \mathbb{R}^d$ and let

$$\tau(\omega) = \inf\{n : \omega_1 + \cdots + \omega_n = 0\}$$

where $\inf \theta = \infty$, and set $\tau(\Delta) = \infty$. If we let $\tau_2(\omega) = \tau(\omega) + \tau(\theta^\tau \omega)$, then on $\{\tau < \infty\}$,

$$\tau(\theta^\tau \omega) = \inf\{n : (\theta^\tau \omega)_1 + \cdots + (\theta^\tau \omega)_n = 0\}$$

$$= \inf\{n : \omega_{\tau+1} + \cdots + \omega_{\tau+n} = 0\}$$

$$\tau(\omega) + \tau(\theta^\tau \omega) = \inf\{m > \tau : \omega_1 + \cdots + \omega_m = 0\}.$$

So τ_2 is the time of the second visit to 0 (and thanks to the conventions $\theta^\infty \omega = \Delta$ and $\tau(\Delta) = \infty$, this is true for all ω). To prepare for developments below we would like the reader to observe that (1.3) implies $P(\tau_2 < \infty) = P(\tau < \infty)^2$. The computations generalize easily to show that if we let

$$\tau_n = \tau_{n-1} + \tau(\theta^{\tau(n-1)}\omega),$$

then τ_n is the time of the nth visit to 0 and

$$P(\tau_n < \infty) = P(\tau < \infty)^n. \tag{1.4}$$

Extending the notation just introduced to an arbitrary stopping time T and letting $t_n = T(\theta^{T(n-1)})$, we can extend (1.3) to:

(1.5) THEOREM   Suppose $P(T < \infty) = 1$. Then the random vectors

$$V_n = (t_n, X_{T_{n-1}+1}, \ldots, X_{T_n})$$

are independent and identically distributed.

Proof   The independence follows from (1.3), since $V_1, \ldots, V_{n-1}$ belong to $\mathscr{F}(T_{n-1})$. Breaking things down according to the value of T_{n-1}, it is easy to show that V_n and V_1 have the same distribution. Writing this out is left as an exercise for the reader.   □

Example 1.4   *Ladder variables.* Let $\alpha(\omega) = \inf\{n : \omega_1 + \cdots + \omega_n > 0\}$ where $\inf \theta = \infty$, and set $\alpha(\Delta) = \infty$. Let $\alpha_0 = 0$ and let $\alpha_k = \alpha_{k-1} + \alpha(\theta^{\alpha(k-1)}\omega)$ for $k \geq 1$. At time α_k the random walk is at a record high value. The next three exercises investigate these times.

Exercise 1.8   (i) If $P(\alpha < \infty) < 1$, then $P(\sup S_n < \infty) = 1$. (ii) If $P(\alpha < \infty) = 1$, then $P(\sup S_n = \infty) = 1$.

Hint   In (ii), $\xi_k = S_{\alpha(k)} - S_{\alpha(k-1)}$ are i.i.d. with $E\xi_k \in (0, \infty]$.

Exercise 1.9   Let $\beta(\omega) = \inf\{n : S_n < 0\}$. Without using (1.1) prove that the four possibilities in (1.2) correspond to the four combinations of $P(\alpha < \infty) = 0$ or 1 and $P(\beta < \infty) = 0$ or 1.

Exercise 1.10   Let $\bar{\beta}(\omega) = \inf\{n \geq 1 : S_n \leq 0\}$. Let

$$A_m^n = \{S_1 < S_m, \ldots, S_{m-1} < S_m, S_m \geq S_{m+1}, \ldots, S_m \geq S_n\}$$

and observe

$$1 = \sum_{m=1}^{n} P(A_m^n) = \sum_{m=1}^{n} P(\alpha \geq m)P(\bar{\beta} > n - m).$$

Let $n \to \infty$ and conclude $E\alpha = 1/P(\bar{\beta} = \infty)$ by proving two inequalities.

Exercise 1.11   Combine the last exercise with the proof of (ii) in Exercise 1.8 to conclude that if $EX_i = 0$, then $P(\bar{\beta} = \infty) = 0$ and if $P(X_i = 0) < 1$, then we are in case (iv) of (1.2).

Our final result about stopping times is:

(1.6) Wald's Equation   Let $X_1, X_2, \ldots$ be i.i.d. with $E|X_i| < \infty$. If N is a stopping time with $EN < \infty$, then $ES_N = EX_1 EN$.

Proof   First suppose the $X_i \geq 0$.

$$ES_N = \int S_N \, dP = \sum_{n=1}^{\infty} \int S_n 1_{\{N=n\}} \, dP$$

$$= \sum_{n=1}^{\infty} \sum_{m=1}^{n} \int X_m 1_{\{N=n\}} \, dP.$$

Since the $X_i \geq 0$, we can interchange the order of summation (i.e., use Fubini's theorem) to conclude that the last expression

$$= \sum_{m=1}^{\infty} \sum_{n=m}^{\infty} \int X_m 1_{\{N=n\}} \, dP$$

$$= \sum_{m=1}^{\infty} \int X_m 1_{\{N \geq m\}} \, dP.$$

Now $\{N \geq m\} = \{N \leq m - 1\}^c \in \mathscr{F}_{m-1}$, so the last expression

$$= \sum_{m=1}^{\infty} EX_m P(N \geq m) = EX_1 EN.$$

To prove the result in general, we run the last argument backward:

$$\infty > \sum_{m=1}^{\infty} E|X_m| P(N \geq m)$$

$$= \sum_{m=1}^{\infty} \sum_{n=m}^{\infty} \int |X_m| 1_{\{N=n\}} \, dP.$$

The last formula shows that the double sum converges absolutely in one order, so Fubini's theorem gives

$$\sum_{m=1}^{\infty} \sum_{n=m}^{\infty} \int X_m 1_{\{N=n\}} \, dP = \sum_{n=1}^{\infty} \sum_{m=1}^{n} \int X_m 1_{\{N=n\}} \, dP,$$

and the result follows.   □

Exercise 1.12   Let $X_1, X_2, \ldots$ be i.i.d. uniform on $(0, 1)$, let $S_n = X_1 + \cdots + X_n$, and let $T = \inf\{n : S_n > 1\}$. Show that $P(T \geq n) = 1/n!$, so $ET = e$ and $ES_T = e/2$.

Example 1.5   *Simple random walk.* Let $X_1, X_2, \ldots$ be i.i.d. with $P(X_i = 1) = 1/2$ and $P(X_i = -1) = 1/2$. Let $a < 0 < b$ be integers and let $N = \inf\{n : S_n \notin (a, b)\}$. To apply (1.6) we have to check that $EN < \infty$. To do this we observe that if $x \in (a, b)$, then

$$P(x + S_{b-a} \notin (a, b)) \geq 2^{-(b-a)},$$

since $b - a + 1$'s in a row will take us out of the interval. Iterating the last inequality, it follows that

$$P(N > n(b - a)) \leq (1 - 2^{-(b-a)})^n,$$

so $EN < \infty$. Applying (1.6) now gives $ES_N = 0$ or

$$bP(S_N = b) + aP(S_N = a) = 0.$$

Since

$$P(S_N = b) + P(S_N = a) = 1,$$

it follows that

$$(b - a)P(S_N = b) = -a,$$

so

$$P(S_N = b) = \frac{-a}{b - a}, \qquad P(S_N = a) = \frac{b}{b - a}.$$

Letting $T_a = \inf\{n : S_n = a\}$, we can write the last conclusion as

$$P(T_a < T_b) = \frac{b}{b - a} \quad \text{for } a < 0 < b. \tag{1.7}$$

Setting $b = M$ and letting $M \to \infty$ give

$$P(T_a < \infty) \geq P(T_a < T_M) \to 1$$

for all $a < 0$. From symmetry it follows that

$$P(T_x < \infty) = 1 \quad \text{for all } x \in \mathbb{Z}. \tag{1.8}$$

Our final fact about T_x is that $ET_x = \infty$ for $x \neq 0$. To see that, note that if $ET_x < \infty$, then (1.6) would imply

$$x = ES_{T_x} = EX_1 ET_x = 0.$$

In Section 3 we will compute the distribution of T_1 and show that

$$P(T_1 > t) \sim Ct^{-1/2}.$$

Exercise 1.13   *Asymmetric simple random walk.* Let $X_1, X_2, \ldots$ be i.i.d. with $P(X_1 = 1) = p > 1/2$ and $P(X_1 = -1) = 1 - p$, and let $S_n = X_1 + \cdots + X_n$.

(i) If $\beta = \inf\{n : S_n < 0\}$, then $P(\beta < \infty) < 1$.
(ii) If $Y = \inf S_n$, then $P(Y \leq -k) = P(\beta < \infty)^k$.
(iii) Let $\alpha = \inf\{m : S_m = 1\}$. Apply Wald's equation to $\alpha \wedge n$ and let $n \to \infty$ to get

$$E\alpha = 1/EX_1 = 1/(2p - 1).$$

Comparing with Exercise 1.10 shows $P(\bar\beta = \infty) = 2p - 1$.

Exercise 1.14   Let $X_n, n \geq 1$, be i.i.d. with $EX_1^+ < \infty$ and let

$$Y_n = \max_{1 \leq m \leq n} X_m - cn.$$

That is, we are looking for a large value of X but we have to pay $c > 0$ for each observation. (i) Let $T = \inf\{n : X_n > a\}$ and $p = P(X_n > a)$, and compute EY_T. (ii) Let α be the unique solution of $E(X_1 - \alpha)^+ = c$. Show that $EY_T = \alpha$ in this case and use the inequality

$$Y_n \leq \alpha + \sum_{m=1}^{n} \left((X_m - \alpha)^+ - c \right)$$

to conclude that if τ is a stopping time with $E\tau < \infty$, then $EY_\tau \leq \alpha$.

Exercise 1.15   Let $X_1, X_2, \ldots$ be i.i.d. with $EX_n = 0$ and $EX_n^2 = \sigma^2 < \infty$. If T is a stopping time with $ET < \infty$, then $ES_T^2 = \sigma^2 ET$.

Hint   Compute $ES_{T \wedge n}^2$ by induction and show that $S_{T \wedge n}$ is a Cauchy sequence in L^2.

A corollary of the last result is that if $X_1, X_2, \ldots$ are i.i.d. with $EX_n = 0$ and $EX_n^2 = 1$, and we let $T_c = \inf\{n \geq 1 : |S_n| > cn^{1/2}\}$, then $ET_c = \infty$ for $c > 1$. To prove that $ET_c < \infty$ for $c < 1$, let $\tau = T_c \wedge n$, observe

$$E\tau = ES_\tau^2 = ES_{\tau-1}^2 + 2E(S_{\tau-1}X_\tau) + EX_\tau^2$$

$$\leq c^2 E\tau + 2c(E\tau EX_\tau^2)^{1/2} + EX_\tau^2,$$

and use the following result:

Exercise 1.16   (Gundy–Siegmund) Let $Y_1, Y_2, \ldots \geq 0$ be independent with

$$n^{-1} \sum_{m=1}^{n} E(Y_m; Y_m > \varepsilon m) \to 0$$

(e.g., Y_m i.i.d. with $EY_m < \infty$). Let T_n be a sequence of stopping times with $ET_n < \infty$ and $ET_n \to \infty$ as $n \to \infty$. Then $EY_{T_n}/ET_n \to 0$.

Hint   Estimates are easy when $Y_{T_n} \leq \varepsilon T_n$ or $T_n \leq N$ with N fixed, and the hypothesis will take care of the rest.

2   Recurrence

In this section we will investigate the question mentioned at the beginning of the chapter. Does the sequence $S_1(\omega), S_2(\omega), \ldots$ return to (or near) 0 infinitely often? The answer to the last question is either yes or no, and the random walk is called recurrent or transient accordingly. We begin with some definitions that formulate the question precisely and a result that establishes a dichotomy between the two cases.

The number $x \in \mathbb{R}^d$ is said to be a *recurrent value* for the random walk S_n if:

For every $\varepsilon > 0$, $P(\|S_n - x\| < \varepsilon \text{ i.o.}) = 1$.

Here $\|x\| = \sup |x_i|$. The Hewitt–Savage 0–1 law (1.1) implies that if the last probability is < 1, it is 0. Our first result shows that to know the set of recurrent values it is enough to check $x = 0$. A number x is said to be a *possible value* of the random walk if for any $\varepsilon > 0$ there is an n so that $P(\|S_n - x\| < \varepsilon) > 0$.

(2.1) THEOREM   The set $\mathscr{V}$ of recurrent values is either $\varnothing$ or a closed subgroup of $\mathbb{R}$. In the second case $\mathscr{V} = \mathscr{U}$, the set of possible values.

Proof   Suppose $\mathscr{V} \neq \varnothing$. It is clear that $\mathscr{V}$ is closed. To prove that $\mathscr{V}$ is a group we will first show that if $x \in \mathscr{U}$ and $y \in \mathscr{V}$, then $y - x \in \mathscr{V}$. This statement has been formulated so that once it is established, (2.1) follows easily. Let

$$p_{\delta,m} = P(\|S_n - z\| \geq \delta \text{ for all } n \geq m).$$

If $y - x \notin \mathscr{V}$, there is an $\varepsilon > 0$ and $m \geq 1$ so that $p_{2\varepsilon,m}(y - x) > 0$. Since $x \in \mathscr{U}$, there is a k so that $P(\|S_k - x\| < \varepsilon) > 0$. Since

$$P(\|S_n - S_k - (y - x)\| \geq 2\varepsilon \text{ for all } n \geq k + m) = p_{2\varepsilon,m}(y - x)$$

and is independent of $\{\|S_k - x\| < \varepsilon\}$, it follows that

$$p_{\varepsilon,m+k}(y) \geq P(\|S_k - x\| < \varepsilon)p_{2\varepsilon,m}(y - x) > 0,$$

contradicting $y \in \mathcal{V}$, so $y - x \in \mathcal{V}$.

To conclude $\mathcal{V}$ is a group, let $q, r \in \mathcal{V}$ and observe: (i) taking $x = y = r$ shows $0 \in \mathcal{V}$, (ii) taking $x = r$, $y = 0$ shows $-r \in \mathcal{V}$, and (iii) taking $x = -r$, $y = q$ shows $q + r \in \mathcal{V}$. To prove that $\mathcal{V} = \mathcal{U}$ now, observe that if $u \in \mathcal{U}$, taking $x = u$, $y = 0$ shows $-u \in \mathcal{V}$, and since $\mathcal{V}$ is a group, it follows that $u \in \mathcal{V}$. $\square$

If $\mathcal{V} = \varnothing$, the random walk is said to be *transient*; otherwise it is called *recurrent*. Before plunging into the technicalities needed to treat a general random walk, we begin by analyzing the special case Polya considered in 1921. Legend has it that Polya thought of this problem while wandering around in a park near Zürich when he noticed that he kept encountering the same young couple. History does not record what the young couple thought.

Example 2.1   *Simple random walk on $\mathbb{Z}^d$*; that is, $P(X_i = e_j) = P(X_i = -e_j) = 1/2d$ for each of the d unit vectors e_j. Let $\tau_0 = 0$ and $\tau_n = \inf\{m > \tau_{n-1} : S_m = 0\}$ be the time of the nth return to 0. From (1.4) it follows that

$$P(\tau_n < \infty) = P(\tau_1 < \infty)^n,$$

a fact that leads easily to:

(2.2) THEOREM   For any random walk, the following are equivalent:

(i) $P(\tau_1 < \infty) = 1$,   (ii) $P(S_m = 0 \text{ i.o.}) = 1$,   and   (iii) $\sum\limits_{m=1}^{\infty} P(S_m = 0) = \infty$.

Proof   If $P(\tau_1 < \infty) = 1$, then $P(\tau_n < \infty) = 1$ for all n and $P(S_m = 0 \text{ i.o.}) = 1$. Let

$$V = \sum_{m=1}^{\infty} 1_{(S_m=0)} = \sum_{n=1}^{\infty} 1_{(\tau_n < \infty)}$$

be the number of visits to 0. It follows from Fubini's theorem that

$$\sum_{m=1}^{\infty} P(S_m = 0) = EV = \sum_{n=1}^{\infty} P(\tau_n < \infty) = \sum_{n=1}^{\infty} P(\tau_1 < \infty)^n,$$

so (ii) implies (iii), and if (i) is false, then (iii) is false (i.e., (iii) implies (i)). $\square$

Let $\rho_d(m) = P(S_m = 0)$, where S_m is a simple random walk on $\mathbb{Z}^d$. $\rho_d(m)$ is 0 if m is odd. From Section 1 in Chapter 2, we know $\rho_1(2n) \sim (\pi n)^{-1/2}$ as $n \to \infty$.

$$\rho_d(2n) \sim 2^{-(d-1)}(\pi n/d)^{-d/2}. \tag{2.3}$$

To explain the last result, let $N_i(n)$ be the number of steps the random walk has taken in the ith direction at time n. The probability all the $N_i(2n)$ are even $\to 2^{-(d-1)}$. By the weak law $N_i(2n) \sim 2n/d$ and if we condition on the $N_i(2n)$, the d-coordinates are independent, so (2.3) follows from the result in one dimension.

Proof of (2.3)   We will prove the result by induction on the dimension. By considering the value of $N_n(1)$,

$$p_{2d}(2n) = \sum_{m=0}^{2n} \binom{2n}{m} \left(\frac{1}{d}\right)^m \left(\frac{d-1}{d}\right)^{2n-m} p_1(m) p_{d-1}(2n-m).$$

By (9.4) in Chapter 1, there are $C, \gamma \in (0, \infty)$ so that

$$P(|N_1(2n) - 2n/d| \geq \varepsilon n) \leq C e^{-\gamma n}, \tag{$*$}$$

and hence the contribution from $|m - 2n/d| \geq \varepsilon n$ is $o(n^{-d/2})$. When $|m - 2n/d| \leq \varepsilon n$ and m is even,

$$p_1(m) p_{d-1}(2n-m) \approx (\pi n/d)^{-1/2} \cdot 2^{-(d-2)} (\pi n/d)^{-(d-1)/2}$$

by induction, so to complete the proof we need to show $h(n) = P(N_1(n) \text{ is even}) \to 1/2$ as $n \to \infty$. $h(0) = 1$ and

$$h(n) = \frac{d-1}{d} h(n-1) + \frac{1}{d}\{1 - h(n-1)\},$$

so

$$h(n) - \frac{1}{2} = \frac{d-1}{d}\left\{h(n-1) - \frac{1}{2}\right\} + \frac{1}{d}\left\{\frac{1}{2} - h(n-1)\right\} = \frac{d-2}{d}\left\{h(n-1) - \frac{1}{2}\right\},$$

and if $d > 1$ it follows that $h(n) \to 1/2$ as $n \to \infty$. $\square$

From (2.2) and (2.3) we see that simple random walk is recurrent in $d \leq 2$ and transient in $d \geq 3$. To steal a joke from Kakutani (U.C.L.A. colloquium talk): "A drunk man will find his way home but a drunk bird may get lost forever." The next two exercises give a proof of the last result without using the large deviations estimate ($*$).

Exercise 2.1   Prove that a simple random walk is recurrent in two dimensions by showing

$$p_2(2n) = 4^{-2n} \sum_{m=0}^{n} 2n!/(m!\,n-m!)^2 = 4^{-2n} \binom{2n}{n}^2$$

and using Stirling's formula to conclude that $\sum p_2(2n) = \infty$.

Remark   The last formula shows us that $\rho_2(2n) = \rho_1(2n)^2$. For a direct proof of this, note that if T_n^1 and T_n^2 are independent one-dimensional random walks, then T_n jumps from x to $x + (1, 1)$, $x + (1, -1)$, $x + (-1, 1)$, and $x + (-1, -1)$ with equal probability, so rotating T_n by 45 degrees and dividing by $\sqrt{2}$ give S_n.

Exercise 2.2   Prove that a simple random walk is transient in three dimensions by showing

$$\rho_3(2n) = 2^{-2n}\binom{2n}{n}\sum_{j,k}\{3^{-n}n!/(j!\,k!\,(n-j-k)!)\}^2$$

$$\leq 2^{-2n}\binom{2n}{n}\max_{j,k} 3^{-n}n!/(j!\,k!\,(n-j-k)!) \leq Cn^{-3/2}.$$

Exercise 2.3   Let $\pi_d = P(S_n = 0$ for some $n \geq 1)$ be the probability a simple random walk on $\mathbb{Z}^d$ returns to 0. Show

$$1/(1 - \pi_d) = \sum_{n=0}^{\infty} P(S_{2n} = 0).$$

In $d = 3$, $P(S_{2n} = 0) \sim Cn^{-3/2}$, so the series converges rather slowly. At the end of the section we will give another formula that leads very easily to accurate results.

Exercise 2.4   Show that $\pi_d = 1/2d + O(d^{-2})$ as $d \to \infty$.

The rest of this section is devoted to proving the following facts about random walk:

(2.7) S_n is recurrent in $d = 1$ if $S_n/n \to 0$ in probability.
(2.8) S_n is recurrent in $d = 2$ if $S_n/n^{1/2} \Rightarrow$ a normal distribution.
(2.12) S_n is transient in $d \geq 3$ if it is "truly three dimensional."

To prove (2.12), we will give a necessary and sufficient condition for recurrence, (2.9), that shows that the conditions in (2.7) and (2.8) are close to the best possible.

The first step in deriving these results is to generalize (2.2):

(2.4) LEMMA   If $\sum_{n=1}^{\infty} P(\|S_n\| < \varepsilon) < \infty$, then $P(\|S_n\| < \varepsilon$ i.o.) $= 0$. If $\sum_{n=1}^{\infty} P(\|S_n\| < \varepsilon) = \infty$, then $P(\|S_n\| < 2\varepsilon$ i.o.) $= 1$.

Proof   The first conclusion follows from the Borel–Cantelli lemma. To prove the second let $F = \{\|S_n\| < \varepsilon$ i.o.$\}^c$. Breaking things down according to the last time $\|S_n\| < \varepsilon$,

$$P(F) = \sum_{m=0}^{\infty} P(\|S_m\| < \varepsilon, \|S_n\| \geq \varepsilon \text{ for all } n \geq m + 1)$$

$$\geq \sum_{m=0}^{\infty} P(\|S_m\| < \varepsilon, \|S_n - S_m\| \geq 2\varepsilon \text{ for all } n \geq m + 1)$$

$$= \sum_{m=0}^{\infty} P(\|S_m\| < \varepsilon)\rho_{2\varepsilon, 1}$$

where $\rho_{\delta,k} = P(\|S_n\| \geq \delta$ for all $n \geq k)$. Since $P(F) \leq 1$ and

$$\sum_{m=0}^{\infty} P(\|S_m\| < \varepsilon) = \infty,$$

it follows that $\rho_{2\varepsilon,1} = 0$. To extend this conclusion to $\rho_{2\varepsilon,k}$ with $k \geq 2$, let

$$A_m = \{\|S_m\| < \varepsilon, \|S_n\| \geq \varepsilon \text{ for all } n \geq m + k\}.$$

Since any ω can be in at most k of the A_m, repeating the argument above gives

$$k \geq \sum_{m=0}^{\infty} P(A_m) \geq \sum_{m=0}^{\infty} P(\|S_m\| < \varepsilon)\rho_{2\varepsilon,k}.$$

So $\rho_{2\varepsilon,k} = P(\|S_n\| \geq 2\varepsilon$ for all $j \geq k) = 0$, and since k is arbitrary, the desired conclusion follows. $\square$

Our second step is to show that the convergence or divergence of the sums in (2.4) is independent of ε.

(2.5) LEMMA   Let m be an integer ≥ 2.

$$\sum_{n=0}^{\infty} P(\|S_n\| < m\varepsilon) \leq (2m)^d \sum_{n=0}^{\infty} P(\|S_n\| < \varepsilon).$$

Proof   We begin by observing

$$\sum_{n=0}^{\infty} P(|S_n| < m\varepsilon) \leq \sum_{n=0}^{\infty} \sum_{k} P(S_n \in k\varepsilon + [0, \varepsilon)^d)$$

where the inner sum is over $k \in \{-m, \ldots, m - 1\}^d$. If we let

$$T_k = \inf\{m \geq 0 : S_m \in k\varepsilon + [0, \varepsilon)^d\},$$

then breaking things down according to the value of T_k and using Fubini's theorem give

$$\sum_{n=0}^{\infty} P(S_n \in k\varepsilon + [0, \varepsilon)^d) = \sum_{m=0}^{\infty} \sum_{n=0}^{\infty} P(S_n \in k\varepsilon + [0, \varepsilon)^d, T_k = m)$$

$$\leq \sum_{m=0}^{\infty} \sum_{n=m}^{\infty} P(\|S_n - S_m\| < \varepsilon, T_k = m)$$

$$= \sum_{m=0}^{\infty} P(T_k = m) \sum_{l=0}^{\infty} P(\|S_l\| < \varepsilon)$$

$$\leq \sum_{l=0}^{\infty} P(\|S_l\| < \varepsilon).$$

Since there are $(2m)^d$ values of k in $\{-m, \ldots, m - 1\}^d$, the proof of (2.5) is complete. $\square$

Combining (2.4) and (2.5) gives

(2.6) COROLLARY   The convergence (resp. divergence) of $\sum_n P(\|S_n\| < \varepsilon)$ for a single value of $\varepsilon > 0$ is sufficient for transience (resp. recurrence).

(2.7) Chung–Fuchs Theorem   Suppose $d = 1$. If the weak law of large numbers holds in the form $S_n/n \to 0$ in probability, then S_n is recurrent.

Remark   (2.7) applies if $EX_i = 0$. If $EX_i = \mu \neq 0$, then the strong law of large numbers implies $S_n/n \to \mu$, so $|S_n| \to \infty$ and S_n is transient.

Proof   Let $u_n(x) = P(|S_n| < x)$ for x real > 0. (2.5) implies

$$\sum_{n=0}^{\infty} u_n(1) \geq \frac{1}{2m} \sum_{n=0}^{\infty} u_n(m) \geq \frac{1}{2m} \sum_{n=0}^{Am} u_n(n/A),$$

since $u_n(x) \geq 0$ and is increasing in x. By hypothesis $u_n(n/A) \to 1$, so letting $m \to \infty$,

$$\sum_{n=0}^{\infty} u_n(1) \geq A/2.$$

Since A is arbitrary, the sum must be ∞ and the conclusion follows from (2.6).   □

(2.8) THEOREM   If S_n is a random walk in $\mathbb{R}^2$ and $S_n/n^{1/2} \Rightarrow$ a nondegenerate normal distribution, then S_n is recurrent.

Remark   The conclusion is also true if the limit is degenerate, but in that case the random walk is essentially one (or zero) dimensional, and the result follows from the Chung–Fuchs theorem.

Proof   Let $u(n, m) = P(\|S_n\| < m)$. (2.5) implies

$$\sum_{n=0}^{\infty} u(n, 1) \geq (4m^2)^{-1} \sum_{n=0}^{\infty} u(n, m).$$

If $m/\sqrt{n} \to c$, then

$$u(n, m) \to \int_{[-c,c]^2} \varkappa(x)\, dx$$

where $\varkappa(x)$ is the density of the limiting normal distribution. If we use $\rho(c)$ to denote the right-hand side and let $n = [\theta m^2]$, it follows that $u([\theta m^2], m) \to \rho(\theta^{-1/2})$. If we write

$$m^{-2} \sum_{n=0}^{\infty} u(n, m) = \int_0^{\infty} u([\theta m^2], m)\, d\theta,$$

let $m \to \infty$, and use Fatou's lemma, we get

$$\liminf_{m \to \infty} (4m^2)^{-1} \sum_{n=0}^{\infty} u(n, m) \geq 4^{-1} \int_0^{\infty} \rho(\theta^{-1/2}) \, d\theta.$$

As $c \to 0$, $\rho(c) \sim n(0)c^2$, so $\rho(\theta^{-1/2}) \sim n(0)/\theta$, and it follows that $\sum_{n=0}^{\infty} u(n, 1) = \infty$, proving (2.8). □

We come now to the promised necessary and sufficient condition for recurrence.

(2.9) THEOREM   Let $\delta > 0$. S_n is recurrent if and only if $\displaystyle\int_{(-\delta, \delta)^d} \mathrm{Re} \frac{1}{1 - \varphi(y)} \, dy = \infty$.

We will prove a weaker result:

(2.9′) THEOREM   Let $\delta > 0$. S_n is recurrent if and only if

$$\sup_{r < 1} \int_{(-\delta, \delta)^d} \mathrm{Re} \frac{1}{1 - r\varphi(y)} \, dy = \infty.$$

Half of the work needed to get (2.9) from (2.9′) is trivial.

$$0 \leq \mathrm{Re} \frac{1}{1 - r\varphi(y)} \to \mathrm{Re} \frac{1}{1 - \varphi(y)} \quad \text{as } r \to 1,$$

so Fatou's lemma shows that if the integral is infinite, the walk is recurrent. The other direction is rather difficult: (2.9′) is in Chung and Fuchs (1951), but a proof of (2.9) had to wait for Ornstein (1969) and Stone (1969) to solve the problem independently. Their proofs use a trick to reduce to the case where the increments have a density and then a second trick to deal with that case, so we will not give the details here. The reader can consult either of the sources cited or Port and Stone (1969), where the result is demonstrated for random walks on abelian groups.

Proof of (2.9′)   The first ingredient in the solution is the:

(2.10) *Parseval Relation*   Let μ and v be probability measures on $\mathbb{R}^d$ with ch.f.'s φ and ψ.

$$\int \psi(t)\mu(dt) = \int \varphi(x)v(dx).$$

Proof   Since e^{itx} is bounded, Fubini's theorem implies

$$\int \psi(t)\mu(dt) = \int\int e^{itx}v(dx)\mu(dt) = \int\int e^{itx}\mu(dt)v(dx) = \int \varphi(x)\mu(dx). \qquad \square$$

Our second ingredient is a little calculus.

(2.11) LEMMA   If $|x| \leq \pi/3$, then $1 - \cos x \geq x^2/4$.

Proof   It suffices to prove the result for $x > 0$. If $0 \leq z < \pi/3$, then $\cos z \geq 1/2$, so

$$\sin y = \int_0^y \cos z \, dz \geq y/2$$

$$1 - \cos x = \int_0^x \sin y \, dy \geq \int_0^x \frac{y}{2} \, dy = x^2/4. \qquad \square$$

From Example 3.3 in Chapter 2, we see that the density

$$\frac{a - |x|}{a^2} \qquad |x| \leq a, \quad 0 \text{ otherwise}$$

has ch.f.

$$\frac{2(1 - \cos at)}{(at)^2}.$$

Let μ_n denote the distribution of S_n. Using (2.11) and then (2.10), we have

$$P(\|S_n\| < 1/\delta) \leq 4^d \int \prod_{i=1}^d \frac{1 - \cos(\delta t_i)}{(\delta t_i)^2} \mu_n(dt)$$

$$= 2^d \int_{(-\delta,\delta)^d} \prod_{i=1}^d \frac{\delta - |x_i|}{\delta^2} \varphi^n(x) \, dx.$$

Our next step is to sum from 0 to ∞. To be able to interchange the sum and the integral, we first multiply by r^n where $r < 1$:

$$\sum_{n=0}^\infty r^n P(\|S_n\| < 1/\delta) \leq 2^d \int_{(-\delta,\delta)^d} \prod_{i=1}^d \frac{\delta - |x_i|}{\delta^2} \frac{1}{1 - r\varphi(x)} \, dx.$$

Symmetry dictates that the integral on the right is real, so we can take the real part without affecting its value. Letting $r \uparrow 1$ and using (2.6) give half of (2.9′).

To prove the other direction, we begin by noting that Example 3.7 from Chapter 2 shows that the density

$$\frac{1 - \cos(x/\delta)}{\pi x^2/\delta}$$

has ch.f.

$1 - |\delta t| \qquad |t| \le 1/\delta, \quad 0 \text{ otherwise.}$

Using (2.10),

$$P(\|S_n\| < 1/\delta) \ge \int_{(-1/\delta,\, 1/\delta)^d} \prod_{i=1}^{d} (1 - |\delta x_i|) \mu_n(dx)$$

$$= \int \prod_{i=1}^{d} \frac{1 - \cos(t_i/\delta)}{\pi t_i^2/\delta} \varphi^n(t)\, dt.$$

Multiplying by r^n and summing give

$$\sum_{n=0}^{\infty} r^n P(\|S_n\| < 1/\delta) \ge \int \prod_{i=1}^{d} \frac{1 - \cos(t_i/\delta)}{\pi t_i^2/\delta} \frac{1}{1 - r\varphi(t)}\, dt.$$

The last integral is real, so its value is unaffected if we integrate only the real part of the integrand. If we do this and apply (2.11) we get

$$\sum_{n=0}^{\infty} r^n P(\|S_n\| < 1/\delta) \ge (4\pi\delta)^{-d} \int_{(-\delta,\, \delta)^d} \mathrm{Re}\, \frac{1}{1 - r\varphi(t)}\, dt.$$

Letting $r \uparrow 1$ and using (2.6) now complete the proof of (2.9′).   □

Exercise 2.5   To see that the imaginary part of $(1 - \varphi(t))$ is indeed important, consider a random walk with mean $\mu \ne 0$ and variance 1. Check that

$$\int_{-\delta}^{\delta} \mathrm{Re}\, \frac{1}{1 - r\varphi(t)}$$

is finite when $r = 1$ and has $\sup_{r<1} < \infty$.

We will now consider some examples. Our goal in $d = 1$ and $d = 2$ is to convince you that the conditions in (2.7) and (2.8) are close to the best possible. In $d = 1$, consider the symmetric stable laws that have ch.f. $\varphi(t) = \exp(-|t|^\alpha)$. To avoid using facts that we have not proved, we will obtain our conclusions from (2.9′). It is not hard to use that form of the criterion in this case, since

$$1 - r\varphi(t) \downarrow 1 - \exp(-|t|^\alpha) \quad \text{as } r \to 1$$

and

$$1 - \exp(-|t|^\alpha) \sim |t|^\alpha \quad \text{as } t \to 0.$$

From this it follows that the corresponding random walk is transient for $\alpha < 1$ and recurrent for $\alpha \ge 1$. The case $\alpha > 1$ is covered by (2.7), since these random walks have mean 0. The result for $\alpha = 1$ is new because the Cauchy distribution does not satisfy $S_n/n \to 0$ in probability. The random walks with $\alpha < 1$ are interesting because (1.2) implies

$$-\infty = \liminf S_n < \limsup S_n = \infty$$

(see Exercise 1.2), but $P(|S_n| < M \text{ i.o.}) = 0$ for any $M < \infty$.

Exercise 2.6   Consider the asymmetric stable laws with ch.f.

$$\exp(-|t|^\alpha(1 + i\kappa \tan(\pi\alpha/2) \text{ sgn}(t)))$$

and $\alpha > 1$. Are the corresponding random walks recurrent or transient? You may use (2.9) if you wish, but if you think this is (not) a mean problem.

Remark   The stable law examples are misleading in one respect. Shepp (1964) has proved that recurrent random walks may have arbitrarily large tails. To be precise, given a function $\varepsilon(x) \downarrow 0$ as $x \uparrow \infty$, there is a recurrent random walk with $P(|X_1| \geq x) \geq \varepsilon(x)$ for large x. Shepp used characteristic functions to prove this. For a "probabilistic proof," see Grey (1989).

In $d = 2$, let $\alpha < 2$ and let $\varphi(t) = \exp(-|t|^\alpha)$ where $|t| = (t_1^2 + t_2^2)^{1/2}$. φ is the characteristic function of a random vector (X_1, X_2) that has two nice properties:

(i) The distribution of (X_1, X_2) is invariant under rotations.
(ii) X_1 and X_2 have symmetric stable laws with index α.

Again $1 - r\varphi(t) \downarrow 1 - \exp(-|t|^\alpha)$ as $r \uparrow 1$ and $1 - \exp(-|t|^\alpha) \sim |t|^\alpha$ as $t \to 0$. Changing to polar coordinates and noticing

$$2\pi \int_0^\delta dx \, xx^{-\alpha} < \infty$$

when $1 - \alpha > -1$ show the random walks with ch.f. $\exp(-|t|^\alpha) \alpha < 2$ are transient. In $d \geq 3$,

$$2\pi \int_0^\delta dx \, x^{d-1}x^{-2} < \infty,$$

so if a random walk is recurrent in $d \geq 3$, its ch.f. must $\to 1$ faster than t^2. In Exercise 3.22 of Chapter 2 we observed that (in one dimension) if $\varphi(r) = 1 + o(r^2)$, then $\varphi(r) \equiv 1$. By considering $\varphi(r\theta)$ where r is real and θ is a fixed vector, the last conclusion generalizes easily to $\mathbb{R}^d$, $d > 1$, and suggests that once we exclude walks that stay on a plane through 0, no three-dimensional random walks are recurrent.

A random walk in $\mathbb{R}^3$ is *truly three dimensional* if the distribution of X_1 has $P(X_1 \cdot \theta \neq 0) > 0$ for all $\theta \neq 0$.

(2.12) THEOREM   No truly three-dimensional random walk is recurrent.

Proof   We will deduce the result from (2.9'). We begin with some arithmetic. If z is complex, the conjugate of $1 - z$ is $1 - \bar{z}$, so

$$\frac{1}{1-z} = \frac{1-\bar{z}}{|1-z|^2} \quad \text{and} \quad \text{Re}\frac{1}{1-z} = \frac{\text{Re}(1-z)}{|1-z|^2}.$$

If $z = a + bi$ with $a \le 1$, then

$$\text{Re}\frac{1}{1-z} = \frac{1-a}{(1-a)^2 + b^2} \le \frac{1}{1-a},$$

so

$$\text{Re}\frac{1}{1-r\varphi(t)} \le \frac{1}{\text{Re}(1-r\varphi(t))} \le \frac{1}{\text{Re}(1-\varphi(t))}. \tag{a}$$

The last calculation shows that it is enough to estimate

$$\text{Re}(1-\varphi(t)) = \int 1 - \cos(x \cdot t)\mu(dx) \ge \int_{|x \cdot t| < \pi/3} |x \cdot t|^2/4\,\mu(dx)$$

by (2.11). Writing $t = \rho\theta$ where $\theta \in S = \{x : |x| = 1\}$ gives

$$\text{Re}(1-\varphi(\rho\theta)) \ge (\rho^2/4)\int_{|x \cdot \theta| < \pi/3\rho} |x \cdot \theta|^2 \mu(dx). \tag{b}$$

Fatou's lemma implies that if we let $\rho \to 0$ and $\theta(\rho) \to \theta$, then

$$\liminf_{\rho \to 0} \int_{|x \cdot \theta(\rho)| < \pi/3\rho} |x \cdot \theta(\rho)|^2 \mu(dx) \ge \int |x \cdot \theta|^2 \mu(dx) > 0. \tag{c}$$

I claim this implies that for $\rho < \rho_0$,

$$\inf_{\theta \in S} \int_{|x \cdot \theta| < \pi/3\rho} |x \cdot \theta|^2 \mu(dx) = C > 0. \tag{d}$$

To get the last conclusion, observe that if it is false, then for $\rho = 1/n$ there is a θ_n so that

$$\int_{|x \cdot \theta_n| < n\pi/3} |x \cdot \theta_n|^2 \mu(dx) \le 1/n.$$

All the θ_n lie in S, a compact set, so if we pick a convergent subsequence we contradict (c). Combining (b) and (d) gives

$$\text{Re}(1 - \varphi(\rho\theta)) \geq C\rho^2/4.$$

Using the last result and (a) in (2.9′) gives the desired conclusion.   □

Remark   The analysis becomes much simpler when we consider random walks on $\mathbb{Z}^d$. The inversion formula given in Exercise 3.2 of Chapter 2 implies

$$P(S_n = 0) = (2\pi)^{-d} \int_{(-\pi, \pi)^d} \varphi^n(t) \, dt.$$

Multiplying by r^n, summing, and letting $r \uparrow 1$ give

$$\sum_{n=0}^{\infty} P(S_n = 0) = (2\pi)^{-d} \int_{(-\pi, \pi)^d} \text{Re}(1 - \varphi(t))^{-1} \, dt. \tag{2.13}$$

The last formula is difficult to prove in general but is not hard to establish in the special case we are interested in.

Exercise 2.7   Prove (2.13) for a simple random walk in $d \geq 3$.

Plugging in the ch.f. for a simple random walk, we see that

$$\sum_{n=0}^{\infty} P(S_n = 0) = (2\pi)^{-3} \int_{(-\pi, \pi)^3} \left(1 - \frac{1}{3} \sum_{m=1}^{3} \cos x_i\right)^{-1} dx.$$

This integral was first evaluated by Watson in 1939 in terms of elliptic integrals, which could be found in tables. Glasser and Zucker (1977) showed that it was

$$(\sqrt{6}/32\pi^3)\Gamma(1/24)\Gamma(5/24)\Gamma(7/24)\Gamma(11/24) = 1.516386059137\ldots,$$

so it follows from Exercise 2.3 that

$$\pi_d = .340537329544\ldots.$$

For numerical results in $4 \leq d \leq 9$, see Kondo and Hara (1987).

Exercise 2.8   Let F and G be two distributions with ch.f.'s φ and ψ.

$$e^{-i\theta t}\varphi(\theta) = \int e^{i\theta(x-t)} \, dF(x),$$

and integrating $dG(\theta)$ gives another version of the Parseval relation (2.10):

$$\int e^{-i\theta t}\varphi(\theta) \, dG(\theta) = \int \psi(x - t) \, dF(x).$$

If G is the normal distribution with mean 0 and variance a^2 and

$$n(x) = (2\pi)^{-1/2} \exp(-x^2/2),$$

then $\psi(\theta) = (2\pi)^{1/2} n(\theta/a)$ and rearranging gives

$$(2\pi)^{-1/2} \int e^{-i\theta t} \varphi(\theta) \exp(-a^2\theta^2/2) \, d\theta = \int \frac{1}{a} n((t-x)/a) \, dF(x).$$

The right-hand side is the density of $F * G$. Use the last equation to show: (i) distinct probability distributions have distinct ch.f.'s, and (ii) the continuity theorem ((3.4) in Chapter 2).

*3 Visits to 0, Arcsine Laws

In the last section we took a broad look at the recurrence of random walks. In this section we will take a deep look at one example: a simple random walk (on $\mathbb{Z}$). To steal a line from Chung (1974): "We shall treat this by combinatorial methods as an antidote to the analytic skulduggery above." The developments here follow Chapter III of Feller, Vol. I (1968). To facilitate discussion we will think of the sequence $S_1, S_2, \ldots, S_n$ as being represented by a polygonal line with segments $(k-1, S_{k-1}) \to (k, S_k)$. A *path* is a polygonal line that is a possible outcome of a simple random walk. To count the number of paths from $(0,0)$ to (n, x), it is convenient to introduce a and b defined by: $a = (n+x)/2$ is the number of positive steps in the path and $b = (n-x)/2$ is the number of negative steps. Notice that $n = a + b$ and $x = a - b$. If $-n \le x \le n$ and $n - x$ is even, the a and b defined above are nonnegative integers, and the number of paths from $(0,0)$ to (n, x) is

$$N_{n,x} = \binom{n}{a}.$$

Otherwise the number of paths is 0.

(3.1) *Reflection Principle*   If $x, y > 0$, then the number of paths from $(0, x)$ to (n, y) that touch the x-axis = the number of paths from $(0, -x)$ to (n, y).

Proof   Suppose $(0, s_0), (1, s_1), \ldots, (n, s_n)$ is a path from $(0, x)$ to (n, y). Let $K = \inf\{k : s_k = 0\}$. Let $s_k' = -s_k$ for $k \le K$ and $s_k' = s_k$ for $K \le k \le n$. Then (k, s_k'), $0 \le k \le n$, is a path from $(0, -x)$ to (n, u). Conversely if $(0, t_0), (1, t_1), \ldots, (n, t_n)$ is a path from $(0, -x)$ to (n, y), then it must cross 0. Let $K = \inf\{k : t_k = 0\}$. Let $t_k' = -t_k$ for $k \le K$ and $t_k' = t_k$ for $K \le k \le n$. Then (k, t_k'), $0 \le k \le n$, is a path from $(0, -x)$ to (n, u) that touches the x-axis at K. The last two observations set up a 1–1 correspondence between the two classes of paths, so their numbers must be equal. $\square$

From (3.1) we get a result first proved in 1878.

(3.2) *The Ballot Theorem*   Suppose that in an election candidate A gets a votes and candidate B gets b votes where $b < a$. The probability that throughout the counting A always leads B is $(a - b)/(a + b)$.

Proof   Let $x = a - b$, $n = a + b$. Clearly there are as many such outcomes as there are paths from $(1, 1)$ to (n, x) that do not touch the axis. The reflection principle implies that the number of paths from $(1, 1)$ to (n, x) that touch the axis = the number of paths from $(1, -1)$ to (n, x), so the number of paths from $(1, 1)$ to (n, x) that do not touch the axis is

$$N_{n-1,x-1} - N_{n-1,x+1} = \binom{n-1}{a-1} - \binom{n-1}{a} = \frac{(n-1)!}{(a-1)!(n-a)!} - \frac{(n-1)!}{a!(n-a-1)!}$$

$$= \frac{a - (n-a)}{n} \frac{n!}{a!(n-a)!} = \frac{a-b}{a+b} N_{n,x},$$

proving (3.2).   □

Using the ballot theorem, we can compute the distribution of the time to hit 0.

(3.3) LEMMA $P(S_1 \neq 0, \ldots, S_{2n} \neq 0) = P(S_{2n} = 0)$.

Proof $P(S_1 > 0, \ldots, S_{2n} > 0) = \sum_{r=1}^{\infty} P(S_1 > 0, \ldots, S_{2n-1} > 0, S_{2n} = 2r)$. From the proof of (3.2) we see that the number of paths from $(0, 0)$ to $(2n, 2r)$ that do not touch the axis at positive times ($=$ the number of paths from $(1, 1)$ to $(2n, 2r)$) is

$$N_{2n-1, 2r-1} - N_{2n-1, 2r+1}.$$

If we let $p_{n,x} = P(S_n = x)$, then

$$P(S_1 > 0, \ldots, S_{2n-1} > 0, S_{2n} = 2r) = \tfrac{1}{2}(p_{2n-1, 2r-1} - p_{2n-1, 2r+1}).$$

Summing from $r = 1$ to ∞ gives

$$P(S_1 > 0, \ldots, S_{2n} > 0) = (1/2)p_{2n-1, 1} = (1/2)P(S_{2n} = 0).$$

Symmetry implies

$$P(S_1 < 0, \ldots, S_{2n} < 0) = (1/2)P(S_{2n} = 0),$$

and the proof is complete.   □

Let $R = \inf\{m \geq 1 : S_m = 0\}$. Combining (3.3) with (1.3) from Chapter 2 gives

$$P(R > 2n) = P(S_{2n} = 0) \sim \pi^{-1/2} n^{-1/2}. \tag{3.4}$$

Since $P(R > x)/P(|R| > x) = 1$, it follows from (7.7) in Chapter 2 that R is in the domain of attraction of the stable law with $\alpha = 1/2$ and $\kappa = 1$. This implies that if R_n is the time of the nth return to 0, then $R_n/n^2 \Rightarrow Y$, the indicated stable law. In Exercise 7.8 in Chapter 2 we considered $\tau = T_1$ where $T_x = \inf\{n \geq 1 : S_n = x\}$. Since $S_1 \in \{-1, 1\}$ and $T_1 \overset{d}{=} T_{-1}$, $R \overset{d}{=} 1 + T_1$, and it follows that $T_n/n^2 \Rightarrow Y$, the same stable law. In Example 6.6 of Chapter 7 we will use this observation to find the density of the limit.

The last remarks complete our discussion of visits to 0. We turn now to the arcsine laws. The first one concerns

$$L_{2n} = \sup\{m \le 2n : S_m = 0\}.$$

It is remarkably easy to compute the distribution of L_{2n}.

(3.5) LEMMA   Let $u_{2m} = P(S_{2m} = 0)$. Then $P(L_{2n} = 2k) = u_{2k}u_{2n-2k}$.

Proof $P(L_{2n} = 2k) = P(S_{2k} = 0, S_{2k+1} \ne 0, \ldots, S_{2n} \ne 0)$, so the desired result follows from (3.3).   □

From the asymptotic formula for u_{2n} it follows that if $k/n \to x$, then

$$nP(L_{2n} = 2k) \to \pi^{-1}(x(1 - x))^{-1/2}.$$

Summing over x leads easily to:

(3.6) *Arcsine Law for the Last Visit to 0*   For $0 < a < b < 1$,

$$P(a \le L_{2n}/2n \le b) \to \int_a^b \pi^{-1}(x(1 - x))^{-1/2}\, dx.$$

To see the reason for the name, substitute $y = x^{1/2}$, $dy = (1/2)x^{-1/2}\, dx$ in the integral to obtain

$$\int_{\sqrt{a}}^{\sqrt{b}} \frac{2}{\pi}(1 - y^2)^{-1/2}\, dy = \frac{2}{\pi}\arcsin(\sqrt{b}) - \frac{2}{\pi}\arcsin(\sqrt{a}).$$

The symmetry of the limit distribution implies

$$P(L_{2n}/2n \le 1/2) \to 1/2.$$

In gambling terms, if two people were to bet \$1 on a coin flip every day of the year, then with probability 1/2 one of the players will be ahead from July 1 to the end of the year, an event that would undoubtedly cause the other player to complain about his bad luck. The next result deals directly with the amount of time one player is ahead.

(3.7) THEOREM   Let π_{2n} be the number of segments $(k - 1, S_{k-1}) \to (k, S_k)$ that lie above the axis (i.e., in $\{(x, y): x \ge 0\}$), and let $u_m = P(S_m = 0)$.

$$P(\pi_{2n} = 2k) = u_{2k}u_{2n-2k}$$

and consequently, if $0 < a < b < 1$,

$$P(a \leq \pi_{2n}/2n \leq b) \to \int_a^b \pi^{-1}(x(1 - x))^{-1/2} \, dx.$$

Remark   Since $\pi_{2n} \stackrel{d}{=} L_{2n}$, the second conclusion follows from the proof of (3.6). The reader should note that the limiting density $\pi^{-1}(x(1 - x))^{-1/2}$ has a minimum at $x = 1/2$ and $\to \infty$ as $x \to 0$ or 1. An equal division of steps between the positive and negative side is therefore the least likely possibility, and one-sided divisions have the highest probability.

Proof   Let $\beta_{2k, 2n}$ denote the probability of interest. We will prove $\beta_{2k, 2n} = u_{2k}u_{2n-2k}$ by induction. When $n = 1$ it is clear that

$$\beta_{0,2} = \beta_{2,2} = 1/2 = u_0 u_2.$$

For a general n, first suppose $k = n$. From the proof of (3.3) we have

$$\tfrac{1}{2}u_{2n} = P(S_1 > 0, \ldots, S_{2n} > 0)$$

$$= P(S_1 = 1, S_2 - S_1 \geq 0, \ldots, S_{2n} - S_1 \geq 0)$$

$$= \tfrac{1}{2}P(S_1 \geq 0, \ldots, S_{2n-1} \geq 0)$$

$$= \tfrac{1}{2}P(S_1 \geq 0, \ldots, S_{2n} \geq 0) = \tfrac{1}{2}\beta_{2n, 2n}.$$

The next to last equality follows from the observation that if $S_{2n-1} \geq 0$, then $S_{2n-1} \geq 1$ and hence $S_{2n} \geq 0$.

The last computation proves the result for $k = n$. Since $\beta_{0, 2n} = \beta_{2n, 2n}$, the result is also true when $k = 0$. Suppose now that $1 \leq k \leq n - 1$. In this case if R is the time of the first return to 0, then $R = 2m$ with $0 < m < n$. Breaking things up according to whether the first excursion was on the positive or negative side gives

$$\beta_{2k, 2n} = \frac{1}{2}\sum_{m=1}^{k} f_{2m}\beta_{2k-2m, 2n-2m} + \frac{1}{2}\sum_{m=1}^{n-k} f_{2m}\beta_{2k, 2n-2m},$$

where $f_{2m} = P(I_0 = 2m)$. Using the induction hypothesis, it follows that

$$\beta_{2k, 2n} = \frac{1}{2}u_{2n-2k}\sum_{m=1}^{k} f_{2m}u_{2k-2m} + \frac{1}{2}u_{2k}\sum_{m=1}^{n-k} f_{2m}u_{2n-2k-2m}.$$

By considering the time of the first return to 0, we see

$$u_{2k} = \sum_{m=1}^{k} f_{2m}u_{2k-2m}, \qquad u_{2n-2k} = \sum_{m=1}^{n-k} f_{2m}u_{2n-2k-2m},$$

and the desired result follows.   □

Our derivation of (3.7) relied heavily on special properties of simple random walk. There is a closely related result due to E. Sparre–Andersen that is valid for very general random walks. However, notice that the hypothesis in (ii) excludes simple random walk.

(3.8) THEOREM   Let $v_n = |\{k : 1 \leq k \leq n, S_k > 0\}|$. Then

(i) $P(v_n = k) = P(v_k = k)P(v_{n-k} = 0)$.

(ii) If the distribution of X_1 is symmetric and $P(S_m = 0) = 0$ for all $m \geq 1$, then

$$P(v_n = k) = u_{2k} u_{2n-k}$$

where $u_{2m} = 2^{-2m} \dbinom{2m}{m}$ is the probability a simple random walk is 0 at time $2m$.

(iii) Under the hypotheses of (ii),

$$P(a \leq v_n/n \leq b) \to \int_a^b \pi^{-1} (x(1-x))^{-1/2} \, dx \quad \text{for } 0 < a < b < 1.$$

Proof   Taking things in reverse order, (iii) is an immediate consequence of (ii) and the proof of (3.6). (ii) follows easily from (i) by induction, so we begin with that proof. When $n = 1$, our assumptions imply $P(v_1 = 0) = 1/2 = u_0 u_2$. If $n > 1$ and $1 < k < n$, then (i) and the induction hypothesis imply

$$P(v_n = k) = u_{2k} u_0 \cdot u_0 u_{2n-2k} = u_{2k} u_{2n-2k},$$

since $u_0 = 1$. To handle the cases $k = 0$ and $k = n$, we note that (3.5) implies

$$\sum_{m=0}^{n} u_{2m} u_{2n-2m} = 1,$$

and our assumptions imply $P(v_n = 0) = P(v_n = n)$, so these probabilities must be equal to $u_0 u_{2n}$.

The proof of (i) is tricky and requires careful definitions, since we are not supposing that X_1 is symmetric or that $P(S_m = 0) = 0$. Let $v_n' = |\{k : 1 \leq k \leq n, S_k \leq 0\}| = n - v_n$.

$$M_n = \max_{0 \leq j \leq n} S_j \qquad l_n = \min\{j : 0 \leq j \leq n, S_j = M_n\}$$

$$M_n' = \min_{0 \leq j \leq n} S_j \qquad l_n' = \max\{j : 0 \leq j \leq n, S_j = M_n'\}$$

The first symmetry is straightforward.   □

(3.9) LEMMA (l_n, S_n) and $(n - l_n', S_n)$ have the same distribution.

Proof   If we let $T_k = S_n - S_{n-k} = X_n + \cdots + X_{n-k+1}$, then T_k, $0 \leq k \leq n$, has the same distribution as S_k, $0 \leq k \leq n$. Clearly

$$\max_{0 \le k \le n} T_k = S_n - \min_{0 \le k \le n} S_{n-k},$$

and the set of k for which the extrema are attained is the same.   □

The second symmetry is much less obvious.

(3.10) LEMMA (l_n, S_n) and (v_n, S_n) have the same distribution. (l'_n, S_n) and (v'_n, S_n) have the same distribution.

Remark   (i) follows from (3.10) and the trivial observation

$$P(l_n = k) = P(l_k = k)P(l_{n-k} = 0),$$

so once (3.10) is established, the proof of (3.8) will be complete.

Proof of (3.10)   When $n = 1, \{l_1 = 0\} = \{S_1 \le 0\} = \{v_1 = 0\}$, and $\{l'_1 = 0\} = \{S_1 > 0\} = \{v'_1 = 0\}$. We shall prove the general case by induction, supposing that both statements have been proved when n is replaced by $n - 1$. Let

$$G(y) = P(l_{n-1} = k, S_{n-1} \le y)$$

$$H(y) = P(v_{n-1} = k, S_{n-1} \le y).$$

On $\{S_n \le 0\}$, we have $l_{n-1} = l_n$ and $v_{n-1} = v_n$, so if $F(y) = P(X_1 \le y)$, then for $x \le 0$,

$$P(l_n = k, S_n \le x) = \int F(x - y) \, dG(y) \tag{*}$$

$$= \int F(x - y) \, dH(y) = P(v_n = k, S_n \le x).$$

On $\{S_n > 0\}$, we have $l'_{n-1} = l'_n$ and $v'_{n-1} = v'_n$, so repeating the last computation shows that for $x \ge 0$,

$$P(l'_n = n - k, S_n > x) = P(v'_n = n - k, S_n > x).$$

Since (l_n, S_n) has the same distribution as $(n - l'_n, S_n)$ and $v'_n = n - v_n$, it follows that for $x \ge 0$,

$$P(l_n = k, S_n > x) = P(v_n = k, S_n > x).$$

Setting $x = 0$ in the last result and in $(*)$ and adding give

$$P(l_n = k) = P(v_n = k).$$

Subtracting the last two equations and combining the result with (∗) give

$$P(l_n = k, S_n \le x) = P(v_n = k, S_n \le x)$$

for all x. Since (l_n, S_n) has the same distribution as $(n - l'_n, S_n)$ and $v'_n = n - v_n$, it follows that

$$P(l'_n = n - k, S_n > x) = P(v'_n = n - k, S_n > x)$$

for all x. This completes the proof of (3.10) and hence of (3.8).   □

4 Renewal Theory

Let $\xi_1, \xi_2, \ldots$ be i.i.d. positive random variables with distribution F and define a sequence of times by $T_0 = 0$ and $T_k = T_{k-1} + \xi_k$ for $k \ge 1$. As explained in Section 8 in Chapter 1, if we think of ξ_i as the lifetime of the ith light bulb, then T_k is the time the kth bulb burns out. A second interpretation from Section 6 in Chapter 2 is that T_k is the time of arrival of the kth customer. To have a neutral terminology we will refer to the T_k as renewals. The term *renewal* refers to the fact that the process "starts afresh" at T_k; that is, $\{T_{k+j} - T_k, j \ge 1\}$ has the same distribution as $\{T_j, j \ge 1\}$.

Departing slightly from the notation in Section 8 of Chapter 1 and Section 6 of Chapter 2, we let $N_t = \inf\{k: T_k > t\}$. N_t is the number of renewals in $[0, t]$, counting the renewal at time 0. In Chapter 1 we showed (see (8.5)):

(4.1) THEOREM   As $t \to \infty$, $N_t/t \to 1/\mu$ a.s. where $\mu = E\xi_i \in (0, \infty]$ and $1/\infty = 0$.

Our first result concerns the asymptotic behavior of $U(t) = EN_t$.

(4.2) THEOREM   As $t \to \infty$, $U(t)/t \to 1/\mu$.

Proof   We will apply Wald's equation to the stopping time N_t. The first step is to show that $P(\xi_i > 0) > 0$ implies $EN_t < \infty$. To do this pick $\delta > 0$ so that $P(\xi > \delta) = \varepsilon > 0$ and pick K so that $K\delta \ge t$. Since K consecutive ξ_i's that are $> \delta$ will make $S_n > t$, we have

$$P(N_t > mK) \le (1 - \varepsilon^K)^m$$

and $EN_t < \infty$. If $\mu < \infty$, applying Wald's equation now gives $\mu EN(t) = ES_{N(t)} \ge t$, so $U(t) \ge t/\mu$. The last inequality is trivial when $\mu = \infty$, so it holds in general.

Turning to the upper bound, we observe that if $P(\xi_i \le c) = 1$, then repeating the last argument shows $\mu EN(t) = ES_{N(t)} \le t + c$ and the result holds for bounded distributions. If we let $\bar{\xi}_i = \xi_i \wedge c$ and define $\bar{T}_n$ and $\bar{N}(t)$ in the obvious way, then

$$EN(t) \le E\bar{N}(t) \le (t + c)/E(\bar{\xi}_i).$$

Letting $t \to \infty$ and then $c \to \infty$ gives $\limsup_{t \to \infty} EN(t)/t \le 1/\mu$, and the proof is complete.   □

Exercise 4.1   Show that $t/E(X_i \wedge t) \le U(t) \le 2t/E(X_i \wedge t)$.

Exercise 4.2   Deduce (4.2) from (4.1) by showing

$$\limsup_{t \to \infty} E(N_t/t)^2 < \infty.$$

To do this pick $\tau > 0$ so that $P(\xi_i \ge \tau) = \eta > 0$. Let $\xi_i' = \tau$ if $\xi_i \ge \tau$, $\xi_i' = 0$ otherwise, and define T_k' and N_t' for the sequence ξ_i' in the obvious fashion. It is clear that $N_t' \ge N_t$. Since T_k' stays constant for a geometric number of trials and then increases by τ, direct calculation shows $\limsup_{t \to \infty} E(N_t'/t)^2 < \infty$.

Exercise 4.3   If $\varphi(\theta) = \int e^{-\theta x} dF(x)$, then $\int e^{-\theta t} dU(t) = 1/(1 - \varphi(\theta))$. Applying Widder's inversion formula (see Example 5.4 in Chapter 1) to the Laplace transform of the probability measure $(1 - \varphi(1))e^{-t} dU(t)$, it follows that each $U(t)$ corresponds to at most one $F(x)$.

Exercise 4.4   Customers arrive at times of a Poisson process with rate 1. If the server is occupied, they leave. (Think of an automated teller or a prostitute.) If not, they enter service and require a service time with mean μ. Show that the times at which customers enter service are a renewal process with mean $\mu + 1$, and use (4.1) to conclude that the asymptotic fraction of customers served is $1/(\mu + 1)$.

To take a closer look at when the renewals occur, we let

$$U(A) = \sum_{n=0}^{\infty} P(T_n \in A).$$

U is called the *renewal measure*. We absorb the old definition, $U(t) = EN_t$, into the new one by regarding $U(t)$ as shorthand for $U([0, t])$. This should not cause problems, since $U(t)$ is the distribution function for the renewal measure. The asymptotic behavior of $U(t)$ depends on whether the distribution F is *arithmetic*— that is, concentrated on $\{\delta, 2\delta, 3\delta, \ldots\}$ for some $\delta > 0$, or nonarithmetic. We will treat the first case in Chapter 5 as an application of Markov chains, so we will restrict our attention to the second case here.

(4.3) Blackwell's Renewal Theorem   If F is nonarithmetic, then $U([t, t + h]) \to h/\mu$ as $t \to \infty$.

We will prove the result in the case $\mu < \infty$ by "coupling" following Lindvall (1977) (and Athreya, McDonald, and Ney (1978)). To set the stage for the proof, we need a definition and some preliminary computations. If $T_0 \ge 0$ is independent of $\xi_1, \xi_2, \ldots$ and has distribution G, then $T_k = T_{k-1} + \xi_k$, $k \ge 1$, defines a *delayed renewal process*, and G is the *delay distribution*. If we let $N_t = \inf\{k: T_k > t\}$ as before and set $V(t) = EN_t$, then breaking things down according to the value of T_0 gives

$$V(t) = \int_0^t U(t - s)\, dG(s).$$ (4.4)

The last integral and all similar expressions below are intended to include the contribution of any mass G at 0. If we let $U(r) = 0$ for $r < 0$, then the last equation can be written as $V = U * G$ where $*$ denotes convolution.

Applying similar reasoning to U gives

$$U(t) = 1 + \int_0^t U(t - s)\, dF(s),$$ (4.5)

or introducing convolution notation,

$$U = 1_{[0, \infty)}(t) + U * F.$$

Convolving each side with G (and recalling $G * U = U * G$) gives

$$V = G * U = G + V * F.$$ (4.6)

We know $U(t) \sim t/\mu$. Our next step is to find a G so that $V(t) = t/\mu$. Plugging what we want into (4.6) gives

$$t/\mu = G(t) + \int_0^t \frac{t - y}{\mu}\, dF(y)$$

or

$$G(t) = t/\mu - \int_0^t \frac{t - y}{\mu}\, dF(y).$$

If we let $H(y) = (y - t)/\mu$ and $K(y) = 1 - F(y)$, integration by parts gives

$$\int_0^t K(y)\, dH(y) = H(t)K(t) - H(0)K(0) - \int_0^t H(y)\, dK(y)$$

or

$$\frac{1}{\mu} \int_0^t 1 - F(y)\, dy = t/\mu - \int_0^t \frac{t - y}{\mu}\, dF(y),$$

so

$$G(t) = \frac{1}{\mu} \int_0^t 1 - F(y)\, dy.$$ (4.7)

It is comforting to note that $\mu = \int_{[0,\infty)} 1 - F(y)\,dy$, so the last formula defines a probability distribution. When the delay distribution G is the one given in (4.7), we call the result the *stationary renewal process*. Something very special happens when $F(t) = 1 - \exp(-\lambda t)$, $t \geq 0$, where $\lambda > 0$ (i.e., the renewal process is a rate λ Poisson process). In this case $G(t) = F(t)$.

Proof of (4.3) for $\mu < \infty$ Let T_n be a renewal process (with $T_0 = 0$) and let T_n' be an independent stationary renewal process. Let $\varepsilon > 0$ and let $K = \inf\{k : |T_k - T_k'| < \varepsilon\}$. Since $S_k = T_k - T_k'$ is a random walk whose steps have mean 0 and a nonarithmetic distribution, $P(K < \infty) = 1$. Let

$$T_n'' = \begin{cases} T_n & \text{if} \quad K \geq n \\ T_K + T_n' - T_K' & \text{if} \quad K < n \end{cases}.$$

In words, the increments $T_{n+1}'' - T_n''$ are the same as $T_{n+1}' - T_n'$ for $n \geq K$. Since K is a stopping time with respect to $\mathscr{F}_n = \sigma(T_m, T_m', m \leq n)$, it is easy to see that T_n and T_n'' have the same distribution. If we let

$$N'[s, t] = |\{n : T_n' \in [s, t]\}| \quad \text{and} \quad N''[s, t] = |\{n : T_n'' \in [s, t]\}|$$

be the number of renewals in $[s, t]$ in the two processes, then on $\{T_K \leq t\}$,

$$N''[t, t + h] = N'[t + T_K' - T_K, t + h + T_K' - T_K] \geq N'[t + \varepsilon, t + h - \varepsilon]$$

and $N''[t, t + h] \leq N'[t - \varepsilon, t + h + \varepsilon]$.

To relate the expected number of renewals in the two processes, we observe that even if we condition on the location of all the renewals in $[0, s]$, the expected number of renewals in $[s, s + t]$ is at most $U(t)$, since the worst thing that could happen is to have a renewal at time s. Combining the last two observations, we see that if $\varepsilon < h/2$,

$$U([t, t + h]) = EN''[t, t + h] \geq E(N'[t + \varepsilon, t + h - \varepsilon]; T_K \leq t)$$

$$\geq (h - 2\varepsilon)/\mu - P(T_K > t)U(h),$$

since $EN'[t + \varepsilon, t + h - \varepsilon] = (h - 2\varepsilon)/\mu$ and $\{T_K > t\}$ is determined by the renewals of T in $[0, t]$ and the renewals of T' in $[0, t + \varepsilon]$. For the other direction we observe

$$U([t, t + h]) \leq E(N'[t - \varepsilon, t + h + \varepsilon]; T_K \leq t) + E(N''[t, t + h]; T_K > t)$$

$$\leq (h + 2\varepsilon)/\mu + P(T_K > t)U(h).$$

The desired result now follows from the fact that $P(T_K > t) \to 0$ and $\varepsilon < h/2$ is arbitrary.   □

Proof of (4.3) for $\mu = \infty$ There is no stationary renewal process, so we have to resort to other methods. Let

$$\beta = \limsup_{t \to \infty} U(t, t+1] = \lim_{k \to \infty} U(t_k, t_k + 1].$$

By considering the location of T_i we get

$$\beta = \lim_{k \to \infty} \int U(t_k - y, t_k + 1 - y] \, dF^{i*}(y).$$

Since β is the limsup, we must have

$$\lim_{k \to \infty} \int_{(j-1,j]} U(t_k - y, t_k + 1 - y] \, dF^{i*}(y) = \beta \cdot P(T_i \in (j - 1, j]).$$

Now if $j - 1 < y \le j$, we have

$$U(t_k - y, t_k + 1 - y] \le U(t_k - j, t_k + 2 - j],$$

so for j with $P(T_i \in (j - 1, j]) > 0$, we must have

$$\liminf_{k \to \infty} U(t_k - j, t_k + 2 - j] \ge \beta.$$

Summing over i, we see that the last conclusion is true when $U(j - 1, j] > 0$.

The support of U is a semigroup. (If x is in the support of F^{m*} and y is in the support of F^{n*}, then $x + y$ is in the support of $F^{(m+n)*}$.) We have assumed F is nonarithmetic, so $U(j - 1, j] > 0$ for $j \ge j_0$. Letting $r_k = t_k - j_0$ and considering the location of the last renewal in $[0, r_k]$ give

$$1 = \int_0^{r_k} (1 - F(r_k - y)) \, dU(y) \ge \sum_{n=1}^{\infty} (1 - F(2n)) U(r_k - 2n, r_k + 2 - 2n].$$

Since

$$\liminf_{k \to \infty} U(r_k - 2n, r_k + 2 - 2n] \ge \beta$$

and

$$\sum_{n=1}^{\infty} (1 - F(2n)) \ge \mu = \infty,$$

β must be 0. This proves the result for $h = 1$, but that is enough to reach the desired conclusion. $\square$

Remark Following Lindvall (1977), we have based the proof for $\mu = \infty$ on part of Feller's (1961) proof the discrete renewal theorem (i.e., for arithmetic distributions). See Freedman (1971b), pp. 22–25, for an account of Feller's proof. Purists can find a proof that does everything by coupling in Thorisson (1987).

Our next topic is the *renewal equation*: $H = h + H * F$. Two cases we have seen in (4.5) and (4.6) are:

Example 4.1 $h \equiv 1$:

$$U(t) = 1 + \int_0^t U(t - s)\, dF(s).$$

Example 4.2 $h(t) = G(t)$:

$$V(t) = G(t) + \int_0^t V(t - s)\, dF(s).$$

The last equation is valid for an arbitrary delay distribution. If we let G be the distribution in (4.7) and subtract the last two equations, we get:

Example 4.3 $H(t) = U(t) - t/\mu$ satisfies the renewal equation with $h(t) = (1/\mu) \int_t^\infty 1 - F(s)\, ds$.

Last but not least, we have an example that is a typical application of the renewal equation.

Example 4.4   Let $x > 0$ be fixed and let $H(t) = P(T_{N(t)} - t > x)$. By considering the value of T_1, we get

$$H(t) = (1 - F(t + x)) + \int_0^t H(t - s)\, dF(s).$$

The examples above should provide motivation for:

(4.8) THEOREM   If h is bounded, then the function

$$H(t) = \int_0^t h(t - s)\, dU(s)$$

is the unique solution of the renewal equation that is bounded on bounded intervals.

Proof   Let $U_n(A) = \sum_{m=0}^n P(T_m \in A)$ and

$$H_n(t) = \int_0^t h(t - s)\, dU_n(S) = \sum_{m=0}^n h * F^{m*}.$$

Here, F^{m*} is the distribution of T_m, and we have extended the definition of h by setting $h(r) = 0$ for $r < 0$. From the last expression it should be clear that

$$H_{n+1} = h + H_n * F.$$

The fact that $U(t) < \infty$ implies $U(t) - U_n(t) \to 0$. Since h is bounded,

$$|H_n(t) - H(t)| \leq \|h\|_\infty |U(t) - U_n(t)|,$$

and $H_n(t) \to H(t)$ uniformly on bounded intervals. From this it follows easily that H is a solution of the renewal equation which is bounded on bounded intervals.

To prove uniqueness, we observe that if H_1 and H_2 are two solutions, then $K = H_1 - H_2$ satisfies $K = K * F$. If K is bounded on bounded intervals, iterating gives $K = K * F^{n*} \to 0$ as $n \to \infty$, so $H_1 = H_2$. □

The proof of (4.8) is valid when $F(\infty) < 1$. In this case we have a *terminating renewal process*. After a geometric number of trials with mean $1/(1 - F(\infty))$, $T_n = \infty$. This "trivial case" has some interesting applications.

Example 4.5   *Cramér's estimates of ruin.* Consider an insurance company that collects money at rate c and experiences i.i.d. claims at the arrival times of a Poisson process N_t with rate 1. If its initial capital is x, its wealth at time t is

$$W_x(t) = x + ct - \sum_{m=1}^{N_t} Y_i.$$

Here $Y_1, Y_2, \ldots$ are i.i.d. with distribution G and mean μ. Let

$$R(x) = P(W_x(t) \geq 0 \text{ for all } t)$$

be the probability of never going bankrupt starting with capital x. By considering the time and size of the first claim:

$$R(x) = \int_0^\infty e^{-s} \int_0^{x+cs} R(x + cs - y)\, dG(y)\, ds.$$

This does not look much like a renewal equation, but with some ingenuity it can be transformed into one. Changing variables $t = x + cs$,

$$R(x)e^{-x/c} = \int_x^\infty e^{-t/c} \int_0^t R(t - y)\, dG(y)\, dt/c.$$

Differentiating with respect to x and then multiplying by $e^{x/c}$,

$$R'(x) = \frac{1}{c}R(x) - \frac{1}{c}\int_0^x R(x - y)\, dG(y).$$

Integrating x from 0 to w,

$$R(w) - R(0) = \frac{1}{c}\int_0^w R(x)\, dx - \frac{1}{c}\int_0^w \int_0^x R(x - y)\, dG(y)\, dx.$$

Interchanging the order of integration in the double integral, letting

$$S(w) = \int_0^w R(x)\,dx,$$

and then integrating by parts,

$$\int_0^w \int_y^w R(x-y)\,dx\,dG(y) = \int_0^w S(w-y)\,dG(y)$$

$$= -\int_0^w S(w-y)\,d(1-G)(y)$$

$$= S(w) - \int_0^w R(w-y)(1-G(y))\,dy.$$

Plugging this into the preceding equation, we finally have a renewal equation:

$$R(w) = R(0) + \int_0^w R(w-y)(1-G(y))/c\,dy.$$

The last equation took a lot of work to derive but is easy to analyze.

$$\int_0^\infty (1-G(y))/c\,dy = \mu/c.$$

If $\mu > c$,

$$\frac{1}{t}\left(ct - \sum_{m=1}^{N_t} Y_i\right) \to c - \mu < 0 \text{ a.s.},$$

so $R(x) \equiv 0$. When $\mu < c$,

$$F(x) = \int_0^x (1-G(y))/c\,dy$$

is a defective probability distribution and (4.8) tells us

$$R(w) = R(0)U(w).$$

(The function $h \equiv R(0)$!) To complete the solution we have to compute the constant $R(0)$. Letting $w \to \infty$ and noticing $R(w) \to 1$, $U(w) \to (1-F(\infty))^{-1} = (1-\mu/c)^{-1}$, we have

$$R(0) = 1 - \mu/c.$$

Exercise 4.5   Consider a terminating renewal process and let M be the last finite value in the sequence $0, T_1, T_2, \ldots$. (i) Show that $V(t) = P(M \le t)$ satisfies:

$$V(t) = 1 - F(\infty) + \int_0^t V(t - y)\, dF(y)$$

and hence $P(M \le t) = (1 - F(\infty))U(t)$. (ii) Show that $EM = \int x\, dF(x)/(1 - F(\infty))$.

Example 4.6   A pedestrian wants to cross a road on which the traffic is a Poisson process with rate λ. She needs one unit of time with no arrival to safely cross the road. Let $M = \inf\{t \ge 0$: there are no arrivals in $(t, t + 1]\}$ be the waiting time until she starts to cross the street. $V(t) = P(M \le t)$ satisfies

$$V(t) = e^{-1} + \int_0^1 V(t - y)e^{-y}\, dy.$$

This is a renewal equation in which $F(t) = 1 - e^{-t}$ for $0 \le t \le 1$ and $F(t) = F(1)$ for $t > 1$.

$$\int_0^1 x\lambda e^{-\lambda x}\, dx = \int_0^\infty - \int_1^\infty = \frac{1}{\lambda} - \left(1 + \frac{1}{\lambda}\right)e^{-\lambda}$$

by the lack of memory property of the exponential, so from part (ii) of Exercise 4.5,

$$EM = e^{\lambda}/\lambda - \left(1 + \frac{1}{\lambda}\right) = \sum_{n=2}^{\infty} \lambda^{n-1}/n!.$$

We will return to this example in Exercise 4.14.

The basic fact about solutions of the renewal equation (in the nonterminating case) is:

(4.9) The Renewal Theorem   If F is nonarithmetic and h is directly Riemann integrable, then

$$H(t) \to \frac{1}{\mu} \int_0^\infty h(s)\, ds \quad \text{as } t \to \infty.$$

We will define the new phrase in a minute. We will start doing the proof and then figure out what we need to assume.

Proof   Suppose

$$h(s) = \sum_{k=0}^{\infty} a_k 1_{[k\delta, (k+1)\delta)}(s)$$

where $\sum_{k=0}^{\infty} |a_k| < \infty$. Since $U([t, t + \delta]) \le U([0, \delta]) < \infty$, it follows easily from (4.3) that

$$\int_0^t h(t-s)\,dU(s) = \sum_{k=0}^{\infty} a_k U((t-(k+1)\delta, t-k\delta]) \to \frac{1}{\mu}\sum_{k=0}^{\infty} a_k\delta.$$

(Pick K so that $\sum_{k \geq K}|a_k| \leq \varepsilon/2U([0,\delta])$ and then T so that

$$|U((t-(k+1)\delta, t-k\delta]) - \delta/\mu| \leq \varepsilon/2K$$

for $t \geq T$ and $0 \leq k < K$.) If h is an arbitrary function on $[0, \infty)$, we let

$$I^\delta = \sum_{k=0}^{\infty} \delta \sup\{h(x): x \in [k\delta, (k+1)\delta)\}$$

and

$$I_\delta = \sum_{k=0}^{\infty} \delta \inf\{h(x): x \in [k\delta, (k+1)\delta)\}$$

be upper and lower Riemann sums approximating the integral of h over $[0, \infty)$. Comparing h with the obvious upper and lower bounds that are constant on $[k\delta, (k+1)\delta)$ and using the result for the special case,

$$I_\delta \leq \liminf_{t\to\infty} \int_0^t h(t-s)\,dU(s) \leq \limsup_{t\to\infty} \int_0^t h(t-s)\,dU(s) \leq I^\delta.$$

If I^δ and I_δ both approach the same finite limit I as $\delta \to 0$, then h is said to be *directly Riemann integrable*, and it follows that

$$\int_0^t h(t-s)\,dU(y) \to I/\mu. \qquad \square$$

Remark   The word "direct" in the name refers to the fact that while the Riemann integral over $[0, \infty)$ is usually defined as the limit of integrals over $[0, a]$, we are approximating the integral over $[0, \infty)$ directly.

In checking the new hypothesis in (4.9), the following result is useful.

(4.10) LEMMA   If $h(x) \geq 0$ is decreasing with $h(0) < \infty$ and $\int_0^\infty h(x)\,dx < \infty$, then h is directly Riemann integrable.

Proof   Because h is decreasing, $I^\delta = \sum_{k=0}^\infty \delta h(k\delta)$ and $I_\delta = \sum_{k=0}^\infty \delta h((k+1)\delta)$. So

$$I^\delta \geq \int_0^\infty h(x)\,dx \geq I_\delta = I^\delta - h(0)\delta,$$

proving the desired result. $\qquad \square$

The last result suffices for all our applications, so we leave it to the reader to do:

Exercise 4.6 If $h \geq 0$ is continuous, then h is directly Riemann integrable if and only if $I^\delta < \infty$ for some $\delta > 0$ (and hence for all $\delta > 0$).

Returning now to our examples, we skip the first two because in those cases $h(t) \to 1$ as $t \to \infty$. In Example 4.3, $h(t) = (1/\mu) \int_{[t,\infty)} 1 - F(s) \, ds$. h is decreasing, $h(0) = 1$, and

$$\int_0^\infty h(t) \, dt = \int_0^\infty \int_t^\infty 1 - F(s) \, ds \, dt$$

$$= \int_0^\infty \int_0^s 1 - F(s) \, dt \, ds = \int_0^\infty s(1 - F(s)) \, ds = E(\xi_i^2/2).$$

So if $\nu \equiv E(\xi_i^2) < \infty$, it follows from (4.10) and (4.9) that

$$0 \leq U(t) - t/\mu \to \nu/2\mu^2 \quad \text{as } t \to \infty.$$

When the renewal process is a rate λ Poisson process—that is, $P(\xi_i > t) = e^{-\lambda t}$, $N(t) - 1$ has a Poisson distribution with mean λt, so $U(t) = 1 + \lambda t$. According to Feller (1971), p. 385, if the ξ_i are uniform on $(0, 1)$, then

$$U(t) = \sum_{k=0}^n (-1)^k e^{t-k}(t - k)^k/k! \quad \text{for } n \leq t \leq n + 1.$$

As he says, the exact expression "reveals little about the nature of U. The asymptotic formula $0 \leq U(t) - 2t \to 2/3$ is much more interesting."

Exercise 4.7 $E(N_t^2) = \sum_{k=0}^\infty (2k + 1)P(N_t \geq k + 1) = \sum_{k=0}^\infty (2k + 1)F^{k*}(t) = 2(U * U) - U$. Use the fact that $0 \leq U(t) - t/\mu \to (\mu^2 + \sigma^2)/\mu^2$ to show

$$E(N_t^2) = t^2/\mu^2 + t\sigma^2/\mu^3 + o(t)$$

in accordance with the central limit theorem in Exercise 4.13 of Chapter 2.

In Example 4.4, $h(t) = 1 - F(t + x)$. Again h is decreasing, but this time $h(0) \leq 1$ and the integral of h is finite when $\mu = E(\xi_i) < \infty$. Applying (4.10) and (4.9) now gives

$$P(T_{N(t)} - t > x) \to \frac{1}{\mu} \int_0^\infty h(s) \, ds = \frac{1}{\mu} \int_x^\infty 1 - F(t) \, dt,$$

so (when $\mu < \infty$) the distribution of the *residual waiting time* $T_{N(t)} - t$ converges to the delay distribution that produces the stationary renewal process. This fact also follows from our proof of (4.3).

Using the method employed to study Example 4.4, one can analyze various other aspects of the asymptotic behavior of renewal processes. To avoid repeating ourselves:

We assume throughout that F is nonarithmetic.

Exercise 4.8   Let $A_t = t - T_{N(t)-1}$ be the "age" at time t—that is, the amount of time since the last renewal. If we fix $x > 0$, then $H(t) = P(A_t > x)$ satisfies the renewal equation

$$H(t) = (1 - F(t)) \cdot 1_{(x,\infty)}(t) + \int_0^t H(t - s)\, dF(s),$$

so $P(A_t > x) \to (1/\mu) \int_{(x,\infty)} (1 - F(t))\, dt$, which is the limit distribution for the residual lifetime $B_t = T_{N(t)} - t$. The last result can be derived from Example 4.4 by noting that if $t > x$, then $P(A_t \geq x) = P(B_{t-x} > x) = P(\text{no renewal in } (t - x, t])$. To check the placement of the strict inequality, recall $N_t = \inf\{k: T_k > t\}$, so we always have $A_s \geq 0$ and $B_s > 0$.

Exercise 4.9   Use the renewal equation in the last exercise and (4.8) to conclude that if T is a rate λ Poisson process, A_t has the same distribution as $\xi_i \wedge t$.

Exercise 4.10   Let $A_t = t - T_{N(t)-1}$ and $B_t = T_{N(t)} - t$. Show that

$$P(A_t > x, B_t > y) \to \frac{1}{\mu} \int_{x+y}^{\infty} (1 - F(t))\, dt.$$

Exercise 4.11   Write a renewal equation for $q(t) = P$ (number of renewals in $[0, t]$ is odd) and use the renewal theorem to show that $q(t) \to 1/2$.

Exercise 4.12   *Alternating renewal process.* Let $\xi_1, \xi_2, \ldots$ be i.i.d. with distribution F_1 and let $\eta_1, \eta_2, \ldots$ be i.i.d. with distribution F_2. Let $T_0 = 0$, and for $k \geq 1$, let $S_k = T_{k-1} + \xi_k$ and $T_k = S_k + \eta_k$. In words, we have a machine that works for an amount of time ξ_k, breaks down, and then requires η_k units of time to be repaired. Let $F = F_1 * F_2$ and let $H(t)$ be the probability the machine is working at time t. Show that H satisfies

$$H(t) = (1 - F_1(t)) + \int_0^t H(t - s)\, dF(s),$$

and if F is nonarithmetic, then as $t \to \infty$,

$$H(t) \to \mu_1/(\mu_1 + \mu_2)$$

where μ_1 is the mean of F_i.

Exercise 4.13   *Renewal densities.* Show that if $F(t)$ has a density function $f(t)$, then the renewal measure U has a density u that satisfies

$$u(t) = f(t) + \int_0^t u(t - s)\,dF(s).$$

Use the renewal theorem to conclude that if f is directly Riemann integrable, then $u(t) \to 1/\mu$ as $t \to \infty$.

Exercise 4.14 *Pedestrian delay continued.* Consider the renewal equation of Example 4.6:

$$V(t) = 1 - F(\infty) + \int_0^t V(t - y)\,dF(y)$$

and rewrite it as

$$1 - V(t) = F(\infty) - F(t) + \int_0^t (1 - V(t - y))\,dF(y).$$

Let θ be chosen so that $\int e^{\theta y}\,dF(y) = 1$ and let $G(x) = \int_{[0,x]} e^{\theta y}\,dF(y)$. Multiplying the last equation by $e^{\theta t}$ gives

$$e^{\theta t}(1 - V(t)) = e^{\theta t}(F(\infty) - F(t)) + \int_0^t e^{\theta(t-y)}(1 - V(t - y))\,dG(y).$$

Use the renewal theorem (4.9) to conclude that $e^{\theta t} P(M > t) \to$ a limit $\in (0, \infty)$.

Exercise 4.15 *Patterns in coin tossing.* Let X_n, $n \geq 1$, take values H and T with probability $1/2$ each. Let $T_0 = 0$ and $T_k = \inf\{n > T_{k-1} : (X_n, \ldots, X_{n+k-1}) = (i_1, \ldots, i_k)\}$ where $(i_1, \ldots, i_k)$ is some pattern of heads and tails. Show that the T_j form a delayed renewal process; that is, $t_j = T_j - T_{j-1}$ are independent for $j \geq 1$ and identically distributed for $j \geq 2$. To see that the distribution of t_1 may be different, let $(i_1, i_2, i_3) = (H, H, H)$. In this case $P(t_1 = 1) = 1/8$, $P(t_2 = 1) = 1/2$.

Exercise 4.16 Continue to use the notation introduced in the last exercise. (i) Show that for any pattern of length k, $Et_j = 2^k$ for $j \geq 2$. (ii) Compute Et_1 when the pattern is HH and when it is HT.

Hint For HH observe

$$Et_1 = P(HH) + P(HT)E(t_1 + 2) + P(T)E(t_1 + 1).$$

Exercise 4.17 *Infinite mean.* Suppose $1 - F(x) = x^{-\alpha}$ where $0 < \alpha < 1$. Let $N_t = \sup\{n : S_n \leq t\}$. Show that $N_t/t^\alpha \Rightarrow a$ nondegenerate limit. Let $Y_t = t - S_{N(t)}$, $Z_t = S_{N(t)+1} - t$. Show that

$$Y_t/t \Rightarrow \pi^{-1} \sin(\pi\alpha) x^{-\alpha}(1 - x)^{\alpha-1}, \qquad x \in [0, 1]$$

$$Z_t/t \Rightarrow \pi^{-1} \sin(\pi\alpha) x^{-\alpha}(1 + x)^{-1}, \qquad x \in [0, \infty),$$

where the last two statements mean that the indicated random variables converge in distribution to distributions with densities given by the formulas above. This result is due to Dynkin (1961).

4 Martingales

A martingale X_n can be thought of as the fortune at time n of a player who is betting on a fair game; submartingales (supermartingales) as the outcome of betting on a favorable (unfavorable) game. There are two basic facts about martingales. The first is that you cannot make money betting on them (see (2.7)), and in particular if you choose to stop playing at some bounded time N, then your expected winnings EX_N are equal to your initial fortune X_0. (We are supposing for the moment that this is not random.) Our second fact, (2.10), concerns submartingales. To use a heuristic we learned from Mike Brennan, "they are the stochastic analogues of nondecreasing sequences and so if they are bounded above (to be precise, $\sup_n EX_n^+ < \infty$), they converge almost surely." As the material in Section 3 shows, this result has diverse applications. Later sections give sufficient conditions for martingales to converge in $L^p, p > 1$ (Section 4), and in L^1 (Section 5); consider martingales indexed by $n \leq 0$ (Section 6); and give sufficient conditions for $EX_N = EX_0$ to hold for unbounded stopping times (Section 7). The last result is quite useful for studying the behavior of random walks and other systems.

1 Conditional Expectation

We begin with a definition that is important for this chapter and the next one. After giving the definition, we will consider five examples to explain it. Given are a probability space $(\Omega, \mathcal{F}_0, P)$, a σ-field $\mathcal{F} \subset \mathcal{F}_0$, and a random variable $X \in \mathcal{F}_0$ with $E|X| < \infty$. We define the *conditional expectation of X given $\mathcal{F}$*, $E(X|\mathcal{F})$, to be any random variable Y that has

(i) $Y \in \mathcal{F}$, and
(ii) for all $A \in \mathcal{F}$, $\int_A X \, dP = \int_A Y \, dP$.

Any Y satisfying (i) and (ii) is said to be a *version* of $E(X|\mathcal{F})$. Putting aside the question of existence of the conditional expectation, it is easy to see that Y is unique (up to modifications on sets of measure 0). If Y' also satisfies (i) and (ii), then

$$\int_A Y \, dP = \int_A Y' \, dP \quad \text{for all } \alpha \in \mathcal{F},$$

so taking $A = \{Y - Y' \geq \varepsilon > 0\}$, we see $P(A) = 0$ and $Y = Y'$ a.s. Technically, all equalities such as $Y = E(X|\mathscr{F})$ should be written as $Y = E(X|\mathscr{F})$ a.s., but we have ignored this point in previous chapters and will continue to do so.

Exercise 1.1   Generalize the last argument to show that if $X_1 = X_2$ on $B \in \mathscr{F}$, then

$$E(X_1|\mathscr{F}) = E(X_2|\mathscr{F}) \text{ a.s. on } B.$$

Intuitively, we think of $\mathscr{F}$ as describing the information we have at our disposal—for each $A \in \mathscr{F}$, we know whether or not A has occurred. $E(X|\mathscr{F})$ is then our "best guess" of the value of X given the information we have. Some examples should help to clarify this and connect $E(X|\mathscr{F})$ with other definitions of conditional expectation.

Example 1.1   If $X \in \mathscr{F}$, then $E(X|\mathscr{F}) = X$; that is, if we know X, then our "best guess" is X itself. Since X always satisfies (ii), the only thing that can keep X from being $E(X|\mathscr{F})$ is condition (i). A special case of this example is $X = c$ where c is a constant.

Example 1.2   At the other extreme from perfect information is no information. Suppose X is independent of $\mathscr{F}$; that is, for all $B \in \mathscr{R}$ and $A \in \mathscr{F}$,

$$P(\{X \in B\} \cap A) = P(X \in B)P(A).$$

In this case $E(X|\mathscr{F}) = EX$. To check this note that $EX \in \mathscr{F}$ and if $A \in \mathscr{F}$, then

$$\int_A X \, dP = E(X1_A) = EX \, E1_A = \int_A EX \, dP,$$

so (ii) holds. The reader should note that here and in what follows the game is "guess and verify." We come up with a formula for the conditional expectation and then check that it satisfies (i) and (ii).

Example 1.3   In this example we relate the new definition of conditional expectation to the first one taught in an undergraduate probability course. Suppose $\Omega_1, \Omega_2, \ldots$ is a finite or infinite partition of Ω into disjoint sets each of which has positive probability, and let $\mathscr{F} = \sigma(\Omega_1, \Omega_2, \ldots)$ be the σ-field generated by these sets. Then

$$E(X|\mathscr{F}) = E(X; \Omega_i)/P(\Omega_i) \quad \text{on } \Omega_i.$$

In words, the information in Ω_i tells us which element of the partition our outcome lies in, and given this information the best guess for X is the average value of X over Ω_i. To prove our guess is correct, observe that the proposed formula is constant on each Ω_i, so it is measurable with respect to $\mathscr{F}$. To verify (ii), it is enough to check the equality for $A = \Omega_i$, but this is trivial:

$$\int_{\Omega_i} E(X;\Omega_i)/P(\Omega_i)\, dP = E(X;\Omega_i) = \int_{\Omega_i} X\, dP.$$

A degenerate but important special case is $\mathscr{F} = \{\phi, \Omega\}$, the trivial σ-field. In this case $E(X|\mathscr{F}) = EX$.

To continue the connection with undergraduate notions, let

$$P(A|\mathscr{G}) = E(1_A|\mathscr{G})$$

$$P(A|B) = P(A \cap B)/P(B),$$

and observe that in the last example $P(A|\mathscr{F}) = P(A|\Omega_i)$ on Ω_i.

Exercise 1.2   *Bayes formula.* Let $G \in \mathscr{G}$ and show that

$$P(G|A) = \int_G P(A|\mathscr{G})\, dP \Big/ \int_\Omega P(A|\mathscr{G})\, dP.$$

When $\mathscr{G}$ is the σ-field generated by a partition, this reduces to the usual Bayes formula

$$P(G_i|A) = P(A|G_i)P(G_i) \Big/ \sum_j P(A|G_j)P(G_j).$$

A typical application of Bayes formula is:

Exercise 1.3   Let X be uniformly distributed on $\{1, 2, \ldots, 6\}$ (i.e., roll a die) and Y be the number of heads when X coins are tossed. Compute $P(X = i| Y = 0)$.

The definition of conditional expectation given a σ-field contains conditioning on a random variable as a special case. We define

$$E(X|Y) = E(X|\sigma(Y))$$

where $\sigma(Y)$ is the σ-field generated by Y.

Exercise 1.4   Let X_1 and X_2 be independent and have a Poisson distribution with mean λ, and let $Y = X_1 + X_2$. Compute $P(X_1 = i| Y)$.

Example 1.4   To continue making connection with definitions of conditional expectation from undergraduate probability, suppose X and Y have joint density $f(x, y)$; that is,

$$P((X, Y) \in B) = \int_B f(x, y)\, dx\, dy \quad \text{for } B \in \mathscr{R}^2,$$

and suppose for simplicity that $\int f(x, y)\, dx > 0$. In this case $E(g(X)|Y) = h(Y)$ where

$$h(y) = \int g(x)f(x, y)\, dx \Big/ \int f(x, y)\, dx.$$

To "derive" this formula (ignoring technicalities), note

$$P(X = x|Y = y) = P(X = x, Y = y)/P(Y = y) = f(x, y)\Big/ \int f(x, y)\, dx$$

(here $P(Y = y)$, etc. denote probability densities), so

$$E(g(X)|Y = y) = \int g(x)P(X = x|Y = y)\, dx.$$

To check the proposed formula, observe $h(Y) \in \sigma(Y)$, so (i) holds. To check (ii), observe that if $A \in \sigma(Y)$, then $A = \{\omega: Y(\omega) \in B\}$ for some $B \in \mathcal{R}$, so

$$E(h(Y); A) = \int_B \int h(y)f(x, y)\, dx\, dy = \int_B \int g(x)f(x, y)\, dx\, dy$$

$$= E(g(X)1_B(Y)) = E(g(X); A).$$

Remark   To drop the assumption that $\int f(x, y)\, dx > 0$, define h by

$$h(y) \int f(x, y)\, dx = \int g(x)f(x, y)\, dx$$

(i.e., h can be anything where $\int f(x, y)\, dx = 0$) and observe this is enough for the proof.

Exercise 1.5   Let X_1 and X_2 be independent with $P(X_i > t) = e^{-t}$ for $t \geq 0$, and let $Y = X_1 + X_2$. Compute $E(X_1|Y)$ and $P(X_1 < 3|Y)$.

Exercise 1.6   *Borel's paradox.* Let X be a randomly chosen point on the earth. Deviating slightly from the traditional practice, let $\theta \in [0, 2\pi)$ be its longitude and $\varphi \in (-\pi/2, \pi/2)$ be its latitude. Compute the conditional distribution of φ given θ and come to the "shocking" conclusion that it is not uniform. This result sounds more paradoxical if we restrict $\theta \in [0, \pi)$—that is, the longitude specifies the great circle on which the point lies, and we compensate for this change by having $\varphi \in (-\pi, \pi)$. In this case conditioning our uniformly distributed point to lie on a particular great circle results in a point not uniformly distributed on that circle. The paradox completely evaporates once we realize that in either formulation φ is independent of θ, so the conditional distribution is the unconditional one, which is not uniform since there is more land near the equator than near the North Pole.

Example 1.5   Having been somewhat casual in Example 1.4, we finish by giving an example that shows caution is needed in applying the intuitive definitions. Let $\Omega = (0, 1)$, $\mathcal{F}_0 = $ the Borel subsets, $P = $ Lebesgue measure, $X(\omega) = \cos(\pi\omega)$, and $\mathcal{F} = \{A \subset (0, 1): A \text{ or } A^c \text{ is countable}\}$. Arguing carelessly we might conclude that "for each $x \in (0, 1)$ we know whether or not $\{x\}$ has occurred so we know ω and hence $X(\omega)$—that is, $E(X|\mathcal{F}) = X$." The last answer cannot be correct, since $X \notin \mathcal{F}$. The correct answer is $E(X|\mathcal{F}) = 0$, since the right-hand side satisfies (i) and (ii). The right intuition is that if we are told the answer to "$\omega \in A_i$?" for a finite number of $A_i \in \mathcal{F}$, we know nothing about X.

Having considered five examples, we turn our attention now to proving that conditional expectations exist. To do this we recall v is said to be *absolutely continuous with respect to μ* (abbreviated $v \ll \mu$) if $\mu(A) = 0$ implies $v(A) = 0$, and we use the:

Radon–Nikodym Theorem   Let μ and v be σ-finite measures on $(\Omega, \mathcal{F})$. If $v \ll \mu$, there is a function $f \in \mathcal{F}$ so that for all $A \in \mathcal{F}$,

$$\int_A f \, d\mu = v(A).$$

f is usually denoted $dv/d\mu$ and called the *Radon–Nikodym derivative*.

Proof   See (8.6) in the Appendix.   □

The last theorem easily gives the existence of conditional expectation. Suppose $X \geq 0$. (To treat the general case, write $X = X^+ - X^-$.) Let $\mu = P$ and

$$v(A) = \int_A X \, dP \quad \text{for } A \in \mathcal{F}.$$

The dominated convergence theorem implies v is a measure (see Exercise 3.9 in Chapter 1) and the definition of the integral implies $v \ll \mu$. The Radon–Nikodym derivative $dv/d\mu$ is $E(X|\mathcal{F})$.

Conditional expectation has many of the same properties that ordinary expectation does.

(1.1a) *Linearity* $E(aX + Y|\mathcal{F}) = aE(X|\mathcal{F}) + E(Y|\mathcal{F})$.

Proof   We need to check that the right-hand side is a version of the left. It clearly is $\mathcal{F}$-measurable. To check (ii) we observe that if $A \in \mathcal{F}$, then by linearity of the integral and the defining properties of $E(X|\mathcal{F})$ and $E(Y|\mathcal{F})$,

$$\int_A aE(X|\mathscr{F}) + E(Y|\mathscr{F})\, dP = a \int_A E(X|\mathscr{F})\, dP + \int_A E(Y|\mathscr{F})\, dP$$

$$= a \int_A X\, dP + \int_A Y\, dP = \int_A aX + Y\, dP. \qquad \square$$

(1.1b) Monotonicity   If $X \le Y$, then $E(X|\mathscr{F}) \le E(Y|\mathscr{F})$.

Proof

$$\int_A E(X|\mathscr{F})\, dP = \int_A X\, dP \le \int_A Y\, dP = \int_A E(Y|\mathscr{F})\, dP.$$

Letting $A = \{E(X|\mathscr{F}) - E(Y|\mathscr{F}) \ge \varepsilon > 0\}$, we see that the indicated set has probability 0 for all $\varepsilon > 0$. $\qquad\square$

Exercise 1.7   Prove *Chebyshev's inequality*. If $a > 0$,

$$P(|X| \ge a|\mathscr{F}) \le a^{-2}E(X^2|\mathscr{F}).$$

(1.1c) Monotone Convergence Theorem   If $X_n \ge 0$ and $X_n \uparrow X$ with $EX < \infty$, then

$$E(X_n|\mathscr{F}) \uparrow E(X|\mathscr{F}).$$

Proof   Let $Y_n = X - X_n$. It suffices to show that $E(Y_n|\mathscr{F}) \downarrow 0$. Since $Y_n \downarrow$, (1.1b) implies $Z_n \equiv E(Y_n|\mathscr{F}) \downarrow$ a limit Z_∞. If $A \in \mathscr{F}$,

$$\int_A Z_n\, dP = \int_A Y_n\, dP.$$

Letting $n \to \infty$ and using the dominated convergence theorem give $\int_A Z_\infty\, dP = 0$ for all $A \in \mathscr{F}$, so $Z_\infty \equiv 0$. $\qquad\square$

Exercise 1.8   Suppose $X \ge 0$ and $EX = \infty$. (There is nothing to prove when $EX < \infty$.) Show there is a unique $\mathscr{F}$-measurable Y with $0 \le Y \le \infty$, so that

$$\int_A X\, dP = \int_A Y\, dP \quad \text{for all } A \in \mathscr{F}.$$

Give an example to show that we can have $0 < P(Y = \infty) < 1$.

(1.1d) Jensen's Inequality   If φ is convex and $E|X|$, $E|\varphi(X)| < \infty$, then

$$\varphi(E(X|\mathscr{F})) \le E(\varphi(X)|\mathscr{F}).$$

Proof   If φ is linear, the result is trivial so we will suppose φ is not linear. We do this so that if we let $S = \{(a, b): a,\, b \in \mathbb{Q},\, ax + b \le \varphi(x) \text{ for all } x\}$, then $\varphi(x) =$

$\sup\{ax + b : (a, b) \in S\}$. (See the proof of (3.2) in Chapter 1 for more details.) If $\varphi(x) \geq ax + b$, then (1.1b) and (1.1a) imply

$$E(\varphi(X)|\mathscr{F}) \geq aE(X|\mathscr{F}) + b \text{ a.s.}$$

Taking the sup over $(a, b) \in S$ gives

$$E(\varphi(X)|\mathscr{F}) \geq \varphi(E(X|\mathscr{F})) \text{ a.s.} \qquad \square$$

Remark Here we have written a.s. by the inequalities to stress that there is an exceptional set for each a, b so we have to take the sup over a countable set.

Exercise 1.9 Prove the conditional Cauchy–Schwarz inequality:

$$E(XY|\mathscr{G})^2 \leq E(X^2|\mathscr{G})E(Y^2|\mathscr{G}).$$

Hint $E((X + \theta Y)^2|\mathscr{G}) \geq 0$ and if $a\theta^2 + b\theta + c \geq 0$ for all rational θ, then $b^2 - 4ac \leq 0$.

(1.1e) Conditional expectation is a contraction in L^p, $p \geq 1$.

Proof (1.1d) implies $|E(X|\mathscr{F})|^p \leq E(|X|^p|\mathscr{F})$. Taking expected values gives

$$E(|E(X|\mathscr{F})|^p) \leq E(E(|X|^p|\mathscr{F})) = E|X|^p. \qquad \square$$

In the last equality we have used an identity that is an immediate consequence of the definition:

$$E(E(Y|\mathscr{F})) = E(Y). \tag{1.1f}$$

(Use property (ii) with $A = \Omega$.)

Exercise 1.10 Suppose X and Y are independent. Let φ be a function with $E|\varphi(X, Y)| < \infty$ and let $g(x) = E(\varphi(x, Y))$. Show that $E(\varphi(X, Y)|X) = g(X)$.

Hint If $\varphi(x, y) = 1_A(x)1_B(y)$, this would be easy. To facilitate passage to limits, note that we want to show

$$\int_C \varphi(X, Y) \, dP = \int_C g(X) \, dP \quad \text{for } C = \{\omega: Y(\omega) \in D\}.$$

Conditional expectation also has properties (like (1.1f)) that have no analogue for "ordinary" expectation.

(1.2) THEOREM If $\mathscr{F}_1 \subset \mathscr{F}_2$, then:

(i) $E(E(X|\mathscr{F}_1)|\mathscr{F}_2) = E(X|\mathscr{F}_1)$.
(ii) $E(E(X|\mathscr{F}_2)|\mathscr{F}_1) = E(X|\mathscr{F}_1)$.

In words, the smaller σ-field always wins.

Proof Once we notice that $E(X|\mathscr{F}_1) \in \mathscr{F}_2$, (i) follows from Example 1.1. To prove (ii) notice that $E(X|\mathscr{F}_1) \in \mathscr{F}_1$ and if $A \in \mathscr{F}_1 \subset \mathscr{F}_2$, then

$$\int_A E(X|\mathscr{F}_1)\, dP = \int_A X\, dP = \int_A E(X|\mathscr{F}_2)\, dP. \qquad \square$$

Exercise 1.11 Give an example on $\Omega = \{a, b, c\}$ in which

$$E(E(X|\mathscr{F}_1)|\mathscr{F}_2) \neq E(E(X|\mathscr{F}_2)|\mathscr{F}_1).$$

The next result shows that for conditional expectation w.r.t. $\mathscr{F}$, r.v.'s $X \in \mathscr{F}$ are like constants. They can be brought outside the "integral." This property in conjunction with (1.2) is useful for computing conditional expectations.

(1.3) THEOREM If $X \in \mathscr{F}$ and $E|Y|, E|XY| < \infty$, then $E(XY|\mathscr{F}) = XE(Y|\mathscr{F})$.

Proof The right-hand side $\in \mathscr{F}$, so we have to check (ii). To do this we use the usual four-step procedure. First, suppose $X = 1_B$ with $B \in \mathscr{F}$. In this case if $A \in \mathscr{F}$,

$$\int_A 1_B E(Y|\mathscr{F})\, dP = \int_{A \cap B} E(Y|\mathscr{F})\, dP = \int_{A \cap B} Y\, dP = \int_A 1_B Y\, dP,$$

so (ii) holds. The last result extends to simple X by linearity. If $X, Y \geq 0$, let X_n be simple random variables that $\uparrow X$, and use the monotone convergence theorem to conclude that

$$\int_A XE(Y|\mathscr{F})\, dP = \int_A XY\, dP.$$

To prove the result in general, split X and Y into their positive and negative parts. $\qquad \square$

Exercise 1.12 Show that each statement implies the next one and give examples with $X, Y \in \{-1, 0, 1\}$ a.s. that show the reverse implications are false: (i) X and Y are independent, (ii) $E(Y|X) = EY$, (iii) $E(XY) = EX\, EY$.

As an application of (1.3) we get the following result, which gives a "geometric interpretation" of $E(X|\mathscr{F})$.

(1.4) THEOREM Suppose $EX^2 < \infty$. $L^2(\mathscr{F}_0) = \{Y \in \mathscr{F}_0 : EY^2 < \infty\}$ is a Hilbert space, and $L^2(\mathscr{F})$ is a closed subspace. In this case $E(X|\mathscr{F})$ is the projection of X onto $L^2(\mathscr{F})$. In statistical terms, $E(X|\mathscr{F})$ is the variable $Y \in \mathscr{F}$ that minimizes the "mean square error" $E(X - Y)^2$. In particular $E[(X - E(X|\mathscr{F}))^2] \leq E[(X - EX)^2]$.

Proof We prove only the statistical interpretation, leaving it to the reader to show that this implies the geometric one. We begin by observing that if $Z \in L^2(\mathscr{F})$, then (1.3) implies

$$ZE(X|\mathcal{F}) = E(ZX|\mathcal{F}).$$

$(E|XZ| < \infty$ by the Cauchy–Schwarz inequality.) Taking expected values gives

$$E(ZE(X|\mathcal{F})) = E(E(ZX|\mathcal{F})) = E(ZX),$$

or rearranging,

$$E[Z(X - E(X|\mathcal{F}))] = 0 \quad \text{for } Z \in L^2(\mathcal{F}). \tag{1.5}$$

If $Y \in L^2(\mathcal{F})$ and $Z = Y - E(X|\mathcal{F})$, then

$$E((X - Y)^2) = E((X - E(X|\mathcal{F}) - Z)^2) = E((X - E(X|\mathcal{F}))^2) + EZ^2.$$

So $E(X - Y)^2$ is minimized when $Z = 0$.   □

Remark   If one wants to avoid using the Radon–Nikodym theorem, (1.4) can be used to define conditional expectation for $X \in L^2$. Monotone limits and (1.1c) then extend the definition to $X \geq 0$, and taking positive and negative parts gives the general definition. Neveu (1975) uses this approach; see Section I-2.

Exercise 1.13   Show that if $\mathcal{G} \subset \mathcal{H}$ and $EX^2 < \infty$, then

$$E(\{X - E(X|\mathcal{H})\}^2) + E(\{E(X|\mathcal{H}) - E(X|\mathcal{G})\}^2) = E(\{X - E(X|\mathcal{G})\}^2).$$

Dropping the second term on the left, we get an inequality that says geometrically, the larger the subspace, the closer the projection is, or statistically, more information means a smaller mean square error.

Exercise 1.14   Let $\text{var}(X|\mathcal{F}) = E(X^2|\mathcal{F}) - E(X|\mathcal{F})^2$. Show that

$$\text{var}(X) = E(\text{var}(X|\mathcal{F})) + \text{var}(E(X|\mathcal{F})).$$

For a concrete example let $Y_1, Y_2, \ldots$ be i.i.d., N an independent integer-valued r.v., and $X = Y_1 + \cdots + Y_N$. In this case $\text{var}(X) = EN\,\text{var}(Y) + EY\,\text{var}(N)$.

Exercise 1.15   Show that if X and Y are random variables with $E(Y|\mathcal{G}) = X$ and $EY^2 = EX^2$, then $X = Y$ a.s.

Exercise 1.16   The result in the last exercise implies that if $EY^2 < \infty$ and $E(Y|\mathcal{G}) \overset{d}{=} Y$, then $E(Y|\mathcal{G}) = Y$ a.s. Prove this under the assumption $E|Y| < \infty$.

Hint   The trick is to prove that if $E|E(X|\mathcal{G})| = E|X|$, then $\text{sign}(X) = \text{sign}(E(X|\mathcal{G}))$ a.s., and then apply the lemma to $X = Y - c$.

Regular conditional probabilities   Let $(\Omega, \mathcal{F}, P)$ be a probability space, $X: (\Omega, \mathcal{F}) \to (S, \mathcal{S})$ a measurable map, and $\mathcal{G}$ a σ-field $\subset \mathcal{F}$. $\mu: \Omega \times \mathcal{S} \to [0, 1]$ is said to be a *regular conditional distribution* (r.c.d.) for X given $\mathcal{G}$ if:

(i) For each A, $\omega \to \mu(\omega, A)$ is a version of $P(X \in A|\mathcal{G})$.
(ii) For almost every ω, $A \to \mu(\omega, A)$ is a probability measure on $(S, \mathcal{S})$.

When $S = \Omega$ and X is the identity map, μ is called a *regular conditional probability*.

Exercise 1.17   *Continuation of Example 1.4.* Suppose X and Y have a joint density $f(x, y) > 0$. Let

$$\mu(y, A) = \int_A f(x, y)\, dx \Big/ \int f(x, y)\, dx.$$

Show that $\mu(Y(\omega), A)$ is a r.c.d. for X given $\sigma(Y)$.

Regular conditional distributions are useful because they allow us to compute the conditional expectation of all functions of X and to generalize properties of ordinary expectation in a more straightforward way.

Exercise 1.18   Let $\mu(\omega, A)$ be a r.c.d. for X given $\mathcal{F}$, and let $f : (S, \mathcal{S}) \to (\mathbb{R}, \mathcal{R})$ have $E|f(X)| < \infty$. Start with simple functions and show that

$$E(f(X)|\mathcal{F}) = \int \mu(\omega, dx) f(x) \text{ a.s.}$$

Exercise 1.19   Use regular conditional distributions to get the conditional Hölder inequality from the unconditional one; that is, show that if p, $q \in (1, \infty)$ with $1/p + 1/q = 1$, then

$$E(|XY| \,|\mathcal{G}) \le E(|X|^p|\mathcal{G})^{1/p} E(|Y|^q|\mathcal{G})^{1/q}.$$

Unfortunately, r.c.d.'s do not always exist. The first example was due to Dieudonné (1948). See Doob (1953), p. 624, or Faden (1985) for more recent developments. Without going into the details of the example, it is easy to see the source of the problem. If $A_1, A_2, \ldots$ are disjoint, then (1.1a) and (1.1c) imply

$$P\left(X \in \bigcup_n A_n \,\middle|\, \mathcal{G}\right) = \sum_n P(X \in A_n|\mathcal{G}) \text{ a.s.},$$

but if $\mathcal{F}$ contains enough countable collections of disjoint sets, the exceptional sets may pile up. Fortunately,

(1.6) THEOREM   r.c.d.'s exist if $(S, \mathcal{S})$ is nice.

Proof   By definition there is a 1–1 map $\varphi : S \to \mathbb{R}$ so that φ and φ^{-1} are measurable. Using monotonicity (1.1b) and throwing away a countable collection of null sets, we find there is a set Ω_0 with $P(\Omega_0) = 1$ and a family of random variables $G(q, \omega)$, $q \in \mathbb{Q}$, so that $q \to G(q, \omega)$ is nondecreasing and $\omega \to G(\omega, q)$ is a version of $P(\varphi(X) \le q|\mathcal{G})$. Let $F(x, \omega) = \inf\{G(\theta, \omega) : q > x\}$. The notation may remind the reader of the proof of (2.4) in Chapter 2. The argument given there shows F is a distribution function. The monotone convergence theorem (1.1c) implies that $F(x, \omega)$ is a version of $P(\varphi(X) \le x|\mathcal{G})$.

Now for each $\omega \in \Omega_0$, there is a unique measure $\nu(\omega, \cdot)$ on $(\mathbb{R}, \mathcal{R})$ so that $\mu(\omega, (-\infty, x]) = F(\omega, x)$. To check that for each $B \in \mathcal{R}$, $\nu(\omega, B)$ is a version of $P(\varphi(X) \in B|\mathcal{G})$, we observe that the class of B for which this statement is true (this includes the measurability of $\omega \to \nu(\omega, B)$) is a λ-system that contains all sets of the form $(a_1, b_1] \cup \cdots \cup (a_k, b_k]$ where $-\infty \le a_i < b_i \le \infty$, so the desired result follows from the $\pi - \lambda$ theorem. To extract the desired r.c.d. notice that if $A \in \mathcal{S}$ and $B = \varphi(A)$, then $B = (\varphi^{-1})^{-1}(A) \in \mathcal{R}$, and set $\mu(\omega, A) = \nu(\omega, B)$. ☐

The following generalization of (1.6) will be needed in Section 5.1.

Exercise 1.20 Suppose X and Y take values in a nice space $(S, \mathcal{S})$ and $\mathcal{G} = \sigma(Y)$. Imitate the proof of (1.6) to show that there is a function $\mu: S \times \mathcal{S} \to [0, 1]$ so that

(i) for each A, $\mu(Y(\omega), A)$ is a version of $P(X \in A|\mathcal{G})$ and
(ii) for a.e. ω, $A \to \mu(Y(\omega), A)$ is a probability measure on $(S, \mathcal{S})$.

2   Martingales, Almost Sure Convergence

In this section we will give some basic definitions and prove the two basic facts, (2.7) and (2.10), mentioned in the introduction to the chapter. Let $\mathcal{F}_n$ be a *filtration*—that is, an increasing sequence of σ-fields. A sequence X_n is said to be *adapted* to $\mathcal{F}_n$ if $X_n \in \mathcal{F}_n$ for all n. If a sequence X_n with $E|X_n| < \infty$ is adapted to $\mathcal{F}_n$ and has $E(X_{n+1}|\mathcal{F}_n) = X_n$ for all n, then X is said to be a *martingale* (with respect to $\mathcal{F}_n$). If in the last definition = is replaced by $\le$ or $\ge$, then X is said to be a *supermartingale* or *submartingale*, respectively.

To give an example of a martingale, consider the successive tosses of a fair coin and let $\xi_n = 1$ if the nth toss is heads and $\xi_n = -1$ if the nth toss is tails. Let $X_n = \xi_1 + \cdots + \xi_n$ and $\mathcal{F}_n = \sigma(\xi_1, \ldots, \xi_n)$ for $n \ge 1$, $X_0 = 0$, and $\mathcal{F}_0 = \{\varnothing, \Omega\}$. I claim that X_n, $n \ge 0$, is a martingale with respect to $\mathcal{F}_n$. To prove this, we observe that $X_n \in \mathcal{F}_n$, $E|X_n| < \infty$, and ξ_{n+1} is independent of $\mathcal{F}_n$, so

$$E(X_{n+1}|\mathcal{F}_n) = E(X_n + \xi_{n+1}|\mathcal{F}_n) = E(X_n|\mathcal{F}_n) + E(\xi_{n+1}|\mathcal{F}_n) = X_n + E\xi_{n+1} = X_n.$$

Note that in this example $\mathcal{F}_n = \sigma(X_1, \ldots, X_n)$ and $\mathcal{F}_n$ is the smallest filtration that X_n is adapted to. In what follows, when the filtration is not mentioned, we will take $\mathcal{F}_n = \sigma(X_1, \ldots, X_n)$.

Exercise 2.1 Suppose $X_n, n \ge 1$, is a martingale w.r.t. $\mathcal{G}_n$ and let $\mathcal{F}_n = \sigma(X_1, \ldots, X_n)$. Then $\mathcal{G}_n \supset \mathcal{F}_n$ and X_n is a martingale w.r.t. $\mathcal{F}_n$.

If the coin tosses considered above have $P(\xi_n = 1) < 1/2$, then the computation just completed shows $E(X_{n+1}|\mathcal{F}_n) \le X_n$; that is, X_n is a supermartingale. In this case X_n corresponds to betting on an unfavorable game, so there is nothing "super" about a supermartingale. The name comes from the fact that if f is superharmonic (i.e., $\partial^2 f/\partial x_1^2 + \cdots + \partial^2 f/\partial x_d^2 \leq 0$), then

$$f(x) \geq \int_{\partial B(x,r)} f(y)\, d\pi(y)$$

where $\partial B(x, r) = \{y: |x - y| = r\}$ is the boundary of the ball of radius r, and π is surface measure normalized to be a probability measure.

Exercise 2.2 Suppose $f \geq 0$ is superharmonic on $\mathbb{R}^d$. Let $\xi_1, \xi_2, \ldots$ be i.i.d. uniform on $\partial B(0, 1)$, and define S_n by $S_n = x + S_{n-1}$ for $n \geq 1$ and $S_0 = x$. Show that $X_n = f(S_n)$ is a supermartingale.

Exercise 2.3 If $\xi_1, \xi_2, \ldots$ are independent and have $E\xi_i = 0$, then

$$X_n = \sum_{1 \leq i_1 < \cdots < i_k \leq n} \xi_{i_1} \cdots \xi_{i_k}$$

is a martingale. When $k = 2$ and $S_n = \xi_1 + \cdots + \xi_n$, $2X_n = S_n^2 - \sum_{m \leq n} \xi_m^2$.

Our first result is an immediate consequence of the definition.

(2.1) THEOREM If X_n is a supermartingale, then for $n > m$, $E(X_n|\mathcal{F}_m) \leq X_m$.

Proof The definition gives the result for $n = m + 1$. Suppose $n = m + k$ with $k \geq 2$. By (1.2),

$$E(X_{m+k}|\mathcal{F}_m) = E(E(X_{m+k}|\mathcal{F}_{m+k-1})|\mathcal{F}_m) \leq E(X_{m+k-1}|\mathcal{F}_m)$$

by the definition and (1.1b). The desired result now follows by induction. □

(2.2) COROLLARY (i) If X_n is a submartingale, then for $n > m$, $E(X_n|\mathcal{F}_m) \geq X_m$. (ii) If X_n is a martingale, then for $n > m$, $E(X_n|\mathcal{F}_m) = X_m$.

Proof To prove (i) note that $-X_n$ is a supermartingale. For (ii) observe that X_n is a supermartingale and a submartingale. □

Remark The idea in the proof of (2.2) can be used many times below. To keep from repeating ourselves, we will just state the result for either supermartingales or submartingales and leave it to the reader to translate the result for the other two.

(2.3) THEOREM If X_n is a martingale and φ is a convex function with $E|\varphi(X_n)| < \infty$ for all n, then $\varphi(X_n)$ is a submartingale (with respect to the same filtration).

Proof By Jensen's inequality and the definition,

$$E(\varphi(X_{n+1})|\mathcal{F}_n) \geq \varphi(E(X_{n+1}|\mathcal{F}_n)) = \varphi(X_n).$$ □

(2.4) COROLLARY   Let $p \geq 1$. If X_n is a martingale and $E|X_n|^p < \infty$ for all n, then $|X_n|^p$ is a submartingale.

(2.5) THEOREM   If X_n is a submartingale and φ is an increasing convex function with $E|\varphi(X_n)| < \infty$ for all n, then $\varphi(X_n)$ is a submartingale (with respect to the same filtration).

Proof   By Jensen's inequality and the definition,

$$E(\varphi(X_{n+1})|\mathcal{F}_n) \geq \varphi(E(X_{n+1}|\mathcal{F}_n)) \geq \varphi(X_n). \qquad \square$$

Exercise 2.4   Give an example of a submartingale X_n so that X_n^2 is a supermartingale.

(2.6) COROLLARIES   (i) If X_n is a submartingale, then X_n^+ is a submartingale. (ii) If X_n is a supermartingale, then $X_n \wedge a$ is a supermartingale.

Exercise 2.5   Generalize (2.6) by showing that if X_n and Y_n are submartingales w.r.t. $\mathcal{F}_n$, then $X_n \vee Y_n$ is also.

Exercise 2.6   Show that if X_n and Y_n are submartingales w.r.t. $\mathcal{F}_n$, then so is $X_n + Y_n$. Show that if Z_n is a submartingale w.r.t. $\mathcal{G}_n$, then $X_n + Z_n$ may not be a submartingale w.r.t. any filtration.

Let $\mathcal{F}_n$, $n \geq 0$, be a filtration. H_n, $n \geq 1$, is said to be a *predictable sequence* if $H_n \in \mathcal{F}_{n-1}$ for all $n \geq 1$. In words, the value of H_n may be predicted (with certainty) from the information available at time $n - 1$. In this section we will be thinking of H_n as the amount of money a gambler will bet at time n. This can be based on the outcomes at times $1, \ldots, n - 1$ but not on the outcome at time n!

Once we start thinking of H_n as a gambling system, it is natural to ask how much money we would make if we used it. For concreteness, let us suppose that the game consists of flipping a coin and that for each \$1 you bet you win \$1 when the coin comes up heads, and you lose your \$1 when the coin comes up tails. Let X_n be the net amount of money you would have won at time n if you had bet \$1 each time. If you bet according to a gambling system H, then your winnings at time n would be

$$(H \cdot X)_n = \sum_{m=1}^{n} H_m(X_m - X_{m-1}),$$

since $X_m - X_{m-1} = +1$ or -1 when the mth toss results in a win or a loss, respectively.

Let $\xi_m = X_m - X_{m-1}$. A famous gambling system called the "martingale" is defined by $H_1 = 1$ and for $n \geq 2$, $H_n = 2H_{n-1}$ if $\xi_{n-1} = -1$ and $H_n = 1$ if $\xi_{n-1} = 1$. In words, we double our bet when we lose, so that if we lose k times and then win, our net winnings will be $-1 - 2 - \cdots - 2^{k-1} + 2^k = 1$. This system seems to provide us with a "sure thing." However, the next result says there is no system for beating an unfavorable game.

(2.7) THEOREM   Let X_n, $n \geq 0$, be a supermartingale. If $H_n \geq 0$ is predictable and each H_n is bounded, then $(H \cdot X)_n$ is a supermartingale.

Proof

$$E((H \cdot X)_{n+1}|\mathcal{F}_n) = (H \cdot X)_n + E(H_{n+1}(X_{n+1} - X_n)|\mathcal{F}_n)$$

$$= (H \cdot X)_n + H_{n+1}E((X_{n+1} - X_n)|\mathcal{F}_n) \leq (H \cdot X)_n,$$

since $E((X_{n+1} - X_n)|\mathcal{F}_n) \leq 0$ and $H_{n+1} \geq 0$. $\qquad\qquad\qquad\qquad\square$

Remark   The same result is obviously true for submartingales and for martingales (in the last case without the restriction $H_n \geq 0$).

The notion of a stopping time, introduced in Section 3.1, is closely related to the concept of a gambling system. Recall that a random variable N is said to be a *stopping time* if $\{N = n\} \in \mathcal{F}_n$ for all $n < \infty$. If you think of N as the time a gambler stops gambling, then the condition above says that the decision to stop at time n must be measurable with respect to the information he has at that time. If we let $H_n = 1_{\{N \geq n\}}$, then $\{N \geq n\} = \{N \leq n - 1\}^c \in \mathcal{F}_{n-1}$, so H_n is predictable, and it follows from (2.7) that $(H \cdot X)_n = X_{N \wedge n} - X_0$ is a submartingale. Since the constant sequence $Y_n = X_0$ is a submartingale and the sum of two submartingales is also, we have:

(2.8) COROLLARY   If N is a stopping time and X_n is a submartingale, then $X_{N \wedge n}$ is a submartingale.

Although you cannot make money with gambling systems, you can prove theorems with them. Suppose X_n, $n \geq 0$, is a submartingale. Let $a < b$, let $N_0 = -1$, and for $k \geq 1$ let

$$N_{2k-1} = \inf\{m > N_{2k-2} : X_m \leq a\}$$

$$N_{2k} = \inf\{m > N_{2k-1} : X_m \geq b\}.$$

The N_j are stopping times and $\{N_{2k-1} < m \leq N_{2k}\} = \{N_{2k-1} \leq m - 1\} \cap \{N_{2k} \leq m - 1\}^c$, so

$$H_m = \begin{cases} 1 & \text{if } N_{2k-1} < m \leq N_{2k} \text{ for some } k \\ 0 & \text{otherwise} \end{cases}$$

defines a predictable sequence. $X(N_{2k-1}) \leq a$ and $X(N_{2k}) \geq b$, so between times N_{2k-1} and N_{2k}, X_m crosses from below a to above b. H_m is a gambling system which tries to take advantage of these "upcrossings." In stock market terms, we buy when $X_m \leq a$ and sell when $X_m \geq b$, so every time an upcrossing is completed, we make a profit of $\geq (b - a)$. Finally, $U_n = \sup\{k : N_{2k} \leq n\}$ is the number of upcrossings completed by time n.

(2.9) The Upcrossing Inequality   If $X_m, m \geq 0$, is a submartingale, then

$$(b - a)EU_n \leq E(X_n - a)^+ - E(X_0 - a)^+.$$

Proof   Let $Y_n = (X_n - a)^+$. By (1.6), Y_n is a submartingale. Clearly it upcrosses $[0, b - a]$ the same number of times that X_n upcrosses $[a, b]$, and we have $(b - a)U_n \leq (H \cdot Y)_n$, since each upcrossing results in a profit $\geq (b - a)$ and a final incomplete upcrossing (if there is one) makes a nonnegative contribution to the right-hand side. Let $K_m = 1 - H_m$. Clearly $Y_n - Y_0 = (H \cdot Y)_n + (K \cdot Y)_n$, and it follows from (2.7) that $E(K \cdot Y)_n \geq E(K \cdot Y)_0 = 0$, so $E(H \cdot Y)_n \leq E(Y_n - Y_0)$, proving (2.9).   □

We have proved the result in its classical form, even though this is a little misleading. The key fact is that $E(K \cdot X)_n \geq 0$; that is, no matter how hard you try you can't lose money betting on a submartingale. From the upcrossing inequality we immediately get:

(2.10) The Martingale Convergence Theorem   If X_n is a submartingale with sup $EX_n^+ < \infty$, then as $n \to \infty$, X_n converges a.s. to a limit X with $E|X| < \infty$.

Proof   Since $(X - a)^+ \leq X^+ + |a|$, (2.9) implies that

$$EU_n \leq (|a| + EX_n^+)/(b - a).$$

As $n \uparrow \infty$, $U_n \uparrow U$, the number of upcrossings of $[a, b]$ by the whole sequence, so if sup $EX_n^+ < \infty$, then $EU < \infty$ and hence $U < \infty$ a.s. Since the last conclusion holds for all rational a and b,

$$\bigcup_{a,b \in \mathbb{Q}} \{\liminf X_n < a < b < \limsup X_n\} \text{ has probability } 0$$

and hence

$$\limsup X_n = \liminf X_n \text{ a.s.;}$$

that is, lim X_n exists a.s. Fatou's lemma guarantees $EX^+ \leq \liminf EX_n^+ < \infty$, so $X < \infty$ a.s. To see $X > -\infty$, we observe that

$$EX_n^- = EX_n^+ - EX_n \leq EX_n^+ - EX_0$$

(since X_n is a submartingale), so another application of Fatou's lemma shows $EX^- \leq \liminf EX_n^- < \infty$ and completes the proof.   □

Remark   To prepare for the proof of (6.1), the reader should note that we have shown that if the number of upcrossings of (a, b) by X_n is finite for all $a, b \in \mathbb{Q}$, then the limit of X_n exists.

An important special case of (2.10) is:

(2.11) COROLLARY If $X_n \geq 0$ is a supermartingale, then as $n \to \infty$, $X_n \to X$ a.s. and $EX \leq EX_0$.

Proof $Y_n = -X_n \leq 0$ is a submartingale with $EY_n^+ = 0$. Since $EX_0 \geq EX_n$, the inequality follows from Fatou's lemma. □

In the next section we will give several applications of the last two results. We close this one by giving some "counterexamples." The first shows that the assumptions of (2.11) (and hence those of (2.10)) do not guarantee convergence in L^1.

Example 2.1 Let S_n be a symmetric simple random walk with $S_0 = 1$; that is, $S_n = S_{n-1} + \xi_n$ where $\xi_1, \xi_2, \ldots$ are i.i.d. with $P(X_i = 1) = P(X_i = -1) = 1/2$. Let $N = \inf\{n: S_n = 0\}$ and let $X_n = S_{N \wedge n}$. (2.8) implies that X_n is a nonnegative martingale. (2.11) implies X_n converges to a limit $X_\infty < \infty$ which must be $\equiv 0$, since convergence to $k > 0$ is impossible. Since $EX_n = EX_0 = 1$ for all n and $X_\infty = 0$, convergence cannot occur in L^1.

The last example is an important counterexample to keep in mind as you read the rest of this chapter. The next two exercises give counterexamples that are just as good.

Exercise 2.7 Let $Y_1, Y_2, \ldots$ be nonnegative i.i.d. random variables with $EY_m = 1$. Show that $X_n = \prod_{m \leq n} Y_m$ defines a martingale and the a.s. limit of X_n is 0 if $P(Y_m = 1) < 1$. This conclusion can also be obtained by applying the strong law to $\log Y_m$.

Exercise 2.8 Define X_n inductively by $X_0 = 1$, and X_n is uniformly distributed on $(0, X_{n-1})$ for $n \geq 1$. Show that $Y_n = 2^n X_n$ is a martingale and $Y_n \to 0$ a.s. The last exercise implies $1/n \log Y_n \to -1$ a.s., but you are asked to prove $Y_n \to 0$ a.s. without using this.

Example 2.2 We will now give an example of a martingale with $X_n \to 0$ in probability but not a.s. Let $X_0 = 0$. When $X_{k-1} = 0$, let $X_k = 1$ or -1 with probability $1/2k$ and $= 0$ with probability $1 - 1/k$. When $X_{k-1} \neq 0$, let $X_k = kX_{k-1}$ with probability $1/k$ and $= 0$ with probability $1 - 1/k$. From the construction $P(X_k = 0) \geq 1 - 1/k$ so $X_k \to 0$ in probability. On the other hand, the second Borel–Cantelli lemma implies $P(X_k = 0$ for $k \geq K) = 0$, and values in $(-1, 1) - \{0\}$ are impossible so X_k does not converge to 0 a.s.

Exercise 2.9 Give an example of a martingale X_n with $X_n \to -\infty$ a.s.

Hint Let $X_n = \xi_1 + \cdots + \xi_n$ where the ξ_i are independent with $E\xi_i = 0$ and $P(\xi_i = -1) = 1 - \varepsilon_i$.

Exercise 2.10 *The switching principle.* Suppose X_n^1 and X_n^2 are supermartingales with respect to $\mathcal{F}_n$ and N is a stopping time so that $X_N^1 \geq X_N^2$. Then

$$Y_n = X_n^1 1_{(N>n)} + X_n^2 1_{(N \leq n)} \text{ is a supermartingale.}$$

Since $N + 1$ is a stopping time, this implies that

$$Z_n = X_n^1 1_{(N \geq n)} + X_n^2 1_{(N < n)} \text{ is a supermartingale.}$$

(2.11) can be proved directly using the next result which improves (2.9).

Exercise 2.11   *Dubins' inequality.* For every positive martingale X_n, $n \geq 0$, the number of upcrossings U of $[a, b]$ satisfies

$$P(U \geq k) \leq \left(\frac{a}{b}\right)^k E \min(X_1/a, 1).$$

Sketch of proof   Let $N_0 = -1$ and for $j \geq 1$ let

$$N_{2j-1} = \inf\{m > N_{2j-2} : X_m \leq a\}$$

$$N_{2j} = \inf\{m > N_{2j-1} : X_m \geq b\}.$$

Show that if we let $Y_n = 1$ for $0 \leq n < N_1$ and for $j \geq 1$,

$$Y_n = (b/a)^{j-1}(X_n/a) \quad \text{for} \quad N_{2j-1} \leq n < N_{2j}$$

$$= (b/a)^j \quad \text{for} \quad N_{2j} \leq n < N_{2j+1}.$$

Then it follows from the switching principle (see Exercise 2.10) and induction that Y_n is a supermartingale. The result then follows from $EY_n \leq EY_0$.

Exercise 2.12   Let X_n, Y_n, Z_n be positive integrable and adapted to $\mathscr{F}_n$. Suppose

$$E(X_{n+1}|\mathscr{F}_n) \leq (1 + Z_n)X_n + Y_n$$

with $\sum Y_n < \infty$ and $\sum Z_n < \infty$ a.s. Prove that X_n converges a.s. to a finite limit.

Our final result is useful in reducing questions about submartingales to questions about martingales.

(2.12) Doob's Decomposition   Any submartingale X_n, $n \geq 0$, can be written in a unique way as $X_n = M_n + A_n$ where M_n is a martingale and A_n is an increasing sequence with $A_n \in \mathscr{F}_{n-1}$ for $n \geq 1$ and $A_0 = 0$.

Proof   We want $X_n = M_n + A_n$, $E(M_n|\mathscr{F}_{n-1}) = M_{n-1}$, and $A_n \in \mathscr{F}_{n-1}$. So we must have

$$E(X_n|\mathscr{F}_{n-1}) = E(M_n|\mathscr{F}_{n-1}) + E(A_n|\mathscr{F}_{n-1}) = M_{n-1} + A_n = X_{n-1} - A_{n-1} + A_n,$$

and it follows that

(a) $A_n - A_{n-1} = E(X_n|\mathscr{F}_{n-1}) - X_{n-1}$
(b) $M_n = X_n - A_n.$

Now $A_0 = 0$ and $M_0 = X_0$ by assumption, so we have A_n and M_n defined for all time and we have proved uniqueness. To check that our recipe works, we observe that $A_n - A_{n-1} \geq 0$, since X_n is a submartingale, and induction shows $A_n \in \mathscr{F}_{n-1}$. To see that M_n is a martingale, we observe

$$E(M_n|\mathscr{F}_{n-1}) = E(X_n - A_n|\mathscr{F}_{n-1}) = E(X_n|\mathscr{F}_{n-1}) - A_n = X_{n-1} - A_{n-1} = M_{n-1}$$

by (a) and (b), so the proof is complete.   □

Exercise 2.13 Let $X_n = \sum_{m \leq n} 1_{B_m}$ and suppose $B_n \in \mathscr{F}_n$. What is the Doob decomposition for X_n?

3 Examples

In this section we will apply the martingale convergence theorem to generalize the second Borel–Cantelli lemma and to study Polya's urn scheme, Radon–Nikodym derivatives, and branching processes. The four topics are independent of each other and are taken up in the order indicated. Our first result shows that martingales with bounded increments either converge or oscillate between $+\infty$ and $-\infty$.

(3.1) THEOREM   Let $X_1, X_2, \ldots$ be a martingale with $X_0 = 0$ and $|X_{n+1} - X_n| \leq M < \infty$. Let

$$C = \left\{ \lim_{n \to \infty} X_n \text{ exists and is finite} \right\}$$

$$D = \left\{ \limsup_{n \to \infty} X_n = +\infty \text{ and } \liminf_{n \to \infty} X_n = -\infty \right\}.$$

Then $P(C \cup D) = 1$.

Proof   Let $N = \inf\{n: X_n \leq -K\}$. $X_{n \wedge N}$ is a martingale with $X_{n \wedge N} \geq -K - M$ a.s., so applying (2.11) to $X_{n \wedge N} + K + M$ shows $\lim X_n$ exists on $\{N = \infty\}$. Letting $K \to \infty$, we see that the limit exists on $\{\liminf X_n > -\infty\}$. Applying the last conclusion to $-X_n$, we see that $\lim X_n$ exists on D^c, and the proof is complete.   □

Exercise 3.1   Let X_n, $n \geq 0$, be a submartingale with $\sup X_n < \infty$. Let $\xi_n = X_n - X_{n-1}$ and suppose $E(\sup \xi_n^+) < \infty$. Show that X_n converges a.s.

Exercise 3.2   Give an example of a martingale X_n with $\sup_n |X_n| < \infty$ and $P(X_n = a$ i.o.$) = 1$ for $a = -1, 0, 1$. With some care you can arrange things so that $P(X_n = 0) \to 1/2$ and $P(X_n = -1)$, $P(X_n = 1) \to 1/4$. This example shows that it is not enough to have $\sup |X_{n+1} - X_n| < \infty$ in (3.1) and that a martingale can converge in distribution without converging a.s. (or in probability).

(3.2) COROLLARY   *Second Borel–Cantelli Lemma, II*   Let $\mathscr{F}_n$, $n \geq 0$, be a filtration with $\mathscr{F}_0 = \{\varnothing, \Omega\}$ and A_n, $n \geq 1$, a sequence of events with $A_n \in \mathscr{F}_n$. Then

$$\{A_n \text{ i.o.}\} = \left\{ \sum_{n=1}^{\infty} P(A_n|\mathscr{F}_{n-1}) = \infty \right\}.$$

Proof If we let $X_0 = 0$ and

$$X_n = \sum_{m=1}^{n} 1_{A(m)} - P(A_m|\mathscr{F}_{m-1}) \quad \text{for } n \geq 1,$$

then X_n is a martingale with $|X_n - X_{n-1}| \leq 1$. Using the notation of (3.1) we have:

On C, $\displaystyle\sum_{n=1}^{\infty} 1_{A(n)} = \infty$ if and only if $\displaystyle\sum_{n=1}^{\infty} P(A_n|\mathscr{F}_{n-1}) = \infty$.

On D, $\displaystyle\sum_{n=1}^{\infty} 1_{A(n)} = \infty$ and $\displaystyle\sum_{n=1}^{\infty} P(A_n|\mathscr{F}_{n-1}) = \infty$.

Since $P(C \cup D) = 1$, the result follows. □

Exercise 3.3 Show that $\sum_{n=2}^{\infty} P(A_n | \bigcap_{m=1}^{n-1} A_m^c) = \infty$ implies $P(\bigcap_{m=1}^{\infty} A_m^c) = 0$.

Polya's urn scheme An urn contains r red and g green balls. At each time we draw a ball out, then replace it, and add c more balls of the color drawn. Let X_n be the fraction of green balls after the nth draw. It is easy to check that X_n is a martingale. Since $X_n \geq 0$, (2.11) implies that $X_n \to X_\infty$ a.s. To compute the distribution of the limit, we observe (a) the probability of getting green on the first m draws and then red on the next $l = n - m$ draws is

$$\frac{g}{g+r} \cdot \frac{g+c}{g+r+c} \cdots \frac{g+(m-1)c}{g+r+(m-1)c} \cdot \frac{r}{g+r+mc} \cdots \frac{r+(l-1)c}{g+r+(n-1)c},$$

and (b) any other outcome of the first n draws with m green balls drawn and l red balls drawn has the same probability, since the denominator remains the same and the numerator is permuted. Consider the special case $c = 1, g = 1, r = 1$. Let G_n be the number of green balls after the nth draw has been completed and the new ball has been added. It follows from (a) and (b) that

$$P(G_n = m + 1) = \binom{n}{m} \frac{m!(n-m)!}{n+1!} = 1/(n+1),$$

so X_∞ has a uniform distribution on $(0, 1)$. In general the distribution of X_∞ has density

$$\frac{\Gamma((g+r)/c)}{\Gamma(g/c)\Gamma(r/c)} (1-x)^{g/c-1} x^{r/c-1}.$$

This is the *beta distribution* with parameters g/c and r/c. In Example 4.2 we will see that the limit behavior changes drastically if we always add one ball of the opposite color.

Radon–Nikodym derivatives   Let μ be a finite measure and v a probability measure on $(\Omega, \mathcal{F})$. Let $\mathcal{F}_n \uparrow \mathcal{F}$ be σ-fields (i.e., $\sigma(\bigcup \mathcal{F}_n) = \mathcal{F}$). Let μ_n and v_n be the restrictions of μ and v to $\mathcal{F}_n$. Suppose $\mu_n \ll v_n$ for all n and let $X_n = d\mu_n/dv_n$.

(3.3) LEMMA X_n (defined on $(\Omega, \mathcal{F}, v)$) is a martingale w.r.t. $\mathcal{F}_n$.

Proof   We observe that by definition $X_n \in \mathcal{F}_n$. Let $A \in \mathcal{F}_n$. Since $X_n \in \mathcal{F}_n$ and v_n is the restriction of v to $\mathcal{F}_n$,

$$\int_A X_n \, dv = \int_A X_n \, dv_n.$$

Using the definition of X_n and Exercise 8.7 in the Appendix,

$$\int_A X_n \, dv_n = \mu_n(A) = \mu(A).$$

If $A \in \mathcal{F}_{m-1} \subset \mathcal{F}_m$, using the last result for $n = m$ and $n = m - 1$ gives

$$\int_A X_m \, dv = \mu(A) = \int_A X_{m-1} \, dv,$$

so $E(X_m | \mathcal{F}_{m-1}) = X_{m-1}$. $\square$

Example 3.1   Suppose $\mathcal{F}_n = \sigma(I_{k,n} : 0 \le k < K_n)$. In this case the condition $\mu_n \ll v_n$ is $v(I_{k,n}) = 0$ implies $\mu(I_{k,n}) = 0$, and the martingale $X_n = \mu(I_{k,n})/v(I_{k,n})$ on $I_{k,n}$ is an approximation to the Radon–Nikodym derivative. For a concrete example consider $\Omega = [0, 1)$, $I_{k,n} = [k2^{-n}, (k + 1)2^{-n})$ for $0 \le k < 2^n$, and $v =$ Lebesgue measure.

Exercise 3.4   (i) Check by direct computation that the X_n in Example 3.1 is a martingale. (ii) Show that if we drop the condition $\mu_n \ll v_n$ and set $X_n = 0$ when $v(I_{k,n}) = 0$, then $E(X_{n+1} | \mathcal{F}_n) \le X_n$.

Returning to the general case, we observe that X_n is a nonnegative martingale, so (2.11) implies that $X_n \to X_\infty$ v a.s. To have the limit defined on the whole space, let $X = \limsup X_n$.

(3.4) THEOREM $\mu(A) = \int_A X \, dv + \mu(A \cap \{X = \infty\})$.

Remark $\mu_r(A) \equiv \int_A X \, dv$ is a measure $\ll v$. Since (2.11) implies $v(X = \infty) = 0$, $\mu_s(A) \equiv \mu(A \cap \{X = \infty\})$ is singular w.r.t. v. Thus $\mu = \mu_r + \mu_s$ gives the Lebesgue decomposition of μ (see (8.5) in the Appendix), and $X_\infty = d\mu_r/dv$, v a.s. Here and in the proof we will need to keep track of the measure to which the a.s. refers.

Proof   Without loss of generality we can suppose μ is a probability measure. Let $\rho = (\mu + \nu)/2$, $\rho_n = (\mu_n + \nu_n)/2 =$ the restriction of ρ to $\mathcal{F}_n$. Let $Y_n = d\mu_n/d\rho_n$, $Z_n = d\nu_n/d\rho_n$. Y_n, $Z_n \geq 0$ and $Y_n + Z_n = 2$ (by Exercise 8.6 in the Appendix), so Y_n and Z_n are bounded martingales with limits Y and Z. As the reader can probably guess,

$$Y = d\mu/d\rho, \qquad Z = d\nu/d\rho. \tag{$*$}$$

It suffices to prove the first equality. From the proof of (3.3), if $A \in \mathcal{F}_m \subset \mathcal{F}_n$,

$$\mu(A) = \int_A Y_n \, d\rho \to \int_A Y \, d\rho$$

by the bounded convergence theorem. The last computation shows that

$$\mu(A) = \int_A Y \, d\rho \quad \text{for all } A \in \mathcal{G} = \bigcup \mathcal{F}_m.$$

$\mathcal{G}$ is a π-system, so the $\pi - \lambda$ theorem implies the equality is valid for all $A \in \mathcal{F} = \sigma(\mathcal{G})$ and $(*)$ is proved.   □

It follows from Exercise 8.8 in the Appendix that $X_n = Y_n/Z_n$ so $X = Y/Z$ ρ a.s. (Recall $X \equiv \limsup X_n$ and $Y + Z = 2\rho$ a.s., so $\rho(Y = 0, Z = 0) = 0$.) Let $W = (1/Z) \cdot 1_{(Z>0)}$. $(*)$ implies

$$\mu(A) = \int_A Y \, d\rho = \int_A YWZ \, d\rho + \int_A 1_{(Z=0)} Y \, d\rho = \int_A X \, d\nu + \int_A 1_{(X=\infty)} \, d\mu,$$

since $d\nu = Z \, d\rho$, $YW = X \nu$ a.s., $d\mu = Y \, d\rho$, and $\{X = \infty\} = \{Z = 0\}$ μ a.s.

Exercise 3.5   Apply (3.2) to Example 3.1 to get a "probabilistic" proof of the Radon–Nikodym theorem. Suppose $\mathcal{F}$ is *countably generated* (i.e., there is a sequence of sets A_n so that $\mathcal{F} = \sigma(A_n : n \geq 1)$). If ν is a σ-finite measure and $\mu \ll \nu$ (μ not necessarily σ-finite), then there is a function g so that $\mu(A) = \int_A g \, d\nu$.

Remark   Before you object to this as circular reasoning (the Radon–Nikodym theorem was used to define conditional expectation!), observe that the conditional expectations that are needed for Example 3.1 have elementary definitions.

Exercise 3.6   Let μ and ν be probability measures satisfying the hypotheses of (3.4) and let $X = \limsup d\mu_n/d\nu_n$. Then

$$\mu \ll \nu \quad \text{iff } \mu(X < \infty) = 1 \quad \text{iff } EX = 1$$
$$\mu \perp \nu \quad \text{iff } \mu(X = \infty) = 1 \quad \text{iff } EX = 0.$$

Here E denotes integration w.r.t. ν, and iff is short for "if and only if."

Let μ and ν be measures on sequence space $(\mathbb{R}^N, \mathscr{R}^N)$ that make the coordinates $X_n(\omega) = \omega_n$ independent. Let $F_n(x) = \mu(X_n \leq x)$, $G_n(x) = \nu(X_n \leq x)$. Suppose $F_n \ll G_n$ and let $q_n = dF_n/dG_n$. Let $\mathscr{F}_n = \sigma(X_m : m \leq n)$ and μ_n, ν_n the restrictions of μ and ν to $\mathscr{F}_n$.

$$X_n = \prod_{m=1}^{n} q_m(\omega_m) \quad \text{so } \{X < \infty\} = \left\{ \sum_{m=1}^{\infty} \log q_m \text{ converges} \right\}.$$

The Kolmogorov 0–1 law implies $\mu(X < \infty) \in \{0, 1\}$, so it follows from Exercise 3.6 that:

(3.5) Kakutani Dichotomy   Either $\mu \ll \nu$ or $\mu \perp \nu$.

Exercise 3.7   Suppose F_n, G_n are concentrated on $\{0, 1\}$ and have $F_n(0) = 1 - \alpha_n$, $G_n(0) = 1 - \beta_n$ with $0 < \varepsilon \leq \alpha_n, \beta_n \leq 1 - \varepsilon < 1$. Use the three series theorem to show that $\mu \ll \nu$ if and only if $\sum (\alpha_n - \beta_n)^2 < \infty$.

Hint $\sum \log x_m$ converges if and only if $\sum (x_m - 1)$ converges.

Exercise 3.8   Show that if $\sum \alpha_n < \infty$ and $\sum \beta_n = \infty$ in Exercise 3.7, then $\mu \perp \nu$. This shows that the condition $\sum (\alpha_n - \beta_n)^2 < \infty$ is not sufficient in general.

(3.6) THEOREM $\mu \ll \nu$ or $\mu \perp \nu$ according as $\prod_{m=1}^{\infty} \int \sqrt{q_m} \, dG_m > 0$ or $= 0$.

Proof   Jensen's inequality implies

$$\left(\int \sqrt{q_m} \, dG_m \right)^2 \leq \int q_m \, dG_m = 1,$$

so the infinite product is well defined. Let $X_n = \prod_{m \leq n} q_m(\omega_m)$, as above, and recall that $X_n \to X$ a.s. If the infinite product in the hypothesis is 0, then

$$\int X_n^{1/2} \, d\nu = \prod_{m=1}^{n} \int \sqrt{q_m} \, dG_m \to 0,$$

Fatou's lemma implies $X_n \to 0$ ν a.s., and (3.4) implies $\mu \perp \nu$. To prove the other direction, let $Y_n = X_n^{1/2}$,

$$E(Y_{n+k} - Y_n)^2 = E(X_{n+k} + X_n - 2X_n^{1/2}X_{n+k}^{1/2}) = 2\left(1 - \prod_{m=n+1}^{n+k} \int \sqrt{q_m} \, dG_m \right).$$

Since $|a - b| = |a^{1/2} - b^{1/2}| \cdot (a^{1/2} + b^{1/2})$,

$$E|X_{n+k} - X_n| = E((Y_{n+k} - Y_n)(Y_{n+k} + Y_n))$$

$$\leq E(Y_{n+k} - Y_n)^2 E(Y_{n+k} + Y_n)^2 \leq 4E(Y_{n+k} - Y_n)^2$$

by Cauchy–Schwarz and the fact $(a + b)^2 \le 4a^2 + 4b^2$. From the last two equations it follows that if the infinite product is > 0, then X_n converges to X in L^1, so $P(X = \infty) = 0$ and the desired result follows from (3.4).   □

Exercise 3.9   Use (3.6) to find a necessary and sufficient condition for $\mu \ll \nu$ in Exercise 3.7.

Branching Processes   Let ξ_i^n, $i, n \ge 0$, be i.i.d. nonnegative integer-valued random variables. Define a sequence Z_n, $n \ge 0$, by $Z_0 = 1$ and

$$Z_{n+1} = \begin{cases} \xi_1^n + \cdots + \xi_{Z_n}^n & \text{if } Z_n > 0 \\ 0 & \text{if } Z_n = 0. \end{cases}$$

Z_n is called a *Galton–Watson process*. The idea behind the definitions is that Z_n is the number of people in the nth generation and each member of the nth generation gives birth independently to an identically distributed number of children. $p_k = P(\xi_i^n = k)$ is called the *offspring distribution*.

(3.7) LEMMA   Let $\mathscr{F}_n = \sigma(\xi_i^m : 1 \le m \le n)$ and $\mu = E\xi_i^m$. Z_n / μ^n is a martingale w.r.t. $\mathscr{F}_n$.

Proof   Clearly $Z_n \in \mathscr{F}_n$.

$$E(Z_{n+1}|\mathscr{F}_n) = \sum_{k=1}^{\infty} E(Z_{n+1} 1_{\{Z_n = k\}}|\mathscr{F}_n)$$

by (1.1.a) and (1.1c). On $\{Z_n = k\}$, $Z_{n+1} = \xi_1^{n+1} + \cdots + \xi_k^{n+1}$, so the sum above is

$$\sum_{k=1}^{\infty} E((\xi_1^{n+1} + \cdots + \xi_k^{n+1}) 1_{\{Z_n = k\}}|\mathscr{F}_n) = \sum_{k=1}^{\infty} 1_{\{Z_n = k\}} E((\xi_1^{n+1} + \cdots + \xi_k^{n+1})|\mathscr{F}_n)$$

by (1.3). Since each ξ_j^{n+1} is independent of $\mathscr{F}_n$, the last expression

$$= \sum_{k=1}^{\infty} 1_{\{Z_n = k\}} k\mu = \mu Z_n.$$

Dividing both sides by μ^{n+1} now gives the desired result.   □

Remark   The reader should notice that in the proof of (3.7) we broke things down according to the value of Z_n to get rid of the random index. A simpler way of doing the last argument (that we will use in the future) is to use Exercise 1.1 to conclude that on $\{Z_n = k\}$,

$$E(Z_{n+1}|\mathscr{F}_n) = E(\xi_1^{n+1} + \cdots + \xi_k^{n+1}|\mathscr{F}_n) = k\mu = \mu Z_n.$$

Z_n / μ^n is a nonnegative martingale, so (2.11) implies $Z_n / \mu^n \to$ a limit a.s. We begin by identifying cases when the limit is trivial.

(3.8) THEOREM   If $\mu < 1$, then $Z_n = 0$ for all n sufficiently large, so $Z_n/\mu^n \to 0$.

Proof $E(Z_n/\mu^n) = E(Z_0) = 1$, so $E(Z_n) = \mu^n$. Now $Z_n \geq 1$ on $\{Z_n > 0\}$, so

$$P(Z_n > 0) \leq E(Z_n; Z_n > 0) = E(Z_n) = \mu^n \to 0$$

exponentially fast if $\mu < 1$. $\square$

The last answer should be intuitive: If each individual on the average gives birth to less than one child, the species will die out. The next result shows that after we exclude a trivial case, the same result holds when $\mu = 1$.

(3.9) THEOREM   If $\mu = 1$ and $P(\xi_i^m = 1) < 1$, then $Z_n = 0$ for all n sufficiently large.

Proof   When $\mu = 1$, Z_n is itself a nonnegative martingale. Since Z_n is integer-valued and by (2.11) converges to an a.s. finite limit Z_∞, we must have $Z_n = Z_\infty$ for large n. If $P(\xi_i^m = 1) < 1$ and $k > 0$, then $P(Z_n = k$ for all $n \geq N) = 0$ for any N, so we must have $Z_\infty \equiv 0$. $\square$

When $\mu \leq 1$, the limit of Z_n/μ^n is 0 because the branching process dies out. Our next step is to show:

(3.10) THEOREM   If $\mu > 1$, then $P(Z_n > 0$ for all $n) > 0$.

Proof   For $s \in [0, 1]$, let $\varphi(s) = \sum_{k \geq 0} p_k s^k$ where $p_k = P(\xi_i^m = k)$. φ is the *generating function* for the offspring distribution p_k. Differentiating and referring to (3.11) in Chapter 1 for the justification give

$$\varphi'(s) = \sum_{k=1}^{\infty} k p_k s^{k-1} \geq 0$$

$$\varphi''(s) = \sum_{k=2}^{\infty} k(k-1) p_k s^{k-2} \geq 0.$$

So φ is increasing and convex, and $\lim_{s \uparrow 1} \varphi'(s) = \sum_{k=1}^{\infty} k p_k = \mu$.

Our interest in φ stems from the following facts.

(a) If $\theta_m = P(Z_m = 0)$, then $\theta_m = \sum_{k=0}^{\infty} p_k (\theta_{m-1})^k$.

Proof of (a)   If $Z_1 = k$, then $Z_m = 0$ if and only if all k families die out in the remaining $m - 1$ units of time.

(b) If $\varphi'(1) = \mu > 1$, there is a unique $\rho < 1$ so that $\varphi(\rho) = \rho$.

Proof of (b) $\varphi(0) \geq 0$. $\varphi(1) = 1$ and $\varphi'(1) > 1$, so $\varphi(1 - \varepsilon) < 1 - \varepsilon$ for small ε. The last two observations imply the existence of a fixed point. To see it is unique observe that $\mu > 1$ implies $p_k > 0$ for some $k > 1$, so $\varphi''(\theta) > 0$ for $\theta > 0$. Since φ is strictly convex, it follows that if $\rho < 1$ is a fixed point, then $\varphi(x) < x$ for $x \in (\rho, 1)$.

(c) As $m \uparrow \infty$, $\theta_m \uparrow \rho$.

Proof of (c) $\theta_0 = 0$, $\varphi(\rho) = \rho$, and φ is increasing, so induction implies θ_m is increasing and $\theta_m \leq \rho$. Let $\theta_\infty = \lim \theta_m$. Taking limits in $\theta_m = \varphi(\theta_{m-1})$, we see $\theta_\infty = \varphi(\theta_\infty)$. Since $\theta_\infty \leq \rho$, it follows that $\theta_\infty = \rho$.

Combining (a)–(c) shows $P(Z_n = 0$ for some $n) = \lim \theta_n = \rho < 1$ and proves (3.10).   □

The last result shows that when $\mu > 1$, the limit of Z_n/μ^n has a chance of being nonzero. The best result on this question is due to Kesten and Stigum:

(3.11) THEOREM $W = \lim Z_n/\mu^n$ is not $\equiv 0$ if and only if $\sum p_k k \log k < \infty$.

For a proof, see Athreya and Ney (1972), pp. 24–29. In the next section we will show that $\sum k^2 p_k < \infty$ is sufficient for a nontrivial limit.

Exercise 3.10   Galton and Watson, who invented the process that bears their names, were interested in the survival of family names. Suppose each family has exactly three children but flips coins to determine their sex. In the 1800s only male children kept the family name, so following the male offspring leads to a branching process with $p_0 = 1/8, p_1 = 3/8, p_2 = 3/8, p_3 = 1/8$. Compute the survival probability.

4 Doob's Inequality, Convergence in L^p, $p > 1$

We begin by proving a consequence of (2.8).

(4.1) THEOREM   If X_n is a submartingale and N is a stopping time with $P(N \leq k) = 1$, then

$$EX_0 \leq EX_N \leq EX_k.$$

Proof   (2.8) implies $X_{N \wedge n}$ is a submartingale, so it follows that

$$EX_0 = EX_{N \wedge 0} \leq EX_{N \wedge k} = EX_N.$$

To prove the other inequality, let $K_n = 1_{\{N < n\}} = 1_{\{N \leq n-1\}}$. K_n is predictable, so (2.7) implies $(K \cdot X)_n = X_n - X_{N \wedge n}$ is a submartingale and it follows that

$$EX_k - EX_N = E(K \cdot X)_k \geq E(K \cdot X)_0 = 0.$$ □

Remark   Let S_n be a simple random walk with $S_0 = 1$ and let $N = \inf\{n: S_n = 0\}$. (See Example 2.1 for more details.) $ES_0 = 1 > 0 = ES_N$ so the first inequality need not hold for unbounded stopping times. In Section 7 we will give conditions that guarantee $EX_0 \leq EX_N$ for unbounded N.

We will see below that (4.1) is very useful. The first indication of this is:

Integrating the inequality in (4.2) gives

(4.3) L^p Maximum Inequality   If X_n is a submartingale and $\bar{X}_n = \max_{0 \le m \le n} X_m^+$ then for $p > 1$,

$$E(\bar{X}_n^p) \le (p/p - 1)^p E((X_n^+)^p).$$

Consequently, if Y_n is a martingale and $Y_n^* = \max_{0 \le m \le n} |Y_m|$,

$$E|Y_n^*|^p \le (p/p - 1)^p E(|Y_n|^p).$$

Proof   The second inequality follows by applying the first to $X_n = |Y_n|$. Using (4.2), Fubini's theorem, and a little calculus gives

$$E(\bar{X}_n^p) = \int_0^\infty p\lambda^{p-1} P(\bar{X}_n > \lambda) \, d\lambda$$

$$\le \int_0^\infty p\lambda^{p-1} \left(\lambda^{-1} \int X_n^+ 1_{\{\bar{X}_n \ge \lambda\}} \, dP \right) d\lambda$$

$$= \int_\Omega X_n^+ \int_0^{\bar{X}_n} p\lambda^{p-2} \, d\lambda \, dP$$

$$= \frac{p}{p-1} \int_\Omega X_n^+ \bar{X}_n^{p-1} \, dP.$$

If we let $q = (p/p - 1)$ be the exponent conjugate to p and apply Hölder's inequality ((3.3) in Chapter 1), we see that the above

$$\le q(E|X_n^+|^p)^{1/p} (E|\bar{X}_n|^p)^{1/q}.$$

If we divide both sides of the last inequality by $(E|\bar{X}_n|^p)^{1/q}$, we get (4.3). Unfortunately the laws of arithmetic do not allow us to divide by something that may be ∞. To remedy this difficulty, we observe that repeating the proof above with $\bar{X}_n$ replaced by $\bar{X}_n \wedge M$ gives

$$E[(\bar{X}_n \wedge M)^p] = \int_0^\infty p\lambda^{p-1} P(\bar{X}_n \wedge M > \lambda) \, d\lambda$$

$$\le \int_0^\infty p\lambda^{p-1} \left(\lambda^{-1} \int X_n^+ 1_{\{\bar{X}_n \wedge M \ge \lambda\}} \, dP \right) d\lambda$$

by (4.2), since $\{\bar{X}_n \wedge M \ge \lambda\} = \{\bar{X}_n \ge \lambda\}$ or $\varnothing$. Continuing to compute as before

(4.2) Doob's Inequality If X_m is a submartingale and $A = \{\max_{0 \le m \le n} X_m \ge \lambda\}$, th

$$\lambda P(A) \le EX_n 1_A \le EX_n^+.$$

Proof Let $N = \inf\{m: X_m \ge \lambda \text{ or } m \ge n\}$. Since $X_N \ge \lambda$ on A,

$$\lambda P(A) \le EX_N 1_A \le EX_n 1_A.$$

The second inequality follows from the fact that (4.1) implies $EX_N \le EX_n$, an have $\lambda \le X_n$ on A^c. The second inequality in (4.2) is trivial.

If we let $S_n = \xi_1 + \cdots + \xi_n$ where the ξ_i are independent and have $E\xi_n$ $\sigma_m^2 = E\xi_m^2 < \infty$, then (2.2) implies $X_n = S_n^2$ is a submartingale. If we let $\lambda = x$ apply (4.2) to X_n, we get Kolmogorov's inequality ((7.2) in Chapter 1):

$$P\left(\max_{1 \le m \le n} |S_n| \ge x\right) \le x^{-2} \, \text{var}(S_n).$$

Using martingales, one can also prove a converse to the last inequality that used instead of the central limit theorem in our proof of the necessity conditions in the three series theorem. (See Example 4.8 in Chapter 2.)

Exercise 4.1 Suppose in addition to the conditions introduced above that $|\xi_n|$ Let

$$A = \left\{\max_{1 \le m \le n} |S_k| \ge x\right\} \quad \text{and} \quad N = \inf\{m: |S_m| \ge x \text{ or } m \ge n\}.$$

Let $s_n^2 = \sum_{m \le n} \sigma_m^2$. Show that $S_n^2 - s_n^2$ is a martingale and use this to conclu

$$P(A^c) \le (x + K)^2/\text{var}(S_n).$$

Exercise 4.2 Let X_n be a martingale with $X_0 = 0$ and $EX_n^2 < \infty$. Show tha

$$P\left(\max_{m \le n} X_m \ge \lambda\right) \le EX_n^2/(EX_n^2 + \lambda^2).$$

Hint Use the fact that $(X_n + c)^2$ is a submartingale and optimize over c.

Exercise 4.3 Let X_n, $n \ge 0$, be a submartingale and $\lambda > 0$. Show that

$$\lambda P\left(\min_{m \le n} X_m \le -\lambda\right) \le E(X_n^+) - EX_0.$$

Exercise 4.4 Consider the special case $c = g = r = 1$ of Polya's urn dis the last section. Use (4.2) to show that $P(X_n \ge 3/4$ for some $n) \le 2/3$. The tion $P(X_2 = 3/4) = 1/3$ gives a lower bound on this probability. Can you either bound?

$$= \int_\Omega X_n^+ \int_0^{\bar{X}_n \wedge M} p\lambda^{p-2} \, d\lambda \, dP$$

$$= \frac{p}{p-1} \int_\Omega X_n^+ (\bar{X}_n \wedge M)^{p-1} \, dP$$

$$\le q(E|X_n^+|^p)^{1/p} (E|\bar{X}_n \wedge M|^p)^{1/q}.$$

$(E|\bar{X}_n \wedge M|^p)^{1/q}$ is finite, so we can divide by this to get

$$(E(\bar{X}_n \wedge M)^p)^{1/p} \le q(E|X_n^+|^p)^{1/p},$$

and letting $M \to \infty$ gives the desired result.   □

The next result is an extension of (4.3) to $p = 1$.

Exercise 4.5   Let X_n be a submartingale; let $\bar{X}_n = \sup_{m \le n} X_m^+$ and $\log^+ x = \max(\log x, 0)$.

$$E\bar{X}_n \le (1 - e^{-1})^{-1} \{1 + E(X_n^+ \log^+(X_n^+))\}.$$

Hint   Begin as in the proof of (4.3) but integrate from 1 to ∞. To deal with $X_n^+ \log \bar{X}_n$, use

$$a \log b \le a \log a + b/e \le a \log^+ a + b/e.$$

Remark   The last result is almost the best possible condition for $\sup|X_n| \in L^1$. Gundy has shown that if X_n is a positive martingale that has $X_{n+1} \le C X_n$ and $E X_0 \log^+ X_0 < \infty$, then $E(\sup X_n) < \infty$ implies $\sup E(X_n \log^+ X_n) < \infty$. For a proof see Neveu (1975), pp. 71–73.

From (4.3) we get the following:

(4.4) L^p Convergence Theorem   If X_n is a martingale with $\sup E|X_n|^p < \infty$ where $p > 1$, then $X_n \to X$ a.s. and in L^p.

Proof   From the martingale convergence theorem (2.10), it follows that $X_n \to X$ a.s. $|X_n|$ is a submartingale, so (4.3) implies $\sup|X_n| \in L^p$, and it follows from the dominated convergence theorem that $E|X_n - X|^p \to 0$.   □

Remark   Example 2.1 shows (4.4) is false when $p = 1$.

The most important special case of (4.4) is $p = 2$. To treat this case, the next two results are useful.

(4.5) *Orthogonality of Martingale Increments*   If X_n is a martingale with $EX_n^2 < \infty$ for all n, then for $l \le m < n$,

$$E((X_n - X_m)X_l) = 0.$$

Proof   Using (1.1f), (1.3), and the definition of a martingale,

$$E((X_n - X_m)X_l) = E[E((X_n - X_m)X_l|\mathscr{F}_m)] = E[X_l E((X_n - X_m)|\mathscr{F}_m)] = 0. \quad \square$$

(4.6) *Conditional Variance Formula*   Under the hypotheses of (4.5),

$$E((X_n - X_m)^2|\mathscr{F}_m) = E(X_n^2|\mathscr{F}_m) - X_m^2.$$

Remark   This is the conditional analogue of $E(X - EX)^2 = EX^2 - (EX)^2$ and is proved in exactly the same way.

Proof

$$E(X_n^2 - 2X_n X_m + X_m^2|\mathscr{F}_m) = E(X_n^2|\mathscr{F}_m) - 2X_m E(X_n|\mathscr{F}_m) + X_m^2$$

$$= E(X_n^2|\mathscr{F}_m) - 2X_m^2 + X_m^2. \quad \square$$

Exercise 4.6   Let X_n and Y_n be martingales with $EX_n^2 < \infty$ and $EY_n^2 < \infty$.

$$EX_n Y_n - EX_0 Y_0 = \sum_{m=1}^{n} E(X_m - X_{m-1})(Y_m - Y_{m-1}).$$

The next two results generalize (7.3) and (8.2) from Chapter 1. Let X_n, $n \ge 0$, be a martingale and let $\xi_n = X_n - X_{n-1}$ for $n \ge 1$.

Exercise 4.7   If $\sum_{m=1}^{\infty} E\xi_m^2 < \infty$, then $X_n \to X_\infty$ a.s. and in L^2.

Exercise 4.8   If $b_m \uparrow \infty$ and $\sum_{m=1}^{\infty} E\xi_m^2/b_m^2 < \infty$, then $X_n/b_n \to 0$ a.s. In particular if $E\xi_n^2 \le K < \infty$, then $X_n/n \to 0$ a.s.

Example 4.1   *Branching processes.* We continue the study begun at the end of the last section. Using the notation introduced there, we suppose $\mu = E(\xi_i^m) > 1$ and var$(\xi_i^m) = \sigma^2 < \infty$. Let $X_n = Z_n/\mu^n$. By (4.6),

$$E(X_n^2|\mathscr{F}_{n-1}) = X_{n-1}^2 + E((X_n - X_{n-1})^2|\mathscr{F}_{n-1}).$$

To compute the second term we observe

$$E((X_n - X_{n-1})^2|\mathscr{F}_{n-1}) = E((Z_n/\mu^n - Z_{n-1}/\mu^{n-1})^2|\mathscr{F}_{n-1})$$

$$= \mu^{-2n} E((Z_n - \mu Z_{n-1})^2|\mathscr{F}_{n-1}).$$

It follows from Exercise 1.1 that on $\{Z_{n-1} = k\}$,

$$E((Z_n - \mu Z_{n-1})^2 | \mathscr{F}_{n-1}) = E\left(\left(\sum_{i=1}^{k} \xi_i^n - \mu k\right)^2 \bigg| \mathscr{F}_{n-1}\right) = k\sigma^2 = Z_{n-1}\sigma^2.$$

Combining the last three equations gives

$$EX_n^2 = EX_{n-1}^2 + E(Z_{n-1}\sigma^2/\mu^{2n}) = EX_{n-1}^2 + \sigma^2/\mu^{n+1},$$

since $E(Z_{n-1}/\mu^{n-1}) = EZ_0 = 1$. Now $EX_0^2 = 1$, so $EX_1^2 = 1 + \sigma^2/\mu^2$, and induction gives

$$EX_n^2 = 1 + \sigma^2 \sum_{k=2}^{n+1} \mu^{-k}.$$

This shows $\sup EX_n^2 < \infty$, so $X_n \to X$ in L^2 and hence $EX_n \to EX$. $EX_n = 1$ for all n, so $EX = 1$ and X is not $\equiv 0$.

$X = 0$ on $\{Z_0 = 0$ for some $n\}$. To prove the converse, let $\theta = P(X = 0)$. By considering what happens at the first step we see

$$\theta = \sum_{k=0}^{\infty} p_k \theta^k$$

where $p_k = P(\xi_i^m = k)$. This shows $\theta = \varphi(\theta)$. Since $\theta < 1$, it follows (see (b) in the proof of (3.10)) that θ is the other fixed point—that is, $P(Z_n = 0$ for some $n)$.

For the rest of this section we will suppose:

X_n is a martingale with $X_0 = 0$ and $EX_n^2 < \infty$ for all n.

(2.4) implies X_n^2 is a submartingale. It follows from Doob's decomposition (2.12) and (4.6) that we can write $X_n^2 = M_n + A_n$ where M_n is a martingale and

$$A_n = \sum_{m=1}^{n} E(X_m^2 | \mathscr{F}_{m-1}) - X_{m-1}^2 = \sum_{m=1}^{n} E((X_m - X_{m-1})^2 | \mathscr{F}_{m-1}).$$

A_n is called the increasing process associated with X_n. A_n can be thought of as indicating the amount of action that has occurred by time n, and $A_\infty = \lim A_n$ as the total action. (4.8) and (4.9) describe the behavior of the martingale on $\{A_n < \infty\}$ and $\{A_n = \infty\}$, respectively. The key to the proof of the first result is the following:

(4.7) THEOREM $E(\sup_m |X_m|^2) \leq 4EA_\infty$.

Proof Applying the L^2 maximum inequality (4.3) to X_n gives

$$E\left(\sup_{0 \leq m \leq n} |X_m|^2\right) \leq 4EX_n^2 = 4EA_n,$$

since $EM_n = EM_0 = 0$. Using the monotone convergence theorem now gives the desired result.   □

(4.8) THEOREM $\lim_{n\to\infty} X_n$ exists and is finite a.s. on $\{A_\infty < \infty\}$.

Proof   Let $a > 0$. Since $A_{n+1} \in \mathcal{F}_n$, $N = \inf\{n: A_{n+1} > a^2\}$ is a stopping time. Applying (4.7) to $X_{N \wedge n}$ gives

$$E\left(\sup_n |X_{N \wedge n}|^2\right) \le 4a^2,$$

so the L^2 convergence theorem (4.4) implies that $\lim X_{N \wedge n}$ exists and is finite a.s. Since a is arbitrary, the desired result follows.   □

The next result is a refinement of the result in Exercise 4.8.

(4.9) THEOREM   Let f be an increasing function with $\int_{[0,\infty)} (1 + f(t))^{-2}\, dt < \infty$. Then $X_n/f(A_n) \to 0$ a.s. on $\{A_\infty = \infty\}$.

Proof $H_m = (1 + f(A_m))^{-1}$ is predictable, so (2.7) implies

$$Y_n \equiv (H \cdot X)_n = \sum_{m=1}^n (X_m - X_{m-1})/(1 + f(A_m)) \text{ is a martingale.}$$

If B_n is the increasing process associated with Y_n, then

$$B_{n+1} - B_n = E((Y_{n+1} - Y_n)^2|\mathcal{F}_n) = E\left(\left.\frac{(X_{n+1} - X_n)^2}{(1 + f(A_{n+1}))^2}\right|\mathcal{F}_n\right) = \frac{(A_{n+1} - A_n)}{(1 + f(A_{n+1}))^2},$$

since $f(A_{n+1}) \in \mathcal{F}_n$. Our hypotheses on f imply that

$$\sum_{n=0}^\infty \frac{(A_{n+1} - A_n)}{(1 + f(A_{n+1}))^2} \le \sum_{n=0}^\infty \int_{[A_n, A_{n+1})} (1 + f(t))^{-2}\, dt < \infty,$$

so it follows from (4.8) that $Y_n \to Y_\infty$, and the desired conclusion follows from Kronecker's lemma, (8.1) in Chapter 1.   □

From (4.9) we get a result due to Dubins and Freedman (1965) that extends (6.6) from Chapter 1 and (3.2) above.

(4.10) THEOREM   Suppose B_n is adapted to $\mathcal{F}_n$ and let $p_n = P(B_n|\mathcal{F}_{n-1})$. Then

$$\sum_{m=1}^n 1_{B(m)} \Bigg/ \sum_{m=1}^n p_m \to 1 \text{ a.s. on } \left\{\sum_{m=1}^n p_m = \infty\right\}.$$

Proof   Define a martingale by $X_0 = 0$ and $X_n - X_{n-1} = 1_{B(n)} - P(B_n|\mathcal{F}_{n-1})$ for $n \ge 1$, so that we have

$$\left(\sum_{m=1}^{n} 1_{B(m)} \Big/ \sum_{m=1}^{n} p_m\right) - 1 = X_n \Big/ \sum_{m=1}^{n} p_m.$$

The increasing process associated with X_n has

$$A_n - A_{n-1} = E((X_n - X_{n-1})^2|\mathscr{F}_{n-1}) = p_n - p_n^2 \le p_n.$$

On $\{A_\infty < \infty\}$, $X_n \to$ a finite limit by (4.8), so on $\{A_\infty < \infty\} \cap \{\sum_m p_m = \infty\}$,

$$X_n \Big/ \sum_{m=1}^{n} p_m \to 0.$$

$\{A_\infty = \infty\} = \{\sum_m p_m(1 - p_m) = \infty\} \subset \{\sum_m p_m = \infty\}$, so on $\{A_\infty = \infty\}$ the desired conclusion follows from (4.9) with $f(t) = t$. □

Exercise 4.9 Give an example with $A_\infty < \infty$ and $\sum p_m = \infty$ a.s.

Example 4.2 *Bernard Friedman's urn.* Consider a variant of Polya's urn (see Section 3) in which we add a balls of the color drawn and b balls of the opposite color where $a, b > 0$. We will show that if we start with g green balls and r red balls where g, $r > 0$, then the fraction of green balls $g_n \to 1/2$. Let G_n and R_n be the number of green and red balls after the nth draw is completed. Let B_n be the event that the nth ball drawn is green, and let D_n be the number of green balls drawn in the first n draws. It follows from (4.10) that

$$D_n \Big/ \sum_{m=1}^{n} g_{m-1} \to 1 \text{ a.s. on } \sum_{m=1}^{n} g_{m-1} = \infty,$$

which always holds since $g_m \ge (g + (a \wedge b)m)/(g + r + (a + b)m) \to (a \wedge b)/(a + b)$ as $m \to \infty$. At this point the argument breaks into three cases.

Case 1: $a = b$. In this case the result is trivial.

Case 2: $a > b$. We begin with the observation

$$g_{n+1} = \frac{G_{n+1}}{G_{n+1} + R_{n+1}} = \frac{g + aD_n + b(n - D_n)}{g + r + n(a + b)}. \qquad (*)$$

If $\limsup_{n\to\infty} g_n \le x$, then $\limsup_{n\to\infty} D_n/n \le x$ and (since $a < b$)

$$\limsup_{n\to\infty} g_{n+1} \le \frac{ax + b(1 - x)}{a + b} = \frac{b + (a - b)x}{a + b}.$$

The right-hand side is a linear function with slope <1 and fixed point at $1/2$, so starting with the trivial upper bound $x = 1$ and iterating, we conclude that $\limsup g_n \le 1/2$. Interchanging the roles of red and green shows $\limsup_{n\to\infty} g_n \ge 1/2$, and the result follows.

Case 3: $a < b$. The result is easier to believe in this case but harder to prove. The trouble is that when $b > a$ and $D_n \leq xn$, the right-hand side of $(*)$ is maximized by taking $D_n = 0$, so we need to also use the fact that if $r_n = $ the fraction of red balls, then

$$r_{n+1} = \frac{R_{n+1}}{G_{n+1} + R_{n+1}} = \frac{r + bD_n + a(n - D_n)}{g + r + n(a + b)}.$$

Combining this with the formula for g_{n+1}, it follows that if

$$\limsup_{n \to \infty} g_n \leq x \quad \text{and} \quad \limsup_{n \to \infty} r_n \leq y,$$

then

$$\limsup_{n \to \infty} g_n \leq \frac{a(1 - y) + by}{a + b} = \frac{a + (b - a)y}{a + b}$$

and

$$\limsup_{n \to \infty} r_n \leq \frac{bx + a(1 - x)}{a + b} = \frac{a + (b - a)x}{a + b}.$$

Starting with the trivial bounds $x = 1$, $y = 1$ and iterating (observe the two upper bounds are always the same), we conclude as in Case 2 that both limsups are $\leq 1/2$.

Remark   B. Friedman (1949) considered a number of different urn models. The result above is due to Freedman (1965), who proved the result by different methods. The proof above is due to Ornstein and comes from a remark in Freedman's paper.

(4.7) came from using (4.3). If we use (4.2) instead, we get a slightly better result.

(4.11) THEOREM $E(\sup_n |X_n|) \leq 3EA_\infty^{1/2}$.

Proof   As in the proof of (4.8) we let $a > 0$ and let $N = \inf\{n : A_{n+1} > a^2\}$.

$$P\left(\sup_m |X_m| > a\right) \leq P(N < \infty) + P\left(\sup_m |X_{N \wedge m}| > a\right).$$

Applying (4.2) to $|X_{N \wedge m}|^2$ gives

$$P\left(\sup_{m \leq n} |X_{N \wedge m}| > a\right) \leq a^{-2} E(X_{N \wedge n}^2) = a^{-2} E(A_{N \wedge n}) \leq a^{-2} E(A_\infty \wedge a^2).$$

Letting $n \to \infty$ in the last inequality, substituting the result in the first one, and integrating give

$$\int_0^\infty P\left(\sup_m |X_m| > a\right) da \leq \int_0^\infty P(A_\infty > a^2) \, da + \int_0^\infty a^{-2} E(A_\infty \wedge a^2) \, da.$$

Since $P(A_\infty > a^2) = P(A_\infty^{1/2} > a)$, the first integral is $EA_\infty^{1/2}$. For the second observe

$$\int_0^\infty a^{-2} E(A_\infty \wedge a^2) \, da = \int_0^\infty a^{-2} \int_0^{a^2} P(A_\infty > b) \, db \, da$$

$$= \int_0^\infty P(A_\infty > b) \int_{\sqrt{b}}^\infty a^{-2} \, da \, db$$

$$= \int_0^\infty b^{-1/2} P(A_\infty > b) \, db = 2EA_\infty^{1/2}$$

by (5.7) in Chapter 1. □

Exercise 4.10   Let $\xi_1, \xi_2, \ldots$ be i.i.d. with $E\xi_i = 0$ and $E\xi_i^2 < \infty$. Let $X_n = \xi_1 + \cdots + \xi_n$. (i) Use (4.11) to conclude that if $EN^{1/2} < \infty$, then $ES_N = 0$. (ii) By considering the case in which $\xi_i \in \{-1, 1\}$ and $T = \inf\{n: X_n = -1\}$, we see that $ET^{1/2} = \infty$. Recall from (3.4) in Chapter 3 that $P(T > t) \sim Ct^{-1/2}$, so the result in (i) is almost the best possible.

5   Uniform Integrability, Convergence in L^1

In this section we will give necessary and sufficient conditions for a martingale to converge in L^1. The key to this is the following definition. A collection of random variables $X_i, i \in I$, is said to be *uniformly integrable* if

$$\lim_{M \to \infty} \left(\sup_{i \in I} E(|X_i|; |X_i| > M)\right) = 0.$$

Our first result shows that uniformly integrable families can be very large.

(5.1) THEOREM   Given a probability space $(\Omega, \mathcal{F}_0, P)$ and an $X \in L^1$, then $\{E(X|\mathcal{F}): \mathcal{F} \text{ is a } \sigma\text{-field} \subset \mathcal{F}_0\}$ is uniformly integrable.

Proof   If A_n is a sequence of sets with $P(A_n) \to 0$, then the dominated convergence theorem implies $E(|X|; A_n) \to 0$. From the last result, it follows that if $\varepsilon > 0$, we can pick $\delta > 0$ so that if $P(A) \leq \delta$, then $E(|X|; A) \leq \varepsilon$. (If not, there are sets A_n with $P(A_n) \leq 1/n$ and $\liminf E(|X|; A_n) > 0$, a contradiction.)

Pick M large enough so that $E|X|/M \leq \delta$. Jensen's inequality and the definition of conditional expectation imply

$$E(|E(X|\mathcal{F})|; |E(X|\mathcal{F})| > M) \leq E(E(|X||\mathcal{F}); E(|X||\mathcal{F}) > M)$$

$$= E(|X|; E(|X||\mathcal{F}) > M),$$

since $\{E(|X|\,|\mathcal{F}) > M\} \in \mathcal{F}$. Using Chebyshev's inequality and recalling the definition of M, we have

$$P(E(|X|\,|\mathcal{F}) > M) \le E[E(|X|\,|\mathcal{F})]/M = E|X|/M \le \delta,$$

so by the choice of δ we have

$$E(|E(X|\mathcal{F})|; |E(X|\mathcal{F})| > M) \le \varepsilon \quad \text{for all } \mathcal{F}.$$

Since ε was arbitrary, the collection is uniformly integrable. $\qquad\qquad\square$

A common way to check uniform integrability is to use:

Exercise 5.1   Let φ be any function with $\varphi(x)/x \to \infty$ as $x \to \infty$; for example, $\varphi(x) = x^p$ with $p > 1$ or $\varphi(x) = x \log^+ x$. If $E\varphi(|X_i|) \le C$ for all $i \in I$, then $\{X_i : i \in I\}$ is uniformly integrable.

The relevance of uniform integrability to convergence in L^1 is explained by:

(5.2) THEOREM   If $X_n \to X$ a.s., then the following are equivalent:

   (i) $\{X_n : n \ge 0\}$ is uniformly integrable.
  (ii) $X_n \to X$ in L^1.
 (iii) $E|X_n| \to E|X| < \infty$.

Proof   (i) *implies* (ii)   Let $\varphi_M(x) = x$ if $|x| \le M$, $\varphi_M(x) = M$ if $x \ge M$, and $\varphi_M(x) = -M$ if $x \le -M$, and observe

$$E|X_n - X| \le E|\varphi_M(X_n) - \varphi_M(X)| + E(|X_n|; |X_n| > M) + E(|X|; |X| > M).$$

As $n \to \infty$, the first term $\to 0$ by the bounded convergence theorem. If $\varepsilon > 0$ and M is large, uniform integrability implies that the second term $\le \varepsilon$. To bound the third term, we observe that uniform integrability implies $\sup E|X_n| < \infty$, so Fatou's lemma implies $E|X| < \infty$, and by making M larger, we can make the third term $\le \varepsilon$. Combining the last three facts shows limsup $E|X_n - X| \le 2\varepsilon$, proving (ii).
   (ii) *implies* (iii) $|E|X_n| - E|X|| \le E||X_n| - |X|| \le E|X_n - X|$.
   (iii) *implies* (i)   Let $\psi_M(x) = x$ on $[0, M-1]$, $\psi_M = 0$ on $[M, \infty)$, and let ψ_M be linear on $[M-1, M]$. If M is large, $E|X| - E\psi_M(|X|) \le \varepsilon/2$. The bounded convergence theorem implies $E\psi_M(|X_n|) \to E\psi_M(|X|)$, so using (iii) we get that if $n \ge n_0$,

$$E(|X_n|; |X_n| > M) \le E|X_n| - E\psi_M(|X_n|) < \varepsilon.$$

By choosing M larger we can make $E(|X_n|; |X_n| > M) \le \varepsilon$ for $0 \le n < n_0$, so X_n is uniformly integrable. $\qquad\qquad\square$

We are now ready to state the main theorems of this section. We have already done all the work, so the proofs are short.

(5.3) THEOREM   For a submartingale the following are equivalent:

(i) It is uniformly integrable.
(ii) It converges in L^1.

Proof   (i) *implies* (ii)   Uniform integrability implies $\sup E|X_n| < \infty$, so the martingale convergence theorem implies $X_n \to X$ a.s., and (5.2) implies $X_n \to X$ in L^1.
   (ii) *implies* (i)   This follows from (5.2).   □

Remark   In the next proof and several times below we will need the following:

If $X_n \to X$ in L^1, then $E|X_n 1_A - EX 1_A| \leq E|X_n - X| \to 0$; that is, $X_n 1_A \to X 1_A$ in L^1.

(5.4) THEOREM   For a martingale the following are equivalent:

(i) It is uniformly integrable.
(ii) It converges in L^1.
(iii) There is an integrable random variable X so that $X_n = E(X|\mathcal{F}_n)$.

Proof   (i) *implies* (ii)   Since martingales are also submartingales, this follows from (5.3).
   (ii) *implies* (iii)   Let $X = \lim X_n$. If $m > n$, $E(X_m|\mathcal{F}_n) = X_n$, so if $A \in \mathcal{F}_n$, $E(X_n; A) = E(X_m; A)$. As $m \to \infty$, $X_m 1_A \to X 1_A$ in L^1, so we have $E(X_n; A) = E(X; A)$ for all $A \in \mathcal{F}_n$; that is, $X_n = E(X|\mathcal{F}_n)$.
   (iii) *implies* (i)   This follows from (5.1).   □

In the proof of (5.4) we saw that any uniformly integrable martingale can be written as $E(X|\mathcal{F}_n)$ where $X = \lim X_n$. The next result goes in the other direction.

(5.5) THEOREM   Suppose $\mathcal{F}_n \uparrow \mathcal{F}_\infty$; that is, $\mathcal{F}_\infty = \sigma(\mathcal{F}_n: n \geq 0)$. As $n \to \infty$,

$$E(X|\mathcal{F}_n) \to E(X|\mathcal{F}_\infty) \text{ a.s. and in } L^1.$$

Proof   (5.4) and the martingale convergence theorem imply that $X_n = E(X|\mathcal{F}_n)$ is uniformly integrable and converges a.s. and in L^1 to a limit X_∞. If $A \in \mathcal{F}_n$, then

$$\int_A X \, dP = \int_A E(X|\mathcal{F}_n) \, dP = \int_A X_n \, dP.$$

As $n \to \infty$, $X_n 1_A \to X_\infty 1_A$ in L^1, so

$$\int_A X \, dP = \int_A X_\infty \, dP \quad \text{for all } A \in \mathcal{F}_n.$$

Since X and X_∞ are integrable and $\bigcup_n \mathcal{F}_n$ is a π-system, the π-λ theorem implies that the last result holds for all $A \in \mathcal{F}_\infty$. Since $X_\infty \in \mathcal{F}_\infty$, it follows that $X_\infty = E(X|\mathcal{F}_\infty)$. $\square$

Exercise 5.2   Let $Z_1, Z_2, \ldots$ be i.i.d. with $E|Z_i| < \infty$, let θ be an independent r.v. with finite mean, and let $Y_i = Z_i + \theta$. If Z_i is normal $(0, 1)$, then in statistical terms we have a sample from a normal population with variance 1 and unknown mean. The distribution of θ is called the *prior distribution*, and $P(\theta \in \cdot\,|\, Y_1, \ldots, Y_n)$ is called the *posterior distribution* after n observations. Show that $E(\theta|\, Y_1, \ldots, Y_n) \to \theta$ a.s.

In the next three exercises $\Omega = [0, 1)$, $I_{k,n} = [k2^{-n}, (k+1)2^{-n})$, $\mathcal{F}_n = \sigma(I_{k,n}: 0 \le k < 2^n)$.

Exercise 5.3 f is said to be *Lipschitz continuous* if $|f(t) - f(s)| \le K|t - s|$ for $0 \le s$, $t < 1$. Show that $X_n = (f((k+1)2^{-n}) - f(k2^{-n}))/2^{-n}$ on $I_{k,n}$ defines a martingale, $X_n \to X_\infty$ a.s. and in L^1, and

$$f(b) - f(a) = \int_a^b X_\infty(\omega)\, d\omega.$$

Exercise 5.4   Suppose f is integrable on $[0, 1)$. $E(f|\mathcal{F}_n)$ is a step function and $\to f$ in L^1. Use this to show that if $\varepsilon > 0$, there is a function g continuous on $[0, 1]$ with $\int |f - g|\, dx < \varepsilon$. This approximation is simpler than the bare-hands approach we used in Exercise 4.3 of the Appendix.

Exercise 5.5   Let v be a probability measure with $v(I_{k,n}) > 0$ for all k, n. Suppose $d\mu = f\, dv$ and let $X_n = \mu(I_{k,n})/v(I_{k,n})$. Show that $X_n \to f$ a.s. and in $L^1(v)$.

An immediate consequence of (5.5) is:

(5.6) Lévy's 0–1 Law   If $A \in \mathcal{F}_\infty = \sigma(\mathcal{F}_n: n \ge 1)$, then $E(1_A|\mathcal{F}_n) \to 1_A$ a.s.

To steal a line from Chung (1974): "The reader is urged to ponder over the meaning of this result and judge for himself whether it is obvious or incredible." We will now argue for the two points of view.

"It is obvious." $1_A \in \mathcal{F}_\infty$ and $\mathcal{F}_n \uparrow \mathcal{F}_\infty$, so our best guess of 1_A given the information in $\mathcal{F}_n$ should approach 1_A (the best guess given $\mathcal{F}_\infty$).

"It is incredible." Let $X_1, X_2, \ldots$ be independent and suppose $A \in \mathcal{T}$, the tail σ-field. For each n, A is independent of $\mathcal{F}_n$, so $E(1_A|\mathcal{F}_n) = P(A)$. As $n \to \infty$, the left-hand side converges to 1_A a.s., so $P(A) \in \{0, 1\}$ and we have proved Kolmogorov's 0–1 law.

The last argument may not show that (5.6) is "too extraordinary and improbable to admit of belief," but this and other applications of (5.6) below show that it is a very useful result.

Exercise 5.6   Let X_n be r.v.'s taking values in $[0, \infty)$, and suppose that $X_n = 0$ implies $X_m = 0$ for all $m \ge n$. Let $D = \{X_n = 0 \text{ for some } n \ge 1\}$ and assume

$$P(D|X_1, \ldots, X_n) \geq \delta(x) > 0 \text{ a.s. on } \{X_n \leq x\}.$$

Use (5.6) to conclude that $P(D \cup \{\lim_{n \to \infty} X_n = \infty\}) = 1$.

Exercise 5.7 Let Z_n be a branching process with offspring distribution p_k (see the end of Section 3 for definitions). Use the last result to show that if $p_1 < 1$, then

$$P\left(\lim_{n \to \infty} Z_n = 0 \text{ or } \infty\right) = 1.$$

Exercise 5.8 Let $X_n \in [0, 1]$ be adapted to $\mathcal{F}_n$. Let $\alpha, \beta > 0$ with $\alpha + \beta = 1$ and suppose

$$P(X_{n+1} = \alpha + \beta X_n | \mathcal{F}_n) = X_n, \qquad P(X_{n+1} = \beta X_n | \mathcal{F}_n) = 1 - X_n.$$

Show $P(\lim_{n \to \infty} X_n = 0 \text{ or } 1) = 1$, and if $X_0 = \theta$, then $P(\lim_{n \to \infty} X_n = 1) = \theta$.

A more technical consequence of (5.5) is:

(5.7) Dominated Convergence Theorem for Conditional Expectations   Suppose $Y_n \to Y$ a.s. and $|Y_n| \leq Z$ for all n where $EZ < \infty$. If $\mathcal{F}_n \uparrow \mathcal{F}_\infty$, then

$$E(Y_n | \mathcal{F}_n) \to E(Y | \mathcal{F}_\infty) \text{ a.s.}$$

Proof   Let $W_N = \sup\{|Y_n - Y_m| : n, m \geq N\}$. $W_N \leq 2Z$, so $EW_N < \infty$. Using monotonicity (1.1b) and applying (5.5) to W_N give

$$\limsup_{n \to \infty} E(|Y_n - Y| | \mathcal{F}_n) \leq \lim_{n \to \infty} E(W_N | \mathcal{F}_n) = E(W_N | \mathcal{F}_\infty) = W_N.$$

The last result is true for all N and $W_N \downarrow 0$ as $N \uparrow \infty$, so Jensen's inequality implies

$$|E(Y_n | \mathcal{F}_n) - E(Y | \mathcal{F}_n)| \leq E(|Y_n - Y| | \mathcal{F}_n) \to 0 \text{ a.s. as } n \to \infty.$$

(5.5) implies $E(Y | \mathcal{F}_n) \to E(Y | \mathcal{F}_\infty)$ a.s., and the desired result follows from the last two conclusions and the triangle inequality.   □

Exercise 5.9 Show that if $\mathcal{F}_n \uparrow \mathcal{F}_\infty$ and $Y_n \to Y$ in L^1, then $E(Y_n | \mathcal{F}_n) \to E(Y | \mathcal{F}_\infty)$ in L^1.

Suppose $X_1, X_2, \ldots$ are uniformly integrable and $\to X$ a.s. (5.2) implies $X_n \to X$ in L^1, and combining this with Exercise 5.9 shows $E(X_n | \mathcal{F}) \to E(X | \mathcal{F})$ in L^1. We will now show that $E(X_n | \mathcal{F})$ need not converge a.s.

Example 5.1 Let $Y_1, Y_2, \ldots$ and $Z_1, Z_2, \ldots$ be independent r.v.'s with

$$P(Y_n = 1) = 1/n, \qquad P(Y_n = 0) = 1 - 1/n,$$

$$P(Z_n = n) = 1/n, \qquad P(Z_n = 0) = 1 - 1/n.$$

Let $X_n = Y_n Z_n$. $P(X_n > 0) = 1/n^2$, so the Borel–Cantelli lemma implies $X_n \to 0$ a.s. $E(X_n; |X_n| \geq 1) = n/n^2$, so X_n is uniformly integrable. Let $\mathcal{F} = \sigma(Y_1, Y_2, \ldots)$.

$$E(X_n|\mathcal{F}) = Y_n E(Z_n|\mathcal{F}) = Y_n E Z_n = Y_n.$$

Since $Y_n \to 0$ in L^1 but not a.s., the same is true for $E(X_n|\mathcal{F})$.

6   Backwards Martingales

A *backwards martingale* (some authors call them reversed) is a martingale indexed by the negative integers; that is, X_n, $n \leq 0$, satisfies

$$E(X_{n+1}|\mathcal{F}_n) = X_n \quad \text{for } n \leq -1,$$

where $\mathcal{F}_n$, $n \leq 0$, is an increasing sequence of σ-fields. Because the σ-fields decrease as $n \downarrow -\infty$, the convergence theory for backwards martingales is particularly simple.

(6.1) THEOREM $\lim_{n \to -\infty} X_n$ exists a.s. and in L^1.

Proof   Let U_n be the number of upcrossings of $[a, b]$ by $X_{-n}, \ldots, X_0$. The upcrossing inequality (2.9) implies $(b-a) EU_n \leq E(X_0 - a)^+$, so $EU_\infty < \infty$, and by the remark after the proof of (2.10) the limit exists a.s. $X_n = E(X_0|\mathcal{F}_n)$, so (5.1) implies X_n is uniformly integrable and (5.2) tells us that the convergence occurs in L^1.   □

Exercise 6.1   Show that if $X_0 \in L^p$, then the convergence in (6.1) occurs in L^p.

The next result identifies the limit in (6.1).

(6.2) THEOREM   If $X_{-\infty} = \lim_{n \to -\infty} X_n$ and $\mathcal{F}_{-\infty} = \bigcap \mathcal{F}_n$, then $X_{-\infty} = E(X_0|\mathcal{F}_{-\infty})$.

Proof   Clearly $X_{-\infty} \in \mathcal{F}_{-\infty}$. If $A \in \mathcal{F}_{-\infty} \subset \mathcal{F}_n$, then $\int_A X_n \, dP = \int_A X_0 \, dP$. (Recall $X_n = E(X_0|\mathcal{F}_n)$.) Since $X_n 1_A \to X_{-\infty} 1_A$ in L^1 (see the remark before (5.4)), it follows that $\int_A X_{-\infty} \, dP = \int_A X_0 \, dP$ for all $A \in \mathcal{F}_{-\infty}$, proving the desired conclusion.   □

The next result is (5.5) backwards.

(6.3) THEOREM   If $\mathcal{F}_n \downarrow \mathcal{F}_{-\infty}$ as $n \downarrow -\infty$ (i.e., $\mathcal{F}_{-\infty} = \bigcap \mathcal{F}_n$), then

$$E(Y|\mathcal{F}_n) \to E(Y|\mathcal{F}_{-\infty}) \quad \text{a.s. and in } L^1.$$

Proof $X_n = E(Y|\mathcal{F}_n)$ is a backwards martingale, so (6.1) and (6.2) imply that as $n \downarrow -\infty$, $X_n \to X_{-\infty}$ a.s. and in L^1, where $X_{-\infty} = E(X_0|\mathcal{F}_{-\infty}) = E(E(Y|\mathcal{F}_0)|\mathcal{F}_{-\infty}) = E(Y|\mathcal{F}_{-\infty})$.   □

Exercise 6.2   Prove the backwards analogue of (5.7). Suppose $Y_n \to Y_{-\infty}$ a.s. and $|Y_n| \leq Z$ a.s. where $EZ < \infty$. If $\mathcal{F}_n \downarrow \mathcal{F}_{-\infty}$, then $E(Y_n|\mathcal{F}_n) \to E(Y_{-\infty}|\mathcal{F}_{-\infty})$ a.s.

Even though the convergence theory for backwards martingales is easy, there are some nice applications.

Example 6.1 *The strong law of large numbers.* Let $\xi_1, \xi_2, \ldots$ be i.i.d. with $E|\xi_i| < \infty$. Let $S_n = \xi_1 + \cdots + \xi_n$, let $X_{-n} = S_n/n$, and let

$$\mathcal{F}_{-n} = \sigma(S_n, S_{n+1}, S_{n+2}, \ldots) = \sigma(S_n, \xi_{n+1}, \xi_{n+2}, \ldots).$$

To compute $E(X_{-n}|\mathcal{F}_{-n-1})$, we observe that if $j, k \le n + 1$, symmetry implies

$$E(\xi_j|\mathcal{F}_{-n-1}) = E(\xi_k|\mathcal{F}_{-n-1}).$$

So

$$E(\xi_{n+1}|\mathcal{F}_{-n-1}) = \frac{1}{n+1} \sum_{k=1}^{n+1} E(\xi_k|\mathcal{F}_{-n-1}) = \frac{1}{n+1} E(S_{n+1}|\mathcal{F}_{-n-1}) = S_{n+1}/(n+1),$$

and it follows that

$$E(X_{-n}|\mathcal{F}_{-n-1}) = E(S_{n+1}/n|\mathcal{F}_{-n-1}) - E(\xi_{n+1}/n|\mathcal{F}_{-n-1})$$

$$= S_{n+1}/n - S_{n+1}/n(n+1) = S_{n+1}/(n+1) = X_{-n-1}.$$

The last computation shows X_{-n} is a backwards martingale, so it follows from (6.1) and (6.2) that $\lim_{n\to\infty} S_n/n = E(X_1|\mathcal{F}_{-\infty})$. Since the events in $\mathcal{F}_{-\infty}$ are permutable, the Hewitt–Savage 0–1 law ((1.1) in Chapter 3) implies $\mathcal{F}_{-\infty}$ is trivial, so we have

$$\lim_{n\to\infty} S_n/n = E(X_1) \text{ a.s.}$$

Example 6.2 *The Ballot theorem.* Let $\{X_j, 1 \le j \le n\}$ be i.i.d. nonnegative integer-valued r.v.'s, and $S_k = X_1 + \cdots + X_k$. Then

$$P(S_j < j, 1 \le j \le n|S_n) \le (1 - S_n/n)^+ \tag{6.4}$$

with equality if $P(X_j \le 2) = 1$.

Proof The result is trivial when $S_n \ge n$, so suppose $S_n < n$. Computations in Example 6.1 show that $X_{-j} = S_j/j$ is a martingale w.r.t. $\mathcal{F}_{-j} = \sigma(S_j, \ldots, S_n)$. Let $T = \inf\{k \ge -n: X_k \ge 1\}$ and set $T = -1$ if the set is $\emptyset$. Let $G = \{S_j < j$ for $1 \le j \le n\} \subset \{T = -1\}$. $X_T = 0$ on G ($S_1 < 1$ implies $S_1 = 0$) and $X_T \ge 1$ on G^c, so

$$P(G|S_n) \le E(X_T|\mathcal{F}_{-n}) = X_{-n} = S_n/n.$$

When $P(X_j \le 2) = 1$, $X_T = 1$ on G^c and the $\le$ is an $=$. $\qquad\square$

To explain the name, let $X_1, X_2, \ldots, X_n$ be i.i.d. and take values 0 or 2 with probability 1/2 each. Interpreting 0's and 2's as votes for candidates A and B, we see that

$$G = \{S_j < j \text{ for } 1 \le j \le n\} = \{A \text{ leads } B \text{ throughout the counting}\},$$

so $P(G | A \text{ gets } r \text{ votes}) = (1 - 2r/n)^+$, the classical result ((3.2) in Chapter 3).

Exercise 6.3   Let f be an integrable function on $[0, 1)$. Extend f to $[0, 2)$ by setting $f(x + 1) = f(x)$, $x \in [0, 1)$. Let $X_0(\omega) = f(\omega)$. For $n \le -1$, let

$$X_n(\omega) = \sum_{m=1}^{2^{-n}} 2^n f(\omega + i2^n), \qquad \omega \in [0, 1)$$

and $\mathcal{F}_n = \sigma(\omega_{-n+j}; j \ge 1)$ where ω_k is the kth digit in the binary expansion of ω. Show that $X_n = E(X_0 | \mathcal{F}_n)$ and conclude that as $n \to -\infty$, $X_n(\omega) \to \int_{[0,1)} f \, dx$ a.s. Intuitively, $\omega \to X_n(\omega)$ is periodic with period $2^n (n \le 0)$, so the limit must be constant.

Example 6.3   *Hewitt–Savage 0–1 law.* Let $\mathcal{E}_n$ be the σ-field generated by random variables of the form $f(X_1, \ldots, X_n) g_1(X_{n+1}) \cdots g_m(X_{n+m})$ where f is symmetric; that is, if π is a permutation of $\{1, \ldots, n\}$, then $f(x_1, \ldots, x_n) = f(x_{\pi(1)}, \ldots, x_{\pi(n)})$. $\mathcal{E} = \bigcap \mathcal{E}_n$ is the exchangeable σ-field. The last definition applies to r.v.'s defined on an arbitrary measurable space $(\Omega, \mathcal{F})$. We leave it to the reader to check that this extends the definition given in Section 1 of Chapter 3 for the special probability space. In that section we proved:

Hewitt–Savage 0–1 Law.   If $X_1, X_2, \ldots$ are i.i.d. and $A \in \mathcal{E}$, then $P(A) \in \{0, 1\}$.

The key to the new proof is:

(6.5) LEMMA   Suppose $X_1, X_2, \ldots$ are i.i.d. and let

$$A_n(\varphi) = \frac{1}{(n)_k} \sum_i \varphi(X_{i_1}, \ldots, X_{i_k})$$

where the sum is over all sequences of distinct integers $1 \le i_1, \ldots, i_k \le n$ and

$$(n)_k = n(n - 1) \cdots (n - k + 1)$$

is the number of such sequences. If φ is bounded,

$$A_n(\varphi) \to E\varphi(X_1, \ldots, X_k) \text{ a.s.}$$

Proof $A_n(\varphi) \in \mathscr{E}_n$, so

$$A_n(\varphi) = E(A_n(\varphi)|\mathscr{E}_n) = E\left(\frac{1}{(n)_k} \sum_i \varphi(X_{i_1}, \ldots, X_{i_k})\middle|\mathscr{E}_n\right)$$

$$= E(\varphi(X_1, \ldots, X_k)|\mathscr{E}_n),$$

since all the terms in the sum are the same. (6.3) implies that

$$E(\varphi(X_1, \ldots, X_k)|\mathscr{E}_n) \to E(\varphi(X_1, \ldots, X_k)|\mathscr{E}).$$

We want to show that the limit is $E(\varphi(X_1, \ldots, X_k))$. The first step is to observe that there are $k(n-1)_{k-1}$ terms in $A_n(\varphi)$ involving X_1 and φ is bounded, so

$$\frac{1}{(n)_k} \sum_{1 \in i} \varphi(X_{i_1}, \ldots, X_{i_k}) \to 0$$

where $1 \in i$ indicates that the sum is over sequences that contain 1. This shows

$$E(\varphi(X_1, \ldots, X_k)|\mathscr{E}) \in \sigma(X_2, X_3, \ldots).$$

Repeating the argument for $2, 3, \ldots, k$ shows

$$E(\varphi(X_1, \ldots, X_k)|\mathscr{S}) \in \sigma(X_{k+1}, X_{k+2}, \ldots).$$

Intuitively if the conditional expectation of a r.v. is independent of the r.v., then

(a) $E(\varphi(X_1, \ldots, X_k)|\mathscr{E}) = E(\varphi(X_1, \ldots, X_k))$.

To show this, we prove:

(b) If $EX^2 < \infty$ and $E(X|\mathscr{G}) \in \mathscr{F}$ with X independent of $\mathscr{F}$, then $E(X|\mathscr{G}) = EX$.

Proof Let $Y = E(X|\mathscr{G})$. By independence $EXY = EX\,EY = (EY)^2$, since $EY = EX$. From the geometric interpretation of conditional expectation, (1.5), $E((X-Y)Y) = 0$, so $EY^2 = EXY = (EY)^2$ and the variance of Y is 0.

(a) holds for all bounded φ, so $\mathscr{E}$ is independent of $\mathscr{F}_k$. Since this holds for all k, $\mathscr{E}$ is independent of $\sigma(\mathscr{F}_k : k \geq 1) \supset \mathscr{E}$, and we get the usual 0–1 law punch line. If $A \in \mathscr{E}$, it is independent of itself and hence $P(A) = P(A \cap A) = P(A)P(A)$; that is, $P(A) \in \{0, 1\}$. $\square$

Exercise 6.4 Generalize the proof of (6.5) to conclude that if $X_1, X_2, \ldots$ are i.i.d. with $EX_i = \mu$ and $\text{var}(X_i) = \sigma^2 < \infty$, then

$$\binom{n}{2}^{-1} \sum_{1 \leq i < j \leq n} (X_i - X_j)^2 \to 2\sigma^2.$$

Example 6.4   *de Finetti's Theorem.* A sequence $X_1, X_2, \ldots$ is *exchangeable* if for each n and permutation π of $\{1, \ldots, n\}$, $(X_1, \ldots, X_n)$ and $(X_{\pi(1)}, \ldots, X_{\pi(n)})$ have the same distribution. Repeating the proof of (6.5) and using the notation introduced there show that for any exchangeable sequence:

$$A_n(\varphi) = E(A_n(\varphi)|\mathscr{E}_n) = E\left(\frac{1}{(n)_k} \sum_i \varphi(X_{i_1}, \ldots, X_{i_k}) \middle| \mathscr{E}_n\right)$$

$$= E(\varphi(X_1, \ldots, X_k)|\mathscr{E}_n),$$

since all the terms in the sum are the same. (6.3) implies that

$$A_n(\varphi) \to E(\varphi(X_1, \ldots, X_k)|\mathscr{E}). \qquad (6.6)$$

This time, however, we cannot hope to show that the limit is $E(\varphi(X_1, \ldots, X_k))$.

Let f and g be bounded functions on $\mathbb{R}^{k-1}$ and $\mathbb{R}$, respectively. If $\varphi(x_1, \ldots, x_k) = f(x_1, \ldots, x_{k-1})g(x_k)$, then

$$A_n(\varphi) = \frac{1}{(n)_k} \sum_i f(X_{i_1}, \ldots, X_{i_{k-1}})g(X_{i_k})$$

$$= \frac{n}{n-k+1} A_n(f)A_n(g) - \frac{1}{n-k+1} \sum_{j=1}^{k-1} A_n(\varphi_j)$$

where $\varphi_j(x_1, \ldots, x_{k-1}) = f(x_1, \ldots, x_{k-1})g(x_j)$. Applying (6.6) to φ, f, g, and all the φ_j gives

$$E(f(X_1, \ldots, X_{k-1})g(X_k)|\mathscr{E}) = E(f(X_1, \ldots, X_{k-1})|\mathscr{E})E(g(X_k)|\mathscr{E}).$$

It follows by induction that

$$E\left(\prod_{j=1}^k f_j(X_j) \middle| \mathscr{E}\right) = \prod_{j=1}^k E(f_j(X_j)|\mathscr{E}),$$

and we have shown:

(6.7) *de Finetti's Theorem*   Conditioned on $\mathscr{E}$, $X_1, X_2, \ldots$ are independent (and identically distributed, since they are exchangeable).

When the X_i take values in a nice space, there is a regular conditional distribution for $(X_1, X_2, \ldots)$ given $\mathscr{E}$, and the sequence can be represented as a mixture of i.i.d. sequences. Hewitt and Savage (1956) call the sequence *presentable* in this case. For the usual measure theoretic problems the last result is not valid when the

X_i take values in an arbitrary measure space. See Dubins and Freedman (1979) and Freedman (1980) for counterexamples.

The simplest special case of (6.7) occurs when the $X_i \in \{0, 1\}$. In this case:

(6.8) THEOREM   There is a probability distribution on $[0, 1]$ so that

$$P(X_1 = 1, \ldots, X_k = 1, X_{k+1} = 0, \ldots, X_n = 0) = \int_0^1 \theta^k (1 - \theta)^{n-k} \, dF(\theta).$$

(6.8) is useful for people concerned about the foundations of statistics (see Section 3.7 of Savage (1972)), since from the palatable assumption of symmetry one gets the powerful conclusion that the sequence is a mixture of i.i.d. sequences. (6.8) has been proved in a variety of different ways. See Feller, Vol. II (1971, pp. 228–229) for a proof that is related to the moment problem. Diaconis and Freedman (1980) have a nice proof that starts with the trivial observation that the distribution of a finite exchangeable sequence X_m, $1 \le m \le n$, has the form $p_0 H_{0,n} + \cdots + p_n H_{n,n}$ where $H_{m,n}$ is "drawing without replacement from an urn with m 1's and $n - m$ 0's." If $m \to \infty$ and $m/n \to p$, then $H_{m,n}$ approaches product measure with density p. (6.8) follows easily from this and one can get bounds on the rate of convergence.

Exercise 6.5   Prove directly from the definition that if $X_1, X_2, \ldots \in \{0, 1\}$ are exchangeable, then

$$P(X_1 = 1, \ldots, X_k = 1 | S_n = m) = \binom{n-k}{n-m} \bigg/ \binom{n}{m}.$$

Exercise 6.6   If $X_1, X_2, \ldots \in \mathbb{R}$ are exchangeable, then $E(X_1 X_2) \ge 0$.

Exercise 6.7 $X_1, X_2, \ldots$ are said to be conditionally i.i.d. given $\mathcal{G}$ (or i.i.d./$\mathcal{G}$) if

$$P(X_1 \le x_1, \ldots, X_n \le x_n | \mathcal{G}) = \prod_{m=1}^n P(X_1 \le x_m | \mathcal{G}).$$

(i) Show that this implies the X_i are exchangeable. (ii) (6.7) says that an exchangeable sequence is i.i.d./$\mathcal{E}$. Show that the limit of $A_n(\varphi)$ is measurable w.r.t. the tail σ-field $\mathcal{T}$, and use this to conclude that an exchangeable sequence is i.i.d./$\mathcal{T}$.

7 Optional Stopping Theorems

In this section we will prove a number of results that allow us to conclude that if X_n is a submartingale and $M \le N$ are stopping times, then $EX_M \le EX_N$. Example 2.1 shows that this is not always true, but (4.1) shows this is true if N is bounded, so our attention will be focused on the case of unbounded N.

(7.1) THEOREM   If X_n is a uniformly integrable submartingale, then for any stopping time N, $X_{N \wedge n}$ is uniformly integrable.

Proof   We begin by observing X_n^+ is a submartingale, so (4.1) implies $EX_{N \wedge n}^+ \leq EX_n^+$. Since X_n^+ is uniformly integrable, (5.3) implies X_n^+ converges in L^1, and it follows that

$$\sup_n EX_{N \wedge n}^+ \leq \sup_n EX_n^+ < \infty.$$

Using the martingale convergence theorem (2.10) now gives $X_{N \wedge n} \to X_N$ a.s. (here $X_\infty = \lim X_n$) and $E|X_N| < \infty$. With this established, the rest is easy, since

$$E(|X_{N \wedge n}|; |X_{N \wedge n}| > K) = E(|X_N|; |X_N| > K, N \leq n) + E(X_n|; |X_n| > K, N > n)$$

$$\leq E(|X_N|; |X_N| > K) + E(|X_n|; |X_n| > K)$$

and X_n is uniformly integrable.   □

From the last computation in the proof of (7.1) we get:

(7.2) COROLLARY   If $E|X_N| < \infty$ and $X_n 1_{(N>n)}$ is uniformly integrable, then $X_{N \wedge n}$ is uniformly integrable.

From (7.1) we immediately get:

(7.3) THEOREM   If X_n is a uniformly integrable submartingale, then for any stopping time $N \leq \infty$, we have $EX_0 \leq EX_N \leq EX_\infty$ where $X_\infty = \lim X_n$.

Proof   (4.1) implies $EX_0 \leq EX_{N \wedge n} \leq EX_n$. Letting $n \to \infty$ and observing that (7.1) and (5.3) imply $X_{N \wedge n} \to X_N$ and $X_n \to X_\infty$ in L^1 give the desired result.   □

From (7.3) we get the following useful corollary.

(7.4) *The Optional Stopping Theorem*   If $L \leq M$ are stopping times and $Y_{M \wedge n}$ is a uniformly integrable submartingale, then $EY_L \leq EY_M$.

Proof   Use the inequality $EX_N \leq EX_\infty$ in (7.3) with $X_n = Y_{M \wedge n}$ and $N = L$.   □

The last result is the one we use the most (usually with $L = 0$). (7.2) is useful in checking the hypothesis. A typical application is the following generalization of Wald's equation, (1.6) in Chapter 3.

(7.5) THEOREM   Suppose X is a submartingale and $E(|X_{n+1} - X_n||\mathscr{F}_n) \leq B$ a.s. If $EN < \infty$, then $X_{N \wedge n}$ is uniformly integrable and hence $EX_N \geq EX_0$.

Remark   As usual, using the last result twice shows that if X is a martingale, then $EX_N = EX_0$. To recover Wald's equation let S_n be a random walk, let $\mu = E(S_n - S_{n-1})$, and apply the martingale result to $X_n = S_n - n\mu$.

Proof   We begin by observing that

$$|X_{N \wedge n}| \leq |X_0| + \sum_{m=0}^{\infty} |X_{m+1} - X_m| 1_{(N>m)}.$$

To prove uniform integrability, it suffices to show that the right-hand side has finite expectation, for then $|X_{N \wedge n}|$ is dominated by an integrable r.v. Now, $\{N > m\} \in \mathcal{F}_m$, so

$$E(|X_{m+1} - X_m|; N > m) = E(E(|X_{m+1} - X_m||\mathcal{F}_m); N > m) \leq BP(N > m)$$

and

$$E \sum_{m=0}^{\infty} |X_{m+1} - X_m| 1_{(N>m)} \leq B \sum_{m=0}^{\infty} P(N > m) = BEN < \infty. \qquad \square$$

Before we delve further into applications, we pause to prove one last stopping theorem that does not require uniform integrability.

(7.6) THEOREM   If X_n is a nonnegative supermartingale and $N \leq \infty$ is a stopping time, then $EX_0 \geq EX_N$ where $X_\infty = \lim X_n$ (which exists by (2.11)).

Proof   By (4.1) $EX_0 \geq EX_{N \wedge n}$. The monotone convergence theorem implies

$$E(X_N; N < \infty) = \lim_{n \to \infty} E(X_N; N \leq n)$$

and Fatou's lemma implies

$$E(X_N; N = \infty) \leq \liminf_{n \to \infty} E(X_N; N > n).$$

Adding the last two lines and using our first observation

$$EX_N \leq \liminf_{n \to \infty} EX_{N \wedge n} \leq EX_0. \qquad \square$$

Exercise 7.1   If $X_n \geq 0$ is a supermartingale, then $P(\sup X_n > \lambda) \leq EX_0/\lambda$.

Exercise 7.2   Suppose X_n is a martingale with $X_0 > 0$ and $\inf X_n < 0$ a.s. Show that $E(\sup X_n) = \infty$.

We will now apply the results above to random walks. In all the exercises below, $\xi_1, \xi_2, \ldots$ are i.i.d., $S_n = \xi_1 + \cdots + \xi_n$, and $\mathcal{F}_n = \sigma(\xi_1, \ldots, \xi_n)$. We will pay special attention to the asymmetric simple random walk in which $P(\xi_i = 1) = p$ and $P(\xi_i = -1) = q \equiv 1 - p$.

Exercise 7.3   Let S_n be an asymmetric simple random walk with $p > 1/2$ and let $\varphi(x) = (1 - p/p)^x$. Show that $\varphi(S_n)$ is a martingale. Use this to conclude that if $T_x = \inf\{n: S_n = x\}$, then $P(T_a < T_b) = (\varphi(b) - \varphi(0))/(\varphi(b) - \varphi(a))$ for $a < 0 < b$. Let $b \to \infty$ to conclude that $P(T_a < \infty) = (1 - p/p)^{-a}$.

Exercise 7.4   Let S_n be an asymmetric simple random walk with $p > 1/2$, and let $T = \inf\{n: S_n = 1\}$. (i) Use the fact that $X_n = S_n - (p - q)n$ is a martingale to show that $ET = 1/(2p - 1)$. (ii) Let $\sigma^2 = 1 - (p - q)^2$. Use the fact that $X_n = (S_n - (p - q)n)^2 - \sigma^2 n$ is a martingale to show that $\text{var}(T) = (1 - (p - q)^2)/(p - q)^3$.

Hint   Use Exercise 7.3 to check that $X_{T \wedge n}$ is dominated by an integrable r.v.

Exercise 7.5   Let S_n be a symmetric simple random walk starting at 0, and let $T = \inf\{n: S_n \notin (-a, a)\}$ where a is an integer. (i) Use the fact that $S_n^2 - n$ is a martingale to show that $ET = a^2$.

Hint   Use the optional stopping theorem at $T \wedge n$ and let $n \to \infty$.
    (ii) Define a sequence of constants by $c_0 = 0$ and $c_{n+1} - c_n = 6n + 5$ for $n \geq 0$. Show that $S_n^4 - 6nS_n^2 + c_n$ is a martingale and use this to compute ET^2.

Exercise 7.6   Suppose $\varphi(\theta) = E \exp(\theta\xi_1) < \infty$ for $\theta \in (-\delta, \delta)$, and let $\psi(\theta) = \log \varphi(\theta)$. (i) $X_n = \exp(\theta S_n - n\psi(\theta))$ is a martingale. (ii) If ξ_i is not constant, φ is strictly convex. Use this result to show that $X_n \to 0$ a.s.
Hint   Use the martingales for θ and for $\theta/2$.

Exercise 7.7   Let $A \in \mathscr{F}_{n-1}$. The result in part (i) of the last exercise implies

$$E(\exp(\theta S_n - n\psi(\theta)); A) = E(\exp(\theta S_{n-1} - (n - 1)\psi(\theta)); A).$$

Differentiating the last equality with respect to θ and setting $\theta = 0$ give other martingales. Do this twice to find martingales that generalize the ones in Exercises 7.4 and 7.5. If you can bear to take four derivatives, you will get a generalization of the one in part (ii) of Exercise 7.5.

Exercise 7.8   Let S_n be an asymmetric simple random walk and suppose $p \geq 1/2$. Let $T = \inf\{n: S_n = 1\}$. We know that $P(T < \infty) = 1$. Use the martingale of Exercise 7.6 to conclude that if $\theta > 0$, then $1 = e^\theta E\varphi(\theta)^{-T}$ where $\varphi(\theta) = pe^\theta + qe^{-\theta}$ and $q = 1 - p$. Then change variables to get $Es^T = (1 - \{1 - 4pqs^2\}^{1/2})/2qs$.

Exercise 7.9   Suppose $\varphi(\theta_o) = E \exp(\theta_o\xi_1) = 1$ for some $\theta_o < 0$. It follows from the result in Exercise 7.6 that $X_n = \exp(\theta_o S_n)$ is a martingale. Let $T = \inf\{n: S_n \notin (a, b)\}$ and $Y_n = X_{n \wedge T}$. Use (7.5) to conclude that $EX_T = 1$ and $P(S_T \leq a) \leq \exp(-\theta_o a)$.

Exercise 7.10   Let $P(\xi_i = k - 1) = p^k(1 - p)$ for $k \geq 0$ where $p > 1/2$, and let $T = \inf\{n: S_n \notin (a, b)\}$ where $a < 0 < b$ are integers. Use the martingale of Exercise 7.9 to compute the distribution of S_T.

Exercise 7.11   Let S_n be the total assets of an insurance company at the end of year n. In year n, premiums totaling $c > 0$ are received and claims ζ_n are paid where ζ_n is normal(μ, σ^2) and $\mu < c$. The company is ruined if its assets drop to 0 or less. Show that

$$P(\text{ruin}) \le \exp(-2(c - \mu)S_0/\sigma^2).$$

Exercise 7.12   Suppose $E\xi_i = -\mu < 0$ and $\text{var}(\xi_i) = \sigma^2 > 0$. Let $\alpha = \mu/\sigma^2$ and

$$f(x) = \begin{cases} (1 + \alpha(z - x))^{-1} & \text{if } x < z \\ 1 & \text{if } x \ge z. \end{cases}$$

Show that $f(S_n)$ is a supermartingale and use this to conclude

$$P(\max S_n \ge z) \le 1/(1 + \alpha z).$$

Hint   For $x < z$, let $g_x(y) = f(x) + f'(x)(y - x) + \alpha f'(x)(y - x)^2$. Notice that $f'(x) = \alpha f(x)^2$, $f''(x) < 2\alpha f'(x)$, $g_x(z) = 1$, and use this to conclude $g_x(y) \ge f(y)$.

5 Markov Chains

The main object of study in this chapter is (temporally homogeneous) Markov chains on a countable state space S. That is, a sequence of r.v.'s X_n, $n \geq 0$, with

$$P(X_{n+1} = j | \mathscr{F}_n) = p(X_n, j)$$

where $\mathscr{F}_n = \sigma(X_0, \dots, X_n)$, $p(i, j) \geq 0$, and $\sum_j p(i, j) = 1$. The theory focuses on the asymptotic behavior of $p^n(i, j) \equiv P(X_n = j | X_0 = i)$. The basic results are that

$$\lim_{n \to \infty} \frac{1}{n} \sum_{m=1}^{n} p^m(i, j) \qquad \text{exists always,} \tag{5.2}$$

and under a mild assumption called aperiodicity,

$$\lim_{n \to \infty} p^n(i, j) \qquad \text{exists.} \tag{5.5}$$

In nice situations (i.e., X_n is irreducible and positive recurrent) the limits in (5.2) and (5.5) are a probability distribution that is independent of the starting state i. In words, the chain converges to equilibrium as $n \to \infty$. One of the attractions of Markov chain theory is that these powerful conclusions come out of assumptions that are satisfied in a large number of cases.

1 Definitions and Examples

We begin with a very general definition and then gradually specialize to the situation described in the introductory paragraph. Let $(S, \mathscr{S})$ be a measurable space. A sequence of random variables taking values in S is said to be a *Markov chain* with respect to a filtration $\mathscr{F}_n$ if $X_n \in \mathscr{F}_n$ and for all $B \in \mathscr{S}$,

$$P(X_{n+1} \in B | \mathscr{F}_n) = P(X_{n+1} \in B | X_n).$$

In words, given the present, the rest of the past is irrelevant for predicting the location of X_{n+1}. As usual we turn to an example to illustrate the definition.

Example 1.1   *Random walk.* Let $X_0, \xi_1, \xi_2, \ldots \in \mathbb{R}^d$ be independent and let $X_n = X_0 + \xi_1 + \cdots + \xi_n$. X_n is a Markov chain with respect to $\mathscr{F}_n = \sigma(X_0, X_1, \ldots, X_n)$. (As in the case of a martingale, this is the smallest filtration we can get away with and the one we will usually use.) To prove our claim we want to show that if μ_j is the distribution of ξ_j, then

$$P(X_{n+1} \in B | \mathscr{F}_n) = \mu_{n+1}(B - X_n) = P(X_{n+1} \in B | X_n). \qquad (*)$$

To prove the first equality in $(*)$, let $A = \{X_0 \in B_0, \xi_1 \in B_1, \ldots, \xi_n \in B_n\}$ and let μ_0 be the distribution of X_0. Now

$$\int_A 1_{(X_{n+1} \in B)} \, dP = P(X_0 \in B_0, \xi_1 \in B_1, \ldots, \xi_n \in B_n, X_{n+1} \in B)$$

$$= \int_{B_0} \mu_0(dy_0) \cdots \int_{B_n} \mu_n(dy_n) \mu_{n+1}(B - (y_0 + \cdots + y_n))$$

$$= \int_A \mu_{n+1}(B - X_n) \, dP.$$

The collection of sets for which the last formula holds is a λ-system, and the collection for which it has been proved is a π-system, so it follows from the $\pi - \lambda$ theorem ((4.2) in Chapter 1) that the first equality in $(*)$ is true for all $A \in \mathscr{F}_n$.

To prove the second equality in $(*)$, we observe:

(1.1) LEMMA   If $\mathscr{F} \subset \mathscr{G}$ and $E(X | \mathscr{G}) \in \mathscr{F}$, then $E(X | \mathscr{F}) = E(X | \mathscr{G})$.

Proof   Since $\mathscr{F} \subset \mathscr{G}$, (1.2) in Chapter 4 implies

$$E(X | \mathscr{F}) = E(E(X | \mathscr{G}) | \mathscr{F}) = E(X | \mathscr{G}),$$

since $E(X | \mathscr{G}) \in \mathscr{F}$.   □

Our second example will turn out to be the general case, so it is a little dishonest to call it an example. Our aim now is to take the rather abstract object defined at the beginning of the section and, without loss of generality, turn it into something more concrete and easier to work with. Once this is done we will be in a position to give five real examples. A function $p: S \times \mathscr{S} \to \mathbb{R}$ is said to be a *transition probability* if:

(i) for each $x \in S$, $p(x, \cdot)$ is a probability measure on $(S, \mathscr{S})$, and
(ii) for each $A \in \mathscr{S}$, $x \to p(x, A)$ is a measurable function.

Example 1.2   Suppose $(S, \mathcal{S})$ is a nice space. Given a sequence of transition probabilities p_n and an *initial distribution* μ on $(S, \mathcal{S})$, we can define a consistent set of finite-dimensional distributions by

$$P(X_j \in B_j, 0 \le j \le n) = \int_{B_0} \mu(dx_0) \int_{B_1} p_1(x_0, dx_1) \cdots \int_{B_n} p_n(x_{n-1}, dx_n). \quad (1.2)$$

Since we have assumed that $(S, \mathcal{S})$ is nice, Kolmogorov's theorem allows us to construct a probability measure P_μ on *sequence space* $(S^{\{0,1,\cdots\}}, \mathcal{S}^{\{0,1,\cdots\}})$ so that the coordinate maps $X_n(\omega) = \omega_n$ have the desired distributions.

   Our next step is to check that X_n is a Markov chain (with respect to $\mathcal{F}_n = \sigma(X_0, X_1, \ldots, X_n)$). To prove this, we let $A = \{X_0 \in B_0, X_1 \in B_1, \ldots, X_n \in B_n\}$, $B_{n+1} = B$, and observe that

$$\int_A 1_{(X_{n+1} \in B)} \, dP = P(X_0 \in B_0, X_1 \in B_1, \ldots, X_n \in B_n, X_{n+1} \in B)$$

$$= \int_{B_0} \mu(dx_0) \int_{B_1} p_1(x_0, dx_1) \cdots \int_{B_{n+1}} p_{n+1}(x_n, dx_{n+1})$$

$$= \int_A p_{n+1}(X_n, B) \, dP.$$

The collection of sets for which the last formula holds is a λ-system, and the collection for which it has been proved is a π-system, so it follows from the $\pi - \lambda$ theorem ((4.2) in Chapter 1) that the equality is true for all $A \in \mathcal{F}_n$. This shows that

$$P(X_{n+1} \in B | \mathcal{F}_n) = p_{n+1}(X_n, B),$$

and it follows from (1.1) that X_n is a Markov chain.

   As mentioned earlier, this "example" is the general case. To prove this observe that if $(S, \mathcal{S})$ is nice, Exercise 1.20 in Chapter 4 shows that a "regular conditional distribution" exists; that is, there is a transition probability p_{n+1} so that

$$P(X_{n+1} \in B | X_n) = p_{n+1}(X_n, B).$$

It should not be hard to believe that the last equation and the Markov property imply that for bounded measurable f,

$$E(f(X_{n+1}) | \mathcal{F}_n) = \int p_{n+1}(X_n, dy) f(y). \quad (1.3)$$

The desired conclusion is a consequence of the next result. Let $\mathcal{H} =$ the collection of bounded functions for which the identity holds.

(1.4) Monotone Class Theorem   Let $\mathcal{A}$ be a π-system and let $\mathcal{H}$ be a collection of real-valued functions that satisfies: (i) If $A \in \mathcal{A}$, then $1_A \in \mathcal{H}$. (ii) If $f, g \in \mathcal{H}$, then $f + g$ and $cf \in \mathcal{H}$ for any real number c. (iii) If $f_n \in \mathcal{H}$ are nonnegative and increase to a bounded function f, then $f \in \mathcal{H}$. Then $\mathcal{H}$ contains all bounded functions measurable with respect to $\sigma(\mathcal{A})$.

Proof   (ii) and (iii) imply that $\mathcal{G} = \{A: 1_A \in \mathcal{H}\}$ is a λ-system, so by (i) and the $\pi - \lambda$ theorem ((4.2) in Chapter 1), $\mathcal{G} \supset \sigma(\mathcal{A})$. (ii) implies $\mathcal{H}$ contains all simple functions, and (iii) implies $\mathcal{H}$ contains all bounded measurable functions. $\square$

Returning to our main topic we observe that (1.3) implies

$$E\left(\prod_{m=0}^{n} f_m(X_m)\right) = EE\left(\prod_{m=0}^{n} f_m(X_m) \middle| \mathcal{F}_{n-1}\right)$$

$$= E\left(\prod_{m=0}^{n-1} f_m(X_m) \int p_n(X_{n-1}, dy) f_n(y)\right).$$

The last integral is a bounded measurable function of X_{n-1}, so it follows by induction that if μ is the distribution of X_0, then

$$E\left(\prod_{m=0}^{n} f_m(X_m)\right) = \int \mu(dx_0) f_0(x_0) \int p_1(x_0, dx_1) f_1(x_1) \cdots \int p_n(x_{n-1}, dx_n) f_n(x_n);$$
(1.5)

that is, the finite-dimensional distributions coincide with those given in (1.2).

Having reduced the general case to Example 1.2, we can and will suppose that the random variables X_n are the coordinate maps ($X_n(\omega) = \omega_n$) on sequence space

$$(\Omega_0, \mathcal{F}) = (S^{\{0,1,\ldots\}}, \mathcal{S}^{\{0,1,\ldots\}}).$$

We choose this representation because it gives us two advantages in investigating the Markov chain: (i) For each initial distribution μ we have a measure P_μ defined by (1.2) that makes X_n a Markov chain with $P_\mu(X_0 \in A) = \mu(A)$. (ii) We can have the shift operators θ_n defined in Section 1 of Chapter 3: $(\theta_n \omega)(m) = \omega_{m+n}$.

One final simplification, this time with loss of generality, is that from now on we will restrict our attention to the *temporally homogeneous* case—that is, the transition probability does not depend on n. Once we do this we can write the Markov property as

$$P(X_{n+1} \in A | \mathcal{F}_n) = p(X_n, A)$$

and the distribution of the chain is specified by giving its transition probability and initial distribution. Having decided on the framework in which we will investigate things, we can finally give some real examples. In each case S is a countable set and $\mathcal{S} = $ all subsets of S. Let $p(i, j) \geq 0$ and suppose $\sum_j p(i, j) = 1$ for all i. Intuitively

$$p(i, j) = P(X_{n+1} = j | X_n = i).$$

From $p(i, j)$ we can define a transition probability by

$$p(i, A) = \sum_{j \in A} p(i, j).$$

We will now give five concrete examples that will be our constant companions as the story unfolds. In each case we will just give the transition probability.

Example 1.3 *Branching processes. $S = \{0, 1, 2, \ldots\}$.*

$$p(i, j) = P\left(\sum_{m=1}^{i} \xi_m = j\right)$$

where $\xi_1, \xi_2, \ldots$ are i.i.d. nonnegative integer-valued random variables. In words, each of the i individuals at time n (or in generation n) gives birth to an independent and identically distributed number of offspring. For a more complete description, see Section 3 in Chapter 4.

Example 1.4 *Renewal chain. $S = \{0, 1, 2, \ldots\}$. Let $f_k \geq 0$ and $\sum_{k=1}^{\infty} f_k = 1$.*

$$p(0, j) = f_{j+1} \qquad \text{for} \quad j \geq 0$$

$$p(i, i - 1) = 1 \qquad \text{for} \quad i \geq 1$$

$$p(i, j) = 0 \qquad \text{otherwise.}$$

To explain the definition, let $\xi_1, \xi_2, \ldots$ be i.i.d. with $P(\xi_m = j) = f_j$, let $T_0 = i_0$, and for $k \geq 1$ let $T_k = T_{k-1} + \xi_k$. T_k is the time of the kth arrival in a renewal process that has its first arrival at time i_0. Let $Y_m = 1$ if $m \in \{T_0, T_1, T_2, \ldots\}$, and let $X_n = \inf\{m - n : m \geq n, Y_m = 1\}$. $Y_m = 1$ if a renewal occurs at time m, and X_n is the amount of time until the first renewal $\geq n$.

Example 1.5 *M/G/1 queue.* In this model customers arrive according to a Poisson process with rate λ. (M is for Markov and refers to the fact that in a Poisson process the number of arrivals in disjoint time intervals is independent.) Each customer requires an independent amount of service with distribution F. (G is for general service distribution. 1 indicates that there is one server.) Let X_n be the number of customers waiting in the queue at the time the nth customer enters service. When $X_0 = x$, the chain starts with x people waiting in line and customer 0 just beginning her service. If we let

$$a_k = \int_0^\infty e^{-\lambda t} \frac{(\lambda t)^k}{k!} \, dF(t)$$

be the probability that k customers arrive during a service time, then it is easy to see that X_n is a Markov chain with transition probability

$$p(0, k) = a_k$$

$$p(j, j - 1 + k) = a_k \quad \text{if } j \geq 1.$$

The formula for a_k is rather complicated and its exact form is not important, so we will simplify things by assuming only that $a_k > 0$ and $\sum_{k \geq 0} a_k = 1$.

Example 1.6   *Ehrenfest chain. $S = \{0, 1, \ldots, r\}$.*

$$p(k, k + 1) = (r - k)/r$$

$$p(k, k - 1) = k/r$$

$$p(i, j) = 0 \quad \text{otherwise.}$$

In words, there is a total of r balls in two urns, k in the first and $r - k$ in the second. We pick one of the r balls at random and move it to the other urn. Ehrenfest used this to model the division of air molecules between two chambers (of equal size and shape) that are connected by a small hole. For an interesting account of this chain, see Kac (1947a).

Example 1.7   *Birth and death chains. $S = \{0, 1, 2, \ldots\}$.* These chains are defined by the restriction $p(i, j) = 0$ when $|i - j| > 1$. The fact that these processes cannot jump over any integers makes it particularly easy to compute things for them.

For a Markov chain on a countable state space, (1.2) says

$$P_\mu(X_k = i_k, 0 \leq k \leq n) = \mu(i_0) \prod_{m=1}^{n} p(i_{m-1}, i_m).$$

When $n = 1$,

$$P_\mu(X_1 = j) = \sum_i \mu(i)p(i, j) = \mu p(j),$$

that is, the product of the row vector μ with the matrix p. When $n = 2$,

$$P_i(X_2 = k) = \sum_j p(i, j)p(j, k) = p^2(i, k),$$

that is, the second power of the matrix p. Combining the two formulas and generalizing,

$$P_\mu(X_n = j) = \sum_i \mu(i)p^n(i, j) = \mu p^n(j).$$

Exercise 1.1   Suppose $S = \{1, 2, 3\}$ and

$$p = \begin{pmatrix} .2 & 0 & .8 \\ .7 & .3 & 0 \\ 0 & .4 & .6 \end{pmatrix}.$$

Compute $p^2(1, 2)$ and $p^3(2, 3)$.

Exercise 1.2   Suppose $S = \{0, 1\}$ and

$$p = \begin{pmatrix} 1 - \alpha & \alpha \\ \beta & 1 - \beta \end{pmatrix}.$$

Use induction to show that

$$P_\mu(X_n = 0) = \beta/(\alpha + \beta) + (1 - \alpha - \beta)^n \{\mu(0) - \beta/(\alpha + \beta)\}.$$

Exercise 1.3   Let

$$p = \begin{pmatrix} 1/2 & 1/2 & 0 \\ 1/2 & 0 & 1/2 \\ 0 & 1/2 & 1/2 \end{pmatrix}, \qquad U = \begin{pmatrix} 1/\sqrt{3} & 1/\sqrt{2} & 1/\sqrt{6} \\ 1/\sqrt{3} & 0 & -2/\sqrt{6} \\ 1/\sqrt{3} & -1/\sqrt{2} & 1/\sqrt{6} \end{pmatrix},$$

$$D = \begin{pmatrix} 1 & 0 & 0 \\ 0 & 1/2 & 0 \\ 0 & 0 & -1/2 \end{pmatrix}.$$

Check that $p = UDU^t$ and $U^tU = I$ where t denotes transpose and I the identity matrix. Use this to compute all the powers of p.

Exercise 1.4   Let $\xi_0, \xi_1, \ldots$ be i.i.d. $\in \{H, T\}$, taking each value with probability $1/2$. Show that $X_n = (\xi_n, \xi_{n+1})$ is a Markov chain and compute its transition probability p. What is p^2?

Exercise 1.5   Let $\xi_1, \xi_2, \ldots$ be i.i.d. $\in \{1, 2, \ldots, N\}$, taking each value with probability $1/N$. Show that $X_n = |\{\xi_1, \ldots, \xi_n\}|$ is a Markov chain and compute its transition probability.

Exercise 1.6   Let $\xi_1, \xi_2, \ldots$ be i.i.d. $\in \{-1, 1\}$, taking each value with probability $1/2$. Let $S_0 = 0, S_n = \xi_1 + \cdots + \xi_n$, and $X_n = \max\{S_m : 0 \le m \le n\}$. Show that X_n is not a Markov chain.

Exercise 1.7   Let $X_n \in \{-1, 0, 1\}$ be a Markov chain and suppose $p(i, j) > 0$ for all i, j. Give a necessary and sufficient condition for $|X_n|$ to be a Markov chain.

Exercise 1.8   Let $\theta, U_1, U_2, \ldots$ be independent and suppose $P(U_i \le x) = x$ for $x \in (0, 1)$. Let $X_i = 1$ if $U_i \le \theta, = -1$ if $U_i > \theta$, and let $S_n = X_1 + \cdots + X_n$. In words, we first pick θ according to some distribution and then flip a coin with probability θ of heads to generate a random walk. Compute $P(X_{n+1} = 1|S_1, \ldots, S_n)$ and conclude S_n is a temporally inhomogeneous Markov chain. This is due to the fact that "S_n is a sufficient statistic for estimating θ."

2 Extensions of the Markov Property

If X_n is a Markov chain with transition probability p, then by definition

$$P(X_{n+1} \in B | \mathcal{F}_n) = p(X_n, B).$$

In this section we will prove two extensions of the last equality in which $\{X_{n+1} \in B\}$ is replaced by a bounded function of the future, $h(X_n, X_{n+1}, \ldots)$, and n is replaced by a stopping time N. These results, especially the second, will be the keys to developing the theory of Markov chain.

As mentioned in Section 1, we can and will suppose that the X_n are the coordinate maps on sequence space

$$(\Omega_0, \mathcal{F}) = (S^{\{0,1,\ldots\}}, \mathcal{S}^{\{0,1,\ldots\}}),$$

$\mathcal{F}_n = \sigma(X_0, X_1, \ldots, X_n)$, and for each initial distribution we have a measure P_μ defined by (1.2) that makes X_n a Markov chain with $P_\mu(X_0 \in A) = \mu(A)$. Define the shift operators $\theta_n: \Omega_0 \to \Omega_0$ by $(\theta_n \omega)(m) = \omega(m + n)$.

(2.1) The Markov Property   Let $Y: \Omega_0 \to \mathbb{R}$ be bounded and measurable.

$$E_\mu(Y \circ \theta_n | \mathcal{F}_n) = E_{X(n)} Y.$$

Here the subscript μ on the left-hand side indicates that the conditional expectation is taken with respect to P_μ. The right-hand side is the function $\varphi(x) = E_x Y$ evaluated at $x = X_n$. To make the connection with the introduction of this section, let

$$Y(\omega) = h(\omega_0, \omega_1, \ldots).$$

We denote the function by Y, a letter usually used for random variables, because that's exactly what Y is, a measurable function defined on our probability space Ω_0.

Proof   We begin by proving the result in a special case and then use the $\pi - \lambda$ and monotone class theorems to get the general result. Let $A = \{\omega: \omega_0 \in A_0, \ldots, \omega_m \in A_m\}$ and $g_0, \ldots, g_n$ be bounded and measurable. Applying (1.5) with $f_k = 1_{A_k}$ for $k < m$, $f_m = 1_{A_m} g_0$, and $f_k = g_{k-m}$ for $m < k \le m + n$ gives

$$E_\mu\left(\prod_{k=0}^{n} g_k(X_{m+k}); A\right) = \int_{A_0} \mu(dx_0) \int_{A_1} p(x_0, dx_1) \cdots \int_{A_m} p(x_{m-1}, dx_m) g_0(x_m)$$

$$\cdot \int p(x_m, dx_{m+1}) g_1(x_{m+1}) \cdots \int p(x_{m+n-1}, dx_{m+n}) g_n(x_{m+n})$$

$$= E_\mu\left(E_{X_m}\left(\prod_{k=0}^{n} g_k(X_k)\right); A\right).$$

The collection of sets for which the last formula holds is a λ-system, and the collection for which it has been proved is a π-system, so using the $\pi - \lambda$ theorem ((4.2) in Chapter 1) shows that the last identity holds for all $A \in \mathscr{F}_m$.

Fix $A \in \mathscr{F}_m$ and let $\mathscr{H}$ be the collection of bounded measurable Y for which

$$E_\mu(Y \circ \theta_m; A) = E_\mu(E_{X_m} Y; A). \tag{$*$}$$

The last computation shows that $(*)$ holds when $Y(\omega) = \prod_{0 \le k \le n} g_k(\omega_k)$. To finish the proof we will apply the monotone class theorem (1.4). Let $\mathscr{A}$ be the collection of sets of the form $\{\omega \colon \omega_0 \in A_0, \dots, \omega_k \in A_k\}$. $\mathscr{A}$ is a π-system, so taking $g_k = 1_{A_k}$ shows (i) of (1.4) holds. $\mathscr{H}$ clearly has properties (ii) and (iii), so (1.4) implies that $\mathscr{H}$ contains the bounded functions measurable w.r.t. $\sigma(\mathscr{A})$, and the proof is complete.

□

Exercise 2.1   Use the Markov property (2.1) to show that if $A \in \sigma(X_0, \dots, X_n)$ and $B \in \sigma(X_n, X_{n+1}, \dots)$, then for any initial distribution μ,

$$P_\mu(A \cap B | X_n) = P_\mu(A | X_n) P_\mu(B | X_n).$$

In words, the past and future are conditionally independent given the present.

Hint   Write the left-hand side as $E_\mu(E_\mu(1_A 1_B | \mathscr{F}_n) | X_n)$.

The next two results illustrate the use of (2.1). We will see many other applications below.

(2.2) *Chapman–Kolmogorov Equation*

$$P_x(X_{m+n} = z) = \sum_y P_x(X_m = y) P_y(X_n = z).$$

Proof $P_x(X_{n+m} = z) = E_x(P_x(X_{n+m} = z | \mathscr{F}_m)) = E_x(P_{X(m)}(X_n = z))$ by the Markov property (2.1), since $1_{(X(n)=z)} \circ \theta_m = 1_{(X(n+m)=z)}$.   □

(2.3) THEOREM   Let X_n be a Markov chain and suppose

$$P\left(\bigcup_{m=n+1}^\infty \{X_m \in B_m\} \,\Big|\, X_n \right) \ge \delta > 0 \quad \text{on } \{X_n \in A_n\}.$$

Then $P(\{X_n \in A_n \text{ i.o.}\} - \{X_n \in B_n \text{ i.o.}\}) = 0$.

To quote Chung (1974), "The intuitive meaning of the preceding theorem has been given by Doeblin as follows: if the chance of a pedestrian's getting run over is greater than $\delta > 0$ each time he crosses a certain street, then he will not be crossing it indefinitely (since he will be killed first)!"

Proof   Let $\Lambda_n = B_{n+1} \cup B_{n+2} \cup \cdots$, $\Lambda = \cap \Lambda_n = \{X_n \in B_n \text{ i.o.}\}$, and $\Gamma = \{X_n \in A_n \text{ i.o.}\}$. Let $\mathscr{F}_n = \sigma(X_0, X_1, \dots, X_n)$ and $\mathscr{F}_\infty = \sigma(\cup \mathscr{F}_n)$. Using the Markov property and the dominated convergence theorem for conditional expectations ((5.7) in Chapter 4),

$$E(1_{\Lambda(n)}|X_n) = E(1_{\Lambda(n)}|\mathscr{F}_n) \to E(1_\Lambda|\mathscr{F}_\infty) = 1_\Lambda.$$

On Γ, the left-hand side is $\geq \delta$ i.o. This is only possible if $\Gamma \supset \Lambda$. $\qquad \square$

Exercise 2.2 A state a is called *absorbing* if $P_a(X_1 = a) = 1$. Let $D = \{X_n = a$ for some $n \geq 1\}$ and let $h(x) = P_x(D)$. Use (2.3) to conclude that $h(X_n) \to 0$ a.s. on D^c. Here a.s. means P_μ a.s. for any initial distribution μ.

We are now ready for our second extension of the Markov property. Recall N is said to be a stopping time if $\{N = n\} \in \mathscr{F}_n$. As in Chapter 3 let

$$\mathscr{F}_N = \{A: A \cap \{N = n\} \in \mathscr{F}_n \text{ for all } n\}$$

be the information known at time N, and let

$$\theta_N \omega = \begin{cases} \theta_n \omega & \text{on } \{N < \infty\} \\ \Delta & \text{on } \{N = \infty\} \end{cases}$$

where Δ is an extra point that we add to Ω_0. Generally we will explicitly restrict our attention to $\{N < \infty\}$, so the reader does not have to worry about the second part of the definition of θ_N.

(2.4) Strong Markov Property Suppose that for each n, $Y_n: \Omega \to \mathbb{R}$ is measurable and $|Y_n| \leq M$ for all n. Then

$$E_\mu(Y_N \circ \theta_N|\mathscr{F}_N) = E_{X(N)} Y_N \quad \text{on } \{N < \infty\}$$

where the right-hand side is $\varphi(x, n) = E_x Y_n$ evaluated at $x = Y(N)$, $n = N$.

Proof Let $A \in \mathscr{F}_N$.

$$E_\mu(Y_N \circ \theta_N; A \cap \{N < \infty\}) = \sum_{n=0}^{\infty} E_\mu(Y_n \circ \theta_n; A \cap \{N = n\}).$$

Since $A \cap \{N = n\} \in \mathscr{F}_n$, using (2.1) now converts the right side into

$$\sum_{n=0}^{\infty} E_\mu(E_{X(n)} Y_n; A \cap \{N = n\}) = E_\mu(E_{X(N)} Y_N; A \cap \{N < \infty\}). \qquad \square$$

Remark The reader should notice that the proof is trivial. All we do is break things down according to the value of N, replace N by n, apply the Markov property (2.1), and reverse the process. This is the standard technique for proving results about stopping times.

The next example illustrates the use of (2.4) and explains why we want to allow the Y that we apply to the shifted path to depend on n.

(2.5) Reflection Principle   Let $\xi_1, \xi_2, \ldots$ be independent and have a common distribution that is symmetric about 0. Let $S_n = \xi_1 + \cdots + \xi_n$. If $a > 0$, then

$$P\left(\sup_{m \leq n} S_m > a\right) \leq 2P(S_n > a).$$

We will do the proof in two steps because that is how formulas like this are derived in practice. One first computes intuitively and then figures out how to extract the desired formula from (2.4).

Proof in Words   Since the ξ_i have a symmetric distribution, $P(S_m > 0) = P(S_m < 0)$, and hence $P(S_m \geq 0) \geq 1/2$. If we let $N = \inf\{m \leq n : S_m > a\}$ (with $\inf \emptyset = \infty$), then on $\{N < \infty\}$, $S_n - S_N$ is independent of S_N and has $P(S_n - S_N \geq 0) \geq 1/2$. So

$$P(S_n > a) \geq (1/2)P(N \leq n).$$

Formal Proof   Let $Y_m(\omega) = 1$ if $m \leq n$ and $\omega_{n-m} > a$, $Y_m(\omega) = 0$ otherwise, and note that if $y > a$ then

$$E_y Y_m = P_y(S_{n-m} > a) \geq P_y(S_{n-m} \geq y) \geq 1/2.$$

The definition of Y_m is chosen so that $Y_N \circ \theta_N(\omega) = 1$ if $\omega_n > a$ (and hence $N \leq n$), and $= 0$ otherwise. The strong Markov property implies

$$E_0(Y_N \circ \theta_N | \mathscr{F}_N) = E_{S_N} Y_N \geq 1/2 \quad \text{on } \{N < \infty\} = \{N \leq n\}.$$

Integrating over $\{N \leq n\}$ gives

$$\tfrac{1}{2}P(N \leq n) \leq E_0(E_0(Y_N \circ \theta_N | \mathscr{F}_N); N \leq n) = E_0(Y_N \circ \theta_N; N \leq n),$$

since $\{N \leq n\} \in \mathscr{F}_N$. Recalling that $Y_N \circ \theta_N = 1_{\{S_n > a\}}$, the last quantity

$$= E_0(1_{\{S_n > a\}}; N \leq n) = P_0(S_n > a),$$

since $\{S_n > a\} \subset \{N < n\}$. $\square$

Remark   The proof of (2.5) shows that if $P(S_k \geq 0) \geq \alpha$ for all k, then

$$P\left(\max_{m \leq n} S_m > a\right) \leq \alpha^{-1} P(S_n > a).$$

Notation   The next five exercises concern various hitting times. These come in two varieties:

$$\tau_A = \inf\{n \geq 0 : X_n \in A\}, \qquad T_A = \inf\{n \geq 1 : X_n \in A\}.$$

It will often be important to choose the right definition, so we will use this notation throughout. To keep the two straight, remember τ is smaller than T.

Exercise 2.3 *First entrance decomposition.* Let $T_y = \inf\{n \geq 1 : X_n = y\}$. Show that

$$p^n(x, y) = \sum_{m=1}^{n} P_x(T_y = m)p^{n-m}(y, y).$$

Exercise 2.4 Show that $\sum_{m=0}^{n} P_x(X_m = x) \geq \sum_{m=k}^{n+k} P_x(X_m = x)$.

Exercise 2.5 Let $A \subset S$ and define T_A^k inductively by $T_A^0 = 0$, $T_A^k = \inf\{n > T_A^{k-1} : X_n \in A\}$. Show that $Y_k = X(T_A^k)$ is a Markov chain.

Exercise 2.6 Suppose that $S - C$ is finite and for each $x \in S - C$ there is an $n(x)$ so that $p^{n(x)}(x, C) > 0$. Let $\tau_C = \inf\{n \geq 0 : X_n \in C\}$, $\varepsilon = \inf\{p^{n(x)}(x, C)\}$, and $N = \sup n(x)$. Show $P_y(\tau_C > kN) \leq (1 - \varepsilon)^k$.

Exercise 2.7 Let $\tau_A = \inf\{n \geq 0 : X_n \in A\}$ and $h(x) = P_x(\tau_A < \tau_B)$. Suppose $A \cap B = \varnothing$, $S - (A \cup B)$ is finite, and $P_x(\tau_{A \cup B} < \infty) > 0$ for all $x \in S - (A \cup B)$. (i) Show that

$$h(x) = \sum_y p(x, y)h(y) \quad \text{for } x \notin A \cup B. \tag{*}$$

(ii) Use Exercise 2.6 to show that $P_x(\tau_{A \cup B} < \infty) = 1$. (iii) Show that if h satisfies $(*)$, then $h(X(n \wedge \tau_{A \cup B}))$ is a martingale, and use this to conclude that $h(x) = P_x(\tau_A < \tau_B)$ is the only solution of $(*)$ that is 1 on A and 0 on B.

Exercise 2.8 Let X_n be a Markov chain with $S = \{0, 1, \ldots, N\}$ and suppose that X_n is a martingale. (i) Show that 0 and N are absorbing states; that is, $p(0, 0) = p(N, N) = 1$. (ii) Let $\tau_x = \inf\{n \geq 0 : X_n = x\}$. Show that $P_x(\tau_N < \tau_0) = x/N$.

Exercise 2.9 *Genetics chains.* Suppose $S = \{0, 1, \ldots, N\}$ and consider

(i) $p(i, j) = \binom{N}{j}(i/N)^j(1 - i/N)^{N-j}$

(ii) $p(i, j) = \binom{2i}{j}\binom{2N - 2i}{N - j} \Big/ \binom{2N}{N}$.

Show that these chains satisfy the hypotheses of Exercise 2.8.

Exercise 2.10 In so-called brother–sister mating, two animals are mated and among their direct descendants two individuals of opposite sex are selected at random. These animals are mated and the process continues. Suppose each individual can be one of three genotypes, AA, Aa, aa, and suppose that the type of the offspring is determined by selecting a letter from each parent. With these rules the pair of genotypes in the nth generation is a Markov chain with six states:

$$AA, AA \quad AA, Aa \quad AA, aa \quad Aa, Aa \quad Aa, aa \quad aa, aa.$$

The first and last states are absorbing states. Show that the number of A's in the pair is a martingale and use this to compute the probability of getting absorbed in AA, AA.

Exercise 2.11   Let $\tau_A = \inf\{n \geq 0 : X_n \in A\}$ and $g(x) = E_x \tau_A$. Suppose that $S - A$ is finite and for each $x \in S - A$, $P_x(\tau_A < \infty) > 0$. (i) Use Exercise 2.6 to conclude that $E_x \tau_A < \infty$. (ii) Show that

$$g(x) = 1 + \sum_y p(x, y)g(y) \quad \text{for } x \notin A. \tag{*}$$

(iii) Show that if g satisfies (*), $g(X(n \wedge \tau_A)) - n \wedge \tau_A$ is a martingale and use this to conclude that $g(x) = E_x \tau_A$ is the only solution of (*) that is 0 on A.

Exercise 2.12   Let $\xi_0, \xi_1, \ldots$ be i.i.d. $\in \{H, T\}$, taking each value with probability $1/2$, and let $X_n = (\xi_n, \xi_{n+1})$ be the Markov chain from Exercise 1.4. Let $N_1 = \inf\{n \geq 0 : (\xi_n, \xi_{n+1}) = (H, H)\}$ and $N_2 = \inf\{n \geq 0 : (\xi_n, \xi_{n+1}) = (H, T)\}$. Use the results in the last exercise to compute EN_1 and EN_2. Before solving the problem try to guess which symbol $>$, $=$, $<$ goes in the blank: EN_1____EN_2.

Exercise 2.13   Consider the Markov chain on $\{1, 2, \ldots, N\}$ with $p_{ij} = 1/(i - 1)$ when $j < i$, $p_{11} = 1$ and $p_{ij} = 0$ otherwise. (i) Let $\tau = \inf\{n \geq 0 : X_n = 1\}$. Use Exercise 2.9 to verify that $E_k \tau = 1 + 1/2 + \cdots + 1/k - 1$. (ii) Let $I_j = 1$ if X_n visits j. Show that if $X_0 = k$, $I_2, \ldots, I_k$ are independent.

Hint   Evaluate $P(I_j = 1 | I_{j+1}, \ldots, I_k)$ by considering $n = \min\{i > j : I_i = 1\}$.

3   Recurrence and Transience

In this section and the next two, we will consider only Markov chains on a countable state space. Let $T_y^0 = 0$, and for $k \geq 1$ let

$$T_y^k = \inf\{n > T_y^{k-1} : X_n = y\}.$$

T_y^k is the time of the kth return to y. The reader should note that $T_y^1 > 0$, so any visit at time 0 does not count. We adopt this convention so that if we let $T_y = T_y^1$ and $\rho_{xy} = P_x(T_y < \infty)$, then:

(3.1) THEOREM $P_x(T_y^k < \infty) = \rho_{xy}\rho_{yy}^{k-1}$.

Intuitively, in order to make k visits to y we first have to go from x to y and then return $k - 1$ times to y.

Proof   When $k = 1$ the result is trivial, so we suppose $k \geq 2$. Let $Y(\omega) = 1$ if $\omega_n = y$ for some $n \geq 1$, $Y(\omega) = 0$ otherwise. If $N = T_y^{k-1}$, then $Y \circ \theta_N = 1$ if $T_y^k < \infty$, so the strong Markov property (2.4) implies

$$E_x(Y \circ \theta_N | \mathscr{F}_N) = E_{X_N} Y \quad \text{on } \{N < \infty\}.$$

On $\{N < \infty\}$, $X_N = y$, so the right-hand side is $P_y(T_y < \infty) = \rho_{yy}$, and it follows that

$$P_x(T_y^k < \infty) = E_x(Y \circ \theta_N; N < \infty)$$
$$= E_x(E_x(Y \circ \theta_N | \mathscr{F}_N); N < \infty)$$
$$= E_x(\rho_{yy}; N < \infty) = \rho_{yy} P_x(T_y^{k-1} < \infty).$$

The result now follows by induction. □

A state y is said to be *recurrent* if $\rho_{yy} = 1$ and *transient* if $\rho_{yy} < 1$. If y is recurrent, then (3.1) implies $P_y(T_y^k < \infty) = 1$ for all k, so $P_y(X_n = y \text{ i.o.}) = 1$.

Exercise 3.1   For $k \geq 0$, let $R_k = T_y^k$ be the time of the kth return to y, and for $k \geq 1$, let $r_k = R_k - R_{k-1}$ be the kth interarrival time. Use the strong Markov property to conclude that under P_y, $(r_k, X_{R(k-1)}, \ldots, X_{R(k)-1})$, $k \geq 1$, are i.i.d.

If y is transient and we let

$$N(y) = \sum_{n=1}^{\infty} 1_{(X_n = y)}$$

be the number of visits to y at positive times, then

$$E_x N(y) = \sum_{k=1}^{\infty} P_x(N(y) \geq k) = \sum_{k=1}^{\infty} P_x(T_y^k < \infty) \tag{3.2}$$

$$= \sum_{k=1}^{\infty} \rho_{xy} \rho_{yy}^{k-1} = \frac{\rho_{xy}}{1 - \rho_{yy}} < \infty.$$

Combining the last computation with our result for recurrent states gives a result that generalizes (2.2) from Chapter 3.

(3.3) THEOREM y is recurrent if and only if $E_y N(y) = \infty$.

Exercise 3.2   Let $a \in S$, $f_n = P_a(T_a = n)$, and $u_n = P_a(X_n = a)$. (i) Show that $u_n = \sum_{1 \leq m \leq n} f_m u_{n-m}$. (ii) Let $u(s) = \sum_{n \geq 0} u_n s^n$, $f(s) = \sum_{n \geq 1} f_n s^n$, and show $u(s) = 1/(1 - f(s))$. Setting $s = 1$ gives (3.2) for $x = y = a$.

Exercise 3.3   Consider an asymmetric simple random walk on $\mathbb{Z}$; that is, $p(i, i + 1) = p$, $p(i, i - 1) = q = 1 - p$. In this case

$$p^{2m}(0, 0) = \binom{2m}{m} p^m q^m, \quad p^{2m+1}(0, 0) = 0, \quad \text{and} \quad u(s) = (1 - 4pqs^2)^{-1/2}.$$

Use the last exercise to conclude $f(s) = 1 - (1 - 4pqs^2)^{1/2}$. Setting $s = 1$ gives the probability the random walk will return to 0.

The next result shows that recurrence is contagious.

(3.4) THEOREM   If x is recurrent and $\rho_{xy} > 0$, then y is recurrent and $\rho_{yx} = 1$.

Proof   We will first show $\rho_{yx} = 1$ by showing that if $\rho_{yx} < 1$, then $\rho_{xx} < 1$. Let $K = \inf\{k: p^k(x, y) > 0\}$. There is a sequence $y_1, \ldots, y_{K-1}$ so that

$$p(x, y_1)p(y_1, y_2)\cdots p(y_{K-1}, y) > 0.$$

Since K is minimal, $y_i \neq y$ for $1 \leq i \leq K - 1$. If $\rho_{yx} < 1$, we have

$$P_x(T_x = \infty) \geq p(x, y_1)p(y_1, y_2)\cdots p(y_{K-1}, y)(1 - \rho_{yx}) > 0,$$

a contradiction. So $\rho_{yx} = 1$.

To prove that y is recurrent, observe that $\rho_{yx} > 0$ implies there is an L so that $p^L(y, x) > 0$. Now

$$p^{L+n+K}(y, y) \geq p^L(y, x)p^n(x, x)p^K(x, y).$$

Summing over n, we see

$$\sum_{n=1}^{\infty} p^{L+n+K}(y, y) \geq p^L(y, x)p^K(x, y) \sum_{n=1}^{\infty} p^n(x, x) = \infty,$$

so (3.3) implies y is recurrent.   □

Exercise 3.4   Use the strong Markov property to show that in general $\rho_{xz} \geq \rho_{xy}\rho_{yz}$.

The next fact will help us identify recurrent states in examples. First we need two definitions. C is *closed* if $x \in C$ and $\rho_{xy} > 0$ implies $y \in C$. The name comes from the fact that if C is closed and $x \in C$, then $P_x(X_n \in C) = 1$ for all n. D is *irreducible* if $x, y \in D$ implies $\rho_{xy} > 0$.

(3.5) THEOREM   Let C be a finite closed set. Then C contains a recurrent state. If C is irreducible, then all states in C are recurrent.

Proof   In view of (3.4) it suffices to prove the first claim. Suppose it is false. Then for all $y \in C$, $\rho_{yy} < 1$ and $E_x N(y) = \rho_{xy}/(1 - \rho_{yy})$, but this is ridiculous, since it implies

$$\infty > \sum_{y \in C} E_x N(y) = \sum_{y \in C}\sum_{n=1}^{\infty} p^n(x, y) = \sum_{n=1}^{\infty}\sum_{y \in C} p^n(x, y) = \sum_{n=1}^{\infty} 1.$$

The first inequality follows from the fact that C is finite and the last equality from the fact that C is closed.   □

To illustrate the use of the last result, consider:

Example 3.1   Let X_n be a Markov chain with transition matrix

	1	2	3	4	5	6
1	0	1	0	0	0	0
2	.4	.6	0	0	0	0
3	.3	0	.4	.2	.1	0
4	0	0	0	.3	.7	0
5	0	0	0	.5	0	.5
6	0	0	0	.8	0	.2

Things become a little clearer if we represent the matrix as a graph with an edge from i to j if $i \neq j$ and $p(i, j) > 0$. (See Figure 5.1. To make the graph easier to draw we ignore the $p(i, j)$. They are irrelevant in computing ρ_{xy} for $x \neq y$.) It is easy to see that (when $x \neq y$) $\rho_{xy} > 0$ if and only if we can get from x to y on the graph. Looking at the graph we see that:

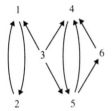

Figure 5.1

(i) $\rho_{34} > 0$ and $\rho_{43} = 0$, so 3 must be transient or we would contradict (3.4).
(ii) $\{1, 2\}$ and $\{4, 5, 6\}$ are irreducible closed sets, so (3.5) implies these states are recurrent.

The last reasoning can be used to identify transient and recurrent states when S is finite, since for $x \in S$, either: (i) there is a y with $\rho_{xy} > 0$ and $\rho_{yx} = 0$ and x must be transient, or (ii) $\rho_{xy} > 0$ implies $\rho_{yx} > 0$. In case (ii), Exercise 3.4 implies $C_x = \{y : \rho_{xy} > 0\}$ is an irreducible closed set. (If $y, z \in C_x$, then $\rho_{yz} \geq \rho_{yx}\rho_{xz} > 0$. If $\rho_{yw} > 0$, then $\rho_{xw} \geq \rho_{xy}\rho_{yw} > 0$, so $w \in C_x$.) So (3.5) implies x is recurrent.

Exercise 3.5   Show that in the Ehrenfest chain (Example 1.6) all states are recurrent.

Example 3.1 motivates the following:

(3.6) Decomposition Theorem   Let $R = \{x : \rho_{xx} = 1\}$ be the recurrent states of a Markov chain. R can be written as $\bigcup_i R_i$ where each R_i is closed and irreducible.

Remark   This result shows that for the study of recurrent states we can, without loss of generality, consider a single irreducible closed set.

Proof   If $x \in R$, let $C_x = \{y : \rho_{xy} > 0\}$. By (3.5) $C_x \subset R$ and if $y \in C_x$, then $\rho_{yx} > 0$. From this it follows easily that either $C_x \cap C_y = \varnothing$ or $C_x = C_y$. To prove the last claim suppose $C_x \cap C_y \neq \varnothing$. If $z \in C_x \cap C_y$, then $\rho_{xy} \geq \rho_{xz}\rho_{zy} > 0$, so if $w \in C_y$ we have $\rho_{xw} \geq \rho_{xy}\rho_{yw} > 0$ and it follows that $C_x \supset C_y$. Interchanging the roles of x and y gives $C_y \supset C_x$, and we have proved our claim. If we let R_i be a listing of the sets which appear as some C_x, we have the desired decomposition.   □

The rest of this section is devoted to examples. Specifically we concentrate on the question: How do we tell whether a state is recurrent or transient? Reasoning based on (3.4) works occasionally when S is infinite.

Example 3.2   *Branching process.* If the probability of no children is positive, then $p_{k0} > 0$ and $\rho_{0k} = 0$ for $k \geq 1$, so (3.4) implies all states $k \geq 1$ are transient. The state 0 has $p(0, 0) = 1$ and is recurrent. It is called an *absorbing state* to reflect the fact that once the chain enters 0, it remains there for all time.

Exercise 3.6   Let Z_n be a branching process with offspring distribution p_k, and let $\varphi(\theta) = \sum p_k \theta^k$. Suppose $\rho < 1$ has $\varphi(\rho) = \rho$. Show that ρ^{Z_n} is a martingale and use this to conclude $P_x(Z_n = 0$ for some $n \geq 1) = \rho^x$.

If S is infinite and irreducible, all that (3.4) tells us is that either all the states are recurrent or all are transient, and we are left to figure out which case occurs.

Example 3.3   *Renewal chain*   Since $p(i, i - 1) = 1$ for $i \geq 1$, it is clear that $\rho_{i0} = 1$ for all $i \geq 1$ and hence also for $i = 0$; that is, 0 is recurrent. If we recall that $p(0, j) = f_{j+1}$ and suppose that $\{k : f_k > 0\}$ is unbounded, then $\rho_{0i} > 0$ for all i and all states are recurrent. If $K = \sup\{k : f_k > 0\} < \infty$, then $\{0, 1, \ldots, K - 1\}$ is an irreducible closed set of recurrent states and all states $k \geq K$ are transient.

Example 3.4   *Birth and death chains on* $\{0, 1, 2, \ldots\}$. Let

$$p(i, i + 1) = p_i, \qquad p(i, i - 1) = q_i, \qquad p(i, i) = r_i$$

where $q_0 = 0$. Let $N = \inf\{n : X_n = 0\}$. To analyze this example, we are going to define a function φ so that $\varphi(X_{N \wedge n})$ is a martingale. We start by setting $\varphi(0) = 0$ and $\varphi(1) = 1$. For the martingale property to hold when $X_n = k \geq 1$, we must have

$$\varphi(k) = p_k \varphi(k + 1) + r_k \varphi(k) + q_k \varphi(k - 1).$$

Using $p_k + q_k + r_k = 1$, we can rewrite the last equation as

$$q_k(\varphi(k) - \varphi(k - 1)) = p_k(\varphi(k + 1) - \varphi(k))$$

or

$$\varphi(k+1) - \varphi(k) = (q_k/p_k)[\varphi(k) - \varphi(k-1)].$$

Here and in what follows, we suppose that $p_k, q_k > 0$ for $k \geq 1$. Otherwise the chain is not irreducible. Since $\varphi(1) - \varphi(0) = 1$, iterating the last result gives

$$\varphi(m+1) - \varphi(m) = \prod_{j=1}^{m} (q_j/p_j) \quad \text{for } m \geq 1$$

and

$$\varphi(n) = \sum_{m=0}^{n-1} \prod_{j=1}^{m} (q_j/p_j) \quad \text{for } n \geq 1,$$

if we interpret the product as 1 when $m = 0$. Let $T_c = \inf\{n \geq 1 : X_n = c\}$. Now I claim that:

(3.7) THEOREM   If $a < x < b$, then $P_x(T_a < T_b) = (\varphi(b) - \varphi(x))/(\varphi(b) - \varphi(a))$.

Proof   If we let $T = T_a \wedge T_b$, then $\varphi(X_{n \wedge T})$ is a bounded martingale and $T < \infty$ a.s., so $\varphi(x) = E_x \varphi(X_T)$. Since $X_T \in \{a, b\}$ a.s.,

$$\varphi(x) = \varphi(a) P_x(T_a < T_b) + \varphi(b)[1 - P_x(T_a < T_b)],$$

and solving gives the indicated formula.   □

Remark   The answer and the proof should remind the reader of Example 1.5 in Chapter 3 and Exercise 7.3 in Chapter 4. To help remember the formula observe that for any α and β, if we let $\psi(x) = \alpha\varphi(x) + \beta$, then $\psi(X_{n \wedge T})$ is also a martingale and the answer we get using ψ must be the same. The last observation explains why the answer is a ratio of differences. To help remember which one, observe that the answer is 1 if $x = a$ and 0 if $x = b$.

Exercise 3.7   A gambler is playing roulette and betting $1 on black each time. The probability she wins $1 is 18/38 and the probability she loses $1 is 20/38. (i) Calculate the probability that, starting with $20, she reaches $40 before losing her money. (ii) Use the fact that $X_n - 2n/38$ is a martingale to calculate $E(T_{40} \wedge T_0)$.

Letting $a = 0$ and $b = M$ in (3.7) gives

$$P_x(T_0 > T_M) = \varphi(x)/\varphi(M).$$

Letting $M \to \infty$ and observing that $T_M \geq M - x$, P_x a.s., we have proved:

(3.8) THEOREM   0 is recurrent if and only if $\varphi(M) \to \infty$ as $M \to \infty$; that is,

$$\varphi(\infty) \equiv \sum_{m=0}^{\infty} \prod_{j=1}^{m} (q_j/p_j) = \infty.$$

If $\varphi(\infty) < \infty$, then $P_x(T_0 = \infty) = \varphi(x)/\varphi(\infty)$.

We will now see what (3.8) says about some concrete cases. Suppose $p_j = p$ and $q_j = 1 - p$ for $j \geq 1$. From (3.8) it follows that 0 is recurrent if and only if $p \leq 1/2$, and if $p > 1/2$, then $P_x(T_0 < \infty) = (\varphi(\infty) - \varphi(x))/\varphi(\infty) = (1 - p/p)^x$. To probe the boundary between recurrence and transience, suppose $p_j = 1/2 + \varepsilon_j$ where $\varepsilon_j \sim Cj^{-\alpha}$ as $j \to \infty$, and $q_j = 1 - p_j$. A little arithmetic shows

$$q_j/p_j = 1 - 2\varepsilon_j/(1/2 + \varepsilon_j) \approx 1 - 4Cj^{-\alpha} \quad \text{for large } j.$$

Case 1: $\alpha > 1$. It is easy to show that for $0 < \delta_j < 1$, we have $\prod_j (1 - \delta_j) > 0$ if and only if $\sum_j \delta_j < \infty$, so if $\alpha > 1$, $\prod_{j \leq k} (q_j/p_j) \downarrow$ a positive limit and 0 is recurrent.

Case 2: $\alpha < 1$. Using the fact that $\log(1 - \delta) \sim -\delta$ as $\delta \to 0$, we see that if $\alpha < 1$,

$$\log\left(\prod_{j=1}^{k} (q_j/p_j)\right) \sim -\sum_{j=1}^{k} 4Cj^{-\alpha} \sim \frac{-4C}{1 - \alpha} k^{1-\alpha} \quad \text{as } k \to \infty,$$

so for $k \geq K$,

$$\prod_{j=1}^{k} (q_j/p_j) \leq \exp\left(\frac{-2C}{1 - \alpha} k^{1-\alpha}\right)$$

and

$$\sum_{k=0}^{\infty} \prod_{j=1}^{k} (q_j/p_j) < \infty,$$

and hence 0 is transient.

Case 3: $\alpha = 1$. Repeating the argument for Case 2 shows

$$\log\left(\prod_{j=1}^{k} (q_j/p_j)\right) \sim -4C \log k.$$

So if $C > 1/4$, 0 is transient, and if $C < 1/4$, 0 is recurrent. The case $C = 1/4$ can go either way.

Example 3.5   *M/G/1 queue.* Let $\mu = \sum ka_k$ be the mean number of customers who arrive during one service time. We will now show that if $\mu > 1$, the chain is transient (i.e., all states are), but if $\mu \leq 1$, it is recurrent. For the case $\mu > 1$, we observe that if $\xi_1, \xi_2, \ldots$ are i.i.d. with $P(\xi_m = j) = a_{j+1}$ and $S_n = \xi_1 + \cdots + \xi_n$, then $X_0 + S_n$ and X_n behave the same until time $N = \inf\{n : X_0 + S_n = 0\}$. When $\mu > 1$, $E\xi_m = \mu - 1 > 0$, so $S_n \to \infty$ a.s. and $\inf S_n > -\infty$ a.s. It follows from the last observation that if x is large, $P_x(N < \infty) < 1$ and the chain is transient.

To deal with the case $\mu \leq 1$ we observe that it follows from arguments in the last paragraph that $X_{n \wedge N}$ is a supermartingale. Let $T = \inf\{n : X_n \geq M\}$. Since $X_{n \wedge N}$

is a nonnegative supermartingale, using the optional stopping theorem ((7.4) in Chapter 4) at time $\tau = T \wedge N$ and observing $X_\tau \geq M$ on $\{T < N\}$, $X_\tau = 0$ on $\{N < T\}$ gives

$$x \geq M P_x(T < N).$$

Letting $M \to \infty$ shows $P_x(N < \infty) = 1$, so the chain is recurrent.

Exercise 3.8   There is another way of seeing that the M/G/1 queue is transient when $\mu > 1$. Show that if we consider the customers who arrive during a person's service time to be her children, then we get a branching process. Use results in Section 3 in Chapter 4 (or Exercise 3.6) to conclude that the chain is recurrent when $\mu \leq 1$ and when $\mu > 1$, $P_x(T_o < \infty) = \rho^x$, where ρ is the smallest fixed point of the function

$$\varphi(\theta) = \sum_{k=0}^\infty a_k \theta^k.$$

Exercise 3.9   Generalize the results for birth and death chains and the M/G/1 queue by showing that if S is irreducible and φ is a nonnegative function with $E_x \varphi(X_1) \leq \varphi(x)$ for $x \notin F$, a finite set, and $\varphi(x) \to \infty$ as $x \to \infty$—that is, $\{x : \varphi(x) \leq M\}$ is finite for any $M < \infty$, then the chain is recurrent. Prove this by showing that if $T = \inf\{n > 0 : X_n \in F\}$, then $P_x(T < \infty) = 1$ for all $x \notin F$, so $P_y(X_n \in F \text{ i.o.}) = 1$ for all $y \in S$, and since F is finite, $P_y(X_n = z \text{ i.o.}) = 1$ for some $z \in F$.

Exercise 3.10   Show that if we replace "$\varphi(x) \to \infty$" by "$\varphi(x) \to 0$" in the last exercise, then we can conclude that the chain is transient.

Exercise 3.11   Let X_n be a birth and death chain with $p_j - 1/2 \sim C/j$ as $j \to \infty$ and $q_j = 1 - p_j$. Show that if we take $C < 1/4$, then we can pick $p > 0$ so that $\varphi(x) = x^p$ satisfies the hypotheses of Exercise 3.9. When $C > 1/4$ we can take $p < 0$ and apply Exercise 3.10. The nice thing about this approach is that it applies if we assume $P_x(|X_1 - x| \leq M) = 1$ and $E_x(X_1 - x) \sim 2C/x$.

Exercise 3.12 f is said to be *superharmonic* if $f(x) \geq \sum_y p(x, y)f(y)$, or equivalently $f(X_n)$ is a supermartingale. Suppose p is irreducible. Show that p is recurrent if and only if every nonnegative superharmonic function is constant.

Exercise 3.13   *M/M/∞ queue.* Consider a telephone system with an infinite number of lines. Let X_n = the number of lines in use at time n, and suppose

$$X_{n+1} = \sum_{m=1}^{X_n} \xi_{n,m} + Y_{n+1}$$

where the $\xi_{n,m}$ are i.i.d. with $P(\xi_{n,m} = 1) = p$ and $P(\xi_{n,m} = 0) = 1 - p$, and Y_n is an independent i.i.d. sequence of Poisson mean λ r.v.'s In words, for each conversation we flip a coin with probability p of heads to see if it continues for another minute.

Meanwhile a Poisson mean λ number of conversations start between time n and $n + 1$. (i) Show that the M/M/∞ queue has $E_x X_1 = px + \lambda$ and hence by Exercise 3.9 is recurrent for any $p < 1$. (ii) Use the last formula and induction to conclude

$$E_x X_n = v + p^n(x - v) \quad \text{where } v = \lambda/(1 - p).$$

4   Stationary Measures

A measure μ is said to be a *stationary measure* if

$$\sum_x \mu(x)p(x, y) = \mu(y).$$

The last equation says $P_\mu(X_1 = y) = \mu(y)$. Using the Markov property and induction, it follows that $P_\mu(X_n = y) = \mu(y)$ for all $n \geq 1$. If μ is a probability measure, we call μ a *stationary distribution*, and it represents a possible equilibrium for the chain. That is, if X_0 has distribution μ, then so does X_n for all $n \geq 1$. If we stretch our imagination a little, we can also apply this interpretation when μ is an infinite measure. (When the total mass is finite, we can divide by $\mu(S)$ to get a stationary distribution.) Before getting into the theory, we consider some examples.

Example 4.1   *The Ehrenfest chain.* $S = \{0, 1, \ldots, r\}$. $p(k, k + 1) = (r - k)/r$, $p(k, k - 1) = k/r$. In this case $\mu(x) = 2^{-r}\binom{r}{x}$ is a stationary distribution. One can check this without pencil and paper by observing that $\mu(x)$ corresponds to flipping r coins to determine which urn each ball is to be placed in, and the transitions of the chain correspond to picking a coin at random and turning it over. Alternatively, you can pick up your pencil and check that

$$\mu(k + 1)p(k + 1, k) + \mu(k - 1)p(k - 1, k) = \mu(k).$$

Example 4.2   *Random walk.* $S = \mathbb{Z}^d$. $p(x, y) = f(y - x)$ where $f(z) \geq 0$ and $\sum f(z) = 1$. In this case $\mu(x) \equiv 1$ is a stationary measure, since

$$\sum_x p(x, y) = \sum_x f(y - x) = 1.$$

A transition probability that has

$$\sum_x p(x, y) = 1$$

is called *doubly stochastic*. This is obviously a necessary and sufficient condition for $\mu(x) \equiv 1$ to be a stationary measure.

Example 4.3   *Asymmetric simple random walk.* $S = \mathbb{Z}$. $p(x, x + 1) = p, p(x, x - 1) = q = 1 - p$. By the last example $\mu(x) \equiv 1$ is a stationary measure. When $p \neq q$, $\mu(x) = (p/q)^x$ is a second one. To check this, we observe that

$$\sum_x \mu(x)p(x, y) = \mu(y + 1)p(y + 1, y) + \mu(y - 1)p(y - 1, y)$$

$$= (p/q)^{y+1}q + (p/q)^{y-1}p = (p/q)^y[p + q] = (p/q)^y.$$

Example 4.4   *Birth and death chains.* $S = \{0, 1, 2, \dots\}.$

$$p(x, x + 1) = p_x, \qquad p(x, x) = r_x, \qquad p(x, x - 1) = q_x$$

with $q_0 = 0$ and $p(i, j) = 0$ otherwise. In this case there is a measure

$$\mu(x) = \prod_{k=1}^x (p_{k-1}/q_k)$$

that has

$$\mu(x)p(x, y) = \mu(y)p(y, x). \tag{4.1}$$

(Notice it is enough to check this for $y = x + 1$.) Summing over x gives

$$\sum_x \mu(x)p(x, y) = \mu(y),$$

so (4.1) is stronger than being a stationary measure. (4.1) asserts that the amount of mass that moves from x to y in one jump is exactly the same as the amount that moves from y to x. A measure μ that satisfies (4.1) is said to be a *reversible measure*. The next exercise explains the name "reversible."

Exercise 4.1   Let μ be a stationary measure and suppose X_0 has "distribution" μ. Then $Y_m = X_{n-m}, 0 \le m \le n$, is a Markov chain with initial measure μ and transition probability

$$q(x, y) = \mu(y)p(y, x)/\mu(x).$$

q is called the *dual transition probability.* If μ is a reversible measure, then $q = p$.

Reviewing the last four examples might convince you that most chains have reversible measures:

1. $\mu(x) = 2^{-r}\binom{r}{x}$ is a reversible measure for the Ehrenfest chain.
2. $\mu(x) \equiv 1$ is a reversible measure if and only if $p(x, y) = p(y, x)$.
3. $\mu(x) = (q/p)^x$ is a reversible measure for a simple random walk.
4. All birth and death chains have reversible measures.

This is a false impression. The M/G/1 queue has no reversible measures because if $x > y + 1, p(x, y) > 0$ but $p(y, x) = 0$. The renewal chain has similar problems.
    Necessary and sufficient conditions for the existence of a reversible measure are that (i) $p(x, y) > 0$ implies $p(y, x) > 0$ and (ii) for any sequence $x_0, x_1, \dots, x_n$ with $x_n = x_0$ and $\prod_{1 \le i \le n} p(x_i, x_{i-1}) > 0$,

$$\prod_{i=1}^{n} \frac{p(x_{i-1}, x_i)}{p(x_i, x_{i-1})} = 1.$$

(i) and (ii) when taken together are a necessary and sufficient condition. This *cycle condition*, due to Kolmogorov, is clearly necessary, since (4.1) implies that for the sequences considered above,

$$\prod_{i=1}^{n} \frac{p(x_{i-1}, x_i)}{p(x_i, x_{i-1})} = \prod_{i=1}^{n} \frac{\mu(x_i)}{\mu(x_{i-1})} = 1.$$

To prove sufficiency, fix $a \in S$, set $\mu(a) = 1$, and if $x_0 = a, x_1, \ldots, x_n = x$ is a sequence with $\prod_{1 \le i \le n} p(x_i, x_{i-1}) > 0$, let

$$\mu(x) = \prod_{i=1}^{n} \frac{p(x_{i-1}, x_i)}{p(x_i, x_{i-1})}.$$

The cycle condition guarantees that the last definition is independent of the path, and it follows easily that (4.1) holds.

Example 4.5   *Random walks on graphs.* A graph is described by giving a countable set of vertices S and an adjacency matrix a_{ij} that has $a_{ij} = 1$ if i and j are adjacent and 0 otherwise. To have an undirected graph with no loops, we suppose $a_{ij} = a_{ji}$ and $a_{ii} = 0$. If we suppose that

$$\mu(i) = \sum_{j} a_{ij} < \infty \quad \text{and let} \quad p(i, j) = a_{ij}/\mu(i),$$

then p is a transition probability that corresponds to picking an edge at random and jumping to the other end. It is clear from the definition that

$$\mu(i)p(i, j) = a_{ij} = a_{ji} = \mu(j)p(j, i),$$

so p is reversible with respect to μ. A little thought reveals that if we assume only that

$$a_{ij} = a_{ji} \ge 0, \quad \mu(i) = \sum_{j} a_{ij} < \infty, \quad \text{and} \quad p(i, j) = a_{ij}/\mu(i), \tag{4.2}$$

the same conclusion is valid. This is the most general example because if p is reversible with respect to μ, we can let $a_{ij} = \mu(i)p(i, j)$.

In the next four exercises we will suppose (4.2) holds and that p is irreducible. Let

$$\tau_\Lambda = \inf\{n \ge 0 : X_n \in \Lambda\}, \qquad \tau_0 = \tau_{\{0\}}$$

$$T_\Lambda = \inf\{n \ge 1 : X_n \in \Lambda\}, \qquad T_0 = T_{\{0\}}.$$

Exercise 4.2 *Dirichlet principle.* Suppose $S - \Lambda$ is finite and let $0 \in S - \Lambda$. The function $h_i^* = P_i(\tau_0 < \tau_\Lambda)$ minimizes

$$\mathscr{D}(h) = (1/2) \sum_{j,k} a_{jk}(h_j - h_k)^2$$

among all the functions that are $= 0$ on Λ and have $h_0 = 1$. Furthermore,

$$\mathscr{D}(h^*) = \mu(0)P_0(T_0 > \tau_\Lambda).$$

Hint Let g be a minimizing function (which exists since $\mathscr{D}(h^n) \to \infty$ if $h_i^n \to \infty$) and look at $\partial \mathscr{D}(g)/\partial g_j$ for $j \notin \Lambda \cup \{0\}$.

Exercise 4.3 *Electric circuits.* If we consider $r_{ij} = 1/a_{ij}$ to be the resistance of the edge from i to j, then a_{ij} is its capacitance. Let v_i be the voltage at i when Λ is grounded (i.e., $v_k = 0$ for $k \in \Lambda$) and a current is introduced at $0 \in S - \Lambda$ to make $v_0 = 1$. Ohm's law says that the current flowing from j to k is $(v_j - v_k)a_{jk}$, and Kirckhoff's law implies

$$\sum_k (v_j - v_k)a_{jk} = 0 \quad \text{for } j \neq 0.$$

Show that if $S - \Lambda$ is finite, $v_i = P_i(\tau_0 < \tau_\Lambda)$ and $\mathscr{D}(v) = $ the current flowing out of 0. Using Ohm's law again, $1/\mathscr{D}(v)$ is the resistance of the circuit connecting 0 to Λ.

Remark Since p is recurrent if and only if $P_0(T_0 > \tau_\Lambda) \downarrow 0$ as $\Lambda \downarrow \varnothing$, the last result says that p is recurrent if and only if the network has infinite resistance. This interpretation gives two interesting results.

Exercise 4.4 *Comparison principle.* Use the last exercise to show that if $a_{ij} \leq \bar{a}_{ij}$ and the $\bar{p}$ associated with $\bar{a}$ is recurrent, then the p associated with a is also. A corollary of the last result is that a random walk on a subgraph of $\mathbb{Z}^2$ is recurrent.

Exercise 4.5 *Nash–Williams criterion.* Suppose $S = \Lambda_0 \cup \Lambda_1 \cup \Lambda_2 \cup \cdots$ where the Λ_k are disjoint finite sets so that if $i \in \Lambda_k$ and $a_{ij} > 0$, then $j \in \Lambda_{k-1} \cup \Lambda_k \cup \Lambda_{k+1}$ $(\Lambda_{-1} = \varnothing)$. Let

$$\alpha_k = \sum_{i \in \Lambda_{k-1}} \sum_{j \in \Lambda_k} a_{ij}.$$

Then

$$P_0(\tau_{\Lambda_n} < T_0) \leq (\alpha_0 \sigma_n)^{-1}$$

where

$$\sigma_n = \sum_{m=1}^n \alpha_k^{-1}.$$

Thus X_n is recurrent provided $\sum \alpha_k^{-1} = \infty$.
   Sketch: Let $a_{ij}^L = a_{ij} + Lb_{ij}$ where $b_{ij} = 1$ if i, j are in the same Λ_k. By Exercise 4.2 the probability of interest increases with L. Let h_x^L be the probability of hitting

Λ_n before 0 starting from x in the chain associated with a_{ij}^L. If x, $y \in \Lambda_k$, then $|h_x^L - h_y^L| \to 0$ as $L \to \infty$ and any subsequential limit satisfies

$$\hat{h}_{k+1} - \hat{h}_k = (\alpha_k / \alpha_{k+1})(\hat{h}_k - \hat{h}_{k-1})$$

where $\hat{h}_k$ is the common value of $\lim h_x^{L_j}$ for $x \in \Lambda_k$.

Remark   Doyle and Snell (1984) call this "Rayleigh's short cut method." Intuitively we set $a_{ij} = \infty$ if i, j are in the same Λ_k (i.e., introduce wires with no resistance). This reduces the network to a linear circuit, and the process to a birth and death chain. The formula for σ_n reflects the fact that resistances in series add, while if we have resistances r_i in parallel, the net resistance is $(\sum r_i^{-1})^{-1}$.

Exercise 4.6   Use the Nash Williams criterion to give a proof that a random walk on a subgraph of $\mathbb{Z}^2$ is recurrent.

Remark   Our treatment of Exercises 4.2–4.6 follows Griffeath and Liggett (1982). For more on this topic, see Lyons (1983) or Doyle and Snell (1984).

Examples 4.1–4.5 show that many Markov chains have stationary measures. The next result proves this.

(4.3) THEOREM   Let x be a recurrent state, and let $T = \inf\{n \geq 1 : X_n = x\}$. Then

$$\mu(y) = E_x \left(\sum_{n=0}^{T-1} 1_{\{X(n)=y\}} \right) = \sum_{n=0}^{\infty} P_x(X_n = y, T > n)$$

defines a stationary measure.

Proof   This is called the "cycle trick." The proof in words is simple. $\mu(y)$ is the expected number of visits to y in $\{0, \ldots, T - 1\}$. $\mu p(x) \equiv \sum \mu(x) p(x, y)$ is the expected number of visits to y in $\{1, \ldots, T\}$, which is $= \mu(y)$, since $X_T = X_0 = x$. To translate this intuition into a proof, let $\bar{p}_n(x, y) = P_x(X_n = y, T > n)$ and use Fubini's theorem to get

$$\sum_y \mu(y) p(y, z) = \sum_{n=0}^{\infty} \sum_y \bar{p}_n(x, y) p(y, z).$$

Case 1: $z \neq x$.

$$\sum_y \bar{p}_n(x, y) p(y, z) = \sum_y P_x(X_n = y, T > n, X_{n+1} = z)$$

$$= P_x(T > n + 1, X_{n+1} = z) = \bar{p}_{n+1}(x, z),$$

so

$$\sum_{n=0}^{\infty} \sum_{y} \bar{p}_n(x, y)p(y, z) = \sum_{n=0}^{\infty} \bar{p}_{n+1}(x, z) = \mu(z),$$

since $\bar{p}_0(x, z) = 0$.

   Case 2: $z = x$.

$$\sum_{y} \bar{p}_n(x, y)p(y, x) = \sum_{y} P_x(X_n = y, T > n, X_{n+1} = x) = P_x(T = n + 1),$$

so

$$\sum_{n=0}^{\infty} \sum_{y} \bar{p}_n(x, y)p(y, x) = \sum_{n=0}^{\infty} P_x(T = n + 1) = 1 = \mu(x),$$

since $P_x(T = 0) = 0$. □

Remark   If x is transient, then we have $\mu p(z) \leq \mu(z)$ with equality for all $z \neq x$.

Technical Note   To show that we are not cheating, we should prove that $\mu(y) < \infty$ for all y. First observe that $\mu p = \mu$ implies $\mu p^n = \mu$ for all $n \geq 1$, and $\mu(x) = 1$, so if $p^n(y, x) > 0$, then $\mu(y) < \infty$. Since the last result is true for all n, we see that $\mu(y) < \infty$ whenever $\rho_{yx} > 0$, but this is good enough. By (3.3), when x is recurrent, $\rho_{xy} > 0$ implies $\rho_{yx} > 0$, and it follows from the argument above that $\mu(y) < \infty$. If $\rho_{xy} = 0$, then $\mu(y) = 0$.

Exercise 4.7   (i) Use the construction in the proof of (4.3) to show that $\mu(j) = \sum_{k \geq j} f_{k+1}$ defines a stationary measure for the renewal chain (Example 1.4). (ii) Show that in this case the dual Markov chain defined in Exercise 4.1 represents the age of the item at use at time n; that is, the amount of time since the last renewal $\leq n$.

   (4.3) allows us to construct a stationary measure for each closed set of recurrent states. Conversely, we have:

(4.4) THEOREM   If p is irreducible and recurrent (i.e., all states are), then the stationary measure is unique up to constant multiples.

Proof   Let v be a stationary measure and let $a \in S$.

$$v(z) = \sum_{y} v(y)p(y, z) = v(a)p(a, z) + \sum_{y \neq a} v(y)p(y, z).$$

Using the last identity to replace $v(y)$ on the right-hand side,

$$v(z) = v(a)p(a, z) + \sum_{y \neq a} v(a)p(a, y)p(y, z) + \sum_{x \neq a} \sum_{y \neq a} v(x)p(x, y)p(y, z)$$

$$= v(a)P_a(X_1 = z) + v(a)P_a(X_1 \neq a, X_2 = z)$$

$$+ P_v(X_0 \neq a, X_1 \neq a, X_2 = z).$$

Continuing in the obvious way, we get

$$v(z) = v(a) \sum_{m=1}^{n} P_a(X_k \neq a, 1 \leq k < m, X_m = z)$$

$$+ P_v(X_j \neq a, 0 \leq j < n, X_n = z).$$

The last term is ≥ 0. Letting $n \to \infty$ gives

$$v(z) \geq v(a)\mu(z) \quad \text{where } \mu(z) = \sum_{m=1}^{\infty} P_a(X_k \neq a, 1 \leq k < m, X_m = z).$$

It follows from (4.3) that μ is a stationary distribution with $\mu(a) = 1$. (Here we are summing from 1 to T rather than from 0 to $T - 1$.) To turn the $\geq$ in the last equation into $=$, we observe

$$v(a) = \sum_{x} v(x)p^n(x, a) \geq v(a) \sum_{x} \mu(x)p^n(x, a) = v(a)\mu(a) = v(a).$$

Since $v(x) \geq v(a)\mu(x)$ and the left- and right-hand sides are equal, we must have $v(x) = v(a)\mu(x)$ whenever $p^n(x, a) > 0$. Since p is irreducible, it follows that $v(x) = v(a)\mu(x)$ for all $x \in S$, and the proof is complete. □

(4.3) and (4.4) make a good team. (4.3) gives us a formula for a stationary distribution and (4.4) shows it is unique up to constant multiples. Together they allow us to derive a lot of formulas.

Exercise 4.8   (i) Use (4.3) and (4.4) to show that for a simple random walk, the expected number of visits to k between successive visits to 0 is 1 for all k. (ii) Derive the last result by computing $P_1(T_k < T_0)$ and $P_k(T_k < T_0)$.

Exercise 4.9   Suppose X_n is irreducible recurrent and has stationary measure μ. Let $A \subset S$ and define T_A^k inductively by $T_A^0 = 0$, $T_A^k = \inf\{n > T_A^{k-1} : X_n \in A\}$. Exercise 2.5 implies that $Y_k = X(T_A^k)$ is a Markov chain. Use (4.3) and (4.4) to conclude that the restriction of μ to A is a stationary measure for Y.

Exercise 4.10   Show that if p is irreducible and recurrent and we let

$$\mu_x(y) = \sum_{n=0}^{\infty} P_x(X_n = y, T_x > n)$$

be the stationary measure that results when we use the cycle trick at x, then

$$\mu_x(y)\mu_y(z) = \mu_x(z).$$

Exercise 4.11   Let $w_{xy} = P_x(T_y < T_x)$ and μ_x be as in the last exercise. Show that

$$\mu_x(y) = w_{xy}/w_{yx}.$$

Exercise 4.12   *Another proof of* (4.4). Suppose p is irreducible and recurrent and let μ be the stationary measure constructed in (4.3). $\mu(x) > 0$ for all x and

$$q(x, y) = \mu(y)p(y, x)/\mu(x) \geq 0$$

defines a "dual" transition probability. (See Exercise 4.1.) (i) Show $q^n(x, y) = \mu(y)p^n(y, x)/\mu(x)$ and conclude q is irreducible and recurrent. (ii) Suppose $v(y) \geq \sum v(x)p(x, y)$ (i.e., v is an *excessive measure*) and let $h(x) = v(x)/\mu(x)$. Verify that $h(y) \geq \sum q(y, x)h(x)$ and use Exercise 3.12 to conclude that h is constant; that is, $v = c\mu$. The last result is stronger than (4.4). The remark after the proof of (4.3) shows that if p is irreducible and transient, there is an excessive measure for each $x \in S$.

Having examined the existence and uniqueness of stationary measures, we turn our attention now to *stationary distributions*—that is, probability measures π with $\pi p = \pi$. Stationary measures may exist for transient chains (e.g., random walks in $d \geq 3$), but:

(4.5) THEOREM   If there is a stationary distribution, then all states with $\pi(y) > 0$ are recurrent.

Proof   Since $\pi p^n = \pi$, Fubini's theorem implies

$$\sum_x \pi(x) \sum_{n=1}^{\infty} p^n(x, y) = \sum_{n=1}^{\infty} \pi(y) = \infty$$

when $\pi(y) > 0$. Using (3.2) now gives

$$\infty = \sum_x \pi(x)\rho_{xy}/(1 - \rho_{yy}) \leq 1/(1 - \rho_{yy}),$$

since $\rho_{xy} \leq 1$ and π is a probability measure. So $\rho_{yy} = 1$.   □

(4.6) THEOREM   If p is irreducible and has stationary distribution π, then

$$\pi(x) = 1/E_x T_x.$$

Proof   Irreducibility implies $\pi(x) > 0$, so all states are recurrent by (4.5). From (4.3)

$$\mu(y) = \sum_{n=0}^{\infty} P_x(X_n = y, T_x > n)$$

defines a stationary measure with $\mu(x) = 1$ and Fubini's theorem implies

$$\sum_y \mu(y) = \sum_{n=0}^{\infty} P_x(T_x > n) = E_x T_x.$$

By (4.4), the stationary measure is unique up to constant multiples, so $\pi(x) = \mu(x)/E_x T_x$.   □

If a state x has $E_x T_x < \infty$, it is said to be *positive recurrent*. A recurrent state with $E_x T_x = \infty$ is said to be *null recurrent*. (5.1) will explain these names. The next result helps us identify positive recurrent states.

(4.7) THEOREM   If p is irreducible, then the following are equivalent:

(i) Some x is positive recurrent.
(ii) There is a stationary distribution.
(iii) All states are positive recurrent.

Proof *(i) implies (ii):*   If x is positive recurrent, then

$$\pi(y) = \sum_{n=0}^{\infty} P_x(X_n = y, T_x > n)/E_x T_x$$

defines a stationary distribution. *(ii) implies (iii):*   (4.4) implies $\pi(y) = 1/E_y T_y$ and irreducibility tells us $\pi(y) > 0$, so $E_y T_y < \infty$. *(iii) implies (i)* is trivial.   □

Exercise 4.13   *Renewal chain.* Show that an irreducible renewal chain (Example 1.4) is positive recurrent (i.e., all the states are) if and only if $\mu = \sum k f_k < \infty$.

Exercise 4.14   Suppose p is irreducible and positive recurrent. Then $E_x T_y < \infty$ for all x, y.

(4.7) shows that being positive recurrent is a *class property*. If it holds for one state in an irreducible set, then it is true for all. Turning to our examples, the Ehrenfest chain (Example 4.1) is positive recurrent. The state space is finite, so there is a stationary distribution and the conclusion follows from (4.5). Random walks (Examples 4.2 and 4.3) are never positive recurrent. Since $\mu(x) \equiv 1$ is a stationary measure, $E_x T_x = \infty$.

Birth and death chains (Example 4.4) have a stationary distribution if and only if

$$\sum_x \prod_{k=1}^{x} (p_{k-1}/q_k) < \infty.$$

By (3.8) the chain is recurrent if and only if

$$\sum_{m=0}^{\infty} \prod_{j=1}^{m} (q_j/p_j) = \infty.$$

When $p_j = p$ and $q_j = (1 - p)$ for $j \geq 1$, there is a stationary distribution if and only if $p < 1/2$ and the chain is transient when $p > 1/2$. In Section 3 we probed the boundary between recurrence and transience by looking at examples with $p_j = 1/2 + \varepsilon_j$ where $\varepsilon_j \sim Cj^{-\alpha}$ as $j \to \infty$ and $C, \alpha \in (0, \infty)$. Since $\varepsilon_j \geq 0$ and hence $p_{j-1}/q_j \geq 1$ for large j, none of these chains have stationary distributions. If we look at chains with $p_j = 1/2 - \varepsilon_j$, then all we have done is interchange the roles of p and q, and results from the last section imply that the chain is positive recurrent when $\alpha < 1$, or $\alpha = 1$ and $C > 1/4$.

Random walks on graphs (Example 4.5) are irreducible if and only if the graph is connected. Since $\mu(i) \geq 1$ in the connected case, we have positive recurrence if and only if the graph is finite.

Exercise 4.15 Compute the expected number of moves it takes a knight to return to its initial position if it starts in a corner of the chessboard, assuming there are no other pieces on the board, and each time it chooses a move at random from its legal moves. (*Note:* A chessboard is $\{0, 1, \ldots, 7\}^2$. A knight's move is L-shaped: two steps in one direction followed by one step in a perpendicular direction.)

Example 4.6 *M/G/1 queue.* Let $\mu = \sum k a_k$ be the mean number of customers who arrive during one service time. In Example 3.5 we showed that the chain is recurrent if and only if $\mu \leq 1$. We will now show that the chain is positive recurrent if and only if $\mu < 1$. First suppose that $\mu < 1$. When $X_n > 0$, the chain behaves like a random walk that has jumps with mean $\mu - 1$, so if $N = \inf\{n \geq 0 : X_n = 0\}$, then $X_{N \wedge n} - (\mu - 1)(N \wedge n)$ is a martingale. If $X_0 = x > 0$, then the martingale property implies

$$x = E_x X_{N \wedge n} + (1 - \mu)E_x(N \wedge n) \geq (1 - \mu)E_x(N \wedge n),$$

since $X_{N \wedge n} \geq 0$ and it follows that $E_x N \leq x/(1 - \mu)$.

To prove that there is equality, observe that X_n decreases by at most 1 each time and for $x \geq 1$, $E_x T_{x-1} = E_1 T_0$, so $E_x N = cx$. To identify the constant observe that

$$E_1 N = 1 + \sum_{k=0}^{\infty} a_k E_k N,$$

so $c = 1 + \mu c$ and $c = 1/(1 - \mu)$. If $X_0 = 0$, then by considering what happens on the first jump we see that

$$E_0 T_0 = 1 + \sum_{k=0}^{\infty} a_k k/(1 - \mu) = 1 + \mu/(1 - \mu) = 1/(1 - \mu) < \infty.$$

This shows that the chain is positive recurrent if $\mu < 1$. To prove the converse observe that the arguments above show that if $E_0 T_0 < \infty$, then $E_k N < \infty$, $E_k N = ck$, and $c = 1/(1 - \mu)$, which is impossible if $\mu \geq 1$.

The last result when combined with (4.4) allows us to conclude that the stationary distribution has $\pi(0) = 1 - \mu$. This may not seem like much, but the equations in $\pi p = \pi$ are:

$$\pi(0) = \pi(0)a_0 + \pi(1)a_0$$

$$\pi(1) = \pi(0)a_1 + \pi(1)a_1 + \pi(2)a_0$$

$$\pi(2) = \pi(0)a_2 + \pi(1)a_2 + \pi(2)a_1 + \pi(3)a_0$$

or, in general,

$$\pi(j) = \pi(0)a_j + \sum_{i=1}^{j+1} \pi(i)a_{j+1-i}.$$

The equations have a "triangular" form, so knowing $\pi(0)$, we can solve for $\pi(1)$, $\pi(2)$, ... The first expression,

$$\pi(1) = \pi(0)(1 - a_0)/a_0,$$

is simple but the formulas get progressively messier, and there is no nice closed form solution except in one case: the M/M/1 queue.

In the M/M/1 queue the service times have an exponential distribution, which we assume to have mean 1. A little calculus shows that in this case

$$a_k = \int_0^\infty e^{-\lambda t} \frac{(\lambda t)^k}{k!} e^{-t} \, dt = \left(\frac{\lambda}{\lambda + 1}\right)^k \frac{1}{\lambda + 1},$$

but if the reader thinks for a moment about the Poisson process and the lack of memory property of the exponential distribution, she can see the result without computation. The mean $\sum k a_k = \lambda$, so if $\lambda < 1$ the chain is positive recurrent and $\pi(0) = \lambda$. The first equation yields $\pi(1) = (1 - \lambda)\lambda$. If we notice $a_{k+1} = a_k \lambda/(\lambda + 1)$ and use the equation for $\pi(j - 1)$, then the equation for $\pi(j)$ can be written as

$$\pi(j) = \pi(j - 1)\lambda/(\lambda + 1) + \pi(j + 1)/(1 + \lambda),$$

so

$$\pi(j + 1) = (\lambda + 1)\pi(j) - \lambda\pi(j - 1)$$

and one can guess and verify that $\pi(j) = (1 - \lambda)\lambda^j$.

Example 4.7 $M/M/\infty$ *queue.* In this chain, introduced in Exercise 3.9,

$$X_{n+1} = \sum_{m=1}^{X_n} \xi_{n,m} + Y_{n+1}$$

where $\xi_{n,m}$ are i.i.d. Bernoulli with mean p and Y_{n+1} is an independent Poisson mean λ. It follows from properties of the Poisson distribution that if X_n is Poisson with mean μ, then X_{n+1} is Poisson with mean $\mu p + \lambda$. Setting $\mu = \mu p + \lambda$, we find that a Poisson distribution with mean $\mu = \lambda/(1 - p)$ is a stationary distribution.

Exercise 4.16   There is a general result that handles Examples 4.6 and 4.7 and is useful in a number of other situations. Let $X_n \geq 0$ be a Markov chain and suppose $E_x X_1 \leq x - \varepsilon$ for $x \geq K$ where $\varepsilon > 0$. Let $Y_n = X_n + n\varepsilon$ and $\tau = \inf\{n : X_n \leq K\}$.

$Y_{n \wedge \tau}$ is a positive supermartingale and the optional stopping theorem implies $E\tau \leq X_0/\varepsilon$.

To close the section we will give a self-contained proof of:

(4.8) THEOREM If p is irreducible and has a stationary distribution π, then any other stationary measure is a multiple of π.

Remark This result is a consequence of (4.5) and (4.4), but we find the method of proof amusing.

Proof Since p is irreducible, $\pi(x) > 0$ for all x. Let φ be a concave function that is bounded on $(0, \infty)$; for example, $\varphi(x) = x/(x + 1)$. Define the *entropy* of μ by

$$\mathscr{E}(\mu) = \sum_y \varphi\left(\frac{\mu(y)}{\pi(y)}\right)\pi(y).$$

The reason for the name will become clear during the proof.

$$\mathscr{E}(\mu p) = \sum_y \varphi\left(\sum_x \frac{\mu(x)p(x, y)}{\pi(y)}\right)\pi(y) = \sum_y \varphi\left(\sum_x \frac{\mu(x)}{\pi(x)}\frac{\pi(x)p(x, y)}{\pi(y)}\right)\pi(y)$$

$$\geq \sum_y \sum_x \varphi\left(\frac{\mu(x)}{\pi(x)}\right)\frac{\pi(x)p(x, y)}{\pi(y)}\pi(y),$$

since φ is concave and $v(x) = \pi(x)p(x, y)/\pi(y)$ is a probability distribution. Since the $\pi(y)$'s cancel and $\sum_y p(x, y) = 1$, the last expression $= \mathscr{E}(\mu)$ and we have shown $\mathscr{E}(\mu p) \leq \mathscr{E}(\mu)$; that is, the entropy of an arbitrary initial measure μ is decreased by an application of p.

If $p(x, y) > 0$ for all x and y, and $\mu p = \mu$, it follows that $\mu(x)/\pi(x)$ must be constant, for otherwise there would be strict inequality in the application of Jensen's inequality. To get from the last special case to the general result, observe that if p is irreducible,

$$\bar{p}(x, y) = \sum_{n=1}^{\infty} 2^{-n}p^n(x, y) > 0 \quad \text{for all } x, y$$

and $\mu p = \mu$ implies $\mu\bar{p} = \mu$. $\square$

5 Asymptotic Behavior

In this section we will investigate the asymptotic behavior of X_n and $p^n(x, y)$. If y is transient, $\sum_n p^n(x, y) < \infty$, so $p^n(x, y) \to 0$ as $n \to \infty$. To deal with the recurrent states, we let

$$N_n(y) = \sum_{m=1}^{n} 1_{\{X(m)=y\}}$$

be the number of visits to y by time n.

(5.1) THEOREM   Suppose y is recurrent. For any $x \in S$, as $n \to \infty$,

$$N_n(y)/n \to \frac{1}{E_y T_y} 1_{\{T_y < \infty\}} \quad P_x \text{ a.s.}$$

Here $1/\infty = 0$.

Proof   Suppose that we start at y. Let $R(k) = \min\{n \geq 1 : N_n(y) = k\}$ = the time of the kth return to y. Let $t_k = R(k) - R(k-1)$ where $R(0) = 0$. Since we have assumed $X_0 = y$, $t_1, t_2, \ldots$ are i.i.d. and the strong law of large numbers implies

$$R(k)/k \to E_y T_y \quad \text{a.s.}$$

Since $R(N_n(y)) \leq n < R(N_n(y) + 1)$,

$$\frac{R(N_n(y))}{N_n(y)} \leq \frac{n}{N_n(y)} < \frac{R(N_n(y) + 1)}{N_n(y) + 1} \cdot \frac{N_n(y) + 1}{N_n(y)}.$$

Letting $n \to \infty$ and recalling $N_n(y) \to \infty$ a.s. since y is recurrent, we have

$$\frac{n}{N_n(y)} \to E_y T_y \quad \text{a.s.}$$

To generalize now to $x \neq y$, observe that if $T_y = \infty$, then $N_n(y) = 0$ for all n and hence

$$N_n(y)/n \to 0 \quad \text{on } \{T_y = \infty\}.$$

The strong Markov property implies that conditional on $\{T_y < \infty\}$, $t_2, t_3, \ldots$ are i.i.d. and have $P_x(t_k = n) = P_y(T_y = n)$, so

$$R(k)/k = t_1/k + (t_2 + \cdots + t_k)/k \to 0 + E_y T_y \quad \text{a.s.}$$

Repeating the last proof for the case $x = y$ shows

$$N_n(y)/n \to 1/E_y T_y \quad \text{a.s. on } \{T_y < \infty\},$$

and combining this with the result for $\{T_y = \infty\}$ completes the proof. $\square$

Remark   (5.1) should help explain the terms "positive" and "null recurrent." If we start from x, then in the first case the asymptotic fraction of time spent at x is positive and in the second case it is 0.

Since $0 \le N_n(y)/n \le 1$, it follows from the bounded convergence theorem that

$$E_x N_n(y)/n \to E_x(1_{\{T_y < \infty\}}/E_y T_y),$$

so

$$\frac{1}{n} \sum_{m=1}^{n} p^m(x, y) \to \rho_{xy}/E_y T_y. \qquad (5.2)$$

The last result was proved for recurrent y but also holds for transient y, since in that case $E_y T_y = \infty$ and the limit is 0, since $\sum_m p^m(x, y) < \infty$.

Exercise 5.1   *M/G/1 queue.* Let $\xi_1, \xi_2, \ldots$ be i.i.d. with $P(\xi_m = k) = a_{k+1}$ for $k \ge -1$. Let $S_n = x + \xi_1 + \cdots + \xi_n$ where $x \ge 0$, and let

$$X_n = S_n - \min_{m \le n} S_m.$$

Show that X_n has the same distribution as the M/G/1 queue (Example 1.5) starting from $X_0 = x$, and use this to conclude that if $\mu = \sum k a_k < 1$, then as $n \to \infty$,

$$\frac{1}{n}|\{m \le n : X_m = 0\}| \to (1 - \mu) \quad \text{a.s.}$$

This gives a roundabout way of getting the result $\pi(0) = 1 - \mu$ proved in Example 4.6.

Exercise 5.2   *Strong law for additive functionals.* Suppose p is irreducible and has stationary distribution π. Let f be a function with $\sum |f(y)|\pi(y) < \infty$. Let T_x^k be the time of the kth return to x. Exercise 3.1 implies

$$V_k = f(X(T_x^k)) + \cdots + f(X(T_x^{k+1} - 1)), \quad k \ge 1 \quad \text{are i.i.d.},$$

and (4.3) implies $E|V_k| < \infty$. Use this to conclude

$$\frac{1}{n} \sum_{m=1}^{n} f(X_m) \to \sum f(y)\pi(y) \quad P_\mu \text{ a.s.}$$

for any initial distribution μ. *Sketch:*   Let $K_n = \inf\{k : T_x^k \ge n\}$ and show that

$$\frac{1}{n} \sum_{m=1}^{K_n} f(X_m) \to EV_1/E_x T_x^1 = \sum f(y)\pi(y) \quad P_\mu \text{ a.s.}$$

and that $P(|V_n| > \varepsilon n \text{ i.o.}) = 0$ for any $\varepsilon > 0$.

Exercise 5.3   *Ratio limit theorems.* (5.1) does not say much about the null recurrent case. To get a more informative limit theorem, suppose that y is recurrent and μ is the (unique up to constant multiples) stationary measure on $C_y = \{z : \rho_{yz} > 0\}$. (i) Break up the path at successive returns to y and show that $N_n(z)/N_n(y) \to \mu(z)/\mu(y)$ P_x a.s. for all $x, z \in C_y$. (ii) Show that

$$\sum_{m=1}^{n} p^m(x, z) \Big/ \sum_{m=1}^{n} p^m(x, y) \to \mu(z)/\mu(y).$$

Sketch of proof of (ii):   Let $\bar{p}_n(x, z) = P_x(X_n = z, T_y > n)$ and decompose $p^m(x, z)$ according to $\sup\{j \le m : X_j = y\}$ to get

$$\sum_{m=1}^{n} p^m(x, z) = \sum_{m=1}^{n} \bar{p}_m(x, z) + \sum_{j=1}^{n-1} p^j(x, y) \sum_{k=1}^{n-j} \bar{p}_k(y, z).$$

The desired conclusion now follows easily from the observations

$$\sum_{m=1}^{\infty} \bar{p}_m(x, z) < \infty \quad \text{and} \quad \sum_{k=1}^{\infty} \bar{p}_k(y, z) = \mu(z)/\mu(y).$$

(5.2) shows that the sequence $p^n(x, y)$ always converges in the Cesaro sense. The next example shows that $p^n(x, y)$ need not converge.

Example 5.1 $p = \begin{pmatrix} 0 & 1 \\ 1 & 0 \end{pmatrix}$, $p^2 = \begin{pmatrix} 1 & 0 \\ 0 & 1 \end{pmatrix}$, $p^3 = p$, $p^4 = p^2, \ldots$.

A similar problem also occurs in the Ehrenfest chain. In that case if X_0 is even, then X_1 is odd, X_2 is even, ..., so $p^n(x, x) = 0$ unless n is even. It is easy to construct examples with $p^n(x, x) = 0$ unless n is a multiple of 3 or 239 or ....

(5.5) below will show that this "periodicity" is the only thing that can prevent the convergence of the $p^n(x, y)$. First we need a definition and two preliminary results. Let x be a recurrent state, let $I_x = \{n \ge 1 : p^n(x, x) > 0\}$, and let d_x be the greatest common divisor (g.c.d.) of I_x. d_x is called the *period* of x. The first result says that the period is a class property.

(5.3) LEMMA   If $\rho_{xy} > 0$, then $d_y = d_x$.

Proof   Let K and L be such that $p^K(x, y) > 0$ and $p^L(y, x) > 0$. (x is recurrent, so $\rho_{yx} > 0$.)

$$p^{K+L}(y, y) \ge p^L(y, x)p^K(x, y) > 0,$$

so d_y divides $K + L$ (abbreviated $d_y | K + L$). Let n be such that $p^n(x, x) > 0$.

$$p^{K+n+L}(y, y) \ge p^L(y, x)p^n(x, x)p^K(x, y) > 0,$$

so $d_y | K + n + L$ and hence $d_y | n$. Since $n \in I_x$ is arbitrary, $d_y | d_x$. Interchanging the roles of y and x gives $d_x | d_y$ and hence $d_x = d_y$. □

The next result implies that $I_x \supset \{m \cdot d_x : m \geq m_0\}$. (Apply (5.4) to $p^{d(x)}$.)

(5.4) LEMMA   If $d_x = 1$, then $p^m(x, x) > 0$ for $m \geq m_0$.

Proof by Example   Suppose $4, 7 \in I_x$. $p^{m+n}(x, x) \geq p^m(x, x)p^n(x, x)$, so I_x is a semi-group; that is, if $m, n \in I_x$, then $m + n \in I_x$. A little calculation shows that in the example

$$I_x \supset \{4, \quad 7, 8, \quad 11, 12, \quad 14, 15, 16, \quad 18, 19, 20, 21, \quad \ldots\},$$

so the result is true with $m_0 = 18$. (Once I_x contains four consecutive integers, it will contain all the rest.)

Proof   Our first goal is to prove that I_x contains two consecutive integers. Let n_0, $n_0 + k \in I_x$. If $k = 1$, we are done. If not, then since the greatest common divisor of I_x is 1, there is an $n_1 \in I_x$ so that k is not a divisor of n_1. Write $n_1 = mk + r$ with $0 < r < k$. Since I_x is a semigroup, $(m + 1)(n_0 + k) > (m + 1)n_0 + n_1$ are both in I_x. Their difference is

$$k(m + 1) - n_1 = k - r < k.$$

Repeating the last argument (at most k times), we eventually arrive at a pair of consecutive integers $N, N + 1$ in I_x. It is now easy to show that the result holds for $m_0 = N^2$. Let $m \geq N^2$ and write $m - N^2 = kN + r$ with $0 \leq r < N$. Then

$$m = r + N^2 + kN = r(1 + N) + (N - r + k)N \in I_x.$$ □

(5.5) Convergence Theorem   Suppose p is irreducible, aperiodic (i.e., all states have $d_x = 1$), and has stationary distribution π. Then as $n \to \infty$, $p^n(x, y) \to \pi(y)$.

Proof   Let $S^2 = S \times S$. Define a transition probability $\bar{p}$ on $S \times S$ by

$$\bar{p}((x_1, y_1), (x_2, y_2)) = p(x_1, x_2)p(y_1, y_2);$$

that is, each coordinate moves independently. Our first step is to check that $\bar{p}$ is irreducible. This may seem like a silly thing to do first, but this is the only step that requires aperiodicity. Since p is irreducible, there are K, L, so that $p^K(x_1, x_2) > 0$ and $p^L(y_1, y_2) > 0$. From (5.4) it follows that if M is large, $p^{L+M}(x_2, x_2) > 0$ and $p^{K+M}(y_2, y_2) > 0$, so

$$\bar{p}^{K+L+M}((x_1, y_1), (x_2, y_2)) > 0.$$

Our second step is to observe that since the two coordinates are independent, $\bar{\pi}(a, b) = \pi(a)\pi(b)$ defines a stationary distribution for $\bar{p}$, and (4.5) implies that for $\bar{p}$

all states are recurrent. Let (X_n, Y_n) denote the chain on $S \times S$, and let T be the first time that this chain hits the diagonal $\{(y, y) : y \in S\}$. Let $T_{(x,x)}$ be the hitting time of (x, x). Since $\bar{p}$ is irreducible and recurrent, $T_{(x,x)} < \infty$ a.s. and hence $T < \infty$ a.s. The final step is to observe that on $\{T \le n\}$, the two coordinates X_n and Y_n have the same distribution. By considering the time and place of the first intersection and then using the Markov property,

$$P(X_n = y, T \le n) = \sum_{m=1}^{n} \sum_{x} P(T = m, X_m = x, X_n = y)$$

$$= \sum_{m=1}^{n} \sum_{x} P(T = m, X_m = x) P(X_n = y | X_m = x)$$

$$= \sum_{m=1}^{n} \sum_{x} P(T = m, Y_m = x) P(Y_n = y | Y_m = x)$$

$$= P(Y_n = y, T \le n).$$

To finish up we observe that

$$P(X_n = y) = P(Y_n = y, T \le n) + P(X_n = y, T > n)$$

$$\le P(Y_n = y) + P(X_n = y, T > n),$$

and similarly $P(Y_n = y) \le P(X_n = y) + P(Y_n = y, T > n)$. So

$$|P(X_n = y) - P(Y_n = y)| \le P(X_n = y, T > n) + P(Y_n = y, T > n),$$

and summing over y gives

$$\sum_{y} |P(X_n = y) - P(Y_n = y)| \le 2P(T > n).$$

If we let $X_0 = x$ and let Y_0 have the stationary distribution π, then Y_n has distribution π and it follows that

$$\sum_{y} |p^n(x, y) - \pi(y)| \le 2P(T > n) \to 0,$$

proving the desired result. If we recall the definition of the total variation distance given in Section 6 in Chapter 2, the last conclusion can be written as

$$\|p^n(x, \cdot) - \pi(\cdot)\| \le P(T > n) \to 0. \qquad \square$$

At first glance it may seem strange to prove the convergence theorem by running independent copies of the chain. An approach that is slightly more complicated but explains better what is happening is to define

$$q((x_1, y_1), (x_2, y_2)) = \begin{cases} p(x_1, x_2)p(y_1, y_2) & \text{if } x_1 \neq y_1 \\ p(x_1, x_2) & \text{if } x_1 = y_1, x_2 = y_2 \\ 0 & \text{otherwise.} \end{cases}$$

In words, the two coordinates move independently until they hit and then move together. It is easy to see from the definition that each coordinate is a copy of the original process. If T' is the hitting time of the diagonal for the new chain (X'_n, Y'_n), then $X'_n = Y'_n$ on $T' \leq n$, so it is clear that

$$\sum_y |P(X'_n = y) - P(Y'_n = y)| \leq 2P(X'_n \neq Y'_n) = 2P(T' > n).$$

On the other hand, T and T' have the same distribution, so $P(T' > n) \to 0$ and the conclusion follows as before. The technique used in the proof is called *coupling*. Generally this term refers to building two sequences X_n and Y_n on the same space to conclude that X_n converges in distribution by showing $P(X_n \neq Y_n) \to 0$, or more generally that for some metric ρ, $P(\rho(X_n, Y_n) > \varepsilon) \to 0$ for all $\varepsilon > 0$.

Having completed the proof of (5.5) we pause to show that coupling can be fun and it doesn't always happen.

Example 5.2   *A coupling card trick.* The following demonstration used by E. B. Dynkin in his probability class is a variation of a card trick that appeared in *Scientific American.* The instructor asks a student to write 100 random digits from 0 to 9 on the blackboard. Another student chooses one of the first 10 numbers and does not tell the instructor. If that digit is 7, say, she counts 7 places along the list, notes the digit at that location, and continues the process. If the digit is 0, she counts 10. A possible sequence is underlined on the list below:

3 4 <u>7</u> 8 2 3 7 5 6 <u>1</u> <u>6</u> 4 6 5 7 8 <u>3</u> 1 5 <u>3</u> 0 7 <u>9</u> 2 3 ...

The trick is that, without knowing the student's first digit, the instructor can point to her final stopping position. To this end, he picks the first digit and forms his own sequence in the same manner as the student and announces his stopping position. He makes an error if the coupling time is larger than 100. Numerical computations done by one of Dynkin's graduate students show that the probability of error is approximately .026.

Example 5.3   *There is a transition probability that is irreducible, aperiodic, and recurrent but has the property that two independent particles need not meet.* Let S_n be a modified two-dimensional simple random walk that stays where it is with probability 1/5 and jumps to each of its four neighbors with probability 1/5 each. Let $S_0 = (0, 0)$, $T = \inf\{n \geq 1 : S_n = (0, 0)\}$, and $f_k = P(T = k)$. Let p be the transition probability with

$$p(0, j) = f_{j+1}, \qquad p(i, i - 1) = 1, \qquad p(i, j) = 0 \quad \text{otherwise.}$$

p is the renewal chain corresponding to the distribution f, so p is recurrent. $f_k > 0$ for all k, so p is irreducible and aperiodic. Let S_n^1 and S_n^2 be independent copies of the random walk starting from $S_0^1 \neq S_0^2$, and let $X_n^i = \inf\{m - n : m \geq n, S_m^i = (0, 0)\}$. From our discussion of the renewal chain in Example 1.4, it follows that X_n^1 and X_n^2 are independent Markov chains with transition probability p. It is easy to see that if $X_n^1 = X_n^2 = j$, then $X_{n+j}^1 = X_{n+j}^2 = 0$, and at this time the four-dimensional random walk (S_n^1, S_n^2) is at $(0, 0, 0, 0)$. Since

$$P((S_n^1, S_n^2) = (0, 0, 0, 0) \text{ for some } n \geq 1) < 1,$$

we have proved our claim.

(5.5) applies almost immediately to the examples considered in Section 1. The M/G/1 queue has $a_k > 0$, so if $\mu < 1$, $P_x(X_n = y) \to \pi(y)$ for any x, $y \geq 0$. The same result holds for the renewal chain provided the greatest common divisor of $\{k : f_k > 0\}$ is 1. In this case $P_x(X_n = 0) \to \pi(0) = 1/v$ where $v = \sum k f_k$ is the mean time between renewals.

Exercise 5.4   Historically the first chain for which (5.5) was proved was the *Bernoulli–Laplace model of diffusion*. Suppose two urns, which we will call left and right, have m balls each. b balls are black and $2m - b$ are white. At each time we pick one ball from each urn and interchange them. Let the state at time n be the number of black balls in the left urn. Compute the transition probability for this chain, find its stationary distribution, and use (5.5) to conclude that the chain approaches equilibrium as $n \to \infty$.

Exercise 5.5   Consider a Markov chain X_n on $\{0, 1, \ldots, N\}$ with transition probability

$$p(i, j) = \binom{N}{j}(p_i)^j(1 - p_i)^{N-j}$$

where $p_i = (i/N)(1 - \alpha_1) + (1 - i/N)\alpha_2$. i is the number of individuals of type 1 in generation n for a population that is held constant at size N, if $\alpha_i > 0$ is the rate at which individuals of type i mutate to the other type. Use (5.5) to conclude that the chain approaches equilibrium as $n \to \infty$. Can you find the stationary distribution?

Exercise 5.6   Consider the chain with transition probability

$$p = \begin{pmatrix} 1/2 & 1/2 & 0 \\ 1/2 & 0 & 1/2 \\ 0 & 1/2 & 1/2 \end{pmatrix}.$$

Suppose X_n starts at x and Y_n starts with the stationary distribution $(1/3, 1/3, 1/3)$. Show that the coupling time defined in the proof of (5.5) has $P(T > n) = (2/3)(1/2)^n$. To see that this accurately measures the rate of convergence, observe that by results in Exercise 1.3, $p^n(1, 1) = 1/3 + (1/2)^{n+1} + (1/6)(-1/2)^n$.

Exercise 5.7   Show that if S is finite, p is irreducible and aperiodic, and T is the coupling time defined in the proof of (5.5), then $P(T > n) \leq Cr^n$ for some $r < 1$ and $C < \infty$. So the convergence to equilibrium occurs exponentially rapidly in this case.

Hint   First consider the case in which $p(x, y) > 0$ for all x and y, and then reduce the general case to this one by looking at a power of p.

Exercise 5.8   For any transition matrix p, define

$$\alpha_n = \sup_{i,j} (1/2) \sum_k |p^n(i, k) - p^n(j, k)|.$$

(i) Show that $\alpha_{m+n} \leq \alpha_n \alpha_m$. (ii) Use (9.1) in Chapter 1 to conclude that

$$\frac{1}{n} \log \alpha_n \to \inf_{m \geq 1} \frac{1}{m} \log \alpha_m,$$

so if $\alpha_m < 1$ for some m, it approaches 0 exponentially fast.

Hint   It is easier to use coupling than algebra. Note that for any i and j we can define r.v.'s X and Y so that $P(X = k) = p^n(i, k)$, $P(Y = k) = p^n(j, k)$, and

$$P(X \neq Y) = (1/2) \sum_k |p^n(i, k) - p^n(j, k)|.$$

And now for something completely different:

Example 5.4   *Shuffling cards.* The state of a deck of n cards can be represented by a permutation, $\pi(i)$ giving the location of the ith card. Consider the following method of mixing the deck. The top card is removed and inserted under one of the $n - 1$ cards that remain. I claim that by following the bottom card of the deck we can see that it takes about $n \log n$ moves to mix up the deck. This card stays at the bottom until the first time (T_1) a card is inserted below it. It is easy to see that when the kth card is inserted below the original bottom card (at time T_k), all $k!$ arrangements of the cards below are equally likely, so at time $\tau_n = T_{n-1} + 1$, all $n!$ arrangements are equally likely. If we let $T_0 = 0$ and $t_k = T_k - T_{k-1}$ for $1 \leq k \leq n - 1$, then these r.v.'s are independent and t_k has a geometric distribution with success probability $1/n - k$. These waiting times are the same as the ones in the coupon collector's problem (Example 5.10 in Chapter 1), so $\tau_n/(n \log n) \to 1$ in probability as $n \to \infty$. For more on card shuffling, see Aldous and Diaconis (1986).

Problem (Too hard to be an exercise!)   Suppose we shuffle cards by picking i and j from $\{1, \ldots, N\}$ and interchange the cards in the ith and jth positions. Show that the deck reaches equilibrium in $CN \log N$ exchanges. It is not hard to find a coupling that works in about N^2 steps.

Exercise 5.9   *Random walk on the hypercube.* Consider $\{0, 1\}^d$ as a graph with edges connecting each pair of points that differ in only one coordinate. Let X_n be a random walk on $\{0, 1\}^d$ that stays put with probability $1/2$ and jumps to one of its d neighbors with probability $1/2d$ each. Let Y_n be another copy of the chain in which

Y_0 is uniformly distributed on $\{0, 1\}^d$. We construct a coupling of X_n and Y_n by letting $U_1, U_2, \ldots$ be uniform on $\{1, 2, \ldots, 2d\}$. The jth coordinate of X and Y is set equal to 1 at time n if $U_n = 2j - 1$, and set equal to 0 if $U_n = 2j$. Let $T_d = \inf\{m : \{U_1, \ldots, U_m\} = \{1, 2, \ldots, 2d\}\}$. Show that for $n \geq T_d$, $X_n = Y_n$. Results for the coupon collector's problem (Example 5.10 in Chapter 1) show that $T_d/(d \log d) \to 1$ in probability as $d \to \infty$.

We turn now to the periodic case.

(5.6) LEMMA   Suppose p is irreducible, recurrent, and all states have period d. Fix $x \in S$ and for each $y \in S$, let $K_y = \{n \geq 1 : p^n(x, y) > 0\}$. (i) There is an $r_y \in \{0, 1, \ldots, d - 1\}$ so that if $n \in K_y$, then $n = r_y$ mod d; that is, the difference $n - r_y$ is a multiple of d. (ii) Let $S_r = \{y : r_y = r\}$ for $0 \leq r < d$. If $y \in S_i$, $z \in S_j$, and $p^n(y, z) > 0$, then $n = (j - i)$ mod d. (iii) $S_0, S_1, \ldots, S_{d-1}$ are irreducible classes for p^d, and all states have period 1.

Proof   (i) Let $m(y)$ be such that $p^{m(y)}(y, x) > 0$. If $n \in K_y$, then $p^{n+m(y)}(x, x) > 0$, so $d|(n + m(y))$. Let $r_y = (d - m(y))$ mod d. (ii) Let m, n be such that $p^m(y, z)$, $p^n(x, y) > 0$. Since $p^{n+m}(x, z) > 0$, it follows from (i) that $n + m = j$ mod d. Since $m = i$ mod d, the result follows. The irreducibility in (iii) follows immediately from (ii). The aperiodicity follows from the definition of the period as the g.c.d. $\{x : p^n(x, x) > 0\}$.   □

A partition of the state space $S_0, S_1, \ldots, S_d$ satisfying (ii) in (5.6) is called a *cyclic decomposition* of the state space. Except for the choice of the set to put first, it is unique. (Pick an $x \in S$. It lies in some S_j, but once the value of j is known, irreducibility and (ii) allow us to calculate all the sets.)

(5.7) *Convergence Theorem, Periodic Case*   Suppose p is irreducible, has a stationary distribution π, and all states have period d. Let $x \in S$, and let $S_0, S_1, \ldots, S_d$ be the cyclic decomposition of the state space with $x \in S_0$. If $y \in S_r$, then

$$\lim_{m \to \infty} p^{md+r}(x, y) = \pi(y)d.$$

Proof   If $y \in S_0$, then using (iii) in (5.6) and applying (5.5) to p^d show

$$\lim_{m \to \infty} p^{md}(x, y) = \pi(y)d.$$

To identify the limit, use Exercise 4.9 or note that (5.2) implies

$$\frac{1}{n} \sum_{m=1}^{n} p^m(x, y) \to \pi(x),$$

and (ii) of (5.6) implies $p^m(x, y) = 0$ unless $d|m$. If $y \in S_r$ with $1 \leq r < d$, then

$$p^{md+r}(x, y) = \sum_{z \in S_r} p^r(x, z)p^{md}(z, y).$$

Since $y, z \in S_r$, it follows from the first case in the proof that $p^{md}(z, y) \to \pi(y)d$ as $m \to \infty$. $p^{md}(z, y) \leq 1$ and $\sum_z p^r(x, z) = 1$, so (5.7) follows from the dominated convergence theorem. $\qquad \square$

Exercise 5.10   *Central limit theorem for additive functionals.* Suppose p is irreducible and has stationary distribution π. Let f be a function with $\sum f(y)\, \pi(y) = 0$. Let T_x^k be the time of the kth return to x. Exercise 3.1 implies

$$V_k = f(X(T_x^k)) + \cdots + f(X(T_x^{k+1} - 1)), \quad k \geq 1, \quad \text{are i.i.d.}$$

Suppose $\sum f(x)\, \pi(x) = 0$ and $E_x V_k^2 < \infty$. Let $K_n = \inf\{k : T_x^k > n\}$. (i) Show that if $a_n \to \infty$, then $(\sum_{n < m \leq K_n} f(X_m))/a_n \to 0$ in probability. (In the aperiodic case $\sum_{n < m \leq K_n} f(X_m)$ has a limit.) (ii) Use (i) and the random index central limit theorem (Exercise 4.12 in Chapter 2) to conclude that for any initial distribution μ,

$$\sum_{m=1}^n f(X_m)/\sqrt{n} \Rightarrow c\chi \quad \text{under } P_\mu.$$

Let $\mathscr{F}_n' = \sigma(X_{n+1}, X_{n+2}, \ldots)$ and $\mathscr{T} = \bigcap_n \mathscr{F}_n'$ be the tail σ-field. The next result is due to Orey. The proof we give is from Blackwell and Freedman (1964).

(5.8) THEOREM   Suppose p is irreducible, recurrent, and all states have period d.

$$\mathscr{T} = \sigma(\{X_0 \in S_r\} : 0 \leq r < d).$$

Remark   To be precise, if μ is any initial distribution and $A \in \mathscr{T}$, then there is an r so that $A = \{X_0 \in S_r\}$ P_μ a.s.

Proof   We build up to the general result in three steps.
    Case 1:   Suppose $P(X_0 = x) = 1$. Let $T_0 = 0$ and for $n \geq 1$, let $T_n = \inf\{m > T_{n-1} : X_m = x\}$ be the time of the nth return to x. Let

$$V_n = (X(T_{n-1}), \ldots, X(T_n - 1)).$$

The vectors V_n are i.i.d. by Exercise 3.1, and the tail σ-field is contained in the exchangeable field of the V_n, so the Hewitt–Savage 0–1 law ((1.1) in Chapter 3, proved there for r.v.'s taking values in a general measurable space) implies that $\mathscr{T}$ is trivial in this case.
    Case 2:   Suppose that the initial distribution is concentrated on one cyclic class—say, S_0. If $A \in \mathscr{T}$, then $P_x(A) \in \{0, 1\}$ for each x by Case 1. If $P_x(A) = 0$ for all $x \in S_0$, then $P_\mu(A) = 0$. Suppose $P_y(A) > 0$ and hence $= 1$ for some $y \in S_0$. Let $z \in S_0$. Since p^d is irreducible and aperiodic on S_0, there is an n so that $p^n(z, y) > 0$ and $p^n(y, y) > 0$. If we write $1_A = 1_B \circ \theta_n$, then the Markov property implies

$$1 = P_y(A) = E_y(E_y(1_B \circ \theta_n | \mathscr{F}_n)) = E_y(E_{X(n)} 1_B),$$

so $P_y(B) = 1$. Another application of the Markov property gives

$$P_z(A) = E_z(E_{X(n)}1_B) \geq p^n(z, y) > 0,$$

so $P_z(A) = 1$, and since $z \in S_0$ is arbitrary, $P_\mu(A) = 1$.

   General case: From Case 2 we see that $P(A|X_0 = y) \equiv 1$ or $\equiv 0$ on each cyclic class. This implies that either $\{X_0 \in S_r\} \subset A$ or $\{X_0 \in S_r\} \cap A = \varnothing$ P_μ a.s. Conversely it is clear that $\{X_0 \in S_r\} = \{X_{nd} \in S_r \text{ i.o.}\} \in \mathcal{T}$, and the proof is complete.

□

The next result will help us identify the tail σ-field in transient examples.

(5.9) THEOREM   Suppose X_0 has initial distribution μ. The equations $h(X_n, n) = E_\mu(Z|\mathcal{F}_n)$ and $Z = \lim_{n\to\infty} h(X_n, n)$ set up a 1–1 correspondence between bounded $Z \in \mathcal{T}$ and bounded *space–time harmonic functions*—that is, bounded $h : S \times \{0, 1, \dots\} \to \mathbb{R}$, so that $h(X_n, n)$ is a martingale.

Proof   Let $Z \in \mathcal{T}$, write $Z = Y_n \circ \theta_n$, and let $h(x, n) = E_x Y_n$.

$$E_\mu(Z|\mathcal{F}_n) = E_\mu(Y_n \circ \theta_n|\mathcal{F}_n) = h(X_n, n)$$

by the Markov property, so $h(X_n, n)$ is a martingale. Conversely if $h(X_n, n)$ is a bounded martingale, using (2.10) and (5.4) from Chapter 4 shows $h(X_n, n) \to Z \in \mathcal{T}$ as $n \to \infty$, and $h(X_n, n) = E_\mu(Z|\mathcal{F}_n)$.

□

Exercise 5.11   A random variable Z with $Z = Z \circ \theta$, and hence $= Z \circ \theta_n$ for all n, is called *invariant*. Show there is a 1–1 correspondence between bounded invariant random variables and bounded harmonic functions. We will have more to say about invariant r.v.'s in Section 1 of Chapter 6.

Example 5.5   *Simple random walk in d dimensions.* We begin by constructing a coupling for this process. Let $i_1, i_2, \dots$ be i.i.d. uniform on $\{1, \dots, d\}$. Let $\xi_1, \xi_2, \dots$ and $\eta_1, \eta_2, \dots$ be i.i.d. uniform on $\{-1, 1\}$. Let e_j be the jth unit vector. Construct a coupled pair of d-dimensional simple random walks by

$$X_n = X_{n-1} + e(i_n)\xi_n$$

$$Y_n = \begin{cases} Y_{n-1} + e(i_n)\xi_n & \text{if } X_{n-1}^i = Y_{n-1}^i \\ Y_{n-1} + e(i_n)\eta_n & \text{if } X_{n-1}^i \neq Y_{n-1}^i. \end{cases}$$

In words, the coordinate that changes is always the same in the two walks, and once they agree in one coordinate, future movements in that direction are the same. It is easy to see that if $X_0^i - Y_0^i$ is even for $1 \leq i \leq d$, then the two random walks will hit with probability 1.

   Let $L_0 = \{z \in \mathbb{Z}^d : z^1 + \cdots + z^d \text{ is even}\}$ and $L_1 = \mathbb{Z}^d - L_0$. Although we have only defined the notion for the recurrent case, it should be clear that L_0, L_1 is the cyclic decomposition of the state space for a simple random walk. If $S_n \in L_i$, then $S_{n+1} \in L_{1-i}$ and p^2 is irreducible on each L_i. To couple two random walks starting from x, $y \in L_i$, let them run independently until the first time all the coordinate differences are even, and then use the last coupling. In the remaining case $x \in L_0$, $y \in L_1$ coupling is impossible.

The next result should explain our interest in coupling two d-dimensional simple random walks.

(5.10) THEOREM   For a d-dimensional simple random walk, $\mathcal{T} = \sigma(\{X_0 \in L_i\}$, $i = 1, 2)$.

Proof   Let x, $y \in L_i$, and let X_n, Y_n be a realization of the coupling defined above for $X_0 = x$ and $Y_0 = y$. Let $h(x, n)$ be a bounded space–time harmonic function. The martingale property implies $h(x, 0) = E_x h(X_n, n)$. If $|h| \leq C$, it follows from the coupling that

$$|h(x, 0) - h(y, 0)| = |Eh(n, X_n) - Eh(n, Y_n)| \leq CP(X_n \neq Y_n) \to 0,$$

so $h(x, 0)$ is constant on L_0 and L_1. Applying the last result to $h'(x, m) = h(x, n + m)$, we see that $h(x, n) = a_n^i$ on L_i. The martingale property implies $a_n^i = a_{n+1}^{1-i}$, and the desired result follows from (5.9).   □

Exercise 5.12   Let $p(x, y) = f(y - x)$ be the transition probability for an irreducible aperiodic random walk on $\mathbb{Z}$. Prove that the tail σ-field is trivial by using *Ornstein's coupling*. Pick M large enough so that the random walk generated by the probability distribution $f_M(x)$ with $f_M(x) = c_M f(x)$ for $|x| \leq M$ and $f_M(x) = 0$ for $|x| > M$ is irreducible and aperiodic. Let $Z_1, Z_2, \ldots$ be i.i.d. with distribution f and let $W_1, W_2, \ldots$ be i.i.d. with distribution f_M. Let $X_n = X_{n-1} + Z_n$ for $n \geq 1$. If $|Z_n| > M$, let $Y_n = Y_{n-1} + Z_n$. If $|Z_n| \leq M$, let $Y_n = Y_{n-1} + W_n$. In words, the big jumps are taken in parallel and the small jumps are independent. Use the recurrence of one-dimensional random walks with mean 0 to conclude $P(X_n \neq Y_n) \to 0$ and then (5.9) to show $\mathcal{T}$ is trivial.

The tail σ-field in (5.10) is essentially the same as in (5.8). To get a more interesting $\mathcal{T}$, we look at:

Example 5.6   *Random walk on a tree.* To facilitate definitions, we will consider the system as a random walk on a group with three generators, a, b, c, which have $a^2 = b^2 = c^2 = e$, the identity element. To form the random walk, let $\xi_1, \xi_2, \ldots$ be i.i.d. with $P(\xi_n = x) = 1/3$ for $x = a, b, c$, and let $X_n = X_{n-1}\xi_n$. (This is equivalent to a random walk on the tree in which each vertex has degree 3, but the algebraic formulation is convenient for computations.) Let L_n be the length of the word X_n when it has been reduced as much as possible, with $L_n = 0$ if $X_n = e$. The reduction can be done as we go along. If the last letter of X_{n-1} is the same as ξ_n, we erase it; otherwise we add the new letter. It is easy to see that L_n is a Markov chain with a transition probability that has $p(0, 1) = 1$ and

$$p(j, j - 1) = 1/3, \qquad p(j, j + 1) = 2/3 \quad \text{for } j \geq 1.$$

As $n \to \infty$, $L_n \to \infty$. From this it follows easily that the word X_n has a limit in the sense that the ith letter X_n^i stays the same for large n. Let X_∞ be the limiting word; that is, $X_\infty^i = \lim X_n^i$. $\mathcal{T} \supset \sigma(X_\infty^i, i \geq 1)$, but it is easy to see that this is not all. If $S_0 =$ the words of even length and $S_1 = S_0^c$, then $X_n \in S_i$ implies $X_{n+1} \in S_{1-i}$, so $\{X_0 \in S_0\} \in \mathcal{T}$. Can the reader prove that we have now found all of $\mathcal{T}$? As Fermat once said, " I have a proof but it won't fit in the margin."

*6 General State Space

In this section we will generalize the results from the last three sections to a collection of Markov chains with uncountable state space called *Harris chains*. The developments here are motivated by three ideas. First, the proofs in the last two sections work if there is one point in the state space that the chain hits with probability 1. (Think, for example, about the construction of the stationary measure in (4.3).) Second, a recurrent Harris chain can be modified to contain such a point. Third, the collection of Harris chains is a comfortable level of generality—broad enough to contain a large number of interesting examples, yet restrictive enough to allow for a rich theory.

We say that a Markov chain X_n is a *Harris chain* if we can find sets $A, B \in \mathcal{S}$, a function q with $q(x, y) \geq \varepsilon > 0$ for $x \in A$, $y \in B$, and a probability measure ρ concentrated on B so that:

(i) If $\tau_A = \inf\{n \geq 0 : X_n \in A\}$, then $P_z(\tau_A < \infty) > 0$ for all $z \in S$.
(ii) If $x \in A$ and $C \subset B$, then $p(x, C) \geq \int_C q(x, y)\rho(dy)$.

To explain the definition we turn to some examples.

Example 6.1 *Countable state space.* If S is countable and there is a point a with $\rho_{xa} > 0$ for all x (a condition slightly weaker than irreducibility), then we can take $A = \{a\}$, $B = \{b\}$ where b is any state with $p(a, b) > 0$, $\mu = $ a point mass at b, and $q(a, b) = p(a, b)$.

Example 6.2 *Chains with continuous densities.* Suppose $X_n \in \mathbb{R}^d$ is a Markov chain with a transition probability that has $p(x, dy) = p(x, y)\, dy$ where $(x, y) \to p(x, y)$ is continuous. Pick (x_0, y_0) so that $p(x_0, y_0) > 0$. Let A and B be open sets around x_0 and y_0 that are small enough so that $p(x, y) \geq \varepsilon > 0$ on $A \times B$. If we let $\rho(C) = |B \cap C|/|B|$ where $|B|$ is the Lebesgue measure of B, then (ii) holds. If (i) holds, then X_n is a Harris chain.

For concrete examples, consider

(a) *Diffusion processes* are a large class of examples that lie outside the scope of this book but are too important to ignore. Specifically, if the generator of X_t has Hölder continuous coefficients satisfying suitable growth conditions (see Dynkin (1965), Appendix), then $P(X_1 \in dy) = p(x, y)\, dy$ and p satisfies the conditions above.

(b) *ARMAP's.* Let $\xi_1, \xi_2, \ldots$ be i.i.d. and $V_n = \theta V_{n-1} + \xi_n$. V_n is called an *autoregressive moving average process* or *armap* for short. We call V_n a *smooth armap* if the distribution of ξ_n has a continuous density. In this case $p(x, dy) = p(x, y)\, dy$ with $(x, y) \to p(x, y)$ continuous.

(c) The *discrete Ornstein Uhlenbeck process* is a special case of (a) and (b). Let $\xi_1, \xi_2, \dots$ be i.i.d. standard normals and let $V_n = \theta V_{n-1} + \xi_n$. The Ornstein Uhlenbeck process is a diffusion process $\{V_t; t \in [0, \infty)\}$ that models the velocity of a particle suspended in a liquid. (See, e.g., Breiman (1968), Section 16.1.) Looking at V_t at integer times (and dividing by a constant to make the variance 1) gives a Markov chain with the indicated distributions.

Example 6.3   *GI/G/1 queue, or storage model.* Let $\xi_1, \xi_2, \dots$ be i.i.d. and define W_n inductively by $W_n = (W_{n-1} + \xi_n)^+$. If $P(\xi_n < 0) > 0$, then we can take $A = B = \{0\}$ and (i) and (ii) hold. To explain the first name in the title, consider a queueing system in which customers arrive at times of a renewal process—that is, at times $0 = T_0 < T_1 < T_2 < \cdots$ with $\zeta_n = T_n - T_{n-1}, n \geq 1$, i.i.d. Let $\eta_n, n \geq 0$, be the amount of service time the nth customer requires and let $\xi_n = \eta_{n-1} - \zeta_n$. I claim that W_n is the amount of time the nth customer has to wait to enter service. To see this notice that the $(n-1)$th customer adds η_{n-1} to the server's workload, and if the server is busy at all times in $[T_{n-1}, T_n)$, she reduces her workload by η_n. If $W_{n-1} + \eta_{n-1} < \zeta_n$, then the server has enough time to finish her work and the next arriving customer will find an empty queue.

The second name in the title refers to the fact that W_n can be used to model the contents of a storage facility. For an intuitive description consider water reservoirs. We assume that rainstorms occur at times of a renewal process $\{T_n : n \geq 1\}$, that the nth rainstorm contributes an amount of water η_n, and that water is consumed at constant rate c. If we let $\zeta_n = T_n - T_{n-1}$ as before, and $\xi_n = \eta_{n-1} - c\zeta_n$, then W_n gives the amount of water in the reservoir just before the nth rainstorm.

History Lesson   Doeblin was the first to prove results for Markov chains on general state space. He supposed that there was an n so that $p^n(x, C) \geq \varepsilon \rho(C)$ for all $x \in S$ and $C \subset S$. See Doob (1953), Section V.5, for an account of his results. Harris (1956) generalized Doeblin's result by observing that it was enough to have a set A so that (i) holds, and the chain viewed on A ($Y_k = X(T_A^k)$ where $T_A^k = \inf\{n > T_A^{k-1} : X_n \in A\}$ and $T_A^0 = 0$) satisfies Doeblin's condition. Our formulation as well as most of the proofs in this section follows Athreya and Ney (1978). For a nice description of the "traditional approach," see Revuz (1984).

Given a Harris chain on $(S, \mathscr{S})$, we will construct a Markov chain $\bar{X}_n$ with transition probability $\bar{p}$ on $(\bar{S}, \bar{\mathscr{S}})$ where $\bar{S} = S \cup \{\alpha\}$ and $\bar{\mathscr{S}} = \{B, B \cup \{\alpha\} : B \in \mathscr{S}\}$. The aim, as advertised earlier, is to manufacture a point α that the process hits with probability 1 in the recurrent case.

If $x \in S - A$, $\bar{p}(x, C) = p(x, C)$ for $C \in \mathscr{S}$.
If $x \in A$, $\bar{p}(x, \{\alpha\}) = \varepsilon$
   $\bar{p}(x, C) = p(x, C) - \varepsilon \rho(C)$ for $C \in \mathscr{S}$.
If $x = \alpha$, $\bar{p}(\alpha, D) = \int \rho(dx)\bar{p}(x, D)$ for $D \in \mathscr{S}$.

Intuitively, $\bar{X}_n = \alpha$ corresponds to X_n being distributed on B according to ρ. Here and in what follows we will reserve A and B for the special sets that occur in the definition and use C and D for generic elements of $\mathscr{S}$. We will often simplify notation by writing $\bar{p}(x, \alpha)$ instead of $\bar{p}(x, \{\alpha\})$, $\mu(\alpha)$ instead of $\mu(\{\alpha\})$, and so on.

Our next step is to prove three technical lemmas that will help us carry out the proofs below. Define a transition probability v by

$$v(x, \{x\}) = 1 \quad \text{if } x \in S, \qquad v(\alpha, C) = \rho(C).$$

(6.1) LEMMA $v\bar{p} = \bar{p}$ and $\bar{p}v = p$.

Proof   Before giving the proof we would like to remind the reader that measures multiply the transition probability on the left; that is, in the first case we want to show $\mu v\bar{p} = \mu\bar{p}$. If we first make a transition according to v and then one according to $\bar{p}$, this amounts to one transition according to $\bar{p}$, since only mass at α is affected by v and

$$\bar{p}(\alpha, D) = \int \rho(dx)\bar{p}(x, D).$$

The second equality also follows easily from the definition. In words, if $\bar{p}$ acts first and then v, then v returns the mass at α to where it came from.   □

From (6.1) it follows easily that we have:

(6.2) LEMMA   Let Y_n be an inhomogeneous Markov chain with $p_{2k} = v$ and $p_{2k+1} = \bar{p}$. Then $\bar{X}_n = Y_{2n}$ is a Markov chain with transition probability $\bar{p}$ and $X_n = Y_{2n+1}$ is a Markov chain with transition probability p.

(6.2) shows that there is an intimate relationship between the asymptotic behaviors of X_n and of $\bar{X}_n$. To quantify this we need a definition. If f is a bounded measurable function on S, let $\bar{f} = vf$; that is, $\bar{f}(x) = f(x)$ for $x \in S$ and $\bar{f}(\alpha) = \int f \, d\rho$.

(6.3) LEMMA   If μ is a probability measure on $(S, \mathscr{S})$, then $E_\mu f(X_n) = E_\mu \bar{f}(\bar{X}_n)$.

Proof   Observe that if X_n and $\bar{X}_n$ are constructed as in (6.2), and $P(\bar{X}_0 \in S) = 1$, then $X_0 = \bar{X}_0$ and X_n is obtained from $\bar{X}_n$ by making a transition according to v.   □

(6.1)–(6.3) will allow us to obtain results for X_n from those for $\bar{X}_n$. We turn now to the task of generalizing the results of Sections 3–5 to $\bar{X}_n$. To facilitate comparison with the results for countable state space, we will break this section into four subsections, the first three of which correspond to Sections 3–5. In the fourth subsection we take an in-depth look at Example 6.3. Before developing the theory we will give one last example that explains why some of the statements are messy.

Example 6.4   *Perverted O.U. process.* Take the discrete Ornstein Uhlenbeck (O.U.) process of Example 6.2 and modify the transition probability on a set $F \subset (1, \infty)$

that has $|F| = 0$. Do anything you want there but keep $p(x, [0, 1]) > 0$, and the result will be a Harris chain. For a concrete nightmare let $G \subset (1, 2)$ have $|G| = 0$, let $F = \bigcup_{n \geq 0} (n + G)$ where $n + G = \{n + x : x \in G\}$, and if $x \in G$, let $p(x, \{x - 1\}) = 1/3$, $p(x, \{x + 1\}) = 2/3$. When $\theta < 1$, the chain will be recurrent (see Exercise 6.4) but the points in G will be transient.

a   Recurrence and Transience

We begin with the dichotomy between recurrence and transience. Let $R = \inf\{n \geq 1 : \bar{X}_n = \alpha\}$. If $P_\alpha(R < \infty) = 1$, then we call the chain *recurrent*; otherwise we call it *transient*. Let $R_1 = R$ and for $k \geq 2$, let $R_k = \inf\{n > R_{k-1} : \bar{X}_n = \alpha\}$ be the time of the kth return to α. The strong Markov property implies $P_\alpha(R_k < \infty) = P_\alpha(R < \infty)^k$, so $P_\alpha(\bar{X}_n = \alpha$ i.o.$) = 1$ in the recurrent case and $= 0$ in the transient case. From this the next result follows easily.

(6.4) THEOREM   In the recurrent case if $p^n(\alpha, C) > 0$ for some n, then $P_\alpha(\bar{X}_n \in C$ i.o.$) = 1$. Let $\lambda(C) = \sum_n 2^{-n} p^n(\alpha, C)$. For λ a.e. x, $P_x(R < \infty) = 1$.

Proof   The first conclusion follows from (2.3). For the second let $D = \{x : P_x(R < \infty) < 1\}$ and observe that if $p^n(\alpha, D) > 0$ for some n, then

$$P_\alpha(X_n = \alpha \text{ i.o.}) \leq \int_D p^n(\alpha, dx) P_x(R < \infty) < 1. \qquad \square$$

Remark   Example 6.4 shows that we cannot expect to have $P_x(R < \infty) = 1$ for all x. To see that, even when the state space is countable, we need not hit every point starting from α, do:

Exercise 6.1   If X_n is a recurrent Harris chain on a countable state space, then S can only have one irreducible set of recurrent states but may have a nonempty set of transient states.

Exercise 6.2   Suppose X_n is a recurrent Harris chain. If (A', B') is another pair satisfying the conditions of the definition, then (6.4) implies $P_\alpha(\bar{X}_n \in A'$ i.o.$) = 1$, so the recurrence or transience does not depend on the choice of (A, B).

As in Section 3, we need special methods to determine whether our examples are recurrent or transient.

Exercise 6.3   In the GI/G/1 queue, the waiting time W_n and the random walk $S_n = X_0 + \xi_1 + \cdots + \xi_n$ agree until $N = \inf\{n : S_n < 0\}$, and at this time $W_N = 0$. Use this observation to show that Example 6.3 is recurrent when $E\xi_n \leq 0$ and transient when $E\xi_n > 0$.

Exercise 6.4   Let V_n be a smooth armap with $E|\xi_i| < \infty$. If $\theta < 1$, then $E_x|X_1| \leq |x|$ for $|x| \geq M$. Use this and ideas from Exercise 3.9 to show that the chain is recurrent in this case. To show transience for $\theta > 1$, let $\varphi(x) = 1/(|x| \vee 1)$ and use Exercise 3.10. In the case $\theta = 1$ the chain is a random walk with mean 0 and hence recurrent.

Exercise 6.5   Let X_n be as in the last exercise and suppose $\theta > 1$. Let $\gamma \in (1, \theta)$ and observe that $P_x(X_1 < \gamma x) \le C/((\theta - \gamma)x)^{-1}$, so if x is large, $P_x(X_n \ge \gamma^n x$ for all $n) > 0$.

Exercise 6.6   In the discrete O.U. process X_{n+1} is normal with mean θX_n and variance 1. What happens to the recurrence and transience if instead Y_{n+1} is normal with mean 0 and variance $\beta|Y_n|$?

b   Stationary Measure

(6.5) THEOREM   In the recurrent case, there is a stationary measure.

Proof   Let $R = \inf\{n \ge 1 : X_n = \alpha\}$, and let

$$\bar{\mu}(C) = E_\alpha\left(\sum_{n=0}^{R-1} 1_{\{X_n \in C\}}\right) = \sum_{n=0}^{\infty} \bar{p}^n(\alpha, C).$$

Repeating the proof of (4.3) shows that $\bar{\mu}\bar{p} = \bar{\mu}$. If we let $\mu = \bar{\mu}v$, then it follows from (6.1) that $\bar{\mu}vp = \bar{\mu}\bar{p}v = \bar{\mu}v$, so $\mu p = \mu$.   □

Exercise 6.7   Let $G_{k,\delta} = \{x : \bar{p}^k(x, \alpha) \ge \delta\}$. Show that $\bar{\mu}(G_{k,\delta}) \le k/\delta$ and use this to conclude that $\bar{\mu}$ and hence μ are σ-finite.

Exercise 6.8   Let λ be the measure defined in (6.4). Show that $\bar{\mu} \ll \lambda$ and $\lambda \ll \bar{\mu}$.

Exercise 6.9   Let V_n be an armap (not necessarily smooth) with $\theta < 1$ and $E\log^+|\xi_n| < \infty$. Show that $\sum_{m\ge0} \theta^m\xi_m$ converges a.s. and defines a stationary distribution for V_n. In the discrete O.U. process, the sum has a normal distribution with mean 0 and variance $(1 - \theta^2)^{-1}$. Find the distribution when $P(\xi_n = 1) = P(\xi_n = -1) = 1/2$ and $\theta = 1/2, 1/3$.

Exercise 6.10   Use the method of Exercise 4.16 to show that the GI/G/1 queue has a stationary distribution when $E\xi_i < 0$.

To investigate uniqueness of the stationary measure we begin with:

(6.6) LEMMA   If v is a σ-finite stationary measure for p, then $v(A) < \infty$ and $\bar{v} = v\bar{p}$ is a stationary measure for $\bar{p}$ with $\bar{v}(\alpha) < \infty$.

Proof   We will first show that $v(A) < \infty$. If $v(A) = \infty$, then part (ii) of the definition implies $v(C) = \infty$ for all sets C with $\rho(C) > 0$. If $B = \bigcup_i B_i$ with $v(B_i) < \infty$, then $\rho(B_i) = 0$ by the last observation and $\rho(B) = 0$ by countable subadditivity, a contradiction. So $v(A) < \infty$ and $\bar{v}(\alpha) = v\bar{p}(\alpha) = \varepsilon v(A) < \infty$. Using the fact that $vp = v$, we find

$$v\bar{p}(C) = v(C) - \varepsilon v(A)\rho(B \cap C),$$

the last subtraction being well defined, since $v(A) < \infty$, and it follows that $\bar{v}v = v$. To check $\bar{v}\bar{p} = \bar{v}$, we observe that (6.1) and the last result imply $\bar{v}\bar{p} = \bar{v}v\bar{p} = v\bar{p} = \bar{v}$. $\square$

(6.7) THEOREM Suppose p is recurrent. If v is a σ-finite stationary measure, then $v = v(\alpha)\mu$ where μ is the measure constructed in the proof of (6.5).

Proof By (6.6) it suffices to prove that if $\bar{v}$ is a stationary measure for $\bar{p}$ with $\bar{v}(\alpha) < \infty$, then $\bar{v} = \bar{v}(\alpha)\bar{\mu}$. Repeating the proof of (4.4) with $a = \alpha$, it is easy to show that $\bar{v}(C) \geq \bar{v}(\alpha)\bar{\mu}(C)$. Continuing to compute as in that proof:

$$\bar{v}(\alpha) = \int \bar{v}(dx)\bar{p}^n(x, \alpha) \geq \bar{v}(\alpha) \int \bar{\mu}(dx)\bar{p}^n(x, \alpha) = \bar{v}(\alpha)\bar{\mu}(\alpha) = \bar{v}(\alpha).$$

Let $S_n = \{x : p^n(x, \alpha) > 0\}$. By assumption $\bigcup_n S_n = S$. If $\bar{v}(D) > \bar{v}(\alpha)\bar{\mu}(D)$ for some D, then $\bar{v}(D \cap S_n) > \bar{v}(\alpha)\bar{\mu}(D \cap S_n)$, and it follows that $\bar{v}(\alpha) > \bar{v}(\alpha)$, a contradiction. $\square$

c Convergence Theorem

We say that a recurrent Harris chain X_n is *aperiodic* if the g.c.d. $\{n \geq 1 : p^n(\alpha, \alpha) > 0\} = 1$. This occurs, for example, if we can take $A = B$ in the definition, for then $p(\alpha, \alpha) > 0$.

(6.8) THEOREM Let X_n be an aperiodic recurrent Harris chain with stationary distribution π. If $P_x(R < \infty) = 1$, then as $n \to \infty$,

$$\|p^n(x, \cdot) - \pi(\cdot)\| \to 0.$$

Note Here $\| \ \|$ denotes the total variation distance between the measures. (6.4) guarantees that π a.e. x satisfies the hypothesis.

Proof In view of (6.3) it suffices to prove the result for $\bar{p}$. We begin by observing that the existence of a stationary probability and the uniqueness result in (6.7) imply that the measure constructed in (6.5) has $E_\alpha R = \bar{\mu}(S) < \infty$. As in the proof of (5.5) we let X_n and Y_n be independent copies of the chain with initial distributions δ_x and π, respectively, and let $\tau = \inf\{n \geq 0 : X_n = Y_n = \alpha\}$. Let S_m (resp T_m), $m \geq 0$, be the times at which X_n (resp Y_n) visit α for the $(m+1)$th time. $S_m - T_m$ is a random walk with mean 0 steps, so $M = \inf\{m \geq 1 : S_m = T_m < \infty\}$ a.s. and it follows that this is true for τ as well. The computations in the proof of (5.5) show $|P(X_n \in C) - P(Y_n \in C)| \leq P(\tau > n)$. Since this is true for all C, $\|p^n(x, \cdot) - \pi(\cdot)\| \leq P(\tau > n)$, and the proof is complete. $\square$

Exercise 6.11 Imitate the proof of (4.5) to show that a Harris chain with a stationary distribution must be recurrent.

Hint We have recurrence if and only if $\sum_n \bar{p}^n(\alpha, \alpha) = \infty$.

Exercise 6.12   Show that an armap with $\theta < 1$ and $E \log^+ |\xi_n| < \infty$ converges in distribution as $n \to \infty$. If the distribution of ξ_n is discrete, $\|p^n(x, \cdot) - \pi(\cdot)\| = 1$ for all n.

Hint   Recall the construction of π in Exercise 6.9.

d   GI-G/1 Queue

For the rest of the section we will concentrate on the GI/G/1 queue. Let $\xi_1, \xi_2, \ldots$ be i.i.d., let $W_n = (W_{n-1} + \xi_n)^+$, and let $S_n = \xi_1 + \cdots + \xi_n$. Recall $\xi_n = \eta_{n-1} - \zeta_n$ where the η's are service times, ζ's are the interarrival times, and suppose $E\xi_n < 0$ so that Exercise 6.10 implies there is a stationary distribution.

Exercise 6.13   Let $m_n = \min(0, S_1, \ldots, S_n)$. (i) Show that $S_n - m_n \overset{d}{=} W_n$. (ii) Let $\xi'_m = \xi_{n+1-m}$ for $1 \leq m \leq n$. Show that $S_n - m_n = \max(0, S'_1, \ldots, S'_n)$. (iii) Use the last result and (6.8) to show that $M \equiv \max(0, S_1, S_2, \ldots)$ is the stationary distribution for W_n. Can you verify this by direct computation?

Explicit formulas for the distribution of M are in general difficult to obtain. However, this can be done if either the arrival or service distribution is exponential. One reason for this is:

Exercise 6.14   Suppose $X, Y \geq 0$ are independent and $P(X > x) = e^{-\lambda x}$. Then the lack of memory property of the exponential distribution implies $P(X - Y > x) = ae^{-\lambda x}$ where $a = P(X - Y > 0)$.

Exercise 6.15   *Exponential service time.* Suppose $P(\eta_n > x) = e^{-\beta x}$ and $E\zeta_n > E\eta_n$. (i) Let $T = \inf\{n : S_n > 0\}$ and $L = S_T$, setting $L = -\infty$ if $T = \infty$. Use the lack of memory property of the exponential distribution to conclude that $P(L > x) = re^{-\beta x}$ where $r = P(T < \infty)$. (ii) Let $L_1, L_2, \ldots$ be independent and $\overset{d}{=} L$. Let $T_n = L_1 + \cdots + L_n$. Show $M \overset{d}{=} \sup T_n$ and use this to conclude that for $x > 0$, the density function

$$P(M = x) = \sum_{k=1}^{\infty} r^k(1-r)e^{-\beta x}\beta^k x^{k-1}/(k-1)! = \lambda r(1-r)e^{-\beta x}.$$

(iii) Compute r by finding $\theta > 0$ so that $\exp(\theta S_n)$ is a martingale.

The derivation of the result for Poisson arrivals (i.e., $P(\zeta_n > x) = e^{-\alpha x}$) is too tricky to be left as an exercise. Let $\bar{S}_n = -S_n$. Reversing time as in (ii) of Exercise 6.13, we see (for $n \geq 1$)

$$P\left(\max_{0 \leq k < n} \bar{S}_k < \bar{S}_n \in A \right) = P\left(\min_{1 \leq k \leq n} \bar{S}_k > 0, \bar{S}_n \in A \right). \tag{6.9}$$

Let $\psi_n(A)$ be the common value of the last two expressions and let $\psi(A) = \sum_{n \geq 0} \psi_n(A)$. $\psi_n(A)$ is the probability the random walk reaches a new maximum (or ladder height, see Example 1.4 in Chapter 3) in A at time n, so $\psi(A)$ is the number of ladder points in A with $\psi(\{0\}) = 1$. Letting the random walk take one more step,

$$P\left(\min_{1 \leq k \leq n} \bar{S}_k > 0, \bar{S}_{n+1} \leq x\right) = \int F(x - z)\, d\psi_n(z).$$

The last identity is valid for $n = 0$ if we interpret the left-hand side as $F(x)$. Let $\tau = \inf\{n \geq 1 : \bar{S}_n \leq 0\}$ and $x \leq 0$. Integrating by parts on the right-hand side and then summing over $n \geq 0$ give

$$P(\bar{S}_\tau \leq x) = \sum_{n=0}^{\infty} P\left(\min_{1 \leq k \leq n} \bar{S}_k > 0, \bar{S}_{n+1} \leq x\right) = \int_{y \leq x} \psi[0, x - y]\, dF(y).$$

$$(6.10)$$

The limit $y \leq x$ comes from the fact that $\psi((-\infty, 0)) = 0$.

Let $\bar{\xi}_n = \bar{S}_n - \bar{S}_{n-1} = -\xi_n$. Exercise 6.15 implies $P(\bar{\xi}_n > x) = ae^{-\alpha x}$. Let $\bar{T} = \inf\{n : \bar{S}_n > 0\}$. $E\bar{\xi}_n > 0$, so $P(\bar{T} < \infty) = 1$. Let $J = \bar{S}_{\bar{T}}$. As in part (i) of Exercise 6.15, $P(J > x) = e^{-\alpha x}$. Let $V_n = J_1 + \cdots + J_n$. V_n is a rate α Poisson process, so $\psi[0, x - y] = 1 + \alpha(x - y)$ for $x - y \geq 0$. Using (6.10) now and integrating by parts give

$$P(\bar{S}_\tau \leq x) = \int_{y \leq x} (1 + \alpha(x - y))\, dF(y) = F(x) + \alpha \int_{-\infty}^{x} F(y)\, dy \quad \text{for } x \leq 0.$$

$$(6.11)$$

Since $P(\bar{S}_n = 0) = 0$ for $n \geq 1$, $-\bar{S}_\tau$ has the same distribution as S_T where $T = \inf\{n : S_n > 0\}$. Combining this with part (ii) of Exercise 6.13 gives a "formula" for $P(M > x)$. Straightforward but somewhat tedious calculations show that if $B(s) = E\exp(-s\eta_n)$, then

$$E\exp(-sM) = \frac{(1 - \alpha \cdot E\eta)s}{s - \alpha + \alpha B(s)},$$

a result known as the *Pollaczek–Khintchine formula*. The computations we omitted can be found in Billingsley (1979), p. 277, or several times in Feller, Vol. II (1971).

Exercise 6.16 One thing is easy to compute for Poisson arrivals. Let $\tau = \inf\{n : S_n < 0\}$. Show that if $P(\zeta_n > x) = e^{-\alpha x}$ and $\mu = E\zeta_n - E\eta_n > 0$, then $E\tau = 1/\alpha\mu$, so the stationary distribution has mass $\alpha\mu$ at 0 and $P(M = 0) = \alpha\mu$.

Note It is tempting to use the martingale $S_n - \mu n$ to reach this conclusion, but Exercise 6.17 shows that if $E(\xi_n^+)^2 = \infty$, then $E(\sup_{n < \tau} S_n) = \infty$. To get around this problem use the martingale $\exp(\theta S_n)/\varphi(\theta)^n$ with $\theta < 0$ and observe that τ and S_τ are independent.

Exercise 6.17 (i) If $E\xi_n < 0$ and $E(\xi_n^+)^\beta = \infty$, then $EM^{\beta-1} = \infty$. (ii) Show that $P(M > x) = P(T = \infty)\psi(x, \infty)$ and use (6.9) to conclude $E(\sup_{n < \tau} S_n^{\beta-1}) = \infty$.

Sketch of proof of (i): Let Y_1, Y_2, ... be i.i.d. and $\overset{d}{=} (\xi_n | \xi_n > 0)$. Let $N_n = |\{m \leq n : \xi_m \leq 0\}|$, $P(\xi_m \leq 0) < a < 1$. Notice that $P(N_n \leq an) \to 1$ and use this to conclude that if $v = -E(\xi_n | \xi_n \leq 0)$ and $\varepsilon > 0$, then for large n,

$$P(M \geq n) \geq (1 - \varepsilon)P\left(\max_{1 \leq m \leq (1-a)n} Y_m \geq (2va + 1)n \right)$$

$$\geq (1 - 2\varepsilon)(1 - a)nP(Y_m \geq (2va + 1)n).$$

6 Ergodic Theorems

X_n, $n \geq 0$, is said to be a stationary sequence if for each $k \geq 1$, it has the same distribution as the shifted sequence X_{n+k}, $n \geq 0$. The basic fact about these sequences, called the *ergodic theorem*, is that if $E|f(X_0)| < \infty$, then

$$\lim_{n \to \infty} \frac{1}{n} \sum_{m=0}^{n-1} f(X_m) \quad \text{exists a.s.} \tag{2.1}$$

If X_n is ergodic (intuitively, it is not a mixture of two other stationary sequences), then the limit is $Ef(X_0)$. Sections 1 and 2 develop the theory needed to prove the ergodic theorem. The remaining five sections of this chapter develop various complements and, with the exception of Section 7, which gives applications of a result proved in Section 6, can be read in any order. In Section 3 we apply the ergodic theorem to study the recurrence of stationary sequences. In Section 4 we study "mixing," an asymptotic independence property stronger than ergodicity. In Section 5 we discuss entropy and give a proof of the Shannon–McMillan–Breiman theorem for X_n's taking values in a finite set. In Section 6 we prove the subadditive ergodic theorem. As five examples in Section 6 and four applications in Section 7 should indicate, this result is a useful generalization of the ergodic theorem.

1 Definitions and Examples

X_0, X_1, ... is said to be a *stationary sequence* if for every k, the sequence X_k, X_{k+1}, ... has the same distribution as the original; that is, for each n, $(X_0, \ldots, X_n)$ and $(X_k, \ldots, X_{k+n})$ have the same distribution. We begin by giving four examples that will be our constant companions.

Example 1.1 X_0, X_1, ... are i.i.d.

Example 1.2 Let X_n be a Markov chain with transition probability $p(x, A)$ and stationary distribution π; that is, $\pi(A) = \int \pi(dx) p(x, A)$. If X_0 has distribution π, then X_0, X_1, ... is a stationary sequence. A special case to keep in mind for counterexamples is the chain with state space $S = \{0, 1\}$ and transition probability $p(x, \{1 - x\}) = 1$. In this case the stationary distribution has $\pi(0) = \pi(1) = 1/2$ and $(X_0, X_1, \ldots) = (0, 1, 0, 1, \ldots)$ or $(1, 0, 1, 0, \ldots)$ with probability $1/2$ each.

Example 1.3   *Rotation of the circle.* Let $\Omega = [0, 1)$, $\mathscr{F} = $ Borel subsets, $P = $ Lebesgue measure. Let $\theta \in (0, 1)$ and for $n \geq 0$, let $X_n(\omega) = (\omega + n\theta) \bmod 1$, where $x \bmod 1 = x - [x]$, $[x]$ being the greatest integer $\leq x$. To see the reason for the name, map $[0, 1)$ into $\mathbb{C}$ by $x \to \exp(2\pi i x)$. This example is a special case of the last one. Let $p(x, \{y\}) = 1$ if $y = (x + \theta) \bmod 1$.

To make new examples from old we can use:

(1.1) THEOREM   If $X_0, X_1, \ldots$ is a stationary sequence and $g: \mathbb{R}^{\{0, 1, \ldots\}} \to \mathbb{R}$ is measurable, then $Y_k = g(X_k, X_{k+1}, \ldots)$ is a stationary sequence.

Proof   If $x \in \mathbb{R}^{\{0, 1, \ldots\}}$, let $g_k(x) = g(x_k, x_{k+1}, \ldots)$, and if $B \in \mathscr{R}^{\{0, 1, \ldots\}}$, let

$$A = \{x: (g_1(x), g_2(x), \ldots) \in B\}.$$

To check stationarity now we observe:

$$P(\omega: (Y_0, Y_1, \ldots) \in B) = P(\omega: (X_0, X_1, \ldots) \in A)$$

$$= P(\omega: (X_k, X_{k+1}, \ldots) \in A) = P(\omega: (Y_k, Y_{k+1}, \ldots) \in B).$$

$\square$

Example 1.4   *Bernoulli shift.* $\Omega = [0, 1)$, $\mathscr{F} = $ Borel subsets, $P = $ Lebesgue measure. $Y_0(\omega) = \omega$ and for $n \geq 1$, let $Y_n(\omega) = (2Y_{n-1}(\omega)) \bmod 1$. This example is a special case of (1.1). Let $X_0, X_1, \ldots$ be i.i.d. with $P(X_i = 0) = P(X_i = 1) = 1/2$, and let

$$g(x) = \sum_{i=0}^{\infty} x_i 2^{-(i+1)}.$$

The name comes from the fact that multiplying by 2 shifts the X's to the left. This example is also a special case of Example 1.2. Let $p(x, \{y\}) = 1$ if $y = (2x) \bmod 1$.

Examples 1.3 and 1.4 are special cases of the following situation.

Example 1.5   Let $(\Omega, \mathscr{F}, P)$ be a probability space. A measurable map $\varphi: \Omega \to \Omega$ is said to be *measure preserving* if $P(\varphi^{-1}A) = P(A)$ for all $A \in \mathscr{F}$. If $X \in \mathscr{F}$, then $X_n(\omega) = X(\varphi^n \omega)$ defines a stationary sequence. To check this, let $B \in \mathscr{R}^{n+1}$ and $A = \{\omega: (X_0(\omega), \ldots, X_n(\omega)) \in B\}$. Then

$$P((X_k, \ldots, X_{k+n}) \in B) = P(\varphi^k \omega \in A) = P(\omega \in A) = P((X_0, \ldots, X_n) \in B).$$

The last example is more than an important example. In fact, it is the only example! If $Y_0, Y_1, \ldots$ is a stationary sequence taking values in a nice space,

Kolmogorov's extension theorem ((7.1) in the Appendix) allows us to construct a measure P on sequence space $(S^{\{0,1,\dots\}}, \mathcal{S}^{\{0,1,\dots\}})$ so that $X_n(\omega) = \omega_n$ has the desired distributions. If we let φ be the shift operator—that is, $\varphi(\omega_0, \omega_1, \dots) = (\omega_1, \omega_2, \dots)$, and let $X(\omega) = \omega_0$, then φ is measure preserving and $X_n(\omega) = X(\varphi^n \omega)$.

In some situations (e.g., in the proof of (3.3) below) it is useful to observe:

(1.2) THEOREM   Any stationary sequence $\{X_n, n \geq 0\}$ can be embedded in a two-sided stationary sequence $\{Y_n : n \in \mathbb{Z}\}$.

Proof   We observe that

$$P(Y_{-m} \in A_0, \dots, Y_n \in A_{m+n}) = P(X_0 \in A_0, \dots, X_{m+n} \in A_{m+n})$$

is a consistent set of finite-dimensional distributions, so a trivial generalization of the Kolmogorov extension theorem implies there is a measure P on $(S^{\mathbb{Z}}, \mathcal{S}^{\mathbb{Z}})$ so that the variables $Y_n(\omega) = \omega_n$ have the desired distributions. $\qquad \square$

In view of the observations above, it suffices to give our definitions and prove our results in the setting of Example 1.5. A set $A \in \mathcal{F}$ is said to be *invariant* if $\varphi^{-1}A = A$. (Here, as usual, two sets are considered to be equal if their symmetric difference has probability 0.)

Exercise 1.1   Show that the class of invariant events $\mathcal{I}$ is a σ-field, and $X \in \mathcal{I}$ if and only if $X \circ \varphi = X$ a.s.

Exercise 1.2   Some authors call A *almost invariant* if $\mu(A \triangle \varphi^{-1}(A)) = 0$. We call such sets *invariant* and call B *invariant in the strict sense* if $B = \varphi^{-1}(B)$. Show that A is almost invariant if and only if there is a B invariant in the strict sense with $\mu(A \triangle B) = 0$.

A measure-preserving transformation on $(\Omega, \mathcal{F}, P)$ is said to be *ergodic* if $\mathcal{I}$ is trivial; that is, for every $A \in \mathcal{I}$, $P(A) \in \{0, 1\}$. If φ is not ergodic, then the space can be split into two sets A and A^c, each having positive measure, so that $\varphi(A) = A$ and $\varphi(A^c) = A^c$. In words, φ is not "irreducible."

To investigate further the meaning of ergodicity we turn to our examples. Before we start, we would like to observe that if $\Omega = \mathbb{R}^{\{0,1,\dots\}}$ and φ is the shift operator, then an invariant set A has $\{\omega : \omega \in A\} = \{\omega : \varphi\omega \in A\} \in \sigma(X_1, X_2, \dots)$. Iterating gives

$$A \in \bigcap_{n=1}^{\infty} \sigma(X_n, X_{n+1}, \dots) = \mathcal{T}, \quad \text{the tail } \sigma\text{-field},$$

so $\mathcal{I} \subset \mathcal{T}$. For an i.i.d. sequence (Example 1.1), Kolmogorov's 0–1 law implies $\mathcal{T}$ is trivial, so $\mathcal{I}$ is trivial and the sequence is ergodic (i.e., when the corresponding measure is put on sequence space the shift is). Turning to Markov chains (Example 1.2), suppose the state space S is countable and the stationary distribution has $\pi(x) > 0$ for all $x \in S$. By (4.5) and (3.6) in

Chapter 5, all states are recurrent, and we can write $S = \bigcup R_i$, where the R_i are disjoint irreducible closed sets. If $X_0 \in R_i$, then with probability 1, $X_n \in R_i$ for all $n \geq 1$, so $\{\omega: X_0(\omega) \in R_i\} \in \mathscr{I}$. The last observation shows that if the Markov chain is not irreducible, then the sequence is not ergodic. To prove the converse, observe that if $A \in \mathscr{I}$, $1_A \circ \theta_n = 1_A$ (where $\theta_n(\omega_0, \omega_1, \ldots) = (\omega_n, \omega_{n+1}, \ldots)$). So if we let $\mathscr{F}_n = \sigma(X_0, \ldots, X_n)$, the shift invariance of 1_A and the Markov property imply

$$E_\pi(1_A|\mathscr{F}_n) = E_\pi(1_A \circ \theta_n|\mathscr{F}_n) = h(X_n)$$

where $h(x) = E_x 1_A$. Lévy's 0–1 law implies that the left-hand side converges to 1_A as $n \to \infty$. If X_n is irreducible and recurrent, then for any $y \in S$, the right-hand side $= h(y)$ i.o., so either $h(x) \equiv 0$ or $h(x) \equiv 1$, and $P_\pi(A) \in \{0, 1\}$. This example also shows that $\mathscr{I}$ and $\mathscr{T}$ may be different. When p is irreducible, $\mathscr{I}$ is trivial, but if all the states have period $d > 1$, $\mathscr{T}$ is not. In (5.8) of Chapter 5 we showed that if $S_0, \ldots, S_{d-1}$ is the cyclic decomposition of S, then $\mathscr{T} = \sigma(\{X_0 \in S_r\} : 0 \leq r < d)$.

Exercise 1.3   Give an example of an ergodic measure-preserving transformation T on $(\Omega, \mathscr{F}, P)$ so that T^2 is not ergodic.

Rotation of the circle (Example 1.3) is not ergodic if $\theta = m/n$ where $1 \leq m < n$. If B is a Borel subset of $[0, 1/n)$ and

$$A = \bigcup_{k=0}^{n-1} (B + k/n),$$

then A is invariant. Conversely, if θ is irrational, then φ is ergodic. To prove this we need a fact from Fourier analysis. If f is a measurable function on $[0, 1)$ with $\int f^2(x)\,dx < \infty$, then f can be written as

$$f(x) = \sum_k c_k e^{2\pi i k x}$$

where the last equality means that

$$\sum_{k=-K}^{K} c_k e^{2\pi i k x} \to f(x) \quad \text{in } L^2[0, 1),$$

and this is possible for only one choice of the coefficients c_k:

$$c_k = \int_0^1 f(x) e^{-2\pi i k x}\,dx.$$

Now

$$f(\varphi(x)) = \sum_k c_k e^{2\pi i k(x+\theta)} = \sum_k (c_k e^{2\pi i k \theta}) e^{2\pi i k x}.$$

The uniqueness of the coefficients c_k implies that $f(\varphi(x)) = f(x)$ if and only if

$$c_k(e^{2\pi i k\theta} - 1) = 0.$$

If θ is irrational, this implies $c_k = 0$ for $k \neq 0$, so f is constant. Applying the last result to $f = 1_A$ with $A \in \mathcal{I}$ shows that $A = \varnothing$ or $[0, 1)$ a.s.

Exercise 1.4   *A direct proof of ergodicity.* (i) Show that if θ is irrational, $x_n = n\theta \mod 1$ is dense in $[0, 1)$.

Hint   All the x_n are distinct, so for any $\varepsilon > 0$, $|x_n - x_m| < \varepsilon$ for some $m < n$. (ii) Use Exercise 3.1 in the Appendix to show that if A is a Borel set with $|A| > 0$, then for any $\delta > 0$ there is an interval $J = (a, b)$ so that $|A \cap J| > (1 - \delta)|J|$. Combine this with (i) to conclude $\mu(A) = 1$.

Finally, the Bernoulli shift (Example 1.4) is ergodic. To prove this, we recall that the stationary sequence $Y_n(\omega) = \varphi^n(\omega)$ can be represented as

$$Y_n = \sum_{m=0}^{\infty} 2^{-(m+1)} X_{n+m}$$

where $X_0, X_1, \ldots$ are i.i.d. with $P(X_k = 1) = P(X_k = 0) = 1/2$, and use the following fact:

(1.3) THEOREM   If $X_0, X_1, \ldots$ is an ergodic stationary sequence and $g: \mathbb{R}^{\{0, 1, \ldots\}} \to \mathbb{R}$ is measurable, then $Y_k = g(X_k, X_{k+1}, \ldots)$ is ergodic.

Proof   Suppose $X_0, X_1, \ldots$ is defined on sequence space with $X_n(\omega) = \omega_n$. If B has $\{\omega: (Y_0, Y_1, \ldots) \in B\} = \{\omega: (Y_1, Y_2, \ldots) \in B\}$, then $A = \{\omega: (Y_0, Y_1, \ldots) \in B\}$ is shift invariant.   □

Remark   The proofs of (1.1) and (1.3) generalize easily to functions $g: \mathbb{R}^{\mathbb{Z}} \to \mathbb{R}$ of a two-sided stationary sequence X_n, $n \in \mathbb{Z}$.

Exercise 1.5   Use Fourier analysis as in Example 1.3 to prove that Example 1.4 is ergodic.

Exercise 1.6   Let ξ_n, $n \in \mathbb{Z}$, be i.i.d. with $E\xi_n = 0$ and $E\xi_n^2 < \infty$. Let c_n, $n \in \mathbb{Z}$, be constants with $\sum c_n^2 < \infty$. Show that $X_n = \sum_m c_{n-m}\xi_m$, $n \geq 0$, is stationary and ergodic.

Exercise 1.7   *Baker's transformation.* $\Omega = [0, 1)^2$, $\mathcal{F} = $ Borel sets, $P = $ Lebesgue measure.

$$\varphi(x, y) = \begin{cases} (2x, y/2) & \text{if } 0 \leq x < 1/2 \\ (2x - 1, (y + 1)/2) & \text{if } 1/2 \leq x < 1. \end{cases}$$

The name comes from imagining the unit square to be bread dough that is stretched in the x-direction until it is twice as long and half as high and then cut along $x = 1$ to make two loaves. Relate the action of φ on Ω to the shift transformation on $\{0, 1\}^{\mathbb{Z}}$ and imitate our treatment of Example 1.4 to show φ is ergodic.

Exercise 1.8   *Continued fractions.* Let $\varphi(x) = 1/x - [1/x]$ for $x \in [0, 1)$ and $A(x) = [1/x]$ where $[1/x] =$ the largest integer $\leq 1/x$. $a_n = A(\varphi^n x), n = 0, 1, 2, \ldots$, gives the continued fraction representation of x; that is,

$$x = 1/(a_0 + 1/(a_1 + 1/(a_2 + 1/\cdots))).$$

Let

$$\mu(A) = \frac{1}{\log 2} \int_A \frac{dx}{1 + x} \quad \text{for } A \subset [0, 1).$$

Show that φ preserves μ. In his 1959 monograph Kac claimed that it was "entirely trivial" to check that φ is ergodic but retracted his claim in a later footnote. We leave it to the reader to construct a proof or look up the answer in Ryll–Nardzewski (1951). Chapter 9 of Lévy (1937) is devoted to this topic and is still interesting reading today.

Exercise 1.9   *Independent blocks.* Let $X_1, X_2, \ldots$ be a stationary sequence. Let $n < \infty$ and let $Y_1, Y_2, \ldots$ be a sequence in which $(Y_{k+1}, \ldots, Y_{n(k+1)}), k \geq 0$, are i.i.d. and $(Y_1, \ldots, Y_n) = (X_1, \ldots, X_n)$. Finally let ν be uniformly distributed on $\{1, 2, \ldots, n\}$ and let $Z_m = Y_{\nu+m}$ for $m \geq 1$. Show that Z is stationary and ergodic. The reader will see the reason for our interest in this example in Exercise 2.10.

2   Birkhoff's Ergodic Theorem

Throughout this section φ is a measure-preserving transformation on $(\Omega, \mathscr{F}, P)$. We begin by proving a result that is usually referred to as:

(2.1) *The Ergodic Theorem*   For any $X \in L^1$,

$$\frac{1}{n} \sum_{m=0}^{n-1} X(\varphi^m \omega) \to E(X | \mathscr{I}) \quad \text{a.s. and in } L^1.$$

This result due to Birkhoff (1931) is sometimes called the pointwise or individual ergodic theorem because of the a.s. convergence in the conclusion. The proof we give is based on an odd integration inequality due to Yosida and Kakutani (1939).

(2.2) *Maximal Ergodic Lemma*   Let $X_j(\omega) = X(\varphi^j \omega), S_k(\omega) = X_0(\omega) + \cdots + X_{k-1}(\omega)$, and $M_k(\omega) = \max(0, S_1(\omega), \ldots, S_k(\omega))$. Then $E(X; M_k > 0) \geq 0$.

Proof   We follow Garsia (1965). The proof is not intuitive but none of the steps are difficult. If $j \leq k$, then $M_k(\varphi \omega) \geq S_j(\varphi \omega)$, so adding $X(\omega)$ gives

$$X(\omega) + M_k(\varphi\omega) \ge X(\omega) + S_j(\varphi\omega) = S_{j+1}(\omega),$$

and rearranging we have

$$X(\omega) \ge S_{j+1}(\omega) - M_k(\varphi\omega) \quad \text{for } j = 1, \dots, k.$$

Trivially $X(\omega) \ge S_1(\omega) - M_k(\varphi\omega)$, since $S_1(\omega) = X(\omega)$ and $M_k(\varphi\omega) \ge 0$. Therefore

$$E(X(\omega); M_k > 0) \ge \int_{\{M_k > 0\}} \max(S_1(\omega), \dots, S_k(\omega)) - M_k(\varphi\omega) \, dP$$

$$= \int_{\{M_k > 0\}} M_k(\omega) - M_k(\varphi\omega) \, dP.$$

Now $M_k(\omega) = 0$ and $M_k(\varphi\omega) \ge 0$ on $\{M_k > 0\}^c$, so the last expression is

$$\ge \int M_k(\omega) - M_k(\varphi\omega) \, dP = 0,$$

since φ is measure preserving. $\square$

Proof of (2.1) $E(X|\mathcal{I})$ is invariant under φ (see Exercise 1.1), so letting $X' = X - E(X|\mathcal{I})$, we can assume without loss of generality that $E(X|\mathcal{I}) = 0$. Let $\bar{X} = \limsup S_n/n$, let $\varepsilon > 0$, and let $D = \{\omega: \bar{X}(\omega) > \varepsilon\}$. Our goal is to prove that $P(D) = 0$. $\bar{X}(\varphi\omega) = \bar{X}(\omega)$, so $D \in \mathcal{I}$. Let

$$X^*(\omega) = (X(\omega) - \varepsilon)1_D(\omega)$$

$$S_n^*(\omega) = X^*(\omega) + \cdots + X^*(\varphi^{n-1}\omega)$$

$$M_n^*(\omega) = \max(0, S_1^*(\omega), \dots, S_n^*(\omega))$$

$$F_n = \{M_n^* > 0\}$$

$$F = \bigcup_n F_n = \left\{\sup_{k \ge 1} S_k^*/k > 0\right\}.$$

Since $X^*(\omega) = (X(\omega) - \varepsilon)1_D(\omega)$ and $D = \{\limsup S_k/k > \varepsilon\}$, it follows that

$$F = \left\{\sup_{k \ge 1} S_k/k > \varepsilon\right\} \cap D = D.$$

(2.2) implies that $E(X^*; F_n) \ge 0$. Since $E|X^*| \le E|X| + \varepsilon < \infty$, the dominated convergence theorem implies $E(X^*; F_n) \to E(X^*; F)$, and it follows that $E(X^*; F) \ge 0$. The last conclusion looks innocent, but $F = D \in \mathcal{I}$, so it implies

$$0 \le E(X^*; D) = E(X - \varepsilon; D) = E(E(X|\mathscr{I}); D) - \varepsilon P(D) = -\varepsilon P(D),$$

since $E(X|\mathscr{I}) = 0$. The last inequality implies that

$$0 = P(D) = P(\limsup S_n/n > \varepsilon),$$

and since $\varepsilon > 0$ is arbitrary, it follows that $\limsup S_n/n \le 0$. Applying the last result to $-X$ shows that $S_n/n \to 0$ a.s.

To prove that convergence occurs in L^1, let

$$X'(\omega) = X(\omega)1_{(|X(\omega)| \le M)} \quad \text{and} \quad X''(\omega) = X(\omega) - X'(\omega).$$

The part of the ergodic theorem we have proved implies

$$\frac{1}{n} \sum_{m=0}^{n-1} X'(\varphi^m \omega) \to E(X'|\mathscr{I}) \quad \text{a.s.}$$

Since X' is bounded, the bounded convergence theorem implies

$$E \left| \frac{1}{n} \sum_{m=0}^{n-1} X'(\varphi^m \omega) - E(X'|\mathscr{I}) \right| \to 0.$$

To handle X'' we observe

$$E \left| \frac{1}{n} \sum_{m=0}^{n-1} X''(\varphi^m \omega) \right| \le \frac{1}{n} \sum_{m=0}^{n-1} E|X''(\varphi^m \omega)| = E|X''|$$

and $E|E(X''|\mathscr{I})| \le EE(|X''||\mathscr{I}) = E|X''|$. So

$$E \left| \frac{1}{n} \sum_{m=0}^{n-1} X''(\varphi^m \omega) - E(X''|\mathscr{I}) \right| \le 2E|X''|,$$

and it follows that

$$\limsup_{n \to \infty} E \left| \frac{1}{n} \sum_{m=0}^{n-1} X(\varphi^m \omega) - E(X|\mathscr{I}) \right| \le 2E|X''|.$$

As $M \to \infty$, $E|X''| \to 0$ by the dominated convergence theorem, so we have completed the proof of (2.1).   □

Exercise 2.1   Show that if $X \in L^p$ with $p > 1$, then the convergence in (2.1) occurs in L^p.

Exercise 2.2   Suppose $g_n(\omega) \to g(\omega)$ a.s. and $E(\sup_k |g_k(\omega)|) < \infty$. Then

$$\lim_{n\to\infty} \frac{1}{n} \sum_{m=0}^{n-1} g_m(T^m\omega) = E(g|\mathscr{I}) \quad \text{a.s.}$$

If we suppose only that $g_n \to g$ in L^1, we get L^1 convergence.

Before turning to examples, we would like to prove a useful result that is a simple consequence of (2.2).

(2.3) Wiener's Maximal Inequality   Let $X_j(\omega) = X(\varphi^j\omega)$, $S_k(\omega) = X_0(\omega) + \cdots + X_{k-1}(\omega)$, $A_k(\omega) = S_k(\omega)/k$, and $D_k = \max(A_1, \ldots, A_k)$. If $\alpha > 0$, then

$$P(D_k > \alpha) \le \alpha^{-1} E|X|.$$

Proof   Let $B = \{D_k > \alpha\}$. Applying (2.2) to $X' = X - \alpha 1_B$ with $X'_j(\omega) = X'(\varphi^j\omega)$, $S'_k = X'_0(\omega) + \cdots + X'_{k-1}$, and $M'_k = \max(0, S'_1, \ldots, S'_k)$ gives $E(X'; M'_k > 0) \ge 0$. Since $\{M'_k > 0\} = \{D_k > \alpha\} \equiv B$, it follows that

$$E|X| \ge \int_B X \, dP \ge \alpha \int_B 1_B \, dP = \alpha P(B). \qquad \square$$

Exercise 2.3   Use (2.3) and the truncation argument at the end of the proof of (2.1) to conclude that if (2.1) holds for bounded r.v.'s, then it holds whenever $E|X| < \infty$.

Our next step is to see what (2.1) says about our examples.

Example 2.1   *i.i.d. sequences.* Since $\mathscr{I}$ is trivial, the ergodic theorem implies that

$$\frac{1}{n} \sum_{m=0}^{n-1} X_m \to EX_0 \quad \text{a.s. and in } L^1.$$

The a.s. convergence is the strong law of large numbers.

Exercise 2.4   Prove the L^1 convergence in the law of large numbers without invoking ergodic theorem.

Sketch   Apply the strong law to X_m^+ and X_m^-; then use (5.2) in Chapter 4.

Example 2.2   *Markov chains.* Let X_n be an irreducible Markov chain on a countable state space which has a stationary distribution π. Let f be a function with

$$\sum_x |f(x)|\pi(x) < \infty.$$

In Section 1 we showed that $\mathscr{I}$ is trivial, so applying the ergodic theorem to $f(X_0(\omega))$

gives

$$\frac{1}{n}\sum_{m=0}^{n-1} f(X_m) \to \sum_x f(x)\pi(x) \quad \text{a.s. and in } L^1.$$

For another proof, see Exercise 5.2 in Chapter 5.

Example 2.3   *Rotation of the circle.* $\Omega = [0, 1)$, $\varphi(\omega) = (\omega + \theta)$ mod 1. Suppose that $\theta \in (0, 1)$ is irrational, so Example 1.3 implies that $\mathscr{I}$ is trivial. If we set $X(\omega) = 1_A(\omega)$, with A a Borel subset of $[0, 1)$, then the ergodic theorem implies

$$\frac{1}{n}\sum_{m=0}^{n-1} 1_{(\varphi^m\omega \in A)} \to |A| \quad \text{a.s.}$$

where $|A|$ denotes the Lebesgue measure of A. The last result for $\omega = 0$ is usually called *Weyl's equidistribution theorem*, although Bohl and Sierpinski should also get credit. For the history and a nonprobabilistic proof, see Hardy and Wright (1959), pp. 390–393.

To recover the number theoretic result we will now show that if $A = [a, b)$, then the exceptional set is $\varnothing$. Let $A_k = [a + 1/k, b - 1/k)$. If $b - a > 2/k$, the last result implies

$$\frac{1}{n}\sum_{m=0}^{n-1} 1_{A_k}(\varphi^m\omega) \to b - a - 2/k$$

for $\omega \in \Omega_k$ with $P(\Omega_k) = 1$. Let $G = \bigcap \Omega_k$ where the intersection is over integers k with $b - a > 2/k$. $P(G) = 1$, so G is dense in $[0, 1)$. If $x \in [0, 1)$ and $\omega \in G$ with $|\omega - x| < 1/k$, then $\varphi^m\omega \in A_k$ implies $\varphi^m x \in A$, so

$$\liminf_{n\to\infty}\frac{1}{n}\sum_{m=0}^{n-1} 1_A(\varphi^m x) \geq b - a - 2/k$$

for all large enough k. Applying similar reasoning to A^c shows

$$\frac{1}{n}\sum_{m=0}^{n-1} 1_A(\varphi^m x) \to b - a.$$

As Gelfand first observed, the equidistribution theorem says something interesting about 2^m. Let $\theta = \log_{10} 2$, $1 \leq k \leq 9$, and $A = [\log_{10} k, \log_{10}(k + 1))$ where $\log_{10} y$ is the logarithm of y to the base 10. Taking $x = 0$ in the last result, we have

$$\frac{1}{n}\sum_{m=0}^{n-1} 1_A(\varphi^m 0) \to \log_{10}\left(\frac{k + 1}{k}\right).$$

A little thought reveals that the first digit of 2^m is k if and only if $m\theta \bmod 1 \in A$. Taking $k = 1$, for example, we have shown that the asymptotic fraction of time 1 is the first digit of 2^m is $\log_{10} 2 = .3010$.

The limit distribution on $\{1, \ldots, 9\}$ is called Benford's (1938) law, although it was discovered by Newcomb (1881). As Raimi (1976) explains, in many tables the observed frequency in which k appears as a first digit is approximately $\log_{10}(k + 1/k)$. He mentions powers of 2 as an example. Two other data sets that fit this very well are (i) the street addresses of the first 342 persons in *American Men of Science* (1938) and (ii) the kilowatt-hours on 1243 electric bills for October 1969 from Honiara in the British Solomon Islands. We leave it to the reader to figure out why Benford's law should appear in the last two situations.

Example 2.4   *Bernoulli shift.* $\Omega = [0, 1)$, $\varphi(\omega) = (2\omega) \bmod 1$. Let $i_1, \ldots, i_k \in \{0, 1\}$, let $r = \sum_{1 \le m \le k} i_m 2^{-m}$, and let $X(\omega) = 1$ if $r \le \omega < r + 2^{-k}$. In words, $X(\omega) = 1$ if the first k digits of the binary expansion of ω are $i_1, \ldots, i_k$. The ergodic theorem implies that

$$\frac{1}{n} \sum_{m=0}^{n-1} X(\varphi^m \omega) \to 2^{-k} \quad \text{a.s.;}$$

that is, in almost every $\omega \in [0, 1)$ the pattern $i_1, \ldots, i_k$ occurs with its expected frequency. Since there are only a countable number of patterns (of finite length), it follows that almost every $\omega \in [0, 1)$ is *normal*; that is, all patterns occur with their expected frequency. This is the binary version of Borel's (1909) normal number theorem.

Exercise 2.5   *Tent map.* Consider the transformation T on $[0, 1]$ defined by

$$T(y) = \begin{cases} 2y & \text{if } y \le 1/2 \\ 2(1 - y) & \text{if } y \ge 1/2. \end{cases}$$

φ preserves Lebesgue measure and is equivalent to a shift and flip map on $\{0, 1\}^{\{0, 1, 2, \ldots\}}$.

$$\tau(\omega_0, \omega_1, \omega_2, \ldots) = \begin{cases} (\omega_1, \omega_2, \ldots) & \text{if } \omega_0 = 0 \\ (1 - \omega_1, 1 - \omega_2, \ldots) & \text{if } \omega_0 = 1. \end{cases}$$

(i) Show that if $X(y) = 1_{[0, 1/2)}(y)$, then $X(y), X(Ty), \ldots$ are i.i.d.
(ii) Generalize (i) to conclude that if $X_{i,j}(y) = 1_{[i/2^j, (i+1)/2^j)}(y)$, then

$$\frac{1}{n} \sum_{m=0}^{n-1} X_{i,j}(T^m y) \to 2^{-j} \quad \text{a.s.}$$

(iii) The result in part (ii) implies that if $X = \sum_i a_i X_{i,j}$, then

$$\frac{1}{n} \sum_{m=0}^{n-1} X(T^m y) \to EX \quad \text{in } L^1.$$

The approximation argument used at the end of the proof of (2.1) implies that the last result holds for all $X \in L^1$, so T is ergodic.

Exercise 2.6   Let X_n be a stationary process. Prove that X_n is ergodic if there is a π-system $\mathcal{A}$ with $\sigma(\mathcal{A}) = \mathcal{R}$ so that whenever $A_1, \ldots, A_k \in \mathcal{A}$,

$$\frac{1}{n} \sum_{m=1}^{n} 1_{(X_{m+1} \in A_1, \ldots, X_{m+k} \in A_k)} \to P(X_1 \in A_1, \ldots, X_k \in A_k).$$

Exercise 2.7 $\varphi(x) = 4x(1 - x)$ maps $[0, 1]$ into itself. Let

$$g(x) = \frac{2}{\pi} \arcsin \sqrt{x}$$

$$h(y) = (\sin \pi y/2)^2.$$

Show that if T is the transformation in Exercise 2.5 then $Ty = g(\varphi(h(y)))$. Use this to show that the measure with density $\pi^{-1}(x(1 - x))^{-1/2}$ is preserved by φ, and φ is ergodic.

Research Problem   Describe the set of $\theta \in (0, 4)$ for which $\varphi(x) = \theta x(1 - x)$ has a stationary distribution on $[0, 1]$ that is absolutely continuous with respect to Lebesgue measure.

Exercise 2.8   Let F be a distribution function that is strictly increasing, continuous, and has $F(0) = 0$, $F(1) = 1$. Show that $F^{-1}(2F(x) \bmod 1)$ is an ergodic transformation that preserves F.

Exercise 2.9   Let $\mathcal{M}$ be the set of all probability measures P so that T is measure preserving on $(\Omega, \mathcal{F}, P)$. (i) Show $\mathcal{M}$ is convex. (ii) Show that if $\mu, \nu \in \mathcal{M}$, $\mu \ll \nu$, and ν is ergodic (i.e., T is ergodic on $(\Omega, \mathcal{F}, \nu)$), then $\mu = \nu$. (iii) Use (ii) to conclude that the extreme points of $\mathcal{M}$ are the ergodic measures.

Exercise 2.10   Let $\Omega = \{0, 1\}^{\{0, 1, \ldots\}}$ and let θ be the shift on Ω. The collection $\mathcal{M}$ defined in the last exercise is the set of stationary sequences. Use the examples in Exercise 1.9 to show that the extreme points are dense in $\mathcal{M}$ in the weak topology (i.e., convergence of finite-dimensional distributions).

Exercise 2.11   Let $p(x, dy)$ be the transition probability of a Markov chain on $(S, \mathcal{S})$. Let Π be the set of probability distributions with $\pi p = \pi$. (i) Show that Π is convex and π is an extreme point if and only if the shift is ergodic under P_π. (ii) Show that π is an extreme point of Π if and only if $\mu \in \Pi$ and $\mu \ll \pi$ implies $\mu = \pi$.

Exercise 2.12   Let p be a transition probability on $(S, \mathcal{S})$ and μ a σ-finite measure with $\mu p \ll \mu$. (i) Show that if $f \in L^1(\mu)$ and we define a signed measure by $\nu(A) = \int_A f(x)\mu(dx)$, then $\nu p \ll \mu$ and the mapping $Tf = d(\nu p)/d\mu$ is a contraction on $L^1(\mu)$. (ii) Imitate the proof of (2.2) to show that if $f \in L^1(\mu)$ and $E = \bigcup_{n \geq 0} \{f + \cdots +$

$T^n f > 0\}$, then $\int_E f\mu(dx) > 0$. (iii) Show that there is a set $C \in \mathscr{S}$ so that for every $f \geq 0$ in $L^1(\mu)$,

$$\sum_{n=0}^{\infty} T^n f \in \{0, \infty\} \quad \text{a.e. on } C$$

$$\sum_{n=0}^{\infty} T^n f < \infty \quad \text{a.e. on } D = C^c.$$

C is called the *conservative* and D the *dissipative* part of S. Intuitively, points in C are recurrent and those in D are transient.

Exercise 2.13   Continuing the development begun in the last exercise, suppose T is conservative; that is, $C = S$. Let $\mathscr{C} = \{C_f : f \geq 0, f \in L^1(\mu)\}$ where $C_f = \{\sum T^n f = \infty\}$. Show that for a function $f \in L^\infty(\mu)$, the following are equivalent: (i) $ph \leq h$, (ii) $ph = h$, (iii) $h \in \mathscr{C}$. Intuitively, each $A \in \mathscr{C}$ is a closed set of recurrent states.
   Sketch of proof   To see that (i) implies (ii) observe

$$\left\langle \sum_{m=0}^{n-1} T^m f, h - ph \right\rangle = \langle f, h - p^n h \rangle$$

where $\langle g, k \rangle = \int g(x) k(x) \mu(dx)$. To get the other equivalences, let $\mathscr{H} = \{h \in L^\infty(\mu) : h = ph\}$. Show that h is a vector space containing the constant functions and is closed under pointwise limits, and that if $g, h \in \mathscr{H}$, then $g \wedge h \in \mathscr{H}$. This implies $\mathscr{H} = b\mathscr{G}$ (i.e., the bounded $\mathscr{G}$-measurable functions) for some $\mathscr{G}$. Then verify that $\mathscr{G} = \mathscr{C}$. For more on this approach to Markov chains, see Foguel (1969) or Revuz (1984), Chapter 4.

3   Recurrence

In this section we will study the recurrence properties of stationary sequences. Our first result is an application of the ergodic theorem. Let $X_1, X_2, \ldots$ be a stationary sequence taking values in $\mathbb{R}^d$, let $S_k = X_1 + \cdots + X_k$, let $A = \{S_k \neq 0 \text{ for all } k \geq 1\}$, and let $R_n = |\{S_1, \ldots, S_n\}|$ be the number of points visited at time n. Kesten, Spitzer, and Whitman (see Spitzer (1964), p. 40) proved the next result when the X_i are i.i.d. In what case $\mathscr{I}$ is trivial, so the limit is $P(A)$.

(3.1) THEOREM   As $n \to \infty$, $R_n/n \to E(1_A | \mathscr{I})$ a.s.

Proof   Suppose $X_1, X_2, \ldots$ are constructed on $(\mathbb{R}^d)^{\{0, 1, \ldots\}}$ with $X_n(\omega) = \omega_n$, and let φ be the shift operator. It is clear that

$$R_n \geq \sum_{m=0}^{n-1} 1_A(\varphi^m \omega),$$

since the right-hand side $= |\{m: 1 \leq m \leq n, S_l \neq S_m \text{ for all } l > m\}|$. Using the ergodic theorem now gives

$$\liminf_{n \to \infty} R_n/n \geq E(1_A|\mathscr{I}) \quad \text{a.s.}$$

To prove the opposite inequality, let $A_k = \{S_1 \neq 0, S_2 \neq 0, \dots, S_k \neq 0\}$. It is clear that

$$R_n \leq k + \sum_{m=0}^{n-k-1} 1_{A_k}(\varphi^m \omega),$$

since the sum on the right-hand side $= |\{m: 1 \leq m \leq n - k, S_l \neq S_m \text{ for } m < l \leq m + k\}|$. Using the ergodic theorem now gives

$$\limsup_{n \to \infty} R_n/n \leq E(1_{A_k}|\mathscr{I}).$$

As $k \uparrow \infty$, $A_k \downarrow A$, so the monotone convergence theorem for conditional expectations, (1.1c) in Chapter 4, implies

$$E(1_{A_k}|\mathscr{I}) \downarrow E(1_A|\mathscr{I}) \quad \text{as } k \uparrow \infty,$$

and the proof is complete.   □

Exercise 3.1   Let $g_n = P(S_1 \neq 0, \dots, S_n \neq 0)$. Show that $ER_n = \sum_{m=1}^{n} g_{m-1}$.

For a simple random walk in $d = 1$ we know that $g_{2n} \sim \pi^{-1/2} n^{-1/2}$ (see (3.4) in Chapter 3), so Exercise 3.1 implies $ER_n \sim 2\pi^{-1/2} n^{1/2}$. Another way of seeing $R_n = O(n^{1/2})$ in this case is to observe $R_n = \max_{m \leq n} S_m - \min_{m \leq n} S_m + 1$. In Exercise 6.4 of Chapter 7 we will see that $R_n/n^{1/2}$ converges weakly to a nondegenerate limit.

Exercise 3.2   Consider a two-dimensional simple random walk. Let $g_n = P(S_1 \neq 0, \dots, S_n \neq 0)$ and $u_n = P(S_n = 0)$. Clearly $u_{2n-1} = 0$ and $g_{2n} = g_{2n+1}$. (i) Show that $\sum_{0 \leq m \leq n} u_m g_{n-m} = 1$ and use this to conclude that $u_n \sim \pi/\log n$. (ii) Show that $ER_n^2 = \sum_{1 \leq j \leq k \leq n} h_{j,k}$ where $h_{j,k}$ is the probability new points are visited at times j and k. Use the fact that $h_{j,k} \leq g_{j-1} g_{k-j}$ to bound var(R_n) and show $R_n/ER_n \to 1$ in probability. The limit exists a.s. as well, but this takes much more work. See Section 5 of Dvoretsky and Erdös (1950).

Remark   In $d \geq 3$ there is a central limit theorem

$$\chi_n \equiv (R_n - ER_n)/(\text{var } R_n)^{1/2} \Rightarrow \chi$$

where χ is the standard normal distribution. When $d \geq 4$, var $R_n \sim Cn$, but when $d = 3$, var $R_n \sim Cn \log n$. Jain and Orey (1969) proved the result for $d \geq 5$, or more generally for random walks that are *strongly transient*; that is, $L = \sup\{n \geq 0: S_n = 0\}$ has $EL < \infty$. See Jain and Pruitt (1971) for $d = 3$ and 4. They showed var $R_n \sim Cn^2 (\log n)^{-4}$ in $d = 2$, but it was almost 15 years before LeGall (1985,

1986a, 1986b) proved that for $d = 2$, $\chi_n \Rightarrow$ a nonnormal limit defined in terms of "self-intersection local time for Brownian motion."

From (3.1) we get a result about the recurrence of random walks with stationary increments that is (for integer-valued random walks) a generalization of the Chung–Fuchs theorem ((2.7) in Chapter 3).

(3.2) THEOREM   Let $X_1, X_2, \ldots$ be a stationary sequence taking values in $\mathbb{Z}$ with $E|X_i| < \infty$. Let $S_n = X_1 + \cdots + X_n$ and $A = \{S_1 \neq 0, S_2 \neq 0, \ldots\}$. (i) If $E(X_1|\mathscr{I}) = 0$, then $P(A) = 0$. (ii) If $P(A) = 0$, then $P(S_n = 0 \text{ i.o.}) = 1$.

Remark   In words, mean 0 implies recurrence. The condition $E(X_1|\mathscr{I}) = 0$ is needed to rule out trivial examples that have mean 0 but are a combination of a sequence with positive and negative means; for example, $P(X_n = 1 \text{ for all } n) = P(X_n = -1 \text{ for all } n) = 1/2$.

Proof of (i)   If $E(X_1|\mathscr{I}) = 0$, then the ergodic theorem implies $S_n/n \to 0$ a.s. Now

$$\limsup_{n \to \infty} \left(\max_{1 \leq k \leq n} |S_k|/n \right) = \limsup_{n \to \infty} \left(\max_{K \leq k \leq n} |S_k|/n \right) \leq \left(\max_{k \geq K} |S_k|/k \right)$$

for any K, and the right-hand side $\downarrow 0$ as $K \uparrow \infty$. The last conclusion leads easily to

$$\lim_{n \to \infty} \left(\max_{1 \leq k \leq n} S_k \right) \bigg/ n = 0 = \lim_{n \to \infty} \left(\min_{1 \leq k \leq n} S_k \right) \bigg/ n.$$

Since

$$R_n \leq \left| \left\{ \min_{1 \leq k \leq n} S_k, \ldots, \max_{1 \leq k \leq n} S_k \right\} \right|,$$

it follows that $R_n/n \to 0$ and (3.1) implies $P(A) = 0$.

Proof of (ii)   Let $F_j = \{S_i \neq 0 \text{ for } i < j, S_j = 0\}$ and $G_{j,k} = \{S_{j+i} - S_j \neq 0 \text{ for } i < k, S_{j+k} - S_j = 0\}$. $P(A) = 0$ implies that $\Sigma P(F_k) = 1$, and stationarity implies $P(G_{j,k}) = P(F_k)$, so $\bigcup_k G_{j,k} = \Omega$ a.s. It follows that

$$\sum_k P(F_j \cap G_{j,k}) = P(F_j)$$

and

$$\sum_{j,k} P(F_j \cap G_{j,k}) = 1.$$

On $F_j \cap G_{j,k}$, $S_j = 0$ and $S_{j+k} = 0$, so we have shown $P(S_n = 0 \text{ at least twice}) = 1$.

Repeating the last argument shows $P(S_n = 0$ at least k times$) = 1$ for all k, and the proof is complete.   □

Exercise 3.3   Imitate the proof of part (i) in (3.2) to show that if we assume $P(X_i > 1) = 0$ and $EX_i > 0$ in addition to the hypotheses of (3.2), then $P(A) = EX_i$.

You have proved the last result twice for an asymmetric simple random walk (Exercise 1.13 in Chapter 3, Exercise 7.3 in Chapter 4). For more general random walks, the conclusion is new. It is interesting to note that we can use martingale theory to prove a result for random walks that do not skip over integers on the way down.

Exercise 3.4   Suppose $X_1, X_2, \ldots$ are i.i.d. with $P(X_i < -1) = 0$ and $EX_i > 0$. If $P(X_i = -1) > 0$, there is a unique $\theta < 0$ so that $E \exp(\theta X_i) = 1$. Let $S_n = X_1 + \cdots + X_n$ and $N = \inf\{n : S_n < 0\}$. Show that $\exp(\theta S_n)$ is a martingale and use the optional stopping theorem to conclude that $P(N < \infty) = e^\theta$.

Extending the reasoning in the proof of part (ii) of (3.2) gives a result of Kac (1947b). Let $X_0, X_1, \ldots$ be a stationary sequence taking values in $(S, \mathscr{S})$. Let $A \in \mathscr{S}$, let $T_0 = 0$, and for $n \geq 1$, let $T_n = \inf\{m > T_{n-1} : X_m \in A\}$ be the time of the nth return to A.

(3.3) THEOREM   If $P(X_n \in A$ at least once$) = 1$, then under $P(\cdot | X_0 \in A)$, $t_n = T_n - T_{n-1}$ is a stationary sequence with $E(T_1 | X_0 \in A) = 1/P(X_0 \in A)$.

Remark   If X_n is an irreducible Markov chain on a countable state space S starting from its stationary distribution π, and $A = \{x\}$, then (3.3) says $E_x T_x = 1/\pi(x)$, which is (4.6) in Chapter 5. (3.3) extends that result to an arbitrary $A \subset S$ and drops the assumption that X_n is a Markov chain.

Proof   We first show that under $P(\cdot | X_0 \in A)$, $t_1, t_2, \ldots$ is stationary. To cut down on $\ldots$'s, we will only show that

$$P(t_1 = m, t_2 = n | X_0 \in A) = P(t_2 = m, t_3 = n | X_0 \in A).$$

It will be clear that the same proof works for any finite-dimensional distribution. Our first step is to extend $\{X_n, n \geq 0\}$ to a two-sided stationary sequence $\{X_n, n \in \mathbb{Z}\}$ using (1.2). Let $C_k = \{X_{-1} \notin A, \ldots, X_{-k+1} \notin A, X_{-k} \in A\}$.

$$\left(\bigcup_{k=1}^{K} C_k \right)^c = \{X_k \notin A \text{ for } -K \leq k \leq -1\}.$$

The last event has the same probability as $\{X_k \notin A$ for $1 \leq k \leq K\}$, so letting $K \to \infty$, we get

$$P\left(\bigcup_{k=1}^{\infty} C_k\right) = 1.$$

To prove the desired stationarity, we observe that

$$P(t_2 = m, t_3 = n, X_0 \in A) = \sum_{l=1}^{\infty} P(X_0 \in A, t_1 = l, t_2 = m, t_3 = n)$$

$$= \sum_{l=1}^{\infty} P(X_0, X_l, X_{l+m}, X_{l+m+n} \in A, X_i \notin A$$

$$\text{for the other } i \in [0, l + m + n])$$

$$= \sum_{l=1}^{\infty} P(X_{-l}, X_0, X_m, X_{m+n} \in A, X_i \notin A$$

$$\text{for the other } i \in [-l, m + n])$$

$$= \sum_{l=1}^{\infty} P(C_l, X_0 \in A, t_1 = m, t_2 = n).$$

To complete the proof, we compute

$$E(t_1 | X_0 \in A) = \sum_{k=1}^{\infty} P(t_1 \geq k | X_0 \in A) = P(X_0 \in A)^{-1} \sum_{k=1}^{\infty} P(t_1 \geq k, X_0 \in A)$$

$$= P(X_0 \in A)^{-1} \sum_{k=1}^{\infty} P(C_k) = 1/P(X_0 \in A),$$

since the C_k are disjoint and their union has probability 1.   □

In the next two exercises we continue to use the notation of (3.3).

Exercise 3.5   Show that $E(\sum_{1 \leq m \leq T_1} 1_{(X_m \in B)} | X_0 \in A) = P(X_0 \in B)/P(X_0 \in A)$. When $A = \{x\}$ and X_n is a Markov chain, this is the "cycle trick" for defining a stationary measure. (See (4.3) in Chapter 5.)

Exercise 3.6   Consider the special case in which $X_n \in \{0, 1\}$, and let $\bar{P} = P(\cdot | X_0 = 1)$. Show $P(T_1 = n) = \bar{P}(T_1 > n)/\bar{E}T_1$. When $t_1, t_2, \ldots$ are i.i.d., this reduces to the formula for the first waiting time in a stationary renewal process.

In checking the hypotheses of Kac's theorem, a result Poincaré proved in 1899 is useful. First we need a definition. Let $T_A = \inf\{n \geq 1 : \varphi^n(\omega) \in A\}$.

(3.4) THEOREM   Suppose $\varphi: \Omega \to \Omega$ preserves P; that is, $P \circ \varphi^{-1} = P$. (i) $T_A < \infty$ a.s. on A; that is, $P(\omega \in A, T_A = \infty) = 0$. (ii) $\{\varphi^n(\omega) \in A$ i.o.$\} \supset A$. (iii) If φ is ergodic and $P(A) > 0$, then $P(\varphi^n(\omega) \in A$ i.o.$) = 1$.

Remark   Note that in (i) and (ii) we assume only that φ is measure preserving. Extrapolating from Markov chain theory, the conclusions can be "explained" by noting that: (i) the existence of a stationary distribution implies the sequence is recurrent, and (ii) since we start in A, we do not have to assume irreducibility. Conclusion (iii) is, of course, a consequence of the ergodic theorem, but as the self-contained proof below indicates, it is a much simpler fact.

Proof   Let $B = \{\omega \in A, T_A = \infty\}$. A little thought shows that if $\omega \in \varphi^{-m}B$, then $\varphi^m(\omega) \in A$, but $\varphi^n(\omega) \notin A$ for $n > m$, so the $\varphi^{-m}B$ are pairwise disjoint. The fact that φ is measure preserving implies $P(\varphi^{-m}B) = P(B)$, so we must have $P(B) = 0$ (or P would have infinite mass). To prove (ii), note that for any k, φ^k is measure preserving, so (i) implies

$$0 = P(\omega \in A, \varphi^{nk}(\omega) \notin A \text{ for all } n \geq 1)$$

$$\geq P(\omega \in A, \varphi^m(\omega) \notin A \text{ for all } m \geq k).$$

Since the last probability is 0 for all k, (ii) follows. Finally for (iii), note that $B \equiv \{\omega: \varphi^n(\omega) \in A$ i.o.$\}$ is invariant and $\supset A$ by (b), so $P(B) > 0$ and it follows from ergodicity that $P(B) = 1$.   □

4   Mixing

A measure-preserving transformation φ on $(\Omega, \mathscr{F}, P)$ is called *mixing* if

$$\lim_{n \to \infty} P(A \cap \varphi^{-n}B) = P(A)P(B). \tag{4.1}$$

A sequence X_n, $n \geq 0$, is said to be *mixing* if the corresponding shift on sequence space is. To see that mixing implies ergodicity, observe that if A is invariant, then (4.1) implies $P(A) = P(A)^2$; that is, $P(A) \in \{0, 1\}$. Conversely, ergodicity implies

$$\frac{1}{n} \sum_{m=0}^{n-1} 1_B(\varphi^m \omega) \to P(B) \qquad \text{a.s.,}$$

so integrating over A and using the bounded convergence theorem give

$$\frac{1}{n} \sum_{m=0}^{n-1} P(A \cap \varphi^{-m}B) \to P(A)P(B);$$

that is, (4.1) holds in a Cesaro sense.

As usual when we meet a new concept, we turn to our examples to see what it means. To handle Examples 1.1 and 1.2, we use a general result. For this result we suppose that φ is the shift operator on sequence space, and $X_n(\omega) = \omega_n$. Let

$$\mathcal{F}_n' = \sigma(X_n, X_{n+1}, \ldots) \quad \text{and} \quad \mathcal{T} = \bigcap_n \mathcal{F}_n'.$$

(4.2) THEOREM   If $\mathcal{T}$ is trivial, then φ is mixing and

$$\lim_{n\to\infty} \sup_B |P(A \cap \varphi^{-n}B) - P(A)P(B)| = 0.$$

Proof   Let $C = \varphi^{-n}B \in \mathcal{F}_n'$.

$$|P(A \cap C) - P(A)P(C)| = \left| \int_C 1_A - P(A)\, dP \right|$$

$$= \left| \int_C P(A|\mathcal{F}_n') - P(A)\, dP \right|$$

$$\leq \int \left| P(A|\mathcal{F}_n') - P(A) \right| dP \to 0,$$

since $P(A|\mathcal{F}_n') \to P(A)$ in L^1 by (6.3) in Chapter 4. $\square$

Conversely if $A \in \mathcal{T}$ has $P(A) \in (0, 1)$ and $A = \varphi^{-n}B_n$, then

$$|P(A \cap \varphi^{-n}B_n) - P(A)P(B)| = P(A) - P(A)^2,$$

so we have:

$\mathcal{T}$ is trivial if and only if

$$\lim_{n\to\infty} \sup_B |P(A \cap \varphi^{-n}B) - P(A)P(B)| = 0 \quad \text{for all } A.$$

By definition, φ is mixing if and only if

$$\lim_{n\to\infty} |P(A \cap \varphi^{-n}B) - P(A)P(B)| = 0 \quad \text{for all } A, B.$$

φ is ergodic if and only if

$$\lim_{n\to\infty} \frac{1}{n} \sum_{m=0}^{n-1} P(A \cap \varphi^{-m}B) - P(A)P(B) = 0 \quad \text{for all } A, B.$$

Exercise 4.1   Prove the unproved half of the last assertion.

Example 4.1   *i.i.d. sequences* are mixing, since $\mathcal{T}$ is trivial.

Example 4.2   *Markov chains.* Suppose the state space S is countable; p is irreducible and has stationary distribution π. If p is aperiodic, $\mathcal{T}$ is trivial by (5.8) in Chapter 5 and the sequence is mixing. Conversely if p has period d and cyclic decomposition $S_0, S_1, \ldots, S_{d-1}$,

$$\lim_{n\to\infty} P(X_0 \in S_0, X_{nd+1} \in S_0) = 0 \neq \pi(S_0)^2,$$

so the sequence is not mixing.

Example 4.3   *Rotation of the circle.* $\Omega = [0, 1)$, $\varphi(\omega) = (\omega + \theta)$ mod 1, θ is irrational. To prove that this transformation is not mixing, we begin by observing that $n\theta$ mod 1 is dense. (Use Exercise 1.4 or recall Example 2.3 shows that for intervals the exceptional set in the ergodic theorem is $\varnothing$.) Since $n\theta$ mod 1 is dense, there is a sequence $n_k \to \infty$ with $n_k\theta$ mod $1 \to 1/2$. Let $A = B = [0, 1/3)$. If k is large, then $A \cap \varphi^{-n(k)}B = \varnothing$, so $P(A \cap \varphi^{-n(k)}B)$ does not converge to $1/9$ and the transformation is not mixing.

To treat our last example, we will use:

(4.3) LEMMA   Let $\mathcal{A}$ be a π-system. If

$$\lim_{n\to\infty} P(A \cap \varphi^{-n}B) = P(A)P(B) \tag{$*$}$$

holds for $A, B \in \mathcal{A}$, then $(*)$ holds for $A, B \in \sigma(\mathcal{A})$.

Proof   As the reader can probably guess, we are going to use the $\pi - \lambda$ theorem ((4.2) in Chapter 1). Fix $A \in \mathcal{A}$ and let $\mathcal{L}$ be the collection of B for which $(*)$ holds. It is clear that $\Omega \in \mathcal{L}$. Let $B_1 \supset B_2$ be in $\mathcal{L}$, $\varphi^{-n}(B_1 - B_2) = \varphi^{-n}B_1 - \varphi^{-n}B_2$, so

$$P(A \cap \varphi^{-n}(B_1 - B_2)) = P(A \cap \varphi^{-n}B_1) - P(A \cap \varphi^{-n}B_2)$$

and

$$\lim_{n\to\infty} P(A \cap \varphi^{-n}(B_1 - B_2)) = P(A)P(B_1) - P(A)P(B_2) = P(A)P(B_1 - B_2).$$

To finish the proof that $\mathcal{L}$ is a λ-system, let $B_k \in \mathcal{L}$ with $B_k \uparrow B$.

$$|P(A \cap \varphi^{-n}B) - P(A \cap \varphi^{-n}B_k)| \leq |P(\varphi^{-n}B) - P(\varphi^{-n}B_k)| = P(B) - P(B_k),$$

since φ^n preserves P. At this point we have shown that $(*)$ holds for $A \in \mathcal{A}$ and $B \in \sigma(\mathcal{A})$. Fixing $B \in \sigma(\mathcal{A})$, the reader can now repeat the last argument to show that $(*)$ holds for $A, B \in \sigma(\mathcal{A})$. Indeed, the proof is simpler, since you do not have to worry about what φ^{-n} does. $\square$

Example 4.4 *Bernoulli shift.* $\Omega = [0, 1)$, $\varphi(\omega) = 2\omega \mod 1$. Let $\mathscr{F}_n = \sigma([m2^{-n}, (m + 1)2^{-n}); 0 \leq m < 2^n)$ and $\mathscr{A} = \bigcup \mathscr{F}_n$. If $A \in \mathscr{F}_n$, $B \in \mathscr{A}$, and then A and $\varphi^{-m}B$ are independent for $m \geq n$; that is, $P(A \cap \varphi^{-m}B) = P(A)P(\varphi^{-m}B) = P(A)P(B)$, since φ is measure preserving. So (4.3) implies φ is mixing.

Remark A second way of approaching the last example is to let $Y_n(\omega) = \varphi^n(\omega)$, $X_n(\omega) = $ the nth digit in the binary expansion of ω, and observe

$$Y_n(\omega) = \sum_{m=0}^{\infty} 2^{-(m+1)}X_m(\omega). \qquad (4.4)$$

$X_1, X_2, \ldots$ are i.i.d. and the tail σ-field of Y is contained in the tail σ-field of X, so (4.2) implies the sequence Y_n is mixing; that is, the associated shift on sequence space is. The careful reader will notice that the two proofs prove that different transformations are mixing: φ on $[0, 1)$ vs. θ on $[0, 1)^{\{0, 1, \ldots\}}$. The next exercise shows that there is essentially no difference between the two conclusions.

Exercise 4.2 Suppose φ is measure preserving on $(\mathbb{R}, \mathscr{R}, \mu)$ and let $X_n(\omega) = \varphi^n(\omega)$. Then φ is mixing if and only if X_n is.

Sketch Apply (4.3) with $\mathscr{A} = $ the collection of sets of the form $\{X_j(\omega) \in B_j, 0 \leq j \leq k\}$ where $B_j \in \mathscr{R}$.

Up to this point all our examples either have a trivial tail field (Examples 4.1, 4.2, and 4.4) or are not mixing (Example 4.3). To complete the picture we close the section with some examples that are mixing but have nontrivial tail fields.

Example 4.5 Let X_n, $n \in \mathbb{Z}$, be i.i.d. with $P(X_n = 1) = P(X_n = 0) = 1/2$, and let

$$Z_n(\omega) = \sum_{m=0}^{\infty} 2^{-(m+1)}X_{n-m}(\omega).$$

Z_n is mixing but $\mathscr{T} = \sigma(X_n: n \in \mathbb{Z})$.

Proof Since all the X_k, $k \leq n$, can be recovered by expanding Z_n in its binary decimal representation, the claim about $\mathscr{T}$ is clear. To prove that Z_n is mixing, we use (4.3) with $\mathscr{A} = $ sets of the form $\{\omega: Z_i(\omega) \in G_i, 0 \leq i \leq k\}$ where the G_i are open sets. Now

$$Z_n = \sum_{m=0}^{n-k-1} 2^{-(m+1)}X_{n-m} + 2^{-(n-k)}Z_k,$$

so if $\mathscr{T}_k = \sigma(X_j: j \leq k)$ and G_0 is open,

$$P(Z_n \in G_0 | \mathscr{F}_k) = \mu_{n-k-1}(G_0 - 2^{-(n-k)}Z_k)$$

where μ_M is the distribution of $\sum_{0 \leq m \leq M} 2^{-(m+1)}X_{-m}$ and $G_0 - c = \{x - c : x \in G_0\}$. As $M \to \infty$, $\mu_M \Rightarrow \mu_\infty$, a distribution without atoms, so for a.e. ω,

$$\mu_{n-k-1}(G_0 - 2^{-(n-k)}Z_k) \Rightarrow \mu_\infty(G_0),$$

and integrating over $A \in \mathcal{F}_k$,

$$P(Z_n \in G_0, A) \to \mu_\infty(G_0)P(A).$$

Letting $B = \{\omega: Z_0(\omega) \in G_0\}$, we have shown

$$P(A \cap \varphi^{-n}B) \to P(A)P(B) \quad \text{for } A \in \mathcal{F}_k.$$

The last argument generalizes easily to $B = \{\omega: Z_i(\omega) \in G_i, 0 \le i \le k\}$.   □

Exercise 4.3   Show that Baker's transformation, introduced in Exercise 1.7, is mixing but $\mathcal{T}$ is not trivial.

Our last example, borrowed from Lasota and MacKay (1985), is named for its famous cousin: geodesic flow on a compact Riemannian manifold M with negative curvature—for example, the two-hole torus with a suitable metric. For more on these examples, see Anosov (1963, 1967).

Example 4.6   *Anosov map.* $\Omega = [0, 1)^2$, $\varphi(x, y) = (x + y, x + 2y)$. By drawing a picture it is easy to see that φ maps Ω 1–1 onto itself. Since the Jacobian

$$J = \det \begin{pmatrix} 1 & 1 \\ 1 & 2 \end{pmatrix} = 1,$$

φ preserves Lebesgue measure. Since φ is invertible, the entire sequence can be recovered from one term and $\mathcal{T}$ is far from trivial. We will now show that φ is mixing. To begin we observe that

$$\varphi^n(x, y) = (a_{2n-2}x + a_{2n-1}y, a_{2n-1}x + a_{2n}y)$$

where the a_n are the *Fibonacci numbers* given by $a_0 = a_1 = 1$ and $a_{n+1} = a_n + a_{n-1}$ for $n \ge 1$. (To check this notice that $a_{2n-2} + 2a_{2n-1}(+ a_{2n}) = (a_{2n} +)a_{2n+1}$.) Let

$$f(x, y) = \exp(2\pi i(px + qy))$$

$$g(x, y) = \exp(2\pi i(rx + sy)).$$

Since $\int_{[0, 1]} \exp(2\pi ikx)\, dx = 0$ unless $k = 0$, it follows that

$$\int_0^1 \int_0^1 f(x, y)g(\varphi^n(x, y))\, dx\, dy = 0$$

unless

$$ra_{2n-2} + sa_{2n-1} + p = 0$$

$$ra_{2n-1} + sa_{2n} + q = 0.$$

(∗)

Now the difference equation $b_{n+1} - b_n - b_{n-1} = 0$ has a two-parameter family of solutions given by

$$b_n = C_1 \left(\frac{1 + \sqrt{5}}{2} \right)^n + C_2 \left(\frac{1 - \sqrt{5}}{2} \right)^n,$$

so the Fibonacci numbers are given by

$$a_n = \frac{1}{2} \left(\frac{1 + \sqrt{5}}{2} \right)^n + \frac{1}{2} \left(\frac{1 - \sqrt{5}}{2} \right)^n \quad \text{for } n \geq 0.$$

From the last formula we see that

$$\lim_{n \to \infty} a_{2n}/a_{2n-1} = \lim_{n \to \infty} a_{2n-1}/a_{2n-2} = \frac{1 + \sqrt{5}}{2},$$

and by considering several cases one sees that (∗) cannot hold for infinitely many n unless $p = q = r = s = 0$. The last result implies that if

$$f(x, y) = \sum_{j=1}^{k} a_j \exp(2\pi i (p_j x + q_j y))$$

$$g(x, y) = \sum_{j=1}^{l} b_j \exp(2\pi i (r_j x + s_j y)),$$

then as $n \to \infty$,

$$\int_0^1 \int_0^1 f(x, y) g(\varphi^n(x, y)) \, dx \, dy \to 0.$$

(∗∗)

To finish the proof, we observe:

(4.5) LEMMA   Since the f and g for which (∗∗) holds are dense in

$$L_0^2 = \left\{ f : \int_0^1 \int_0^1 f(x, y) \, dx \, dy = 0, \int_0^1 \int_0^1 f^2(x, y) \, dx \, dy < \infty \right\},$$

it follows that φ is mixing.

Proof   Let

$$\langle h, k \rangle = \int_0^1 \int_0^1 h(x, y) k(x, y) \, dx \, dy$$

and $\|h\|_2 = \langle h, h \rangle^{1/2}$. Adding and subtracting $\langle h, g \circ \varphi^n \rangle$ give

$$|\langle f, g \circ \varphi^n \rangle - \langle h, k \circ \varphi^n \rangle| \le |\langle f, g \circ \varphi^n \rangle - \langle h, g \circ \varphi^n \rangle| + |\langle h, g \circ \varphi^n \rangle - \langle h, k \circ \varphi^n \rangle|$$

$$\le \|f - h\|_2 \|g \circ \varphi^n\|_2 + \|h\|_2 \|(g - k) \circ \varphi^n\|_2$$

$$= \|f - h\|_2 \|g\|_2 + \|h\|_2 \|(g - k)\|_2$$

by Cauchy–Schwarz and the fact that φ^n is measure preserving. Suppose now that $h = 1_A - P(A)$, $k = 1_B - P(B) \in L_0^2$; pick f and g so that (∗∗) holds and $\|h - f\|_2$, $\|k - g\|_2$ are $< \varepsilon$. The last inequality implies

$$\limsup_{n \to \infty} |P(A \cap \varphi^{-n}B) - P(A)P(B)| = \limsup_{n \to \infty} |\langle h, k \circ \varphi^n \rangle| \le \varepsilon(\|k\|_2 + \varepsilon) + \|h\|_2 \varepsilon,$$

and since ε is arbitrary, the proof is complete.   □

Exercise 4.4   Does the independent blocks process, defined in Exercise 1.9, always have a trivial tail σ-field? Is it mixing?

Remark   The continued fraction transformation (Exercise 1.8) is mixing. See the end of Section 7 in Chapter 7. Of course, this is more difficult to prove than ergodicity.

*5 Entropy

Throughout this section we will suppose that X_n, $n \in \mathbb{Z}$, is an ergodic stationary sequence taking values in a finite set S. Let

$$p(x_0, \ldots, x_{n-1}) = P(X_0 = x_0, \ldots, X_{n-1} = x_{n-1})$$

and

$$p(x_n | x_{n-1}, \ldots, x_0) = P(X_n = x_n | X_{n-1} = x_{n-1}, \ldots, X_0 = x_0)$$

whenever the conditioning event has positive probability. Define random variables

$$p(X_0, \ldots, X_n) \quad \text{and} \quad p(X_n | X_{n-1}, \ldots, X_0)$$

by setting $x_j = X_j(\omega)$ in the corresponding definitions. Since $P(p(X_0, \ldots, X_n) = 0) = 0$, the conditional probability makes sense a.s.

(5.1) The Shannon–McMillan–Breiman Theorem   asserts that

$$-\frac{1}{n} \log p(X_0, \ldots, X_{n-1}) \to H \qquad \text{a.s.}$$

where $H = \lim_{n \to \infty} E\{-\log p(X_n | X_{n-1}, \ldots, X_0)\}$ is the *entropy rate* of X_n.

Remark The three names indicate the evolution of the theorem: Shannon (1948), McMillan (1953), Breiman (1957). Our proof follows Algoet and Cover (1988).

Proof The first step is to prove that the limit defining H exists. To do this, we observe that if $\mathscr{F}_n = \sigma(X_{-1}, \dots, X_{-n})$ and

$$Y_n = p(x_0 | X_{-1}, \dots, X_{-n}) = P(X_0 = x_0 | \mathscr{F}_n),$$

then Y_n is a bounded martingale, so as $n \to \infty$,

$$Y_n \to Y_\infty \equiv p(x_0 | X_{-1}, X_{-2}, \dots) \text{ a.s. and in } L^1.$$

Since S is finite and $x \log x$ is bounded on $[0, 1]$, it follows that

(a) $\quad H_k \equiv E(-\log p(X_0 | X_{-1}, \dots, X_{-k}))$

$$= E\left(-\sum_x p(x | X_{-1}, \dots, X_{-k}) \log p(x | X_{-1}, \dots, X_{-k}) \right)$$

$$\to E\left(-\sum_x p(x | X_{-1}, X_{-2}, \dots) \log p(x | X_{-1}, X_{-2}, \dots) \right) \equiv H$$

as $k \to \infty$. Now

$$P(X_0, \dots, X_{n-1} | X_{-1}, \dots, X_{-k}) = \prod_{m=0}^{n-1} p(X_m | X_{m-1}, \dots, X_{-k}),$$

so it is easy to see that

$$p(X_0, \dots, X_{n-1} | X_{-1}, X_{-2}, \dots) \equiv \lim_{k \to \infty} p(X_0, \dots, X_{n-1} | X_{-1}, \dots, X_{-k})$$

exists and

$$p(X_0, \dots, X_{n-1} | X_{-1}, X_{-2}, \dots) = \prod_{m=0}^{n-1} p(X_m | X_{m-1}, X_{m-2}, \dots).$$

Taking logs gives

(b) $\quad -\dfrac{1}{n} \log p(X_0, \dots, X_{n-1} | X_{-1}, X_{-2}, \dots) = -\dfrac{1}{n} \sum_{m=0}^{n-1} \log p(X_m | X_{m-1}, X_{m-2}, \dots)$

$$\to E(-\log p(X_0 | X_{-1}, \dots)) = H$$

by the ergodic theorem applied to $F(\omega) = -\log p(\omega_0 | \omega_{-1}, \dots)$.

For a bound in the other direction, let

$$p^k(X_0, \ldots, X_{n-1}) = p(X_0, \ldots, X_{k-1}) \prod_{m=k}^{n} p(X_m | X_{m-1}, \ldots, X_{m-k})$$

for $k \leq n$, and observe that another application of the ergodic theorem gives

$$-\frac{1}{n} \log p^k(X_0, \ldots, X_{n-1}) \to E(-\log p(X_0 | X_{-1}, \ldots, X_{-k})) = H_k. \tag{c}$$

Let

$$A_n^k = p^k(X_0, \ldots, X_{n-1})$$

$$B_n = p(X_0, \ldots, X_{n-1})$$

$$C_n = p(X_0, \ldots, X_{n-1} | X_{-1}, X_{-2}, \ldots)$$

$$W_n^1 = A_n^k / B_n, \qquad W_n^2 = B_n / C_n$$

and note that

$$-\frac{1}{n} \log A_n^k = -\frac{1}{n} \log W_n^1 - \frac{1}{n} \log B_n \tag{d}$$

$$-\frac{1}{n} \log B_n = -\frac{1}{n} \log W_n^2 - \frac{1}{n} \log C_n. \tag{e}$$

To get from (b) and (c) to the desired result we use:

(f) **LEMMA** If $W_n \geq 0$ have $EW_n \leq 1$, then $\limsup_{n \to \infty} n^{-1} \log W_n \leq 0$ a.s.

Proof If $\varepsilon > 0$, then

$$P(n^{-1} \log W_n \geq \varepsilon) \leq P(W_n \geq e^{\varepsilon n}) \leq e^{-\varepsilon n}$$

by Chebyshev's inequality. $\Sigma \exp(-\varepsilon n) < \infty$, so the Borel–Cantelli lemma implies that the limsup $\leq \varepsilon$. ε is arbitrary, so the proof is complete. $\square$

To check that $EW_n^i \leq 1$ we observe that

$$EW_n^1 = \sum_x \frac{p^k(x_0, \ldots, x_{n-1})}{p(x_0, \ldots, x_{n-1})} p(x_0, \ldots, x_{n-1}) = 1.$$

For the second step we observe that

$$E\left(\frac{p(X_0, \ldots, X_{n-1})}{p(X_0, \ldots, X_{n-1} | X_{-1}, \ldots, X_{-k})}\right) = E\left(\frac{p(X_{-k}, \ldots, X_{-1}) p(X_0, \ldots, X_{n-1})}{p(X_{-k}, \ldots, X_{n-1})}\right)$$

$$\sum_x p(x_{-k}, \ldots, x_{-1}) p(x_0, \ldots, x_{n-1}) = 1.$$

Letting $k \to \infty$ and using Fatou's lemma give $EW_n^2 \le 1$. Applying (f) now and using (c), (d), (e), and (b) give

$$H_k \ge \limsup_{n \to \infty} n^{-1} \log p(X_0, \ldots, X_{n-1})$$

$$\ge \liminf_{n \to \infty} n^{-1} \log p(X_0, \ldots, X_{n-1}) \ge H.$$

(a) implies $H_k \to H$ as $k \to \infty$, and the proof of (5.1) is complete. □

Example 5.1 $X_0, X_1, \ldots$ are i.i.d. In this case

$$p(X_0, \ldots, X_{n-1}) = \prod_{m=0}^{n-1} p(X_m),$$

so the strong law of large numbers implies

$$-\frac{1}{n} \log p(X_0, \ldots, X_{n-1}) \to -E \log p(X_0) = -\sum_x p(x) \log p(x).$$

For a concrete example, suppose $P(X_i = 0) = 2/3$ and $P(X_i = 1) = 1/3$. In this case

$$H = \tfrac{2}{3} \log \tfrac{3}{2} + \tfrac{1}{3} \log 3 \approx .6365.$$

Example 5.2 $X_0, X_1, \ldots$ is a Markov chain. In this case $H = H_1$, so

$$H = E(-\log p(X_0|X_{-1}) = \sum_{x,y} \pi(x)p(x, y)(-\log p(x, y))$$

where p is the transition probability and π is the stationary distribution. For a concrete example, suppose

$$p = \begin{pmatrix} .5 & .5 \\ 1 & 0 \end{pmatrix}, \qquad \pi = (2/3, 1/3).$$

In this case, $H = \tfrac{2}{3} \log 2 \approx .4621$.

Comparing the values of H in the two examples, we see that although they have the same marginal distributions, the i.i.d. sequence is "more random" than the Markov chain. To explain what the values of H tell us about the last two examples, we state a result that is merely a reformulation of (5.1).

(5.2) Asymptotic Equipartition Property Let $a = |S|$. Given $\varepsilon > 0$, there is an N_ε so that for $n \ge N_\varepsilon$, the a^n possible outcomes of $(X_0, \ldots, X_{n-1})$ can be divided into two classes: (i) a class B_n with probability $< \varepsilon$, and (ii) a class G_n in which each outcome $(x_0, x_1, \ldots, x_{n-1})$ has

$$|-H - n^{-1} \log p(x_0, \ldots, x_{n-1})| < \varepsilon.$$

Since the outcomes in G_n have probability $\exp(-(H \pm \varepsilon)n)$, it follows that the number of outcomes in G_n is $\exp((H \pm \varepsilon)n)$. A trivial but important special case occurs when X_n is an i.i.d. sequence with $P(X_n = x) = 1/a$ for $x \in S$. In this case $H = \log a$, so we need almost all the outcomes to capture $1 - \varepsilon$ of the probability mass. Since all outcomes have equal probability, the last conclusion should not be too surprising.

Exercise 5.1   *Breiman's proof.* Let $g_n(\omega) = -\log p(\omega_0 | \omega_{-1}, \ldots, \omega_{-n})$ and use Exercise 2.2 to prove (5.1). Checking $E(\sup g_n) < \infty$ is the hard part. Breiman (1957) did this incorrectly and had to submit a correction note, so we will give you a hint. Let

$$A_k = \left\{ g_k > \lambda, \sup_{j \le k} g_j \le \lambda \right\}$$

and observe

$$P(A_j) = \sum_{\omega \in A_j} p(\omega_{-j}, \ldots, \omega_0) \le e^{-\lambda} \sum_{\omega \in A_j} p(\omega_{-j}, \ldots, \omega_{-1}).$$

Chung (1961) used a similar approach to prove a version of the result for infinite alphabets.

Let p be a probability distribution on a countable set I and write p_i for the probability assigned to i. We defined the entropy of p by

$$H(p) = -\sum p_i \log p_i \quad \text{where } 0 \log 0 = 0.$$

The next four exercises show that some common distributions maximize entropy. Their solutions are based on the following simple observation: $\log x \le x - 1$ (since $\log$ is concave), so if p and q are probability distributions on a countable set I,

$$\sum_{i \in I} p_i \log(q_i/p_i) \le \sum_{i \in I} q_i - p_i = 0$$

(our convention implies $0 \log(q_i/0) = 0$) and

$$-\sum p_i \log p_i \le -\sum p_i \log q_i.$$

Since $\log x = x - 1$ only for $x = 1$, we have equality if and only if $p = q$.

Exercise 5.2   Show that of all the probability distributions on $\{1, \ldots, M\}$, the uniform $q_k = 1/M$ has the largest entropy.

Exercise 5.3   Show that of all the probability distributions on $\{1, 2, \ldots\}$ with mean $= \lambda > 1$, the geometric $q_k = (1/\lambda)(1 - 1/\lambda)^{k-1}$ has the largest entropy.

If a distribution on $\mathbb{R}$ has density $p(x)$, we define its entropy by

$$H = -\int p(x) \log p(x) \, dx \quad \text{where } 0 \log 0 = 0.$$

Repeating the argument above shows that if p and q are probability densities,

$$\int p(x) \log(q(x)/p(x)) \le 0.$$

Exercise 5.4   Show that of all the probability densities on $[0, \infty)$ with mean λ, the exponential $q(x) = \lambda^{-1} \exp(-x/\lambda)$ has the largest entropy.

Exercise 5.5   Show that of all the probability densities on $\mathbb{R}$ with $EX^2 = 1$, the standard normal has the largest entropy.

Exercise 5.6   Find the probability density on $\mathbb{R}$ with $EX^{2k} = 1$ that has the largest entropy.

For the last three exercises we restrict our attention to a random variable taking values in a countable set.

Exercise 5.7   Show that if a_{ij} is doubly stochastic and $\bar{p}_i = \sum a_{ij} p_j$, then $H(\bar{p}) \ge H(p)$. Since $a_{ij} = 1/M$ is doubly stochastic, this gives another solution of Exercise 5.2.

Exercise 5.8   We define the entropy of a random variable to be that of its distribution. Show that $H((X, Y)) \le H(X) + H(Y)$ with equality if and only if X and Y are independent.

Exercise 5.9   Let $p(j|i) = P(Y = j | X = i)$. Define $H(Y|X = i) = -\sum_j p(j|i) \log p(j|i)$ (the entropy of the conditional distribution) and $H(Y|X) = \sum_i p_i H(Y|X = i)$ where $p_i = P(X = i)$. Show that $H((X, Y)) = H(X) + H(Y|X)$, so $H(Y|X) \le H(Y)$ with equality if and only if X and Y are independent. The difference $H(Y) - H(Y|X)$ is the *information* about Y contained in X.

*6   A Subadditive Ergodic Theorem

In this section we will prove Liggett's (1985) version of Kingman's (1968):

(6.1) *Subadditive Ergodic Theorem*   Suppose $X_{m,n}$, $0 \le m < n$, satisfy:

(i) $X_{0,m} + X_{m,n} \ge X_{0,n}$.
(ii) For each k, $\{X_{nk,(n+1)k}, n \ge 1\}$ is a stationary sequence.
(iii) The distribution of $\{X_{m,m+k}, k \ge 1\}$ does not depend on m.
(iv) $EX_{0,1}^+ < \infty$ and for each n, $EX_{0,n} \ge \gamma_0 n$ where $\gamma_0 > -\infty$.

Then

(a) $\lim_{n\to\infty} EX_{0,n}/n = \inf_m EX_{0,m}/m \equiv \gamma$.
(b) $X = \lim_{n\to\infty} X_{0,n}/n$ exists a.s. and in L^1, so $EX = \gamma$.
(c) If all the stationary sequences in (ii) are ergodic, then $X = \gamma$ a.s.

Remark   Since (i) implies $X_{0,m}^+ + X_{m,n}^+ \geq X_{0,n}^+$, (iv) implies $E|X_{0,n}| \leq Cn < \infty$. King-man assumed (iv), but instead of (i)–(iii) he assumed that $X_{l,m} + X_{m,n} \geq X_{l,n}$ for all $l < m < n$ and that the distribution of $\{X_{m+k,n+k}, 0 \leq m < n\}$ does not depend on k. In two of the four applications in Section 7 these stronger conditions do not hold.

As usual, we will begin by considering several examples. Since the validity of (ii) and (iii) in each case is clear, we will check only (i) and (iv). The first example shows that (6.1) contains the ergodic theorem (5.1) as a special case.

Example 6.1   Suppose $\xi_1, \xi_2, \ldots$ is a stationary sequence with $E|\xi_k| < \infty$, and let $X_{m,n} = \xi_{m+1} + \cdots + \xi_n$. Then $X_{0,n} = X_{0,m} + X_{m,n}$ and (iv) holds.

Example 6.2   *Range of random walk.* Suppose $\xi_1, \xi_2, \ldots$ is a stationary sequence and let $S_n = \xi_1 + \cdots + \xi_n$. Let $X_{m,n} = |\{S_{m+1}, \ldots, S_n\}|$. It is clear that $X_{0,m} + X_{m,n} \geq X_{0,n}$. $0 \leq X_{0,n} \leq n$, so (iv) holds. Applying (6.1) now gives $X_{0,n}/n \to X$ a.s. and in L^1, but it does not tell us what the limit is.

Exercise 6.1   Suppose $\xi_1, \xi_2, \ldots$ are i.i.d. in Example 6.2. Use (c) and (a) of (6.1) to conclude that $|\{S_1, \ldots, S_n\}|/n \to P$ (no return to 0).

Example 6.3   *Records from "improving populations."* Ballerini and Resnick (1985, 1987). Let $\xi_1, \xi_2, \ldots$ be a stationary sequence and let $\zeta_n = \xi_n + cn$ where $c > 0$. Let

$$X_{m,n} = |\{l: m < l \leq n \text{ and } \zeta_l > \zeta_k \text{ for } m < k < l\}|.$$

It is clear that $X_{0,m} + X_{m,n} \geq X_{0,n}$. $0 \leq X_{0,n} \leq n$, so (iv) holds. Applying (6.1) now gives that $X_{0,n}/n \to X$ a.s. and in L^1. To identify the limit extend ξ_n, $n \geq 0$, to $\{\xi_n: n \in \mathbb{Z}\}$ using (1.2), and let $\zeta_n = \xi_n + cn, n \in \mathbb{Z}$. Let $Y_m = 1$ if $\zeta_m > \zeta_k$ for all $k < m$. An easy extension of (1.1) implies Y_m is a stationary sequence, so the ergodic theorem implies

$$(Y_1 + \cdots + Y_n)/n \to Y \quad \text{a.s. and in } L^1.$$

To see that $X = Y$ observe that $X \geq Y$, but

$$EX_{0,n}/n = \frac{1}{n} \sum_{m=1}^{n} P(\zeta_0 > \zeta_k \text{ for } 0 > k > -m) \to EY$$

as $n \to \infty$, so $EX = EY$ and hence $X = Y$ a.s.

The analysis in the last paragraph could be applied to identify the limit in Example 6.2. If we let $Y_m = 1$ when $S_k \neq S_m$ for all $k > m$, then we get a stationary sequence that has the same limit as $X_{0,n}/n$. (This was the key to the proof in Section 3.) Kingman's original proof shows that we can always identify the limit by this method. That is, we can write

$$X_{m,n} = \sum_{k=m+1}^{n} Y_k + Z_{m,n}$$

where Y_m, $m \geq 1$, is a stationary sequence, and $Z_{m,n} \geq 0$ is a subadditive process with $EZ_{0,n}/n \to 0$. We will prove (6.1) without proving Kingman's "decomposition theorem."

The next example shows that the convergence in (a) of (6.1) may occur arbitrarily slowly.

Example 6.4   Suppose $X_{m,m+k} = f(k) \geq 0$ where $f(k)/k$ is decreasing.

$$X_{0,n} = f(n) = m\frac{f(n)}{n} + (n-m)\frac{f(n)}{n} \leq m\frac{f(m)}{m} + (n-m)\frac{f(n-m)}{n-m}$$

$$= X_{0,m} + X_{m,n}.$$

Exercise 6.2   *Strong law of large numbers in Banach space.* (i) Let $X_1, X_2, \ldots$ be i.i.d. taking values in a Banach space B with norm $\|\cdot\|$, and suppose $E\|X_i\| < \infty$. Show that $\|S_n/n\| \to$ a limit a.s. (ii) Turning the property we need into the definition, we say that X is *Bochner integrable* if for any $\varepsilon > 0$, there is a linear operator θ with finite-dimensional range so that $E\|X - \theta X\| < \varepsilon$. Use (i) to conclude that in this case $S_n/n \to EX$ a.s. This example is from Steele (1989), who gives David Aldous credit for the proof.

Exercise 6.3   *Longest common subsequences.* Let $X_1, X_2, \ldots$ and $Y_1, Y_2, \ldots$ be ergodic stationary sequences. Let $L_n = \max\{K : X_{i_k} = Y_{j_k}$ for $1 \leq k \leq K$ where $1 \leq i_1 < i_2 < \cdots < i_K \leq n$ and $1 \leq j_1 < j_2 < \cdots < j_K \leq n\}$. Show that $L_n/n \to \gamma = \sup_{m \geq 1} E(L_m/m)$.

Exercise 6.4   Suppose that in the last exercise $X_1, X_2, \ldots$ and $Y_1, Y_2, \ldots$ are i.i.d. and take the values 0 and 1 with probability 1/2 each. (i) Compute EL_1 and $EL_2/2$ to get lower bounds on γ. (ii) Show that $\gamma < 1$ by computing the expected number of i and j sequences of length $K = (1 - \varepsilon)n$ with the desired property.

Remark   Chvátal and Sankoff (1975) have shown $.727273 \leq \gamma \leq .866595$.

The examples above should provide enough motivation for now. In Section 7 we will give four more applications of (6.1).

Proof of (6.1)   There are four steps. The first, second, and fourth date back to Kingman (1968). The half-dozen proofs of subadditive ergodic theorems that exist all do the crucial third step in a different way. Here we use the approach of S. Leventhal (1988), who in turn based his proof on Katznelson and Weiss (1982).

Step 1   Let $a_n = EX_{0,n}$. (i) and (iii) imply that

$$a_m + a_{n-m} \geq a_n. \tag{6.2}$$

From this it follows easily that

$$a_n/n \to \inf_{m \geq 1} a_m/m \equiv \gamma. \tag{6.3}$$

To prove this, we observe that the liminf is clearly $\geq \gamma$, so all we have to show is that the limsup $\leq a_m/m$ for any m. The last fact is easy, for if we write $n = km + l$ with $0 \leq l < m$, then repeated use of (6.2) gives

$$a_n \leq ka_m + a_l.$$

Dividing by $n = km + l$ gives

$$\frac{a(n)}{n} \leq \left(\frac{km}{km + l} \right) \frac{a(m)}{m} + \frac{a(l)}{n}.$$

Letting $n \to \infty$ and recalling $0 \leq l < m$ give (6.3) and prove (a) in (6.1).

Remark   Chvátal and Sankoff (1975) attribute (6.3) to Fekete (1923).

Step 2   Making repeated use of (i), we get

$$X_{0,n} \leq X_{0,km} + X_{km,n}$$

$$X_{0,n} \leq X_{0,(k-1)m} + X_{(k-1)m,km} + X_{km,n}$$

and so on until the first term on the right is $X_{0,m}$. Dividing by $n = km + l$ then gives

$$\frac{X(0, n)}{n} \leq \left(\frac{k}{km + l} \right) \frac{X(0, m) + \cdots + X((k - 1)m, km)}{k} + \frac{X(km, n)}{n}. \tag{6.4}$$

Using (ii) and the ergodic theorem now gives that

$$\frac{X(0, m) + \cdots + X((k - 1)m, km)}{k} \to A_m \quad \text{a.s. and in } L^1.$$

If we fix l and let $\varepsilon > 0$, then (iii) implies

$$\sum_{k=1}^{\infty} P(X_{km,km+l} > n\varepsilon) \leq \sum_{k=1}^{\infty} P(X_{0,l} > k\varepsilon) < \infty,$$

since $EX_{0,1}^{+} < \infty$ by the remark after (6.1). The last two observations imply

$$\bar{X} \equiv \limsup_{n\to\infty} X_{0,n}/n \leq A_m/m. \tag{6.5}$$

Taking expected values now gives $E\bar{X} \leq E(X_{0,m}/m)$ and taking the infimum over m, we have $E\bar{X} \leq \gamma$. Note that if all the stationary sequences in (ii) are ergodic, we have $\bar{X} \leq \gamma$.

Exercise 6.5   Show that if (i)–(iii) hold, $EX_{0,1}^{+} < \infty$, and inf $EX_{0,m}/m = -\infty$, then as $n \to \infty$, $X_{0,n}/n \to -\infty$ a.s.

Step 3   The next step is to let

$$\underline{X} = \liminf_{n\to\infty} X_{0,n}/n$$

and show that $E\underline{X} \geq \gamma$. Since $\infty > EX_{0,1} \geq \gamma \geq \gamma_0 > -\infty$ and we have shown in Step 2 that $E\bar{X} \leq \gamma$, it will follow that $\underline{X} = \bar{X}$; that is, the limit of $X_{0,n}/n$ exists a.s. Let

$$\underline{X}_m = \liminf_{n\to\infty} X_{m,m+n}/n.$$

(i) implies

$$X_{0,m+n} \leq X_{0,m} + X_{m,m+n}.$$

Dividing both sides by n and letting $n \to \infty$ give $\underline{X} \leq \underline{X}_m$ a.s. However, (iii) implies that $\underline{X}_m$ and $\underline{X}$ have the same distribution, so $\underline{X} = \underline{X}_m$ a.s.

Let $\varepsilon > 0$ and let $Z = \varepsilon + (\underline{X} \vee -M)$. Since $\underline{X} \leq \bar{X}$ and $E\bar{X} \leq \gamma < \infty$ by Step 2, $E|Z| < \infty$. Let

$$Y_{m,n} = X_{m,n} - (n-m)Z.$$

Y satisfies (i)–(iv) (since $Z_{m,n} = -(n-m)Z$ does) and has

$$\underline{Y} \equiv \liminf_{n\to\infty} Y_{0,n}/n \leq -\varepsilon. \tag{6.6}$$

Let $T_m = \min\{n \geq 1 : Y_{m,m+n} \leq 0\}$. (iii) implies $T_m \overset{d}{=} T_0$ and

$$E(Y_{m,m+1}; T_m > N) = E(Y_{0,1}; T_0 > N).$$

(6.6) implies that $P(T_0 < \infty) = 1$, so we can pick N large enough so that

$$E(Y_{0,1}; T_0 > N) \leq \varepsilon.$$

Let $S_m = T_m$ on $\{T_m \leq N\}$ and $= 1$ on $\{T_m > N\}$. (This is not a stopping time but there is nothing special about stopping times for a stationary sequence!) Let $\xi_m = 0$

on $\{T_m \le N\}$ and $Y_{m,m+1}$ on $\{T_m > N\}$. Since $Y(m, m + T_m) \le 0$ always and we have $S_m = 1$, $Y_{m,m+1} > 0$ on $\{T_m > N\}$, we have $Y(m, m + S_m) \le \xi_m$ and $\xi_m \ge 0$. Let $R_0 = 0$, and for $k \ge 1$ let $R_k = R_{k-1} + S(R_{k-1})$. Let $K = \max\{k: R_k \le n\}$. From (i) it follows that

$$Y(0, n) \le Y(R_0, R_1) + \cdots + Y(R_{K-1}, R_K) + Y(R_K, n).$$

Since $\xi_m \ge 0$ and $n - R_K \le N$, the last quantity is

$$\le \sum_{m=0}^{n-1} \xi_m + \sum_{j=1}^{N} |Y_{n-j,n-j+1}|.$$

(Here we have used (i) on $Y(R_K, n)$.) Dividing both sides by n, taking expected values, and letting $n \to \infty$ give

$$\limsup_{n \to \infty} E Y_{0,n}/n \le E\xi_0 \le E(Y_{0,1}; T_0 > N) \le \varepsilon.$$

It follows from (a) and the definition of $Y_{0,n}$ that

$$\gamma = \limsup_{n \to \infty} EX_{0,n}/n \le 2\varepsilon + E(\underline{X} \vee -M).$$

Since $\varepsilon > 0$ and M are arbitrary, it follows that $E\underline{X} \ge \gamma$ and Step 3 is complete.

Step 4   It only remains to prove convergence in L^1. Let $\Gamma_m = A_m/m$ be the limit in (6.5), recall $E\Gamma_m = E(X_{0,m}/m)$, and let $\Gamma = \inf \Gamma_m$. Observing that $|z| = 2z^+ - z$ (consider two cases $z \ge 0$ and $z < 0$), we can write

$$E|X_{0,n}/n - \Gamma| = 2E(X_{0,n}/n - \Gamma)^+ - E(X_{0,n}/n - \Gamma) \le 2E(X_{0,n}/n - \Gamma)^+,$$

since

$$E(X_{0,n}/n) \ge \gamma = \inf E\Gamma_m \ge E\Gamma.$$

Using the trivial inequality $(x + y)^+ \le x^+ + y^+$ now gives

$$E(X_{0,n}/n - \Gamma)^+ \le E(X_{0,n}/n - \Gamma_m)^+ + E(\Gamma_m - \Gamma).$$

Now $E\Gamma_m \to \gamma$ as $m \to \infty$ and $E\Gamma \ge E\bar{X} \ge E\underline{X} \ge \gamma$ by Steps 2 and 3, so $E\Gamma = \gamma$, and it follows that $E(\Gamma_m - \Gamma)$ is small if m is large. To bound the other term, observe

$$E(X_{0,n}/n - \Gamma_m)^+ \le E\left(\frac{X(0, m) + \cdots + X((k - 1)m, km)}{km + l} - \Gamma_m\right)^+$$
$$+ E\left(\frac{X(km, n)}{n}\right)^+.$$

The second term $= E(X_{0,1}^+/n) \to 0$. For the first we observe that the ergodic theorem implies

$$E\left|\frac{X(0, m) + \cdots + X((k-1)m, km)}{k} - \Gamma_m\right| \to 0,$$

and the proof of (6.1) is complete.   □

The next two exercises develop a generalization of (6.1). The third gives an application of the new result.

Exercise 6.6   If a_n and c_n are sequences that satisfy $a_{m+n} \le a_m + a_n + c_n$ and $\lim c_n/n = 0$, then a_n/n has a limit. Show that the limit may be $>$ inf a_m/m.

Exercise 6.7   *Derriennic's (1983) almost subadditive ergodic theorem.* Let T be a measure-preserving transformation on $(\Omega, \mathcal{F}, P)$ and let f_n be a sequence of integrable functions with

$$f_{n+m} - f_n - f_m \circ T^n \le h_m \circ T^n$$

where $h_m \ge 0$ have $\lim (1/n)\int h_n\, dP = 0$. Show that if $\inf(1/n)\int f_n\, dP > -\infty$, then f_n/n converges in L^1 to $\bar{f} = \limsup f_n/n$.

Note   The hypotheses also imply that the sequence converges a.s., but that part of the proof is too hard to be an exercise.

Exercise 6.8   *Shannon–McMillan–Breiman theorem.* Let X_n, $n \in \mathbb{Z}$, be an ergodic stationary sequence taking values in a finite alphabet S, and suppose X_n is constructed on sequence space $(S^{\mathbb{Z}}, \mathscr{S}^{\mathbb{Z}})$. Let $\mathscr{F}_m^n = \sigma(X_k : m \le k \le n)$ and

$$I_n = \sum_{\text{atoms } A \in \mathscr{F}_0^{n-1}} (-\log P(A)) 1_A.$$

(i) Show that $I_{n+m} \le I_n + I_m \circ \theta^n + (J_m - I_m)^+ \circ \theta^n$ where θ is the shift and

$$J_m = \sup_{k \ge 1} \sum_{\text{atoms } A \in \mathscr{F}_0^{m-1}} (-\log P(A|\mathscr{F}_{-k}^{-1})).$$

(ii) Show that $P(J_m - I_m \ge x) \le e^{-x}$ and apply the result in the last exercise to get (5.1).

*7   Applications

In this section we will give four applications of our subadditive ergodic theorem (6.1). These examples are independent of each other and can be read in any order. In the last two we encounter situations to which Liggett's version applies but Kingman's version does not.

Example 7.1   *Products of random matrices.* Suppose $A_1, A_2, \ldots$ is a stationary sequence of $k \times k$ matrices with positive entries and let $\alpha_{m,n}(i, j) = (A_{m+1} \cdots A_n)(i, j)$

—that is, the entry in row i of column j of the product $A_{m+1} \cdots A_n$. It is clear that

$$\alpha_{0,m}(1, 1)\alpha_{m,n}(1, 1) \le \alpha_{0,n}(1, 1),$$

so if we let $X_{m,n} = -\log \alpha_{m,n}(1, 1)$, then $X_{0,m} + X_{m,n} \ge X_{0,n}$. To check (iv) we observe that

$$\prod_{m=1}^{n} A_m(1, 1) \le \alpha_{0,n}(1, 1) \le k^{n-1} \prod_{m=1}^{n} \left(\sup_{i,j} A_m(i, j) \right),$$

or taking logs,

$$-\sum_{m=1}^{n} \log A_m(1, 1) \ge X_{0,n} \ge -(n \log k) - \sum_{m=1}^{n} \log \left(\sup_{i,j} A_m(i, j) \right).$$

So if $E \log A_m(1, 1) > -\infty$, then $EX_{0,1}^+ < \infty$, and if

$$E \log \left(\sup_{i,j} A_m(i, j) \right) < \infty,$$

then $EX_{0,n}^- \le \gamma_0 n$. If we observe that

$$P \left(\log \left(\sup_{i,j} A_m(i, j) \right) \ge x \right) \le \sum_{i,j} P(\log A_m(i, j) \ge x),$$

we see that it is enough to assume that

$$E|\log A_m(i, j)| < \infty \quad \text{for all } i, j. \tag{*}$$

When (*) holds, applying (6.1) gives $X_{0,n}/n \to X$ a.s. Using the strict positivity of the entries, it is easy to improve that result to

$$\frac{1}{n} \log \alpha_{0,n}(i, j) \to -X \quad \text{a.s. for all } i, j,$$

a result first proved by Furstenberg and Kesten (1960).

The key to the proof above was the fact that $\alpha_{0,n}(1, 1)$ was supermultiplicative. An alternative approach is to let

$$\|A\| = \max_i \sum_j |A(i, j)| = \max\{\|xA\|_1 : \|x\|_1 = 1\}$$

where $(xA)_j = \sum_i x_i A(i, j)$ and $\|x\|_1 = |x_1| + \cdots + |x_k|$. From the second definition it is clear that $\|AB\| \le \|A\| \cdot \|B\|$, so if we let

$$\beta_{m,n} = \|A_{m+1} \cdots A_n\|$$

and $Y_{m,n} = \log \beta_{m,n}$, then $Y_{m,n}$ is subadditive. It is not hard to see that

$$\frac{1}{n} \log\|A_{m+1} \cdots A_n\| \to -X \quad \text{a.s.}$$

where X is the limit of $X_{0,n}/n$. To see the advantage in having two proofs of the same result, we observe that if $A_1, A_2, \ldots$ is an i.i.d. sequence, then X is constant and we can get upper and lower bounds by observing

$$\sup_{m \geq 1} (E \log \alpha_{0,m})/m = -X = \inf_{m \geq 1} (E \log \beta_{0,m})/m.$$

Remark Oseledec (1968) proved a result which gives the asymptotic behavior of all the eigenvalues of A. As Ragunathan (1979) and Ruelle (1979) have observed, this result can also be obtained from (6.1). See Krengel (1985) or the papers cited for details. Furstenberg and Kesten (1960) and later Ishitani (1977) have proved central limit theorems:

$$(\log \alpha_{0,n}(1, 1) - \mu n)/n^{1/2} \to \sigma\chi.$$

The next example shows that σ may be 0.

Exercise 7.1 Let A_n, $n \in \mathbb{Z}$, be strictly positive transition probability matrices; that is, $A_n(i, j) > 0$ and $\sum_j A_n(i, j) = 1$. (i) Show that

$$\max_{i,j} \sum_k |\alpha_{0,n}(i, k) - \alpha_{0,n}(j, k)| \to 0 \quad \text{a.s.}$$

(ii) Show that $\alpha_{-n,0}(i, j)$ converges a.s. as $n \to \infty$. Use this to conclude that $\alpha_{0,n}(i, j)$ converges in distribution.

Remark For more about products of random matrices, including the last example, see Cohen, Kesten, and Newman (1985).

Example 7.2 *Increasing sequences in random permutations.* Let π be a permutation of $\{1, 2, \ldots, n\}$ and let $l(\pi)$ be the length of the longest increasing sequence in π—that is, the largest k for which there are integers $i_1 < i_2 < \cdots < i_k$ so that $\pi(i_1) < \pi(i_2) < \cdots < \pi(i_k)$. Hammersley (1970) attacked this problem by putting a rate 1 Poisson process in the plane and, for $s < t \in [0, \infty)$, letting $Y_{s,t}$ denote the length of the longest increasing path lying in the square $R_{m,n}$ with vertices (s, s), (s, t), (t, t), and (t, s)—that is, the largest k for which there are points (x_i, y_i) in the Poisson process with $s < x_1 < \cdots < x_k < t$ and $s < y_1 < \cdots < y_k < t$. It is clear that $Y_{0,m} + Y_{m,n} \leq Y_{0,n}$. Applying (6.1) to $-Y_{0,n}$ shows

$$Y_{0,n}/n \to \gamma \equiv \sup_{m \geq 1} EY_{0,m}/m \quad \text{a.s.}$$

(For each k, $Y_{nk,(n+1)k}$, $n \geq 0$, is i.i.d., so the limit is constant. We will show that $\gamma < \infty$ in Exercise 7.3.)

To get from the result about the Poisson process back to the random permutation problem, let $\tau(n)$ be the smallest value of t for which there are n points in $R_{0,t}$. Let the n points in $R_{0,\tau(n)}$ be written as (x_i, y_i) where $0 < x_1 < x_2 < \cdots < x_n \leq \tau(n)$ and let π_n be the unique permutation of $\{1, 2, \ldots, n\}$ so that $y_{\pi_n(1)} \leq y_{\pi_n(2)} < \cdots < y_{\pi_n(n)}$. It is clear that $Y_{0,\tau(n)} = l(\pi_n)$. An easy argument shows $\tau(n)/\sqrt{n} \to 1$ a.s. (Let S_n be the number of points in $R_{0,\sqrt{n}}$, observe $S_n - S_{n-1}$ are i.i.d., use the strong law of large numbers, and then argue as in the proof of (8.5) in Chapter 1.) So it follows that

$$n^{-1/2}l(\pi_n) \to \gamma \quad \text{a.s.}$$

Hammersley (1970) has a proof that $\pi/2 \leq \gamma \leq e$, and Kingman (1973) shows that $1.59 < \gamma < 2.49$. (See Exercises 7.4 and 7.5.) Subsequent work on the random permutation problem (see Logan and Shepp (1977) and Vershik and Kerov (1977)) has shown that $\gamma = 2$. Exercises 7.2 and 7.3 are devoted to showing that $0 < \gamma < \infty$. For them it is useful to observe that if we let $X_1, X_2, \ldots$ be i.i.d. uniform on $(0, 1)$ and L_n is the length of the longest increasing subsequence, then $L_n \overset{d}{=} l(\pi_n)$. We begin with a nonrandom result due to Erdös and Szerkeres (1935).

(7.1) THEOREM   Every sequence a_i, $1 \leq i \leq mn + 1$, of distinct real numbers must contain an increasing subsequence of length $m + 1$ or a decreasing subsequence of length $n + 1$.

Proof   Construct increasing subsequences by letting a_1 start the first one and then putting a_j at the end of the first subsequence for which $a_j >$ the last term, starting a new subsequence if a_j is always $<$ the last term. It is easy to check that the last terms of the sequences are always a decreasing sequence, so the desired result follows from the "pigeon hole principle." In words, if we have $mn + 1$ pigeons, then there must be either $m + 1$ holes or $n + 1$ pigeons in some hole, for otherwise we would have $\leq mn$ pigeons.   □

Exercise 7.2   Use (7.1) to show that if $m = n^2 + 1$, then $EL_m \geq (n + 1)/2$ and hence $\gamma \geq 1/2$.

Exercise 7.3   Let J_k^n be the number of subsets of $\{X_1, \ldots, X_n\}$ of size k that form an increasing subsequence. Compute EJ_k^n and use Stirling's formula to show that $\gamma \leq e$.

Exercise 7.4   Improve the upper bound in the last exercise by observing that if $l(\pi_n) \geq l$, then $J_k^n \geq \binom{l}{k}$, so $P(l(\pi_n) \geq l) \leq \binom{n}{k}(k!\binom{l}{k})^{-1}$. Let $l \sim \beta n^{1/2}$ and $k \sim \alpha n^{1/2}$, use Stirling's formula to evaluate the right-hand side, and pick α to optimize the bound to show $\gamma < 2.49$.

Exercise 7.5   Given a rate 1 Poisson process in $[0, \infty) \times [0, \infty)$, let (x_1, y_1) be the point that minimizes $z_1 = x_1 + y_1$. Let (x_2, y_2) be the point in $[x_1, \infty) \times [x_2, \infty)$

that minimizes $z_2 = x_2 + y_2$, and so on. Use this construction to show that $\gamma \geq (8/\pi)^{1/2} > 1.59$.

Example 7.3 *Age-dependent branching processes.* This is a variation of the branching process introduced in Section 3 of Chapter 4 in which each individual lives for an amount of time with distribution F before producing k offspring with probability p_k. The description of the process is completed by supposing that the process starts with one individual in generation 0 who is born at time 0, and when this particle dies, its offspring start independent copies of the original process.

Suppose $p_0 = 0$, let $X_{0,m}$ be the birth time of the first member of generation m, and let $X_{m,n}$ be the time lag necessary for that individual to have an offspring in generation n. (In case of ties, pick an individual at random from those in generation m born at time $X_{0,m}$.) It is clear that $X_{0,n} \leq X_{0,m} + X_{m,n}$. Since $X_{0,n} \geq 0$, (iv) holds if we assume F has finite mean. Notice that the inequality $X_{l,m} + X_{m,n} \geq X_{l,n}$ is false when $l > 0$ because if we call i_m the individual that determines the value of $X_{m,n}$ for $n > m$, then i_m may not be a descendant of i_l.

Applying (6.1) now, it follows that $X_{0,n}/n \to \gamma$ a.s. The limit is constant because the sequences $\{X_{nk,(n+1)k} : n \geq 0\}$ are i.i.d., but as usual one has to use other methods to identify the constant. In the next three exercises, let $t_1, t_2, \ldots$ be i.i.d. with distribution F, let $T_n = t_1 + \cdots + t_n$, and $\mu = \sum k p_k$.

Exercise 7.6 (i) Let $Z_n(an)$ be the number of individuals in generation n born by time an. Show that $EZ_n(an) = \mu^n P(T_n \leq an)$. (ii) Show that if $EZ_n(an) > 1$ for some n, then $\gamma \leq a$. (iii) By results in Section 9 of Chapter 1; $n^{-1} \log P(T_n \leq an) \to -c(a)$ as $n \to \infty$. Combine this with (i) and (ii) to conclude $\gamma = \inf\{a : \log \mu - c(a) > 0\}$.

The last result is from Biggins (1977). See his 1978 and 1979 papers for extensions and refinements. Kingman (1975) has an approach to the problem via martingales:

Exercise 7.7 Let $\varphi(\theta) = E \exp(-\theta t_i)$ and

$$Y_n = (\mu\varphi(\theta))^{-n} \sum_{i=1}^{Z_n} \exp(-\theta T_n(i))$$

where the sum is over individuals in generation n and $T_n(i)$ is the ith person's birth time. Show that Y_n is a nonnegative martingale and use this to conclude that if $\exp(-\theta a)/\mu\varphi(\theta) > 1$, then $P(X_{0,n} \leq an) \to 0$. A little thought reveals that this bound is the same as the answer in the last exercise.

Exercise 7.8 Let Z_t be the number of individuals alive at time t when $Z_0 = 1$, and let $m(t) = EZ_t$. Considering what happens at the first split gives

$$m(t) = (1 - F(t)) + \int_0^t \left(\sum_{k=0}^{\infty} p_k k m(t - s) \right) dF(s)$$

where p_k is the probability of k offspring. If $\mu = \sum k p_k > 1$ and $F(0) = 0$, then there is a unique α so that

$$\mu \int_0^\infty e^{-\alpha t} \, dF(t) = 1.$$

Multiplying through by $e^{-\alpha t}$ gives an equation for $H(t) = e^{-\alpha t} m(t)$:

$$H(t) = h(t) + \int_0^t H(t - s) \, dG(s)$$

where $h(t) = e^{-\alpha t}(1 - F(t))$ and $G(t) = \mu \int_{(0,t]} e^{-\alpha s} \, dF(s)$. Applying the renewal theorem ((4.9) in Chapter 3) now shows that $e^{-\alpha t} E Z_t$ converges to a positive limit as $t \to \infty$. α is called the *Malthusian parameter*, after Malthus who investigated the exponential growth of populations.

Example 7.4 *First passage percolation.* Consider $\mathbb{Z}^d$ as a graph with edges connecting each $x, y \in \mathbb{Z}^d$ with $|x - y| = 1$. Assign an independent nonnegative random variable $\tau(e)$ to each edge that represents the time required to traverse the edge going in either direction. If e is the edge connecting x and y, let $\tau(x, y) = \tau(y, x) = \tau(e)$. If $x_0 = x, x_1, \ldots, x_n = y$ is a path from x to y—that is, a sequence with $|x_m - x_{m-1}| = 1$ for $1 \le m \le n$, we define the *travel time* for the path to be $\tau(x_0, x_1) + \cdots + \tau(x_{n-1}, x_n)$. Define the *passage time* from x to y, $t(x, y) = $ the infimum of the travel times over all paths from x to y. Let $z \in \mathbb{Z}^d$ and let $X_{m,n} = t(mu, nu)$ where $u = (1, 0, \ldots, 0)$.

Clearly $X_{0,m} + X_{m,n} \ge X_{0,n}$. $X_{0,n} \ge 0$, so if $E\tau(x, y) < \infty$, then (iv) holds and (6.1) implies that $X_{0,n}/n \to X$ a.s. To see that the limit is constant, enumerate the edges in some order $e_1, e_2, \ldots$ and observe that X is measurable with respect to the tail σ-field of the i.i.d. sequence $\tau(e_1), \tau(e_2), \ldots$.

Exercise 7.9 It is not hard to see that the assumption of a finite first moment can be weakened. If τ has distribution F with

$$\int_0^\infty (1 - F(x))^{2d} \, dx < \infty, \tag{$*$}$$

that is, the minimum of $2d$ independent copies has finite mean, then by finding $2d$ disjoint paths from 0 to $u = (1, 0, \ldots, 0)$, one concludes that $E\tau(0, u) < \infty$ and (6.1) can be applied. The condition $(*)$ is also necessary for $X_{0,n}/n$ to converge to a finite limit. If $(*)$ fails and Y_n is the minimum of $t(e)$ over all the edges from nu, then

$$\limsup_{n \to \infty} X_{0,n}/n \ge \limsup_{n \to \infty} Y_n/n = \infty \quad \text{a.s.}$$

Above we considered the *point-to-point passage time*. A second object of interest is the *point-to-line passage time*:

$$a_n = \inf\{t(0, x) : x_1 = n\}.$$

Unfortunately it does not seem to be possible to embed this sequence in a subadditive family. To see the difficulty let $\bar{t}(0, x)$ be the infimum of travel times over paths from 0 to x that lie in $\{y : y_1 \ge 0\}$, let

$$\bar{a}_m = \inf\{\bar{t}(0, x) : x_1 = m\},$$

and let x^m be a point at which the infimum is achieved. (We leave to the reader the highly nontrivial task of proving that such a point exists. See Smythe and Wierman (1978) for a proof.) If we let $\bar{a}_{m,n}$ be the infimum of travel times over all paths which start at x^m, stay in $\{y: y_1 \geq m\}$, and end on $\{y: y_1 = n\}$, then $\bar{a}_{m,n}$ is independent of $\bar{a}_m$ and

$$\bar{a}_m + \bar{a}_{m,n} \geq \bar{a}_n.$$

The last inequality is true without the half-space restriction but the independence is not, and without the half-space restriction we cannot get the stationarity properties needed to apply (6.1).

Remark   The family $\bar{a}_{m,n}$ is another example where $\bar{a}_{l,m} + \bar{a}_{m,n} \geq \bar{a}_{l,n}$ need not hold for $l > 0$.

A second approach to limit theorems for a_m is to prove a result about the set of points which can be reached by time t: $\xi_t = \{x: t(0, x) \leq t\}$. Cox and Durrett (1981) have shown:

(7.2) THEOREM   For any passage time distribution F with $F(0) = 0$, there is a convex set A so that for any $\varepsilon > 0$ we have with probability 1

$$\xi_t \subset (1 + \varepsilon)tA \quad \text{for all } t \text{ sufficiently large}$$

and

$$|\xi_t^c \cap (1 - \varepsilon)tA \cap \mathbb{Z}^d|/t^d \to 0, \quad t \to \infty.$$

Ignoring boring details, the last result says $\xi_t/t \to A$ a.s. It implies that $a_n/n \to \gamma$ a.s. where $\gamma = 1/\sup\{x_1: x \in A\}$. (Use the convexity and reflection symmetry of A.) When the distribution has finite mean (or satisfies the condition in Exercise 7.9), γ is the limit of $t(0, nu)/n$. Without any assumptions $t(0, nu)/n \to \gamma$ in probability. For more details see the paper cited above. Kesten (1986, 1987) are good sources for more about first passage percolation.

Exercise 7.10   *Oriented first passage percolation.* Consider a graph with vertices $\{(m, n) \in \mathbb{Z}^2: m + n \text{ is even and } n \leq 0\}$ and oriented edges connecting (m, n) to $(m - 1, n - 1)$ and (m, n) to $(m + 1, n - 1)$. Assign i.i.d. exponential mean 1 r.v.'s to each edge. Thinking of the number on edge e as giving the time it takes water to travel down the edge, define $t(m, n)$ as the time at which the fluid first reaches (m, n) and $a_n = \inf\{t(m, -n)\}$. Show that as $n \to \infty$, a_n/n converges to a limit γ a.s.

Exercise 7.11   Continue with the set-up in the last exercise. (i) Show $\gamma < 1/2$ by considering a_1 and a_2. (ii) Get a positive lower bound on γ by looking at the expected number of paths down to $\{(m, -n) : -n \le m \le n\}$ with passage time $\le an$ and using results from Section 9 in Chapter 1.

Remark   If we replace the graph in Exercise 7.10 by a binary tree, then we get a problem equivalent to the first birth problem (Example 7.3) for $p_2 = 2$, $P(t_i > x) = e^{-x}$. In the first birth problem the lower bound obtained by the methods of part (ii) Exercise 7.11 was sharp, but in oriented first passage percolation it is not.

7 Brownian Motion

Brownian motion is a process of tremendous practical and theoretical significance. It originated (a) as a model of the phenomenon observed by Robert Brown in 1828 that "pollen grains suspended in water perform a continual swarming motion," and (b) in Bachelier's (1900) work as a model of the stock market. These are just two of many systems that Brownian motion has been used to model. On the theoretical side Brownian motion is a Gaussian Markov process with stationary independent increments. It lies in the intersection of three important classes of processes and is a fundamental example in each theory.

The first part of this chapter develops properties of Brownian motion. In Section 1 we define Brownian motion and investigate continuity properties of its paths. In Section 2 we prove the Markov property and a related 0–1 law. In Section 3 we define stopping times and prove the strong Markov property. In Section 4 we take a close look at the zero set of Brownian motion. In Section 5 we introduce some martingales associated with Brownian motion and use them to compute the distribution of T and B_T for some stopping times T.

The second part of this chapter applies Brownian motion to some of the problems considered in Chapters 1 and 2. In Section 6 we embed random walks into Brownian motion to prove Donsker's theorem, a far-reaching generalization of the central limit theorem. In Section 7 we extend Donsker's theorem to martingales satisfying "Lindeberg–Feller conditions" and to weakly dependent stationary sequences. In Section 8 we show that the discrepancy between the empirical distribution and the true distribution when suitably magnified converges to Brownian bridge. In Section 9 we prove laws of the iterated logarithm for Brownian motion and random walks with finite variance. The last three sections depend on Section 6 but are independent of each other and can be read in any order.

1 Definition and Construction

A one-dimensional *Brownian motion* is a real-valued process B_t, $t \geq 0$, that has the following properties:

(a) If $t_0 < t_1 < \cdots < t_n$, then $B(t_0), B(t_1) - B(t_0), \ldots, B(t_n) - B(t_{n-1})$ are independent.
(b) If $s, t \geq 0$, then

$$P(B(s + t) - B(s) \in A) = \int_A (2\pi t)^{-1/2} \exp(-x^2/2t)\, dx.$$

(c) With probability 1, $t \to B_t$ is continuous.

(a) says that B_t has independent increments. (b) says that the increment $B(s + t) - B(s)$ has a normal distribution with mean 0 and variance t. (c) is self-explanatory.

Thinking of Brown's pollen grain, (c) is certainly reasonable. (a) can be justified by noting that the movement of the pollen grain is due to the net effect of the bombardment of millions of water molecules. The next exercise shows that if we accept (c) and (a), (b) is forced on us by the central limit theorem.

Exercise 1.1   Suppose (a) holds, the distribution of $B(t) - B(s)$ depends on only $t - s$, $E(B(t) - B(s)) = 0$, $E(B(t) - B(s))^2 < \infty$, and $nP(B(t + 1/n) - B(t) > \varepsilon) \to 0$ as $n \to \infty$ for any $\varepsilon > 0$. Apply the Lindeberg–Feller theorem ((4.6) in Chapter 2) to $X_{n,k} = B(kt/n) - B((k - 1)t/n)$ and conclude that $B(t) - B(s)$ has a normal distribution with variance $t - s$.

Note   The condition $nP(B(t + 1/n) - B(t) > \varepsilon) \to 0$ is needed to guarantee that B_t has continuous paths. If N_t is a rate 1 Poisson process, then $N_t - t$ has all the properties but the last one.

Two immediate consequences of the definition that will be useful many times are:

(1.1) *Translation Invariance* $\{B_t - B_0, t \geq 0\}$ is independent of B_0 and has the same distribution as a Brownian motion with $B_0 = 0$.

(1.2) *The Brownian Scaling Relation*   For any $t > 0$,

$$\{B_{st}, s \geq 0\} \stackrel{d}{=} \{t^{1/2} B_s, s \geq 0\};$$

that is, the two families of r.v.'s have the same finite-dimensional distributions.

A second equivalent definition of Brownian motion starting from $B_0 = 0$ that we will occasionally find useful is that B_t, $t \geq 0$, is a real-valued process satisfying:

(a') $B(t)$ is a *Gaussian process* (i.e., all its finite-dimensional distributions are multivariate normal),
(b') $EB_s B_t = s \wedge t$, and
(c') with probability 1, $t \to B_t$ is continuous.

It is easy to see that (a) and (b) imply (a'). To get (b') from (a) and (b), suppose $s < t$ and write

$$EB_s B_t = E(B_s^2) + E(B_s(B_t - B_s)) = s.$$

The converse is even easier. (a') and (b') specify the finite-dimensional distributions of B_t, which by the last calculation must agree with the ones defined in (a) and (b).

The first question that must be addressed in any treatment of Brownian motion is, "Is there a process with these properties?" The answer is yes, of course, or this chapter would not exist. For pedagogical reasons we will pursue an approach that leads to a dead end and then retreat a little to rectify the difficulty. Fix an $x \in \mathbb{R}$ and for each $0 < t_1 < \cdots < t_n$ define a measure on $\mathbb{R}^n$ by

$$\mu_{x,t_1,\ldots,t_n}(A_1 \times \cdots \times A_n) = \int_{A_1} dx_1 \cdots \int_{A_n} dx_n \prod_{m=1}^n p_{t_m - t_{m-1}}(x_{m-1}, x_m)$$

where $x_0 = x$, $t_0 = 0$, and

$$p_t(a, b) = (2\pi t)^{-1/2} \exp(-(b-a)^2/2t).$$

From the formula above it is easy to see that for fixed x the family μ is a consistent set of finite-dimensional distributions (f.d.d.'s); that is, if $\{s_1, \ldots, s_{n-1}\} \subset \{t_1, \ldots, t_n\}$ and $t_j \notin \{s_1, \ldots, s_{n-1}\}$, then

$$\mu_{x,s_1,\ldots,s_{n-1}}(A_1 \times \cdots \times A_{n-1}) = \mu_{x,t_1,\ldots,t_n}(A_1 \times \cdots \times A_{j-1} \times \mathbb{R} \times A_j \times \cdots \times A_{n-1}).$$

This is clear when $j = n$. To check the equality when $1 \leq j < n$, it is enough to show that

$$\int p_{t_j - t_{j-1}}(x, y) p_{t_{j+1} - t_j}(y, z)\, dy = p_{t_{j+1} - t_{j-1}}(x, z).$$

By translation invariance, we can without loss of generality assume $x = 0$, but all this says is that the sum of independent normals with mean 0 and variances $t_j - t_{j-1}$ and $t_{j+1} - t_j$ has a normal distribution with mean 0 and variance $t_{j+1} - t_{j-1}$.

With the consistency of f.d.d.'s verified, we get our first construction of Brownian motion:

(1.3) THEOREM Let $\Omega_o = \{$functions $\omega: [0, \infty) \to \mathbb{R}\}$ and $\mathscr{F}_o$ be the σ-field generated by the finite-dimensional sets $\{\omega: \omega(t_i) \in A_i$ for $1 \leq i \leq n\}$ where $A_i \in \mathscr{R}$. For each $x \in \mathbb{R}$, there is a unique probability measure ν_x on $(\Omega_o, \mathscr{F}_o)$ so that $\nu_x\{\omega: \omega(0) = x\} = 1$ and

$$\nu_x(\{\omega: \omega(t_i) \in A_i\}) = \mu_{x,t_1,\ldots,t_n}(A_1 \times \cdots \times A_n) \quad \text{when } 0 < t_1 < \cdots < t_n. \quad (*)$$

This follows from a generalization of Kolmogorov's extension theorem ((7.1) in the Appendix). We will not bother with the details, since at this point we are at the dead end referred to above. If $C = \{\omega: t \to \omega(t)$ is continuous$\}$, then $C \notin \mathscr{F}_o$; that is, C is not a measurable set. The easiest way of proving $C \notin \mathscr{F}_o$ is to do:

Exercise 1.2 $A \in \mathcal{F}_o$ if and only if there is a sequence of times $t_1, t_2, \ldots \in [0, \infty)$ and a $B \in \mathcal{R}^{\{1,2,\ldots\}}$ so that $A = \{\omega : (\omega(t_1), \omega(t_2), \ldots) \in B\}$. In words, all events in $\mathcal{F}_o$ depend on only countably many coordinates.

The above problem is easy to solve. Let $\mathbb{Q}_2 = \{m2^{-n} : m, n \geq 0\}$ be the *dyadic rationals*. If $\Omega_q = \{\omega : \mathbb{Q}_2 \to \mathbb{R}\}$ and $\mathcal{F}_q$ is the σ-field generated by the finite-dimensional sets, then enumerating the rationals $q_1, q_2, \ldots$ and applying Kolmogorov's extension theorem show that we can construct a probability v_x on $(\Omega_q, \mathcal{F}_q)$ so that $v_x \{\omega : \omega(0) = x\} = 1$ and (*) in (1.3) holds when the $t_i \in \mathbb{Q}_2$. To extend B_t to a process defined on $[0, \infty)$ we will show:

(1.4) THEOREM   Let $T < \infty$ and $x \in \mathbb{R}$. v_x assigns probability 1 to paths $\omega : \mathbb{Q}_2 \to \mathbb{R}$ that are uniformly continuous on $\mathbb{Q}_2 \cap [0, T]$.

Remark   It will take quite a bit of work to prove (1.4). Before taking on that task, we will attend to the last measure theoretic detail: We tidy things up by moving our probability measures to $(C, \mathcal{C})$ where $C = \{\text{continuous } \omega : [0, \infty) \to \mathbb{R}\}$ and $\mathcal{C}$ is the σ-field generated by the coordinate maps $t \to \omega(t)$. To do this, we observe that the map ψ that takes a uniformly continuous point in Ω_q to its extension in C is measurable, and we set

$$P_x = v_x \circ \psi^{-1}.$$

Our construction guarantees that $B_t(\omega) = \omega_t$ has the right finite-dimensional distributions for $t \in \mathbb{Q}_2$. Continuity of paths and a simple limiting argument show that this is true when $t \in [0, \infty)$. Finally, the reader should note that, as in the case of Markov chains, we have one set of random variables $B_t(\omega) = \omega(t)$ and a family of probability measures P_x, $x \in \mathbb{R}$, so that under P_x, B_t is a Brownian motion with $P_x(B_t = x) = 1$.

Proof of (1.4)   By (1.1) and (1.2), we can without loss of generality suppose $B_0 = 0$ and prove the result for $T = 1$. In this case, the scaling relation (1.2) implies

$$E|B_t|^4 = Ct^2 \quad \text{where } C = E|B_1|^4 < \infty.$$

From the last observation we get the desired uniform continuity by using a computation due to Kolmogorov. The reader should check as she goes along that all we use is

$$E|B_t - B_s|^4 \leq C(t - s)^2. \tag{1.5}$$

In particular, we do not take advantage of the fact that B_t has independent increments.

Let $\gamma < 1/4$, $\delta > 0$, and observe that $a^4 P(|X| \geq a) \leq E|X|^4$, so

$$P(|B(j2^{-n}) - B(i2^{-n})| > ((j - i)2^{-n})^\gamma \text{ for some } 0 \leq i, j \leq 2^n \text{ with } 0 < j - i \leq 2^{n\delta})$$

$$\leq \sum ((j - i)2^{-n})^{-4\gamma} E|B(j2^{-n}) - B(i2^{-n})|^4$$

where the sum is over the (i, j) satisfying the condition in the event on the left-hand side. Using (1.5) and then noticing the number of (i, j)'s in the sum is $\leq 2^n 2^{n\delta}$, we see that the right-hand side is

$$\leq C \sum ((j - i)2^{-n})^{-4\gamma+2} \leq C2^n 2^{n\delta} (2^{n\delta} 2^{-n})^{-4\gamma+2} = C2^{-n\varepsilon}$$

where $\varepsilon = (1 - \delta)(2 - 4\gamma) - (1 + \delta)$. Since $\gamma < 1/4$, we can pick δ small enough so that $\varepsilon > 0$, and the Borel–Cantelli lemma implies:

(a) For almost every ω there is an $N(\omega)$ so that for all $n \geq N(\omega)$,

$$|B(j2^{-n}) - B(i2^{-n})| \leq ((j - i)2^{-n})^\gamma \quad \text{whenever } 0 < j - i \leq 2^{n\delta}.$$

To get from (a) to (1.4) we have to relate points at different levels. Let q, $r \in \mathbb{Q}_2 \cap [0, 1]$ with $0 < r - q < 2^{-(1-\delta)N(\omega)}$. (The reason for the $1 - \delta$ will become clear in a moment.) Pick $m \geq N(\omega)$ so that

$$2^{-(m+1)(1-\delta)} \leq r - q < 2^{-m(1-\delta)}$$

and write

$$r = j2^{-m} + 2^{-r(1)} + \cdots + 2^{-r(l)}$$

$$q = i2^{-m} - 2^{-q(1)} - \cdots - 2^{-q(k)}$$

where $m < r(1) < \cdots < r(l)$ and $m < q(1) < \cdots < q(k)$. Now $0 < r - q < 2^{-m(1-\delta)}$, so $(j - i) < 2^{m\delta}$ and it follows from (a) that

(b) $|B(i2^{-m}) - B(j2^{-m})| \leq ((2^{m\delta})2^{-m})^\gamma$.

Using (a) and the triangle inequality gives (here and in what follows C is a constant whose value is unimportant and will change from line to line)

(c) $|B(q) - B(i2^{-m})| \leq \sum_{h=1}^{k} (2^{-q(h)})^\gamma \leq \sum_{h=m}^{\infty} (2^{-\gamma})^h \leq C2^{-\gamma m}$.

Repeating the last computation shows

(d) $|B(r) - B(j2^{-m})| \leq C2^{-\gamma m}$.

Combining (b)–(d) gives

$$|B(q) - B(r)| \leq C2^{-\gamma m(1-\delta)} \leq C'|r - q|^\gamma,$$

since $2^{-m(1-\delta)} \leq 2^{1-\delta}|r - q|$, and we have shown B_t is Hölder continuous with exponent γ. $\square$

Exercise 1.3 Show that if we replace (1.5) by $E|X_t - X_s|^\beta \leq C|t - s|^{1+\alpha}$ where $\alpha, \beta > 0$, then (a) holds for any $\gamma < \alpha/\beta$ and hence the paths of X_t are Hölder continuous with exponent γ for any $\gamma < \alpha/\beta$. This is *Kolmogorov's continuity criterion*.

The scaling relation (1.2) implies

$$E|B_t - B_s|^{2m} = C_m|t - s|^m \quad \text{where } C_m = E|B_1|^{2m},$$

so using Exercise 1.3 gives a result of Wiener (1923).

(1.6) THEOREM   Brownian paths are Hölder continuous with exponent γ for any $\gamma < 1/2$.

It is easy to show:

(1.7) THEOREM   With probability 1, Brownian paths are not Lipschitz continuous (and hence not differentiable) at any point.

Remark   The nondifferentiability of Brownian paths was discovered by Paley, Wiener, and Zygmund (1933). Paley died in 1933 at the age of 26 in a skiing accident while the paper was in press. The proof we are about to give is due to Dvoretsky, Erdös, and Kakutani (1961).

Proof   Let $A_n = \{\omega:$ there is an $s \in [0, 1]$ so that $|B_t - B_s| \le C|t - s|$ when $|t - s| \le 3/n\}$.

$$Y_{k,n} = \max\left\{\left|B\left(\frac{k+j}{n}\right) - B\left(\frac{k+j-1}{n}\right)\right| : j = 0, 1, 2\right\} \quad \text{for } 1 \le k \le n - 2$$

$B_n = \{$at least one $Y_{k,n}$ is $\le 5C/n\}$.

The triangle inequality implies $A_n \subset B_n$. The worst case is $s = 1$. We pick $k = n - 2$ and observe

$$\left|B\left(\frac{n-3}{n}\right) - B\left(\frac{n-2}{n}\right)\right| \le \left|B\left(\frac{n-3}{n}\right) - B(1)\right| + \left|B(1) - B\left(\frac{n-2}{n}\right)\right| \le C(3/n + 2/n).$$

Using $A_n \subset B_n$ and the scaling relation (1.2) gives

$$P(A_n) \le P(B_n) \le nP(|B(1/n)| \le 5C/n)^3 = nP(|B(1)| \le 5C/n^{1/2})^3$$

$$\le n\{(10C/n^{1/2}) \cdot (2\pi)^{-1/2}\}^3,$$

since $\exp(-x^2/2) \le 1$. Letting $n \to \infty$ shows $P(A_n) \to 0$. Noticing $n \to A_n$ is increasing shows $P(A_n) = 0$ for all n and proves our claim, since C is arbitrary.   □

Exercise 1.4   Looking at the proof of (1.7) carefully shows that if $\gamma > 5/6$, then B_t is not Hölder continuous with exponent γ at any point in $[0, 1]$. Show by considering k increments instead of three that the last conclusion is true for all $\gamma > 1/2 + 1/k$.

The next result is more evidence that the sample paths of Brownian motion behave locally like $\sqrt{t}$.

Exercise 1.5   Let $\Delta_{m,n} = B(tm2^{-n}) - B(t(m-1)2^{-n})$. Compute $E(\sum_{m \le 2^n} \Delta_{m,n}^2 - t)^2$ and use Borel–Cantelli to conclude that $\sum_{m \le 2^n} \Delta_{m,n}^2 \to t$ a.s. as $n \to \infty$.

Remark   The last result is true if we consider a sequence of partitions $\Pi_1 \subset \Pi_2 \subset \cdots$ with mesh $\to 0$. See Freedman (1971a), pp. 42–46. In Exercise 2.3 we will see that the true quadratic variation, defined as the sup over all partitions, is ∞.

In the rest of this section we will explore several other ways of constructing Brownian motion. Perhaps the simplest is:

Lévy's (1948) *interpolation construction*   We will define inductively random variables so that for each $n \ge 0$, $U_{m,n}$, $1 \le m \le 2^n$, are i.i.d. normal$(0, 2^{-n})$. Let $U_{1,0}$ be normal$(0, 1)$. Assuming that the nth row has been defined, let $V_{m,n}$, $1 \le m \le 2^n$, be i.i.d. normal$(0, 2^{-n})$ that are independent of the $U_{m,n}$ and let

$$U_{2m-1,n+1} = (U_{m,n} + V_{m,n})/2, \qquad U_{2m,n+1} = (U_{m,n} - V_{m,n})/2.$$

To conclude that the $(n+1)$th row has the desired property we observe that

$$E(U_{j,n+1} U_{k,n+1}) = 0 \quad \text{when } j \ne k.$$

(This is clear by induction if $|j - k| > 1$ or $j = 2m$, $k = 2m + 1$. When $j = 2m - 1$, $k = 2m$,

$$E(U_{2m-1,n+1} U_{2m,n+1}) = \tfrac{1}{4}(EU_{m,n}^2 - EV_{m,n}^2) = 0.)$$

To get Brownian motion, let $S_{m,n} = U_{1,n} + \cdots + U_{m,n}$ and define $B_n(t)$ by requiring that $B(m2^{-n}) = S_{m,n}$ and $B_n(t)$ is linear on each interval $[(m-1)2^{-n}, m2^{-n}]$.

$$\sup_{t \in [0,1]} |B_n(t) - B_{n+1}(t)| = \sup_{m \le 2^n} |V_{m,n}|.$$

Scaling and (1.3) in Chapter 1 imply

$$P(|V_{m,n}| > n^2 2^{-n/2}) = P(V_{0,1} > n^2) \le \exp(-n^2/2).$$

So it follows from the Borel–Cantelli lemma that

$$P\left(\sup_m |V_{m,n}| \ge n^2 2^{-n/2} \text{ for infinitely many } n \right) = 0.$$

The last result implies that $B_n(t)$ converges uniformly on $[0, 1]$ to a continuous limit that we call $B(t)$. (a) and (b) of the definition clearly hold when the times are of the form $m2^{-n}$. Path continuity and a limiting argument show that they are true in general.

There are two ways of constructing Brownian motion from random series.

Paley and Wiener (1934)   They showed that if $\xi_0, \xi_1, \ldots$ are i.i.d. standard normals and

$$\xi_0 t + \sum_{n=1}^{\infty} \left(\sum_{k=2^{n-1}}^{2^n-1} \xi_k \sqrt{2}(\sin \pi k t)/\pi k \right),$$

then the series converges uniformly and the limit is Brownian motion. (See Itô and McKean (1965), pp. 21–22, or Hida (1980), pp. 67–72, for proofs.) This definition is reminiscent of the continuous function

$$f(t) = \sum_{n=1}^{\infty} \sin(\pi n^2 t)/n^2$$

conjectured to be nowhere differentiable by Riemann and proved so by Hardy.

P. Lévy (1948), Z. Ciesielski (1961)   The *Haar functions* are defined by $H_0(t) \equiv 1$, and for $n \geq 0$, $2^n \leq k < 2^{n+1}$,

$$H_k(t) = \begin{cases} 2^{n/2} & \text{if} \quad (k - 2^n)/2^n \leq t < (k + .5 - 2^n)/2^n \\ -2^{n/2} & \text{if} \quad (k + .5 - 2^n)/2^n \leq t < (k + 1 - 2^n)/2^n \\ 0 & \text{otherwise.} \end{cases}$$

The *Schauder functions* are the integrals of the Haar functions

$$S_k(t) = \int_0^t H_k(s)\, ds.$$

The $S_k(t)$ are little tents with height $2^{-n/2}/2$, so it is not difficult to show:

Exercise 1.6   Let $\xi_0, \xi_1, \ldots$ be i.i.d. standard normals.

$$C(t) \equiv \sum_{n=0}^{\infty} \xi_n S_n(t) \quad \text{converges uniformly on } [0, 1].$$

It is easy to see that $EC(t) = 0$ and the finite-dimensional distributions are Gaussian. In view of our second definition, to identify the limit as Brownian motion we only have to show

$$\sum_{n=0}^{\infty} S_n(s) S_n(t) = s \wedge t.$$

McKean (1969) settles the matter with incantation "using Parseval's relation for the Haar functions applied to the indicator functions f and g of $r \leq s$ and $r \leq t$." The next exercise gives Karlin and Taylor's (1975) probabilistic derivation of the needed relation.

Exercise 1.7   (i) Let U be uniform on $(0, 1)$ and f a function with $E|f(U)| < \infty$. Let $Y_k = H_k(U)$ and $a_k = E(f(U)H_k(U))$. Show that

$$E(f(U)|Y_1, \ldots, Y_k) = \sum_{m=0}^{n} a_k Y_k \quad \text{and} \quad \rightarrow f(U) \text{ a.s. as } k \rightarrow \infty.$$

(ii) Show that if $E(f(U)^2) < \infty$, then $E(f(U)^2) = \sum_k a_k^2$. (iii) Applying (ii) to $f + g$ and $f - g$, show that if g has $E(g(U)^2) < \infty$ and $b_k = E(g(U)H_k(U))$, then $E(f(U)g(U)) = \sum_k a_k b_k$. When f and g are the functions McKean suggests, $s \wedge t = E(f(U)g(U))$ and $a_k = S_k(s)$, so the desired result follows.

2   Markov Property, Blumenthal's 0–1 Law

Intuitively the Markov property says "if $s \geq 0$, then $B(t + s) - B(s)$, $t \geq 0$, is a Brownian motion that is independent of what happened before time s." The first step in making this into a precise statement is to explain what we mean by "what happened before time s." The first thing that comes to mind is

$$\mathscr{F}_s^\circ = \sigma(B_r \colon r \leq s).$$

For reasons that will become clear as we go along it is convenient to replace $\mathscr{F}_s^\circ$ by

$$\mathscr{F}_s^+ = \bigcap_{t > s} \mathscr{F}_t^\circ.$$

The fields $\mathscr{F}_s^+$ are nicer because they are *right continuous*:

$$\bigcap_{t > s} \mathscr{F}_t^+ = \bigcap_{t > s} \left(\bigcap_{u > t} \mathscr{F}_u^\circ \right) = \bigcap_{u > s} \mathscr{F}_u^\circ = \mathscr{F}_s^+.$$

In words, the $\mathscr{F}_s^+$ allow us an "infinitesimal peek at the future"; that is, $A \in \mathscr{F}_s^+$ if it is in $\mathscr{F}_{s+\varepsilon}^\circ$ for any $\varepsilon > 0$. The random variable

$$\limsup_{t \downarrow s} (B_t - B_s)/f(t - s)$$

is measurable with respect to $\mathscr{F}_s^+$ but not $\mathscr{F}_s^\circ$. We will see below that there are no interesting examples; that is, $\mathscr{F}_s^+$ and $\mathscr{F}_s^\circ$ are the same (up to null sets).

To state the Markov property we need some notation. Recall that we have a family of measures P_x, $x \in \mathbb{R}$, on $(C, \mathscr{C})$ so that under P_x, $B_t(\omega) = \omega(t)$ is a Brownian motion starting at x. For $s \geq 0$ we define the *shift transformation* $\theta_s : C \to C$ by

$$(\theta_s \omega)(t) = \omega(s + t) \quad \text{for } t \geq 0.$$

In words, we cut off the part of the path before time s and then shift the path so that time s becomes time 0.

(2.1) The Markov Property   If $s \geq 0$ and Y is bounded and $\mathscr{C}$-measurable, then for all $x \in \mathbb{R}$,

$$E_x(Y \circ \theta_s | \mathscr{F}_s^+) = E_{B(s)} Y$$

where the right-hand side is the function $\varphi(x) = E_x Y$ evaluated at $x = B(s)$.

Proof   As in the proof of the Markov property in Chapter 5 (see (2.1)), we will prove the result for a carefully chosen special case and then use the monotone class theorem (MCT) to get the general case. Suppose $Y(\omega) = \prod_{1 \leq m \leq n} f_m(\omega(t_m))$ where $0 < t_1 < \cdots < t_n$ and the f_m are bounded and continuous. Let $0 < h < t_1$, let $0 < s_1 < \cdots < s_k \leq s + h$, and let $A = \{\omega : \omega(s_j) \in A_j, 1 \leq j \leq k\}$ where $A_j \in \mathscr{R}$ for $1 \leq j \leq k$. From the definition of Brownian motion it follows that

$$E(Y \circ \theta_s; A) = \int_{A_1} dx_1 \, p_{s_1}(x, x_1) \cdots \int_{A_k} dx_k \, p_{s_k - s_{k-1}}(x_{k-1}, x_k) \int dy \, p_{t+h-s_k}(x_k, y) \varphi(y, h)$$

where

$$\varphi(y, h) = \int dy_1 \, p_{t_1 - h}(y, y_1) f_1(y_1) \cdots \int dy_k \, p_{t_k - t_{k-1}}(y_{k-1}, y_k) f_k(y_k),$$

so we have

$$E(Y \circ \theta_s; A) = E(\varphi(B_{s+h}, h); A).$$

The last equality holds for all finite-dimensional sets A, so the $\pi - \lambda$ theorem ((4.2) in Chapter 1) implies that it is valid for all $A \in \mathscr{F}_{s+h}^o \supset \mathscr{F}_s^+$.

If f is bounded and continuous, the dominated convergence theorem implies

$$x \to \int dy \, p_t(x, y) f(y) \quad \text{is continuous.}$$

The last result and induction imply

$$\varphi(y, h) \quad \text{is continuous.} \tag{2.2}$$

Letting $h \downarrow 0$ and using the bounded convergence theorem show that for the special

φ considered above,

$$E(Y \circ \theta_s; A) = E(\varphi(B_s, 0); A) \quad \text{for all } A \in \mathscr{F}_s^+. \tag{$*$}$$

The desired conclusion now follows from the MCT, (1.4) in Chapter 5. Let $\mathscr{H}$ = the collection of bounded functions for which $(*)$ holds. $\mathscr{H}$ clearly has properties (ii) and (iii) of the MCT. Let $\mathscr{A}$ be the collection of sets of the form $\{\omega : \omega(t_j) \in G_j\}$ where G_j is an open set. If G is open, the function 1_G is an increasing limit of continuous functions $f_k(x) = (1 \wedge k \operatorname{dist}(x, G^c))$, so if $A \in \mathscr{A}$, then $1_A \in \mathscr{H}$. This shows (i) holds and the desired conclusion follows from the MCT.   □

The next two examples give typical applications of the Markov property.

Example 2.1   Let $R = \inf\{t > 1 : B_t = 0\}$. R is for right or return. Breaking things down according to the value at time 1 gives

$$P_x(R > 1 + t) = \int p_1(x, y) P_y(T_0 > t) \, dy \tag{2.3}$$

where $T_0 = \inf\{s > 0 : B_s = 0\}$.

Example 2.2   Let $L = \sup\{t \le 1 : B_t = 0\}$. L is for left or last. Breaking things down according to the value at time $t \in (0, 1]$ gives

$$P_0(L \le t) = \int p_t(0, y) P_y(T_0 > 1 - t) \, dy. \tag{2.4}$$

Exercise 2.1   Show that (2.3) and (2.4) are consequences of (2.1).

We will compute the distribution of T_0 in Section 3 and the distributions of L and R in Section 4.

The reader will see many applications of the Markov property below, so we turn our attention now to a "triviality" that has surprising consequences. Since

$$E_x(Y \circ \theta_s | \mathscr{F}_s^+) = E_{B(s)} Y \in \mathscr{F}_s^o,$$

it follows from (1.1) in Chapter 5 that

$$E_x(Y \circ \theta_s | \mathscr{F}_s^+) = E_x(Y \circ \theta_s | \mathscr{F}_s^o).$$

From the last equation it is a short step to:

(2.5) THEOREM   If $Z \in \mathscr{C}$ is bounded, then for all $s \ge 0$ and $x \in \mathbb{R}$,

$$E_x(Z | \mathscr{F}_s^+) = E_x(Z | \mathscr{F}_s^o)$$

Proof   As in the proof of (2.1) it suffices to prove the result when

$$Z = \prod_{m=1}^{n} f_m(B(t_m))$$

and the f_m are bounded and measurable. In this case Z can be written as $X(Y \circ \theta_s)$ where $X \in \mathscr{F}_s^\circ$ and Y is $\mathscr{C}$-measurable, so

$$E_x(Z|\mathscr{F}_s^+) = XE_x(Y \circ \theta_s|\mathscr{F}_s^+) = XE_{B(s)}Y \in \mathscr{F}_s^\circ,$$

and the proof is complete.   □

If we let $Z \in \mathscr{F}_s^+$, then (1.5) implies $Z = E_x(Z|\mathscr{F}_s^\circ) \in \mathscr{F}_s^\circ$, so the two σ-fields are the same up to null sets. At first glance, this conclusion is not exciting. The fun starts when we take $s = 0$ in (2.5) to get:

(2.6) Blumenthal's 0–1 Law   If $A \in \mathscr{F}_0^+$, then for all $x \in \mathbb{R}$, $P_x(A) \in \{0, 1\}$.

Proof   Using (2.5), the fact that $A \in \mathscr{F}_0^+$ and $\mathscr{F}_0^\circ = \sigma(B_0)$ is trivial under P_x gives

$$1_A = E_x(1_A|\mathscr{F}_0^+) = E_x(1_A|\mathscr{F}_0^\circ) = P_x(A)   P_x \text{ a.s.}$$

This shows that the indicator function 1_A is a.s. equal to the number $P_x(A)$ and the result follows.   □

In words, the last result says that the *germ field*, $\mathscr{F}_0^+$, is trivial. This result is very useful in studying the local behavior of Brownian paths.

(2.7) THEOREM   If $\tau = \inf\{t \geq 0 : B_t > 0\}$, then $P_0(\tau = 0) = 1$.

Proof

$$P_0(\tau \leq t) \geq P_0(B_t > 0) = 1/2,$$

since the normal distribution is symmetric about 0. Letting $t \downarrow 0$, we conclude

$$P_0(\tau = 0) = \lim_{t \downarrow 0} P_0(\tau \leq t) \geq 1/2,$$

so it follows from (2.6) that $P_0(\tau = 0) = 1$.   □

Once Brownian motion must hit $(0, \infty)$ immediately starting from 0, it must also hit $(-\infty, 0)$ immediately. Since $t \to B_t$ is continuous, this forces:

(2.8) THEOREM   If $T_0 = \inf\{t > 0 : B_t = 0\}$, then $P_0(T_0 = 0) = 1$.

A corollary of (2.8) is:

Exercise 2.2   For any $a < b$, $P(\max_{a \leq t \leq b} B_t > B_a$ and $> B_b) = 1$. Use this to conclude that the set of local maxima of B_t is a dense set.

Another typical application of (2.6) is:

Exercise 2.3   Show that $\limsup_{t \downarrow 0} B(t)/t^{1/2} = \infty$ P_0 a.s., so with probability 1 Brownian paths are not Hölder continuous of order 1/2 at 0.

Remark   Let $\mathcal{H}_\gamma(\omega)$ be the set of times at which the path $\omega \in C$ is Hölder continuous of order γ. (1.6) shows that $P(\mathcal{H}_\gamma = [0, \infty)) = 1$ for $\gamma < 1/2$. Exercise 1.4 shows that $P(\mathcal{H}_\gamma = \varnothing) = 1$ for $\gamma > 1/2$. The last exercise shows $P(t \in \mathcal{H}_{1/2}) = 0$ for each t, but B. Davis (1983) has shown $P(\mathcal{H}_{1/2} \neq \varnothing) = 1$.

Exercise 2.4   Let $K < \infty$, cover $[0, 1]$ with intervals (s, t) having $|B(t) - B(s)| \geq K(t - s)^{1/2}$, and conclude that $\sup \sum_i |B(t_i) - B(t_{i-1})|^2 = \infty$ where the sup is taken over all partitions of $[0, 1]$.

Exercise 2.5   Let $t \to f(t)$ be an increasing function with $f(t) > 0$ for all $t > 0$. Use (2.6) to conclude that $\limsup_{t \downarrow 0} B(t)/f(t) = c$ P_0 a.s. where $c \in [0, \infty]$ is a constant. From Exercise 2.3 we see that $c = \infty$ when $f(t) = t^{1/2}$. In Section 9 we will show that $c = 2^{1/2}$ when $f(t) = (t \log \log t)^{1/2}$.

With our discussion of Blumenthal's 0–1 law complete, the distinction between $\mathcal{F}_s^+$ and $\mathcal{F}_s^o$ is no longer important, so we will make one final improvement in our σ-fields and remove the superscripts. Let

$$\mathcal{N} = \{A: P_x(A) = 0 \text{ for all } x \in \mathbb{R}\}$$

$$\mathcal{F}_s = \sigma(\mathcal{F}_s^+ \cup \mathcal{N}) = \sigma(\mathcal{F}_s^o \cup \mathcal{N}).$$

$\mathcal{N}$ is for *null sets*. Since we have a family of measures, a set can be safely ignored only if it has measure 0 under every P_x.

(2.6) concerns the behavior of B_t as $t \to 0$. By using a trick we can use this result to get information about the behavior as $t \to \infty$.

(2.9) THEOREM   If B_t is a Brownian motion starting at 0, then so is the process defined by $X_0 = 0$ and $X_t = tB(1/t)$ for $t > 0$.

Proof   We begin by observing that the strong law of large numbers implies $X_t \to 0$ as $t \to 0$, so X has continuous paths and we only have to check that X has the right f.d.d.'s. By the second definition of Brownian motion, it suffices to show that (i) if $0 < t_1 < \cdots < t_n$, then $(X(t_1), \ldots, X(t_n))$ has a multivariate normal distribution with mean 0, and (ii) if $s < t$, then

$$E(X_s X_t) = st E(B(1/s)B(1/t)) = s. \qquad \square$$

(2.9) allows us to relate the behavior of B_t as $t \to \infty$ and as $t \to 0$. Combining this idea with Blumenthal's 0–1 law leads to a very useful result. Let

$$\mathcal{F}_t' = \sigma(B_s \colon s \geq t) = \text{the future at time } t$$

$$\mathcal{T} = \bigcap_{t \geq 0} \mathcal{F}_t' = \text{the tail } \sigma\text{-field}.$$

(2.10) THEOREM   If $A \in \mathcal{T}$, then either $P_x(A) \equiv 0$ or $P_x(A) \equiv 1$.

Remark   Notice that this is stronger than the conclusion of Blumenthal's 0–1 law (2.6). The examples $A = \{\omega \colon \omega(0) \in B\}$ show that for A in the germ σ-field $\mathcal{F}_0^+$, the value of $P_x(A)$ may depend on x.

Proof   Since the tail σ-field of B is the same as the germ σ-field for X, it follows that $P_0(A) \in \{0, 1\}$. To improve this to the conclusion given observe that $A \in \mathcal{F}_1'$, so A can be written as $1_B \circ \theta_1$. Applying the Markov property gives

$$P_x(A) = E_x(1_B \circ \theta_1) = E_x(E_x(1_B \circ \theta_1 | \mathcal{F}_1)) = E_x(E_{B(1)} 1_B)$$

$$= \int (2\pi)^{-1/2} \exp(-(y - x)^2/2t) P_y(B) \, dy.$$

Taking $x = 0$ we see that if $P_0(A) = 0$, then $P_y(B) = 0$ for a.e. y with respect to Lebesgue measure, and using the formula again shows $P_x(A) = 0$ for all x. To handle the case $P_0(A) = 1$ observe that $A^c \in \mathcal{T}$ and $P_0(A^c) = 0$, so the last result implies $P_x(A^c) = 0$ for all x.   □

Exercise 2.6   Let $\{B_t = 0 \text{ i.o.}\} = \bigcap \{B_t = 0 \text{ for some } t \geq n\}$. Use (2.8), (2.9), and (2.10) to show $P_x(B_t = 0 \text{ i.o.}) = 1$ for all x.

3   Stopping Times, the Strong Markov Property

Generalizing a definition in Section 1 of Chapter 3, we call a random variable S taking values in $[0, \infty]$ a *stopping time* if for all $t \geq 0$, $\{S < t\} \in \mathcal{F}_t$. In the last definition we have obviously made a choice between $\{S < t\}$ and $\{S \leq t\}$. This makes a big difference in discrete time but none in continuous time (for a right continuous filtration $\mathcal{F}_t$):

If $\{S \leq t\} \in \mathcal{F}_t$, then $\{S < t\} = \bigcup_n \{S \leq t - 1/n\} \in \mathcal{F}_t$.

If $\{S < t\} \in \mathcal{F}_t$, then $\{S \leq t\} = \bigcap_n \{S < t + 1/n\} \in \mathcal{F}_t$.

The first conclusion requires only that $t \to \mathcal{F}_t$ is increasing. The second relies on the fact that $t \to \mathcal{F}_t$ is right continuous. (3.2) and (3.3) below show that when checking whether something is a stopping time, it is nice to know that the two definitions are equivalent.

(3.1) THEOREM   If G is an open set and $T = \inf\{t \geq 0 : B_t \in G\}$, then T is a stopping time.

Proof   Since G is open and $t \to B_t$ is continuous, $\{T < t\} = \bigcup_{q<t}\{B_q \in G\}$ where the union is over all rational q, so $\{T < t\} \in \mathcal{F}_t$.   □

(3.2) THEOREM   If T_n is a sequence of stopping times and $T_n \downarrow T$, then T is a stopping time.

Proof $\{T < t\} = \bigcup_n \{T_n < t\}$.   □

(3.3) THEOREM   If T_n is a sequence of stopping times and $T_n \uparrow T$, then T is a stopping time.

Proof $\{T \leq t\} = \bigcap_n \{T_n \leq t\}$.   □

(3.4) THEOREM   If K is a closed set and $T = \inf\{t \geq 0 : B_t \in K\}$, then T is a stopping time.

Proof   Let $G_n = \bigcup\{(x - 1/n, x + 1/n) : x \in K\}$ and let $T_n = \inf\{t \geq 0 : B_t \in G_n\}$. Since G_n is open, it follows from (3.1) that T is a stopping time. I claim that as $n \uparrow \infty$, $T_n \uparrow T$. To prove this notice that $T \geq T_n$ for all n, so $\lim T_n \leq T$. To prove $T \leq \lim T_n$ we can suppose that $T_n \uparrow t < \infty$. Since $B(T_n) \in \bar{G}_n$ for all n and $B(T_n) \to B(t)$, it follows that $B(t) \in K$ and $T \leq t$.   □

(3.1) and (3.4) will take care of all the hitting times we will consider. Our next goal is to state and prove the strong Markov property. To do this, we need to generalize two definitions from Section 1 in Chapter 3.

$$(\theta_S \omega)(t) = \begin{cases} \omega(S(\omega) + t) & \text{on } \{S < \infty\} \\ \Delta & \text{on } \{S = \infty\} \end{cases}$$

where Δ is an extra point we add to C. As in Section 2 of Chapter 5, we will usually explicitly restrict our attention to $\{S < \infty\}$, so the reader does not have to worry about the second half of the definition.

$$\mathcal{F}_S = \{A : A \cap \{S \leq t\} \in \mathcal{F}_t \text{ for all } t \geq 0\}.$$

Exercise 3.1   Since $\mathcal{F}_t$ is right continuous, the last definition is unchanged if we replace $\{S \leq t\}$ by $\{S < t\}$. As in the case of stopping times, it is nice to know that the two definitions are equivalent.

Most of the properties of $\mathcal{F}_N$ derived in Section 1 of Chapter 3 carry over to continuous time. Two that will be useful below are:

(3.5) THEOREM   If $S \leq T$ are stopping times, then $\mathcal{F}_S \subset \mathcal{F}_T$.

Proof   If $A \in \mathcal{F}_S$, then $A \cap \{T \leq t\} = (A \cap \{S \leq t\}) \cap \{T \leq t\} \in \mathcal{F}_t$.   □

(3.6) THEOREM   If $T_n \downarrow T$ are stopping times, then $\mathcal{F}_T = \bigcap \mathcal{F}(T_n)$.

Proof   (3.5) implies $\mathcal{F}(T_n) \supset \mathcal{F}_T$ for all n. To prove the other inclusion, let $A \in \bigcap \mathcal{F}(T_n)$. Since $A \cap \{T_n < t\} \in \mathcal{F}_t$ and $T_n \downarrow T$, it follows that $A \cap \{T < t\} \in \mathcal{F}_t$.   □

We come now to the most important property of stopping times.

(3.7) Strong Markov Property   Let $(s, \omega) \to Y(s, \omega)$ be bounded and $\mathcal{R} \times \mathcal{C}$ measurable. If S is a stopping time, then for all $x \in \mathbb{R}$,

$$E_x(Y_S \circ \theta_S | \mathcal{F}_S) = E_{B(S)} Y_S \quad \text{on } \{S < \infty\}$$

where the right-hand side is the function $\varphi(x, t) = E_x Y_t$ evaluated at $x = B(S)$, $t = S$.

Proof   We first prove the result under the assumption that there is a sequence of times $t_n \uparrow \infty$, so that $P_x(S < \infty) = \sum P_x(S = t_n)$. In this case the proof is basically the same as the proof of (2.4) in Chapter 5. We break things down according to the value of S, apply the Markov property, and put the pieces back together. If we let $Z_n = Y_{t_n}(\omega)$ and $A \in \mathcal{F}_S$, then

$$E_x(Y_S \circ \theta_S; A \cap \{S < \infty\}) = \sum_{n=1}^{\infty} E_x(Z_n \circ \theta_{t_n}; A \cap \{S = t_n\}).$$

Now if $A \in \mathcal{F}_S$, $A \cap \{S = t_n\} = (A \cap \{S \leq t_n\}) - (A \cap \{S \leq t_{n-1}\}) \in \mathcal{F}(T_n)$, so it follows from the Markov property that the above sum is

$$= \sum_{n=1}^{\infty} E_x(E_{B(t_n)} Z_n; A \cap \{S = t_n\}) = E_x(E_{B(S)} Y_S; A \cap \{S < \infty\}).$$

To prove the result in general we let $S_n = ([2^n S] + 1)/2^n$ where $[x] = $ the largest integer $\leq x$; that is,

$$S_n = (m + 1)2^{-n} \quad \text{if } m2^{-n} \leq S < (m + 1)2^{-n}.$$

In words, we stop at the first time of the form $k2^{-n}$ after S (i.e., $> S$). From the verbal description it should be clear that S_n is a stopping time. To check this we observe that if $m2^{-n} < t \leq (m + 1)2^{-n}$, then $\{S_n < t\} = \{S < m2^{-n}\} \in \mathcal{F}_t$. To be able to let $n \to \infty$, we restrict our attention to Y's of the form

$$Y_s(\omega) = f_0(s) \prod_{m=1}^{n} f_m(\omega(t_m)) \tag{*}$$

where $0 < t_1 < \cdots < t_n$ and $f_0, \ldots, f_n$ are bounded and continuous. In this case, as we observed in (2.2), $\varphi(x, s) = E_x Y_s$ is bounded and continuous.

Having assembled the necessary ingredients, we can now complete the proof. Let $A \in \mathcal{F}_S$. Since $S \leq S_n$, (3.5) implies $A \in \mathcal{F}(S_n)$. Applying the special case of (3.7) proved above to S_n and observing that $\{S_n < \infty\} = \{S < \infty\}$ give

$$E_x(Y_{S_n} \circ \theta_{S_n}; A \cap \{S < \infty\}) = E_x(\varphi(B(S_n), S_n); A \cap \{S < \infty\}).$$

Now as $n \to \infty$, $S_n \downarrow S$, $B(S_n) \to B(S)$, $\varphi(B(S_n), S_n) \to \varphi(B(S), S)$, and

$$Y_{S_n} \circ \theta_{S_n} \to Y_S \circ \theta_S,$$

so the bounded convergence theorem implies that (3.7) holds when Y has the form given in (∗). An application of the monotone class theorem as in the proofs of (2.1) and (2.5) now gives the desired result.   ☐

Example 3.1   *Zeros of Brownian motion.* Let $R_t = \inf\{u > t : B_u = 0\}$ and let $T_0 = \inf\{u > 0 : B_u = 0\}$. The strong Markov property and (2.8) imply

$$P_x(T_0 \circ R_t > 0 | \mathscr{F}_{R(t)}) = P_0(T_0 > 0) = 0.$$

(Here we are using $P_x(R_t < \infty) = 1$, a consequence of Exercise 2.6.) From this it follows that if a point $u \in \mathscr{Z}(\omega) \equiv \{t : B_t(\omega) = 0\}$ is isolated on the left (i.e., there is a rational $t < u$ so that $(t, u) \cap \mathscr{Z}(\omega) = \varnothing$), then it is, with probability 1, a decreasing limit of points in $\mathscr{Z}(\omega)$. This shows that the closed set $\mathscr{Z}(\omega)$ has no isolated points and hence must be uncountable. For the last step see Hewitt and Stromberg (1965), p. 72.

If we let $|\mathscr{Z}(\omega)|$ denote the Lebesgue measure of $\mathscr{Z}(\omega)$, then Fubini's theorem implies

$$E_x(|\mathscr{Z}(\omega)| \cap [0, T]) = \int_0^T P_x(B_t = 0)\, dt = 0.$$

So $\mathscr{Z}(\omega)$ is a set of measure 0. Combining this observation with the one in the last paragraph shows that $\mathscr{Z}(\omega)$ is like the classical Cantor set, which is constructed by taking $[0, 1] - (1/3, 2/3)$ and then successively removing the middle third of the intervals that remain.

Exercise 3.2   What is wrong with the following proof that $P_0(B_t = 0$ for all $t \geq 0) = 1$? Let $T = \inf\{t : B_t = 0$ but $B_s \neq 0$ for $s \in (t, t + h)$ for some $h > 0\}$. Since T is a stopping time, it follows that no point of $\mathscr{Z}(\omega)$ is isolated on the right, and since $\mathscr{Z}(\omega)$ is closed, it must be all of $[0, \infty)$.

Our second application shows why we want to allow the function Y that we apply to the shifted path to depend on the stopping time S.

Example 3.2   *Reflection principle.* Let $a > 0$ and let $T_a = \inf\{t : B_t = a\}$. Then

$$P_0(T_a < t) = 2P_0(B_t \geq a). \tag{3.8}$$

Remark   D. André used a symmetry argument in his proof of the ballot theorem (see (3.1) and (3.2) in Chapter 3) that can be viewed as a reflection principle for random walk, so this result is sometimes named after him. Lévy (1939, p. 293) stated the reflection principle for Brownian motion, but a proof had to wait until ones were given independently by Dynkin (1955) and Hunt (1956). Hunt probably had Lévy in mind when he said, "Although mathematicians use this extended Markoff property, at least as a heuristic principle, I have nowhere found it discussed with rigor." We will prove the result in words first and then show how it follows from (3.7).

Intuitive Proof   We observe that if B_s hits a at some time $s < t$, then the strong Markov property implies that $B_t - B(T_a)$ is independent of what happened before time T_a. The symmetry of the normal distribution and $P_0(B_t = a) = 0$ then imply

$$P_0(T_a < t, B_t > a) = \tfrac{1}{2}P_0(T_a < t). \tag{3.9}$$

Since $\{B_t > a\} \subset \{T_a < t\}$, the result follows.

Proof   To make the intuitive proof rigorous we only have to prove (3.9). To extract this from the strong Markov property (3.7), we let

$$Y_s(\omega) = \begin{cases} 1 & \text{if } s < t, \omega(t-s) > a \\ 0 & \text{otherwise.} \end{cases}$$

We do this so that if we let $S = \inf\{s \le t : B_s = a\}$ with $\inf \varnothing = \infty$, then

$$Y_S(\theta_S\omega) = \begin{cases} 1 & \text{if } S < t, B_t > a \\ 0 & \text{otherwise,} \end{cases}$$

and the strong Markov property implies

$$E_0(Y_S \circ \theta_S | \mathscr{F}_S) = \varphi(B_S, S) \quad \text{on } \{S < \infty\} = \{T_a < t\}$$

where $\varphi(x, s) = E_x Y_s$. $B_S = a$ on $\{S < \infty\}$ and $\varphi(x, s) = 1/2$ if $s < t$ and $x = a$, so

$$P_0(T_a < t, B_t \ge a) = E_0(1/2; T_a < \infty),$$

which proves (3.9).   □

Exercise 3.3   Generalize the proof of (3.9) to conclude that if $u < v \le a$, then

$$P_0(T_a < t, u < B_t < v) = P_0(2a - v < B_t < 2a - u). \tag{3.10}$$

Here a picture is worth a hundred words; see Figure 7.1.

Exercise 3.4   Find $\lim_{\varepsilon \to 0} P_\varepsilon(B_1 \ge x | T_0 > 1)$.

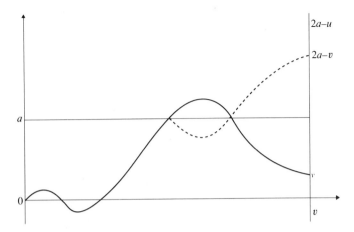

Figure 7.1

Exercise 3.5   Use the strong Markov property to conclude that if $a < b \leq c < d$, then

$$P\left(\max_{a \leq t \leq b} B_t = \max_{c \leq t \leq d} B_t \right) = 0.$$

Exercise 3.6   Dvoretsky, Erdös, and Kakutani (1961) claim that B_t has uncountably many local maxima. Use the last exercise to show this is false.

Remark t is said to be a *point of increase* if there are $s < t < u$ so that $B_r < B_t$ for $r \in (s, t)$ and $B_r > B_t$ for $r \in (t, u)$. By an improper use of the reflection principle one could conclude that such points are as numerous as local maxima, which satisfy $B_r < B_t$ for $r \in (s, t)$ and $B_t > B_r$ for $r \in (t, u)$. However, Dvoretsky, Erdös, and Kakutani (1961) have shown (see Adelman (1985) and Burdzy (1989) for more recent proofs) that with probability 1, there are no points of increase (vs. countably many local maxima).

Let $M_t = \max_{s \leq t} B_s$. (3.8) tells us that if $B_0 = 0$, then $M_t \stackrel{d}{=} |B_t|$. Our first application of this will be to estimate the *modulus of continuity*

$$\mathrm{osc}(\delta) = \sup\{|B_s - B_t| : s, t \in [0, 1], |t - s| < \delta\}.$$

(3.11) THEOREM   With probability 1, $\limsup_{\delta \to 0} \mathrm{osc}(\delta)/(\delta \log(1/\delta))^{1/2} \leq 6$.

Proof   Let $I_{m,n} = [m2^{-n}, (m + 1)2^{-n}]$, $\Delta_{m,n} = \sup\{|B_t - B(m2^{-n})| : t \in I_{m,n}\}$.

$$P(\Delta_{m,n} \geq a2^{-n/2}) \leq 4P(B(2^{-n}) \geq a2^{-n/2}) = 4P(B(1) \geq a) \leq 4\exp(-a^2/2)$$

by (1.3) in Chapter 1 if $a \geq 1$. Let $\varepsilon > 0$, $b = 2(1 + \varepsilon)(\log 2)$, and $a_n = (bn)^{1/2}$. By (3.8)

$$P(\Delta_{m,n} \geq a_n 2^{-n/2} \text{ for some } m \leq 2^n) \leq 2^{-n\varepsilon},$$

so the Borel–Cantelli lemma implies that if $n \geq N(\omega)$, $\Delta_{m,n} \leq (bn)^{1/2} 2^{-n/2}$. Now if $s \in I_{m,n}$, $s < t$, and $|s - t| < 2^{-n}$, then $t \in I_{m,n}$ or $I_{m+1,n}$. In either case the triangle inequality implies

$$|B_t - B_s| \leq 3(bn)^{1/2} 2^{-n/2}.$$

(The worst case is $t \in I_{m+1,n}$, but even in this case

$$|B_t - B_s| \leq |B_t - B((m+1)2^{-n})| + |B((m+1)2^{-n}) - B(m2^{-n})| + |B(m2^{-n}) - B_s|.)$$

It follows from the last estimate that for $2^{-(n+1)} \leq \delta < 2^{-n}$,

$$\text{osc}(\delta) \leq 3(bn)^{1/2} 2^{-n/2} \leq 3(b \log_2(1/\delta))^{1/2} (2\delta)^{1/2} = 6((1 + \varepsilon)\delta \log(1/\delta))^{1/2}.$$

Recall $b = 2(1 + \varepsilon) \log 2$ and observe $\exp((\log 2)(\log_2 1/\delta)) = 1/\delta$.   □

The constant 6 is not the best possible because the end of the proof is sloppy. Lévy (1937) showed

$$\limsup_{\delta \to 0} \text{osc}(\delta)/(\delta \log(1/\delta))^{1/2} = \sqrt{2}.$$

One half of this is easy.

Exercise 3.7   Show that if $\varepsilon > 0$,

$$P(\max \Delta_{m,n} \geq \{2(1 - \varepsilon)(\log 2)n\}^{1/2} 2^{-n/2}) \to 1.$$

The upper bound can be proved by imitating the proof of (1.6). See McKean (1969), pp. 14–16, or Itô and McKean (1965), pp. 36–38, where a sharper result due to Chung, Erdös, and Sirao (1959) is proved.

*4 Maxima and Zeros

We have two interrelated aims in this section. The first is to understand the behavior of the hitting times $T_a = \inf\{t: B_t = a\}$ as the level a varies. The second is to take a closer look at the zero set of Brownian motion and give the reader a glimpse of Brownian local time, a rich and beautiful subject. We do not have the time (and this is not the place) to give a full account of the theory of local time. If our sketchy treatment becomes confusing, the reader can go on to the next section.

We begin by observing that the reflection principle (3.8) implies

$$P(T_a \leq t) = 2P_0(B_t \geq a) = 2 \int_a^\infty (2\pi t)^{-1/2} \exp(-x^2/2t) \, dx.$$

To find the probability density of T_a, we change variables $x = (t^{1/2}a)/s^{1/2}$ to get

$$P_0(T_a \le t) = 2 \int_t^0 (2\pi t)^{-1/2} \exp(-a^2/2s)(-t^{1/2}a/2s^{3/2})\, ds \tag{4.1}$$

$$= \int_0^t (2\pi s^3)^{-1/2}a \exp(-a^2/2s)\, ds.$$

Using the last formula we can compute the distribution of $L = \sup\{t \le 1 : B_t = 0\}$ and $R = \inf\{t \ge 1 : B_t = 0\}$, completing work we started in Examples 2.1 and 2.2. By (2.4),

$$P_0(L \le s) = \int_{-\infty}^\infty p_s(0, x)P_x(T_0 > 1 - s)\, dx$$

$$= 2 \int_0^\infty (2\pi s)^{-1/2} \exp(-x^2/2s) \int_{1-s}^\infty (2\pi r^3)^{-1/2} x \exp(-x^2/2r)\, dr\, dx$$

$$= \frac{1}{\pi} \int_{1-s}^\infty (sr^3)^{-1/2} \int_0^\infty x \exp(-x^2(r + s)/2rs)\, dx\, dr$$

$$= \frac{1}{\pi} \int_{1-s}^\infty (sr^3)^{-1/2}rs/(r + s)\, dr.$$

Our next step is to let $t = s/(r + s)$ to convert the integral over $r \in [1 - s, \infty)$ into one over $t \in [0, s]$. $dt = -s/(r + s)^2\, dr$, so to make the calculations easier, we first rewrite the integral as

$$= \frac{1}{\pi} \int_{1-s}^\infty \left(\frac{(r + s)^2}{rs}\right)^{1/2} \frac{s}{(r + s)^2}\, dr$$

and then change variables to get

$$P_0(L \le s) = \frac{1}{\pi} \int_0^s (t(1 - t))^{-1/2}\, dt = \frac{2}{\pi} \arcsin(\sqrt{s}). \tag{4.2}$$

The arcsin may remind the reader of the limit theorem for $L_{2n} = \sup\{m \le 2n : S_m = 0\}$ given in (3.6) of Chapter 3. We will see in Section 6 that our new result is a consequence of the old one.

The computation for R is much easier and is left to the reader.

Exercise 4.1   Show that the probability density $P(R = 1 + t) = 1/(\pi t^{1/2}(1 + t))$.

Exercise 4.2   Let $\mathcal{Z}(\omega) = \{t : B_t = 0\}$. Using the scaling relation (1.2) and some trivial trigonometry, the result in (4.2) can be rewritten as

$$P_0(\mathscr{Z} \cap [t_1, t_2] \neq \varnothing) = \frac{2}{\pi} \arccos(\sqrt{t_1/t_2}).$$

Let $I_m = [m2^{-n}, (m + 1)2^{-n}]$. Show that $E|\{m \leq 2^n : I_m \cap \mathscr{Z} \neq \varnothing\}| \sim c2^{n/2}$ and use this to conclude that if $\alpha > 1/2$, the α-dimensional Hausdorff measure of $\mathscr{Z} \cap [0, 1]$ is 0. See Exercise 2.5 in the Appendix for the definition of Hausdorff measure.

Formula (4.1) begins the fulfillment of a promise we made in Section 7 of Chapter 2. To explain this, we notice that if $0 < a < b$, then

$$T_{b-a} \circ \theta_{T_a} = T_b - T_a,$$

so the strong Markov property implies that:

(4.3) THEOREM   Under P_0, $\{T_a, a \geq 0\}$ has stationary independent increments.

The scaling relation (1.2) implies

$$T_a \overset{d}{=} a^2 T_1. \tag{4.4}$$

Combining (4.3) and (4.4) and using (7.15) from Chapter 2, we see that T_a has a stable law with index $\alpha = 1/2$. Since $T_a \geq 0$, the skewness parameter $\kappa = 1$. For a derivation that does not rely on the fact that (7.13) gives all the stable laws, combine Example 6.6 below with (7.13) in Chapter 2.

For developments below it will be useful to know the Laplace transform $\varphi_a(\lambda) = E_0 \exp(-\lambda T_a)$. It is easy to determine the form of the answer. To do this, we start by observing that (4.3) implies

$$\varphi_x(\lambda)\varphi_y(\lambda) = \varphi_{x+y}(\lambda). \tag{4.5}$$

Since $a \to \varphi_a(\lambda)$ is decreasing, it follows easily from this that

$$\varphi_a(\lambda) = \exp(-ac(\lambda)).$$

(Let $c(\lambda) = -\log \varphi_1(\lambda)$. Use (4.5) with $x = y = 2^{-m}$ to prove the result first for $a = 2^{-m}$, $m \geq 1$, and then let $x = k2^{-m}$ and $y = 2^{-m}$ to get the result for $a = (k + 1)2^{-m}$, $k \geq 1$.) To identify $c(\lambda)$ we observe that (4.4) implies

$$E \exp(-T_a) = E \exp(-a^2 T_1),$$

so $ac(1) = c(a^2)$; that is, $c(\lambda) = c\sqrt{\lambda}$. In the next section (see Exercise 5.4) we will show that $c = \sqrt{2}$, so

$$E_0(\exp(-\lambda T_a)) = \exp(-a\sqrt{2\lambda}). \tag{4.6}$$

Since T_a has a stable law with $\alpha = 1/2$ and $\kappa = 1$, it follows from the proof of (7.7) in Chapter 2 that T_a is the sum of the points in a Poisson process on $(0, \infty)$

with mean measure $Cal^{-3/2} dl$. (See the end of Section 6 in Chapter 2 for the definition of the Poisson process.) To verify this directly and identify the constant, let N_a be the Poisson process.

$$E\left(\exp\left(-\lambda\int_0^\infty l\,N_a(dl)\right)\right) = \exp\left(-a\int_0^\infty (1 - e^{-\lambda l})Cl^{-3/2}\,dl\right).$$

Integrating by parts and then changing variables $x = \lambda l$,

$$\int_0^\infty (1 - e^{-\lambda l})l^{-3/2}\,dl = \int_0^\infty \lambda e^{-\lambda l}2l^{-1/2}\,dl$$

$$= 2\lambda^{1/2}\int_0^\infty e^{-x}x^{-1/2}\,dx = 2\Gamma(1/2)\lambda^{1/2}.$$

Comparing this with (4.6) shows that if $C = (2\pi)^{-1/2}$, then

$$E\left(\exp\left(-\lambda\int_0^\infty lN_a(dl)\right)\right) = E_0(\exp(-\lambda T_a)).$$

Looking at the graph in Figure 7.2 sideways, we see that the jumps in $a \to T_a$ correspond to flat stretches in the graph of $M_t = \max_{s\le t} B_s$, and $N_a([x, \infty))$ gives the distribution of the number of flat stretches of size $\ge x$ by time T_a.

Our next topic is the joint distribution of B_t and $M_t = \max_{s\le t} B_s$. Writing $P_0(B_t = x)$ for the probability density, (3.10) becomes

$$P_0(M_t > a, B_t = b) = P_0(B_t = 2a - b) = (2\pi t)^{-1/2}e^{-(2a-b)^2/2t}.$$

Differentiating w.r.t. a,

$$P_0(M_t = a, B_t = b) = (2/\pi t^3)^{1/2}(2a - b)e^{-(2a-b)^2/2t}.$$

Setting $b = a - x$ and integrating da,

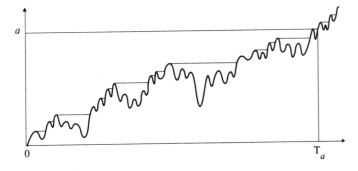

Figure 7.2

$$P_0(M_t - B_t = x) = \int_0^\infty (2/\pi t)^{1/2} \left(\frac{a+x}{t} \right) e^{-(a+x)^2/2t}\, da = (2/\pi t)^{1/2} e^{-x^2/2t}.$$

The last equality shows $M_t - B_t$ has the same distribution as $|B_t|$. A little more thought improves the last result to:

(4.7) THEOREM   Let $A_s = M_s - B_s$. Under P_0, $\{A_s: s \geq 0\}$ and $\{|B_s|: s \geq 0\}$ have the same distribution. (A is for absolute value.)

Exercise 4.3   Prove (4.7) by generalizing the proof of $M_t - B_t \overset{d}{=} |B_t|$ to demonstrate

$$P(A_t = y|\mathscr{F}_s) = p_{t-s}(A_s, y) + p_{t-s}(A_s, -y)$$

and then using this to conclude that the finite-dimensional distributions are the same.

The reason for interest in (4.7) centers on the fact that M_s increases only when A_s is 0. From results in Example 3.1, $\mathscr{Z}(\omega) = \{s: A_s = 0\}$ is a set of measure 0, so $t \to M_t$ is the distribution function of a singular measure supported on $\mathscr{Z}(\omega)$. Noticing

$$M_t - M_s \leq \sup_{s \leq r \leq t} |B_r - B_s|$$

and using our estimate of the modulus of continuity, (3.11) leads easily to an estimate of the Hausdorff dimension of $\mathscr{Z}$.

Exercise 4.4   Let $\mathscr{Z}_t(\omega) = \mathscr{Z}(\omega) \cap [0, t]$. If $\alpha < 1/2$, then with probability 1, the α-dimensional Hausdorff measure $h_\alpha(\mathscr{Z}_t) = \infty$. Combining this with Exercise 4.2 shows that $\mathscr{Z}(\omega)$ has Hausdorff dimension 1/2. See Exercise 2.6 in the Appendix for the definition of Hausdorff dimension.

Exercise 4.5   Compute $\lim_{\varepsilon \to 0} P_0(\mathscr{Z} \cap [\varepsilon, t] = \varnothing | \mathscr{Z} \cap [\varepsilon, s] = \varnothing)$.

Hint   The answer is related to the mean measure of the Poisson process N_a introduced in the discussion after (4.6).

Lévy (1948) proved (4.7). His reason for interest in this was that M_t provides a nontrivial measurement of the amount of time A_s was 0 in the interval $[0, t]$, called the *local time* at 0. There are many ways of defining local time directly for Brownian motion. The easiest to state is $L_t = h_{1/2}(\mathscr{Z}_t)$ where $\mathscr{Z}_t = \mathscr{Z}(\omega) \cap [0, t]$ and $h_{1/2}$ is the 1/2-dimensional Hausdorff measure. The next three exercises give others. We leave it to the interested reader to show that our five definitions are the same (up to constant multiples).

Exercise 4.6   *Downcrossings.* Let $T_x = \inf\{t: |B_t| = x\}$ and define $R_\varepsilon^k, S_\varepsilon^k$ by $S_\varepsilon^0 = T_\varepsilon$, $R_\varepsilon^k = \inf\{t > S_\varepsilon^{k-1} : |B_t| = 0\}$, and $S_\varepsilon^k = \inf\{t > R_\varepsilon^k : |B_t| = \varepsilon\}$. $N(\varepsilon) = \sup\{k: R_\varepsilon^k < T_1\}$ is the number of downcrossings completed by time T_1. Let $Z_n = N(2^{-n})$,

$X_n = N(2^{-n})/2^n$, and suppose $B_0 = 0$. Show that X_n is an L^2 bounded martingale and hence converges a.s. to a limit X_∞.

Hint Z_n is a branching process.

Exercise 4.7   *Occupation times.* Use the last exercise to show that

$$(1/\varepsilon) \int_0^{T_1} 1_{(|B_s| < \varepsilon)} \, ds \to cX_\infty \quad \text{in probability.}$$

Hint   The amounts of time spent in $[0, \varepsilon)$ during $[S_\varepsilon^{k-1}, S_\varepsilon^k)$ are i.i.d. Compute the mean and variance.

By working harder H. Trotter (1958) showed that

$$(1/\varepsilon) \int_0^t 1_{(|B_s - x| < \varepsilon)} \, ds \to L_t(x) \quad \text{a.s.}$$

and $L_t(x)$ is jointly continuous in t and x. For a simple proof, see McKean (1969).

Exercise 4.8   *An intrinsic definition.* For the last definition we return to the representation $A_t = M_t - B_t$ and let $N_a([\delta, \infty))$ be the Poisson process introduced in the discussion after (4.6). Show that $\delta^{1/2} N_a([\delta, \infty)) \to a(2/\pi)^{1/2}$ in probability and use this to get yet another definition of local time:

$$\delta^{1/2}(\# \text{ of components of } [0, t] - \mathscr{Z} \text{ with length} \geq \delta) \to M_t \text{ in probability.}$$

This definition is called intrinsic, since it looks at only the zero set.

5   Martingales

At the end of Section 7 in Chapter 4 we used martingales to study the hitting times of random walks. The same methods can be used on Brownian motion once we prove:

(5.1) THEOREM   Let X_t be a right continuous martingale adapted to a right continuous filtration. If T is a bounded stopping time, then $EX_T = EX_0$.

Proof   Let n be an integer so that $P(T \leq n - 1) = 1$. As in the proof of the strong Markov property, let $T_m = ([2^m T] + 1)/2^m$. $Y_k^m = X(k2^{-m})$ is a martingale with respect to $\mathscr{F}_k^m = \mathscr{F}(k2^{-m})$ and $S_m = 2^m T_m$ is a stopping time for $(Y_k^m, \mathscr{F}_k^m)$, so

$$X(T_m) = E(X_n | \mathscr{F}(T_m)).$$

As $m \uparrow \infty$, $X(T_m) \to X(T)$ by right continuity and $\mathscr{F}(T_m) \downarrow \mathscr{F}(T)$ by (3.6), so it follows from (7.3) in Chapter 4 that

$$X(T) = E(X_n | \mathcal{F}(T)).$$

Taking expected values now gives $EX(T) = EX_n = EX_0$, since X_n is a martingale.

$\square$

(5.2) THEOREM B_t is a martingale w.r.t. the σ-fields $\mathcal{F}_t$ defined in Section 3.

Note   We will use these σ-fields in (5.4), (5.6), and (5.8) but will not mention them explicitly in the statements.

Proof   The Markov property implies that

$$E_x(B_t | \mathcal{F}_s) = E_{B(s)}(B_{t-s}) = B_s,$$

since the symmetry of the normal distribution implies $E_y B_u = y$ for all $u \geq 0$. $\square$

From (5.2) it follows immediately that we have:

(5.3) THEOREM   If $a < x < b$, then $P_x(T_a < T_b) = (b - x)/(b - a)$.

We have already done this argument for simple random walk ((1.7) in Chapter 3), so we leave the details to the reader and then go off on a tangent.

Exercise 5.1   Prove (5.3).

Exercise 5.2   *Optimal doubling in backgammon* (Keeler and Spencer (1975)). In our idealization, backgammon is a Brownian motion starting at $1/2$ run until it hits 1 or 0, and B_t is the probability you will win given the events up to time t. Initially the "doubling cube" sits in the middle of the board and either player can "double" — that is, tell the other player to play on for twice the stakes, or give up and pay the current wager. If a player accepts the double (i.e., decides to play on), she gets possession of the doubling cube and is the only one who can offer the next double. A doubling strategy is given by two numbers $b < 1/2 < a$; that is, offer a double when $B_t \geq a$ and give up if the other player doubles and $B_t < b$. It is not hard to see that for the optimal strategy $b^* = 1 - a^*$ and that when $B_t = b^*$, accepting and giving up must have the same payoff. Use the last two facts and (5.3) to conclude $a^* = 4/5$.

(5.4) THEOREM $B_t^2 - t$ is a martingale.

Proof

$$E_x(B_t^2 | \mathcal{F}_s) = E_x(B_s^2 + 2B_s(B_t - B_s) + (B_t - B_s)^2 | \mathcal{F}_s)$$

$$= B_s^2 + 2B_s E_x(B_t - B_s | \mathcal{F}_s) + E_x((B_t - B_s)^2 | \mathcal{F}_s)$$

$$= B_s^2 + 0 + (t - s),$$

since $B_t - B_s$ is independent of $\mathcal{F}_s$ and has mean 0 and variance $t - s$. $\square$

(5.5) THEOREM   Let $T = \inf\{t: B_t \notin (a, b)\}$ where $a < 0 < b$. $E_0 T = -ab$.

Proof   (5.1) and (5.4) imply $E_0(B^2(T \wedge t)) = E(T \wedge t)$. Letting $t \to \infty$ and using the monotone and bounded convergence theorems give $E(T \wedge t) \uparrow ET$ and

$$EB^2(T \wedge t) \to EB_T^2 = a^2\frac{b}{b-a} + b^2\frac{-a}{b-a} = ab\frac{a-b}{b-a} = -ab. \qquad \square$$

(5.6) THEOREM $\exp(\theta B_t - \theta^2 t/2)$ is a martingale.

Proof $E_x(\exp(\theta B_t)|\mathcal{F}_s) = \exp(\theta B_s)E(\exp(\theta(B_t - B_s))|\mathcal{F}_s) = \exp(\theta B_s)\exp(\theta^2(t - s)/2)$, since $B_t - B_s$ is independent of $\mathcal{F}_s$ and has a normal distribution with mean 0 and variance $t - s$. $\qquad \square$

(5.7) THEOREM   Let $T = \inf\{t: B_t \notin (-a, a)\}$. $E_0 \exp(-\lambda T) = 1/\cosh(a\sqrt{2\lambda})$.

Proof   (5.1) and (5.6) imply that $1 = E_0 \exp(\theta B(T \wedge t) - \theta^2(T \wedge t))$. Letting $t \to \infty$ and using the bounded convergence theorem give

$$1 = E_0 \exp(\theta B_T - \theta^2 T/2).$$

Symmetry implies that $P(B_T = a) = P(B_T = -a) = 1/2$ and that $B(T)$ is independent of T, so

$$1 = \cosh(\theta a)E_0 \exp(-\theta^2 T/2).$$

Setting $\theta = \sqrt{2\lambda}$ now gives the desired result. $\qquad \square$

Exercise 5.3   Derive (5.7) by showing that $\exp(-\theta^2 t/2)\cosh(\theta B_t)$ is a martingale.

Exercise 5.4   Let $\tau = \inf\{t: B_t = a + bt\}$ where $a > 0$. (i) Use the martingale $\exp(\theta B_t - \theta^2 t/2)$ with $\theta = b + (b^2 + 2\lambda)^{1/2}$ to show

$$E_0 \exp(-\lambda\tau) = \exp(-a\{b + (b^2 + 2\lambda)^{1/2}\}).$$

Setting $b = 0$ gives the formula promised in the last section. (ii) Suppose $b > 0$ and use the martingale with $\theta = 2b$ to conclude

$$P_0(\tau < \infty) = \exp(-2ab).$$

Remark   The last exercise gives information about the amount of time it takes Brownian motion with drift $-b$, $X_t \equiv B_t - bt$, to hit level a.

Exercise 5.5   Let $\sigma = \inf\{t: B_t \notin (a, b)\}$ and let $\lambda > 0$. From the strong Markov property it follows that

$$E_x \exp(-\lambda T_a) = E_x(e^{-\lambda\sigma}; T_a < T_b) + E_x(e^{-\lambda\sigma}; T_b < T_a)E_b \exp(-\lambda T_a).$$

Interchange the roles of a and b to get a second equation and solve to get

$$E_x(e^{-\lambda T}; T_a < T_b) = \sinh(\sqrt{2\lambda}(b - x))/\sinh(\sqrt{2\lambda}(b - a))$$

$$E_x(e^{-\lambda T}; T_b < T_a) = \sinh(\sqrt{2\lambda}(x - a))/\sinh(\sqrt{2\lambda}(b - a)).$$

(5.8) THEOREM $B_t^3 - 3tB_t, B_t^4 - 6tB_t^2 + t^2, \ldots$ are martingales.

Proof   The result in (5.6) can be written as

$$E(\exp(\theta B_t - \theta^2 t/2); A) = \exp(\theta B_s - \theta^2 s/2) \quad \text{for } A \in \mathscr{F}_s.$$

Taking k derivatives of $f(\theta) = \exp(\theta x - \theta^2 t/2)$ with respect to θ and setting $\theta = 0$ then will give a function $h_k(\theta)$ so that $h_k(B_t)$ is a martingale. The first two we have seen before:

k	$h_k(\theta)$	Martingale
1	$(x - \theta t)f(\theta)$	B_t
2	$\{(x - \theta t)^2 - t\}f(\theta)$	$B_t^2 - t$
3	$\{(x - \theta t)^3 - 3t(x - \theta t)\}f(\theta)$	$B_t^3 - 3tB_t$
4	$\{(x - \theta t)^4 - 6t(x - \theta t)^2 + 3t^2\}f(\theta)$	$B_t^4 - 6tB_t^2 + 3t^2$

$\square$

(5.9) THEOREM   Let $T = \inf\{t: B_t \notin (-a, a)\}$. $ET^2 = 5a^4/3$.

Proof   (5.1) and (5.8) imply $E(B(T \wedge t)^4 - 6(T \wedge t)B(T \wedge t)^2) = -3E(T \wedge t)^2$. From (5.5) we know that $ET = a^2 < \infty$. Letting $t \to \infty$ and using the dominated convergence theorem on the left-hand side and the monotone convergence theorem on the right give

$$a^4 - 6a^2 ET = -3E(T^2).$$

Plugging in $ET = a^2$ gives the desired result. $\square$

Exercise 5.6   If $U = \inf\{t: B_t \notin (a, b)\}$ where $a < 0 < b$ and $a \neq -b$, then U and B_U^2 are not independent, so we cannot calculate EU^2 as we did in the proof of (5.9). Use the Cauchy–Schwarz inequality to estimate $E(UB_U^2)$ and conclude $EU^2 \leq CE(B_U^4)$ where C is independent of a and b.

Exercise 5.7   Let $u(t, x)$ be a function that satisfies

$$\frac{\partial u}{\partial t} = \frac{1}{2}\frac{\partial^2 u}{\partial x^2} \quad \text{and} \quad \left|\frac{\partial^2 u}{\partial x^2}(t, x)\right| \leq C_T \exp(x^2/(t + \varepsilon)) \quad \text{for } t \leq T. \tag{$*$}$$

Show that $u(t, B_t)$ is a martingale by checking

$$\frac{\partial}{\partial t}p_t(x, y) = \frac{1}{2}\frac{\partial^2}{\partial y^2}p_t(x, y),$$

interchanging $\partial/\partial t$ and $\int$, and then integrating by parts twice to show

$$\frac{\partial}{\partial t} E_x u(t, B_t) = \int \frac{\partial}{\partial t}(p_t(x, y)u(t, y)) \, dy = 0.$$

Examples of functions that satisfy (*) are $\exp(\theta x - \theta^2 t/2)$, x, $x^2 - t$, ...

Exercise 5.8 Find a martingale of the form $B_t^6 - atB_t^4 + bt^2 B_t^2 - ct^3$ and use it to compute the third moment of $T = \inf\{t: B_t \notin (-a, a)\}$.

Note You can differentiate $\exp(\theta x - \theta^2 t/2)$ six times, but it is easier to get a, b, and c from the last exercise.

Exercise 5.9 Show that $(a + t)^{-1/2} \exp(B_t^2/(a + t))$ is a martingale for any $a > 0$ and use this to conclude that $B_t/(t \log t)^{1/2} \to 0$ a.s. as $t \to \infty$.

6 Skorokhod's Representation, Donsker's Theorem

Let X_1, X_2, ... be i.i.d. with $EX = 0$ and $EX^2 = 1$, and let $S_n = X_1 + \cdots + X_n$. In this section we will show that as $n \to \infty$, $S(nt)/n^{1/2}$, $0 \le t \le 1$, converges in distribution to B_t, a Brownian motion starting from $B_0 = 0$. We will say precisely what the last sentence means below. The key to its proof is:

(6.1) Skorokhod's Representation Theorem If $EX = 0$ and $EX^2 < \infty$, then there is a stopping time T for Brownian motion so that $B_T \overset{d}{=} X$.

Remark The Brownian motion in the statement and all the Brownian motions in this section have $B_0 = 0$.

Proof Suppose first that X is supported on $\{a, b\}$ where $a < 0 < b$. Since $EX = 0$, we must have

$$P(X = a) = \frac{b}{b - a}, \qquad P(X = b) = \frac{-a}{b - a}.$$

If we let $T_{a,b} = \inf\{t: B_t \notin (a, b)\}$, then (5.3) implies $B_T \overset{d}{=} X$ and (5.5) tells us that

$$ET = -ab = EB_T^2.$$

To treat the general case we will write $F(x) = P(X \le x)$ as a mixture of two point distributions with mean 0. Let

$$c = \int_{-\infty}^{0} (-u) \, dF(u) = \int_{0}^{\infty} v \, dF(v).$$

If φ is bounded and $\varphi(0) = 0$, then

$$c \int \varphi(x)\, dF(x) = \left(\int_0^\infty \varphi(v)\, dF(v)\right) \int_{-\infty}^0 (-u)\, dF(u) + \left(\int_{-\infty}^0 \varphi(u)\, dF(u)\right) \int_0^\infty v\, dF(v)$$

$$= \int_0^\infty dF(v) \int_{-\infty}^0 dF(u)(v\varphi(u) - u\varphi(v)).$$

So

$$\int \varphi(x)\, dF(x) = c^{-1} \int_0^\infty dF(v) \int_{-\infty}^0 dF(u)(v - u)\left\{\frac{v}{v-u}\varphi(u) + \frac{-u}{v-u}\varphi(v)\right\}.$$

The last equation gives the desired mixture. If we let $(U, V) \in \mathbb{R}^2$ have

$$P((U, V) = (0, 0)) = F(\{0\})$$

$$P((U, V) \in A) = c^{-1} \iint_{(u,v)\in A} dF(u)\, dF(v)(v - u)$$

(6.2)

and define probability measures by $\mu_{0,0}(\{0\}) = 1$ and

$$\mu_{u,v}(\{u\}) = \frac{v}{v-u}, \qquad \mu_{u,v}(\{v\}) = \frac{-u}{v-u} \quad \text{for } u < 0 < v,$$

then

$$\int \varphi(x)\, dF(x) = E\left(\int \varphi(x)\mu_{U,V}(dx)\right).$$

We proved the last formula when $\varphi(0) = 0$, but it is easy to see that it is true in general. Letting $\varphi \equiv 1$ in the last equation shows that the measure defined in (6.2) has total mass 1.   □

From the calculations above it follows that if we have (U, V) with distribution given in (6.2) and an independent Brownian motion defined on the same space, then $B(T_{U,V}) \overset{d}{=} X$. Sticklers for detail will notice that $T_{U,V}$ is not a stopping time for B_t, since (U, V) is independent of the Brownian motion. This is not a serious problem, since if we condition on $U = u$ and $V = v$, then $T_{u,v}$ is a stopping time and this is good enough for all the calculations below. For instance, to compute $E(T_{U,V})$ we observe

$$E(T_{U,V}) = E\{E(T_{U,V}|(U, V))\} = E(-UV)$$

by (5.5). (6.2) implies

$$E(-UV) = \int_{-\infty}^{0} dF(u)(-u) \int_{0}^{\infty} dF(v)v(v - u)c^{-1}$$

$$= \int_{-\infty}^{0} dF(u)(-u)\left\{-u + \int_{0}^{\infty} dF(v)c^{-1}v^2\right\},$$

since

$$c = \int_{0}^{\infty} v \, dF(v) = \int_{-\infty}^{0} (-u) \, dF(u).$$

Using the second expression for c now gives

$$E(T_{U,V}) = \int_{-\infty}^{0} u^2 \, dF(u) + \int_{0}^{\infty} v^2 \, dF(v) = EX^2.$$

Exercise 6.1   Use Exercise 5.6 to conclude that $E(T_{U,V}^2) \le CEX^4$.

Exercise 6.2   Dubins (1968). Let X be a random variable with $EX = 0$ and suppose for simplicity that X has a continuous distribution. Define constants inductively by $a_0^0 = 0$, $a_{-1}^0 = -\infty$, $a_1^0 = \infty$, and for $k \ge 0$,

$$a_{2j}^{k+1} = a_j^k \quad \text{for } |j| \le 2^{k-1}$$

$$a_{2j-1}^{k+1} = E(X | a_{j-1}^k < X < a_j^k) \quad \text{for } -2^{k-1} < j \le 2^{k-1}.$$

To check the indexing, note

$$a_{-2}^1 = a_{-1}^0 = -\infty, \qquad a_0^1 = a_0^0 = 0, \qquad a_2^1 = a_1^0 = \infty$$

$$a_{-1}^1 = E(X | X < 0), \qquad a_1^1 = E(X | X > 0).$$

Let $T_1 = \inf\{t: B_t \notin (a_{-1}^1, a_1^1)\}$ and if $B(T_k) = a_j^k$, let

$$T_{k+1} = \inf\{t > T_k : B_t \notin (a_{2j-1}^{k+1}, a_{2j+1}^{k+1})\}.$$

Show that (i) $B(T_k) \overset{d}{=} E(X | \mathscr{G}_k)$ where $\mathscr{G}_k = \sigma(\{a_{j-1}^k < X < a_j^k\})$ and (ii) $T_k \uparrow T < \infty$ a.s. and conclude that $B(T) \overset{d}{=} X$.

Hint for (ii)   (i) implies $E|B(T_k)| \le E|X| < \infty$, so $B(T_k)$ converges a.s. To see $P(T_k \to \infty) = 0$ note that when $B(T_1) > 0$, $B_t > 0$ for all $t \in [T_1, T)$ and recall the result in Exercise 2.6.

Exercise 6.3   Suppose X takes on the values 2, 1, -1, and -2 with probability 1/4 each. Let $T_1 = \inf\{t: B_t \notin (-3/2, 3/2)\}$ and $\alpha = \inf\{t > T_1 : |B_t - B(T_1)| > 1/2\}$. α is Dubins' stopping time for X. Skorokhod's approach is to let $(U, V) = (-1, 1)$ or $(-2, 2)$ with probability 1/2 each and let $\beta = \inf\{t: B_t \notin (U, V)\}$. Show that $E\alpha = E\beta = EX^2$ but $E\alpha^2 < E\beta^2$.

For other approaches to embedding without randomization, see Root (1969) or Sheu (1986). From (6.1) it is only a small step to:

(6.3) THEOREM   Let $X_1, X_2, \ldots$ be i.i.d. with a distribution F that has mean 0 and variance 1, and let $S_n = (X_1 + \cdots + X_n)$. There is a sequence of stopping times $T_0 = 0, T_1, T_2, \ldots$ such that $S_n = B(T_n)$ and $T_n - T_{n-1}$ are independent and identically distributed.

Proof   Let $(U_1, V_1), (U_2, V_2), \ldots$ be i.i.d. and have distribution given in (6.2) and let B_t be an independent Brownian motion. Let $T_0 = 0$, and for $n \geq 1$ let

$$T_n = \inf\{t \geq T_{n-1} : B_t - B(T_{n-1}) \notin (U_n, V_n)\}. \qquad \square$$

As a corollary of (6.3) we get:

(6.4) *Central Limit Theorem*   Under the hypotheses of (6.3), $S_n/\sqrt{n} \Rightarrow \chi$, where χ has the standard normal distribution.

Proof   If we let $W_n(t) = B(nt)/\sqrt{n} \overset{d}{=} B_t$ by Brownian scaling, then

$$S_n/\sqrt{n} \overset{d}{=} B(T_n)/\sqrt{n} = W_n(T_n/n).$$

Let $\varepsilon > 0$ and pick δ so that

$$P(|B_t - B_1| > \varepsilon \text{ for some } t \in (1 - \delta, 1 + \delta)) < \varepsilon/2.$$

The strong law of large numbers implies that $T_n/n \to 1$ a.s., so we can pick N large enough so that for $n \geq N$, $P(|T_n/n - 1| > \delta) < \varepsilon/2$. The last two estimates imply that for $n \geq N$,

$$P(|W_n(T_n/n) - W_n(1)| > \varepsilon) < \varepsilon,$$

so if $F_n(x) = P(S_n/\sqrt{n} < x) = P(W_n(T_n/n) \leq x)$ and $\mathcal{N}(x) = P(W_n(1) \leq x)$, then

$$\mathcal{N}(x - \varepsilon) - \varepsilon \leq F_n(x) \leq \mathcal{N}(x + \varepsilon) + \varepsilon \quad \text{for } n \geq N.$$

Since ε is arbitrary, the result follows. $\square$

Our next goal is to prove a strengthening of (6.4) that allows us to obtain limit theorems for functionals of $\{S_m : 0 \leq m \leq n\}$—for example, $\max_{0 \leq m \leq n} S_m$ or $|\{m \leq n : S_m > 0\}|$. Let $C[0, 1] = \{\text{continuous } \omega : [0, 1] \to \mathbb{R}\}$. When equipped with the norm $\|\omega\| = \sup\{|\omega(s)| : s \in [0, 1]\}$, $C[0, 1]$ becomes a complete separable metric space. To fit $C[0, 1]$ into the framework of Section 9 in Chapter 2 we want our measures defined on $\mathcal{B} = $ the σ-field generated by the open sets. Fortunately,

(6.5) LEMMA $\mathcal{B}$ is the same as $\mathcal{C}$, the σ-field generated by the finite-dimensional sets $\{\omega : \omega(t_i) \in A_i\}$.

Proof   Observe that if ξ is a given continuous function,

$$\{\omega: \|\omega - \xi\| \le \varepsilon - 1/n\} = \bigcap_q \{\omega: |\omega(q) - \xi(q)| \le \varepsilon - 1/n\}$$

where the intersection is over all rationals in $[0, 1]$. Letting $n \to \infty$ shows $\{\omega: \|\omega - \xi\| < \varepsilon\} \in \mathscr{C}$ and $\mathscr{B} \subset \mathscr{C}$. To prove the reverse inclusion observe that if the A_i are open, the finite-dimensional set $\{\omega: \omega(t_i) \in A_i\}$ is open, so the $\pi - \lambda$ theorem implies $\mathscr{B} \supset \mathscr{C}$. $\square$

A sequence of probability measures μ_n on $C[0, 1]$ is said to *converge weakly* to a limit μ if for all bounded continuous functions $\varphi: C[0, 1] \to \mathbb{R}, \int \varphi \, d\mu_n \to \int \varphi \, d\mu$. Let

$$S(u) = \begin{cases} S_k & \text{if} \quad u = k \\ \text{linear on } [k, k + 1] & \text{for} \quad k \ge 0. \end{cases}$$

We will prove:

(6.6) Donsker's Theorem $S(n \cdot)/\sqrt{n} \Rightarrow B(\cdot)$; that is, the associated measures on $C[0, 1]$ converge weakly.

To motivate ourselves for the proof we will begin by extracting several corollaries. The key to each one is the following consequence of (9.1) in Chapter 2.

(6.7) THEOREM   If $\psi: C[0, 1] \to \mathbb{R}$ has the property that it is continuous P_0 a.e., then

$$\psi(S(n \cdot)/\sqrt{n}) \Rightarrow \psi(B(\cdot)). \tag{$*$}$$

Example 6.1   Let $\psi(\omega) = \omega(1)$. In this case $\psi: C[0, 1] \to \mathbb{R}$ is continuous and $(*)$ is the central limit theorem.

Example 6.2   Let $\psi(\omega) = \max\{\omega(t): 0 \le t \le 1\}$. Again $\psi: C[0, 1] \to \mathbb{R}$ is continuous. This time $(*)$ says

$$\max_{0 \le m \le n} S_m/\sqrt{n} \Rightarrow M_1 \equiv \max_{0 \le t \le 1} B_t.$$

To complete the picture, we observe that by (3.8) the distribution of the right-hand side is

$$P_0(M_1 \ge a) = P_0(T_a \le 1) = 2P_0(B_1 \ge a);$$

that is, $M_1 \overset{d}{=} |B_1|$.

Exercise 6.4   Suppose S_n is a one-dimensional simple random walk and let

$$R_n = 1 + \max_{m \leq n} S_m - \min_{m \leq n} S_m$$

be the number of points visited by time n. Show that $R_n/\sqrt{n} \Rightarrow$ a limit.

Example 6.3   Let $\psi(\omega) = \sup\{t \leq 1 : \omega(t) = 0\}$. This time ψ is not continuous, for if ω_ε has $\omega_\varepsilon(0) = 0$, $\omega_\varepsilon(1/3) = 1$, $\omega_\varepsilon(2/3) = \varepsilon$, $\omega(1) = 2$, and linear on each interval $[j, (j + 1)/3]$, then $\psi(\omega_0) = 2/3$ but $\psi(\omega_\varepsilon) = 0$ for $\varepsilon > 0$. It is easy to see that if $\psi(\omega) < 1$ and $\omega(t)$ has positive and negative values in each interval $(\psi(\omega) - \delta, \psi(\omega))$, then ψ is continuous at ω. By arguments in Example 3.1, the last set has P_0 measure 1. (If the 0 at $\psi(\omega)$ was isolated on the left, it would not be isolated on the right.) It follows that

$$\sup\{m \leq n : S_{m-1} \cdot S_m \leq 0\}/n \Rightarrow L = \sup\{t \leq 1 : B_t = 0\}.$$

The distribution of L is given in (4.2). The last result shows that the arcsine law, (3.6) in Chapter 3, proved for simple random walks holds when the mean is 0 and the variance is finite.

Example 6.4   Let $\psi(\omega) = |\{t \in [0, 1] : \omega(t) > 0\}|$. The point $\omega \equiv 0$ shows that ψ is not continuous, but it is easy to see that ψ is continuous at points with $|\{t \in [0, 1] : \omega(t) = 0\}| = 0$. Fubini's theorem implies that

$$E_0|\{t \in [0, 1] : B_t = 0\}| = \int_0^1 P_0(B_t = 0) \, dt = 0,$$

so ψ is continuous P_0 a.e. Applying (∗) now shows

$$|\{m \leq n : S_m > 0\}|/n \Rightarrow |\{t \in [0, 1] : B_t > 0\}|.$$

To compute the distribution of the right-hand side observe that we proved in Section (3.8) of Chapter 3 that if $S_n \overset{d}{=} -S_n$ and $P(S_m = 0 \text{ for all } m \geq 1) = 0$—for example, the X_i have a symmetric continuous distribution, then the left-hand side converges to the arcsine law, so the right-hand side has that distribution and is the limit for any random walk with mean 0 and finite variance. The last argument uses an idea called the "invariance principle" that originated with Erdös and Kac (1946, 1947): The asymptotic behavior of functionals of S_n should be the same as long as the central limit theorem applies. Our final application is from the original paper of Donsker (1951). Erdös and Kac (1946) give the limit distribution for the case $k = 2$.

Example 6.5   Let $\psi(\omega) = \int_{[0, 1]} \omega(t)^k \, dt$ where $k > 0$ is an integer. ψ is continuous, so applying (6.7) gives

$$n^{-1-(k/2)} \sum_{m=1}^n S_m^k \Rightarrow \int_0^1 B_t^k \, dt.$$

It is remarkable that the last result holds under the assumption that $EX_i = 0$ and $EX_i^2 = 1$; that is, we do not need to assume that $E|X_i^k| < \infty$.

Exercise 6.5   When $k = 1$ the last result says

$$n^{-3/2} \sum_{m=1}^{n} (n + 1 - m)X_m \Rightarrow \int_0^1 B_t \, dt.$$

The right-hand side has a normal distribution with mean 0 and variance 1/2. Deduce this result from the Lindeberg–Feller theorem.

Proof of (6.6)   To simplify the proof and prepare for generalizations in the next section, let $X_{n,m}$, $1 \le m \le n$, be a triangular array of random variables, $S_{n,m} = X_{n,1} + \cdots + X_{n,m}$ and suppose $S_{n,m} = B(\tau_m^n)$. Let

$$S_{n,(u)} = \begin{cases} S_{n,m} & \text{if} \quad u = m \\ \text{linear for } u \in [m, m+1], & m \ge 0. \end{cases} \qquad \square$$

(6.8) LEMMA   If $\tau_{[ns]}^n \to s$ in probability for each $s \in [0, 1]$, then

$$\|S_{n,(n\cdot)} - B(\cdot)\| \to 0 \quad \text{in probability.}$$

To make the connection with the original problem, let $X_{n,m} = X_m/\sqrt{n}$ and define $\tau_1^n, \ldots, \tau_n^n$ so that $(S_{n,1}, \ldots, S_{n,n}) \overset{d}{=} (B(\tau_1^n), \ldots, B(\tau_n^n))$. If $T_1, T_2, \ldots$ are the stopping times defined in (6.3), Brownian scaling implies $\tau_m^n \overset{d}{=} T_m/n$, so the hypothesis of (6.8) is satisfied.

Proof   The fact that B has continuous paths (and hence is uniformly continuous on $[0, 1]$) implies that if $\varepsilon > 0$, then there is a $\delta > 0$ so that $1/\delta$ is an integer and

(a) $P(|B_t - B_s| < \varepsilon \text{ for all } 0 \le s \le 1, |t - s| < 2\delta) > 1 - \varepsilon.$

The hypothesis of (6.8) implies that if $n \ge N_\delta$, then

$$P(|\tau_{[nk\delta]}^n - k\delta| < \delta \text{ for } k = 1, 2, \ldots, 1/\delta) \ge 1 - \varepsilon.$$

Since $m \to \tau_m^n$ is increasing, it follows that if $s \in ((k-1)\delta, k\delta)$,

$$\tau_{[ns]}^n - s \ge \tau_{[n(k-1)\delta]}^n - (k-1)\delta - \delta$$

$$\tau_{[ns]}^n - s \le \tau_{[nk\delta]}^n - k\delta + \delta,$$

so if $n \ge N_\delta$,

(b) $P\left(\sup_{0 \le s \le 1} |\tau_{[ns]}^n - s| < 2\delta \right) \ge 1 - \varepsilon.$

When the events in (a) and (b) occur,

(c) $|S_{n,m} - B_{m/n}| < \varepsilon$ for all $m \leq n$.

To deal with $t = (m + \theta)/n$ with $0 < \theta < 1$ we observe that

$$|S_{n,(nt)} - B_t| \leq (1 - \theta)|S_{n,m} - B_{m/n}| + \theta|S_{n,m+1} - B_{(m+1)/n}|$$
$$+ (1 - \theta)|B_{m/n} - B_t| + \theta|B_{(m+1)/n} - B_t|.$$

Using (c) on the first two terms and (a) on the last two, we see that if $n \geq N_\delta$ and $1/n < 2\delta$, then $\|S_{n,(n\cdot)} - B(\cdot)\| < 2\varepsilon$ with probability $\geq 1 - 2\varepsilon$. Since ε is arbitrary the proof of (6.8) is complete. $\square$

To get (6.6) now we have to show:

(6.9) LEMMA   If φ is bounded and continuous, $E\varphi(S_{n,(n\cdot)}) \to E\varphi(B(\cdot))$.

Proof   For fixed $\varepsilon > 0$ let $G_\delta = \{\omega: \text{if } \|\omega - \omega'\| < \delta, \text{then } |\varphi(\omega) - \varphi(\omega')| < \varepsilon\}$. Since φ is continuous, $G_\delta \uparrow C[0, 1]$ as $\delta \downarrow 0$. Let $\Delta = \|S_{n,(n\cdot)} - B(\cdot)\|$. The desired result now follows from (6.8) and the trivial inequality

$$|E\varphi(S(n\cdot)/\sqrt{n}) - \varphi(B(\cdot))| \leq \varepsilon + (2 \sup|\varphi(\omega)|)\{P(G_\delta^c) + P(\Delta > \delta)\}.$$ $\square$

For a different approach to the end of the proof, do:

Exercise 6.6   Let X_n, $1 \leq n \leq \infty$, be a sequence of r.v.'s taking values in a metric space (S, ρ). Show that $X_n \Rightarrow X_\infty$ if and only if $Ef(X_n) \to Ef(X_\infty)$ for all bounded functions that are uniformly continuous (i.e., for each $\varepsilon > 0$ there is a $\delta > 0$ so that $\rho(x, y) < \delta$ implies $|f(x) - f(y)| < \varepsilon$).

Hint   To prove the nontrivial direction use the proof of (9.1) in Chapter 2 and observe that if K is closed, $(1 - k\rho(x, K))^+$ is uniformly continuous.

Exercise 6.7   Use the last exercise to prove a converging together result for r.v.'s taking values in a metric space. If $X_n \Rightarrow X_\infty$ and $\rho(X_n, Y_n) \to 0$ in probability, then $Y_n \Rightarrow X_\infty$.

To accommodate our final example we need a trivial generalization of (6.6). Let $C[0, \infty) = \{\text{continuous } \omega: [0, \infty) \to \mathbb{R}\}$ and let $\mathscr{C}[0, \infty)$ be the σ-field generated by the finite-dimensional sets. Given a probability measure μ on $C[0, \infty)$, there is a corresponding measure $\pi_M\mu$ on $C[0, M] = \{\text{continuous } \omega: [0, M] \to \mathbb{R}\}$ (with $\mathscr{C}[0, M]$ being the σ-field generated by the finite-dimensional sets) obtained by "cutting off the paths at time M." Let $(\psi_M\omega)(t) = \omega(t)$ for $t \in [0, M]$ and let $\pi_M\mu = \mu \circ \psi_M^{-1}$. We say that a sequence of probability measures μ_n on $C[0, \infty)$ converges weakly to μ if for all M, $\pi_M\mu_n$ converges weakly to $\pi_M\mu$ on $C[0, M]$, the last concept being defined by a trivial extension of the definitions for $M = 1$. With these definitions it is easy to conclude:

(6.10) THEOREM $S(n\cdot)/\sqrt{n} \Rightarrow B(\cdot)$; that is, the associated measures on $C[0, \infty)$ converge weakly.

Proof   By definition all we have to show is that weak convergence occurs on $C[0, M]$ for all $M < \infty$. The proof of (6.6) works in the same way when 1 is replaced by M.   □

Example 6.6   Let $N_n = \inf\{m: S_m \ge \sqrt{n}\}$ and $T_1 = \inf\{t: B_t = 1\}$. Since $\psi(\omega) = T_1(\omega) \wedge 1$ is continuous P_0 a.e. on $C[0, 1]$ and the distribution of T_1 is continuous, it follows from (6.7) that for $0 < t < 1$,

$$P(N_n \le nt) \to P(T_1 \le t).$$

Repeating the last argument with 1 replaced by M and using (6.10) show that the last conclusion holds for all t.

*7 CLT's for Dependent Variables

In this section we will prove central limit theorems for some dependent sequences. First, by embedding martingales in Brownian motion we will prove a Lindeberg–Feller theorem for martingales, (7.3). Then using an idea of Gordin (1969) we will use the result for martingales to get a CLT for stationary sequences, (7.6). The condition in (7.6) may look difficult to check, but we show that it is implied by the usual mixing conditions.

(7.1) THEOREM   If S_n is a martingale with $S_0 = 0$ and B_t is a Brownian motion, there is a sequence of stopping times $0 = T_0 \le T_1 \le T_2 \le \cdots$ for the Brownian motion, so that

$$(S_0, S_1, \ldots, S_k) \overset{d}{=} (B(T_0), B(T_1), \ldots, B(T_k)) \quad \text{for all } k \ge 0.$$

Remark   This is due to Strassen (1967); see Theorem 4.3.

Proof   We include $S_0 = 0 = B(T_0)$ only for the sake of starting the induction argument. Suppose we have $(S_0, \ldots, S_{k-1}) \overset{d}{=} (B(T_0), \ldots, B(T_{k-1}))$ for some $k \ge 1$. The strong Markov property implies that $\{B(T_{k-1} + t) - B(T_{k-1}): t \ge 0\}$ is a Brownian motion that is independent of $\mathscr{F}(T_{k-1})$. Let $\mu_k(s_0, \ldots, s_{k-1}; \cdot)$ be a regular conditional distribution of $S_k - S_{k-1}$ given $S_j = s_j, 0 \le j \le k$; that is, for each Borel set A,

$$P(S_k - S_{k-1} \in A | S_0, \ldots, S_{k-1}) = \mu_k(s_0, \ldots, s_{k-1}; A).$$

By results at the end of Section 1 of Chapter 4, this exists and we have

$$0 = E(S_k - S_{k-1} | S_0, \ldots, S_{k-1}) = \int x \mu_k(s_0, \ldots, s_{k-1}; dx),$$

so the mean of the conditional distribution is 0 almost surely. Using (6.1) now we see that for almost every $\vec{S} \equiv (S_0, \ldots, S_{k-1})$ there is a stopping time $\tau_{\vec{S}}$ (for $\{B(T_{k-1} + t) - B(T_{k-1}) : t \geq 0\}$) so that

$$B(T_{k-1} + \tau_{\vec{S}}) - B(T_{k-1}) \overset{d}{=} \mu_k(S_1, \ldots, S_{k-1}; \cdot).$$

If we let $T_k = T_{k-1} + \tau_{\vec{S}}$, then $(S_0, \ldots, S_k) \overset{d}{=} (B(T_0), \ldots, B(T_k))$ and the result follows by induction.   □

Remark   While the details of the proof are fresh in the reader's mind, we would like to observe that if $E(S_k - S_{k-1})^2 < \infty$, then

$$E(\tau_{\vec{S}} | S_0, \ldots, S_{k-1}) = \int x^2 \mu_k(S_0, \ldots, S_{k-1}; dx),$$

since $B_t^2 - t$ is a martingale and $\tau_{\vec{S}}$ is the exit time from a randomly chosen interval $(S_{k-1} + U_k, S_{k-1} + V_k)$.

Our first step toward the promised Lindeberg–Feller theorem is to prove a result of Freedman (1971a). We say that $X_{n,m}, \mathcal{F}_{n,m}, 1 \leq m \leq n$, is a *martingale difference array* if $X_{n,m} \in \mathcal{F}_{n,m}$ and $E(X_{n,m} | \mathcal{F}_{n,m-1}) = 0$ for $1 \leq m \leq n$. (We set $\mathcal{F}_{n,0} = \{\varnothing, \Omega\}$.)

(7.2) THEOREM   Suppose $\{X_{n,m}, \mathcal{F}_{n,m}\}$ is a martingale difference array. Let $S_{n,m} = X_{n,1} + \cdots + X_{n,m}$ and $V_{n,k} = \sum_{1 \leq m \leq k} E(X_{n,m}^2 | \mathcal{F}_{n,m-1})$. If (i) $|X_{n,m}| \leq \varepsilon_n$ for all m with $\varepsilon_n \to 0$ and (ii) for each t, $V_{n,[nt]} \to t$ in probability, then $S_{n,(n\cdot)} \Rightarrow B(\cdot)$

Here and throughout the section $S_{n,(n\cdot)}$ denotes the linear interpolation of $S_{n,m}$ defined by

$$S_{n,(u)} = \begin{cases} S_{n,k} & \text{if} \quad u = k \\ \text{linear for } u \in [k, k+1] & k \geq 0 \end{cases}$$

and $B(\cdot)$ is a Brownian motion with $B_0 = 0$.

Proof   (ii) implies $V_{n,n} \to 1$ in probability. By stopping each sequence at the first time $V_{n,k} > 2$ and setting the later $X_{n,m} = 0$, we can suppose without loss of generality that $V_{n,n} \leq 2 + \varepsilon_n^2$ for all n. By (7.1) we can find stopping times $T_{n,1}, \ldots, T_{n,n}$ so that $(S_{n,1}, \ldots, S_{n,n}) \overset{d}{=} (B(T_{n,1}), \ldots, B(T_{n,n}))$. By (6.8) it suffices to show that $T_{n,[nt]} \to t$ in probability for each $t \in [0, 1]$. To do this we let $t_{n,m} = T_{n,m} - T_{n,m-1}$ (with $T_{n,0} = 0$) and observe that by the remark after the proof of (7.1), $E(t_{n,m} | \mathcal{F}_{n,m-1}) = E(X_{n,m}^2 | \mathcal{F}_{n,m-1})$. The last observation and hypothesis (ii) imply

$$\sum_{m=1}^{[nt]} E(t_{n,m} | \mathcal{F}_{n,m-1}) \to t \quad \text{in probability.}$$

To get from this to $T_{n,[nt]} \to t$ in probability, we observe

$$E\left(\sum_{m=1}^{[nt]} t_{n,m} - E(t_{n,m}|\mathscr{F}_{n,m-1})\right)^2 = E \sum_{m=1}^{[nt]} \{t_{n,m} - E(t_{n,m}|\mathscr{F}_{n,m-1})\}^2$$

by the orthogonality of martingale increments ((4.5) in Chapter 4). Now

$$E(\{t_{n,m} - E(t_{n,m}|\mathscr{F}_{n,m-1})\}^2|\mathscr{F}_{n,m-1}) \le E(t_{n,m}^2|\mathscr{F}_{n,m-1})$$

$$\le CE(X_{n,m}^4|\mathscr{F}_{n,m-1}) \le C\varepsilon_n^2 E(X_{n,m}^2|\mathscr{F}_{n,m-1})$$

by Exercise 5.6 and assumption (i). Taking expected values, summing over n, and recalling we have assumed $V_{n,n} \le 2 + \varepsilon_n^2$, it follows that

$$E\left(\sum_{m=1}^{[nt]} t_{n,m} - E(t_{n,m}|\mathscr{F}_{n,m-1})\right)^2 \le C\varepsilon_n^2 EV_{n,n} \to 0.$$

Unscrambling the definitions, we have shown $E(T_{n,[nt]} - V_{n,[nt]})^2 \to 0$, so Chebyshev's inequality implies $P(|T_{n,[nt]} - V_{n,[nt]}| > \varepsilon) \to 0$, and using (ii) now completes the proof. □

Exercise 7.1   Get rid of assumption (ii) in (7.2) by putting the martingale on its "natural time scale." Let $X_{n,m}$, $\mathscr{F}_{n,m}$, $1 \le m < \infty$, be a martingale difference array with $|X_{n,m}| \le \varepsilon_n$ and $\varepsilon_n \to 0$. Let $V_{n,k} = \sum_{1 \le m \le k} E(X_{n,m}^2|\mathscr{F}_{n,m-1})$. Suppose that $V_{n,m} \to \infty$ as $m \to \infty$. Define $B_n(t)$ by requiring $B_n(V_{n,m}) = S_{n,m}$ and $B_n(t)$ is linear on each interval $[V_{n,m-1}, V_{n,m}]$. Show that $B_n(\cdot) \Rightarrow B(\cdot)$.

With (7.2) established, a truncation argument gives us:

(7.3) Lindeberg–Feller Theorem for Martingales   Suppose $X_{n,m}$, $\mathscr{F}_{n,m}$, $1 \le m \le n$, is a martingale difference array and let $V_{n,k} = \sum_{m \le k} E(X_{n,m}^2|\mathscr{F}_{n,m-1})$. If

(i) $V_{n,[nt]} \to t$ in probability for all $t \in [0, 1]$, and
(ii) $\sum_{m \le n} E(X_{n,m}^2 1_{(|X_{n,m}| > \varepsilon)}|\mathscr{F}_{n,m-1}) \to 0$ in probability,

then $S_{n,(n\cdot)} \Rightarrow B(\cdot)$.

Remark   Literally dozens of papers have been written proving results like this. See Hall and Heyde (1980) for some history. Here we follow Durrett and Resnick (1978).

Proof   The first step is to truncate so that we can apply (7.2). Let

$$\hat{V}_n(\varepsilon) = \sum_{m=1}^n E(X_{n,m}^2 1_{(|X_{n,m}| > \varepsilon_n)}|\mathscr{F}_{n,m-1}).$$

(a) If $\varepsilon_n \to 0$ slowly enough, then $\varepsilon_n^{-2}\hat{V}_n(\varepsilon_n) \to 0$ in probability.

Remark   The ε_n^{-2} is in front so that we can conclude

$$\sum_{m \leq n} P(|X_{n,m}| > \varepsilon_n | \mathscr{F}_{n,m-1}) \to 0 \quad \text{in probability.}$$

Proof of (a)   Let N_m be chosen so that $P(m^2 \hat{V}_n(1/m) > 1/m) \leq 1/m$ for $n \geq N_m$. Let $\varepsilon_n = 1/m$ for $n \in [N_m, N_{m+1})$ and $\varepsilon_n = 1$ if $n < N_1$. If $\delta > 0$ and $1/m < \delta$, then for $n \in [N_m, N_{m+1})$,

$$P(\varepsilon_n^{-2} \hat{V}_n(\varepsilon_n) > \delta) \leq P(m^2 \hat{V}_n(1/m) > 1/m) \leq 1/m. \qquad \square$$

Let

$$\hat{X}_{n,m} = X_{n,m} 1_{(|X_{n,m}| > \varepsilon_n)}$$

$$\bar{X}_{n,m} = X_{n,m} 1_{(|X_{n,m}| \leq \varepsilon_n)}$$

$$\tilde{X}_{n,m} = \bar{X}_{n,m} - E(\bar{X}_{n,m} | \mathscr{F}_{n,m-1}).$$

Our next step is to show:

(b) If we define $\tilde{S}_{n,(n\cdot)}$ in the obvious way, then (7.2) implies $\tilde{S}_{n,(n\cdot)} \Rightarrow B(\cdot)$.

Proof of (b)   Since $|\tilde{X}_{n,m}| \leq 2\varepsilon_n$ we only have to check (ii) in (7.2). To do this, we observe that the conditional variance formula ((4.6) in Chapter 4) implies

$$E(\tilde{X}_{n,m}^2 | \mathscr{F}_{n,m-1}) = E(\bar{X}_{n,m}^2 | \mathscr{F}_{n,m-1}) - E(\bar{X}_{n,m} | \mathscr{F}_{n,m-1})^2.$$

For the first term we observe

$$E(\bar{X}_{n,m}^2 | \mathscr{F}_{n,m-1}) = E(X_{n,m}^2 | \mathscr{F}_{n,m-1}) - E(\hat{X}_{n,m}^2 | \mathscr{F}_{n,m-1}).$$

For the second we observe that $E(X_{n,m} | \mathscr{F}_{n,m-1}) = 0$ implies

$$E(\bar{X}_{n,m} | \mathscr{F}_{n,m-1})^2 = E(\hat{X}_{n,m} | \mathscr{F}_{n,m-1})^2 \leq E(\hat{X}_{n,m}^2 | \mathscr{F}_{n,m-1})$$

by Jensen's inequality, so it follows from (a) and (i) that

$$\sum_{m=1}^{[nt]} E(\tilde{X}_{n,m}^2 | \mathscr{F}_{n,m-1}) \to t \quad \text{for all } t \in [0, 1]. \qquad \square$$

Having proved (b), it remains to estimate the difference between $S_{n,(n\cdot)}$ and $\tilde{S}_{n,(n\cdot)}$. On $\{|X_{n,m}| \leq \varepsilon_n$ for all $1 \leq m \leq n\}$ we have

(c) $\|S_{n,(n\cdot)} - \tilde{S}_{n,(n\cdot)}\| \leq \sum_{m=1}^{n} |E(\bar{X}_{n,m} | \mathscr{F}_{n,m-1})|.$

$$\sum_{n \geq 1} \|E(X_0|\mathscr{F}_{-n})\|_2 < \infty$$

re $\mathscr{F}_m = \sigma(X_k, k \leq m)$ and $\|Y\|_2 = (EY^2)^{1/2}$. Let $S_n = X_1 + \cdots + X_n$.
$/\sqrt{n} \Rightarrow \sigma B(\cdot)$ where

$$\sigma^2 = EX_0^2 + 2\sum_{n=1}^{\infty} EX_0 X_n$$

nd the series in the definition converges absolutely.

Proof Suppose $X_n, n \in \mathbb{Z}$, is defined on sequence space $(\mathbb{R}^{\mathbb{Z}}, \mathscr{R}^{\mathbb{Z}}, P)$ with $X_n(\omega) =$
and let $(\theta^n \omega)(m) = \omega(m + n)$. Let

$$H_n = \{Y \in \mathscr{F}_n \text{ with } EY^2 < \infty\}$$

$$K_n = \{Y \in H_n \text{ with } E(YZ) = 0 \text{ for all } Z \in H_{n-1}\}.$$

Geometrically, $H_0 \supset H_{-1} \supset H_{-2} \supset \cdots$ is a sequence of subspaces of L^2 and K_n is the orthogonal complement of H_{n-1} in H_n. If Y is a random variable, let $(\theta^n Y)(\omega) = Y(\theta^n \omega)$. Generalizing from the example $Y = f(X_{-j}, \ldots, X_k)$, which has $\theta^n Y = f(X_{n-j}, \ldots, X_{n+\kappa})$, it is easy to see that if $Y \in H_k$, then $\theta^n Y \in H_{k+n}$, and hence if $Y \in K_j$, then $\theta^n Y \in K_{n+j}$.

If X_0 happened to be in K_0, then we would be happy, since then $X_n = \theta^n X_0 \in K_n$ for all n, and taking $Z = 1_A \in H_{n-1}$, we would have $E(X_n 1_A) = 0$ for all $A \in \mathscr{F}_{n-1}$ and hence $E(X_n|\mathscr{F}_{n-1}) = 0$. The next best thing to having $X_0 \in K_0$ is to have

$$X_0 = Y_0 + Z_0 - \theta Z_0$$

with $Y_0 \in K_0$ and $Z_0 \in L^2$, for then if we let   (*)

$$S_n = \sum_{m=1}^{n} X_m = \sum_{m=1}^{n} \theta^m X_0 \quad \text{and} \quad T_n = \sum_{m=1}^{n} \theta^m Y_0,$$

then

$$S_n = T_n + \theta Z_0 - \theta^{n+1} Z_0.$$

The $\theta^m Y_0$ are a stationary ergodic martingale difference sequence (ergodicity follows from (1.3) in Chapter 6), so (7.5) implies

$$T_{(n\cdot)}/\sqrt{n} \Rightarrow \sigma B(\cdot) \quad \text{where } \sigma^2 = EY_0^2.$$

To get rid of the other term we observe $\theta Z_0/\sqrt{n} \to 0$ a.s. and

$$P\left(\sup_{1 \leq m \leq n} \theta^{m+1} Z_0 > \varepsilon\sqrt{n}\right) \leq nP(Z_0 > \varepsilon\sqrt{n}) \leq \varepsilon^{-2} E(Z_0^2; Z_0 > \varepsilon\sqrt{n}) \to 0$$

by dominated convergence.

To handle the right-hand side, we observe

(d) $$\sum_{m=1}^{n} |E(\bar{X}_{n,m}|\mathscr{F}_{n,m-1})| = \sum_{m=1}^{n} |E(\hat{X}_{n,m}|\mathscr{F}_{n,m-1})|$$

$$\leq \sum_{m=1}^{n} E(|\hat{X}_{n,m}||\mathscr{F}_{n,m-1})$$

$$\leq \varepsilon_n^{-1} \sum_{m=1}^{n} E(\hat{X}_{n,m}^2|\mathscr{F}_{n,m-1}) \to 0$$

in probability by (a). To complete the proof now, it suffices to show

(e) $$P(|X_{n,m}| > \varepsilon_n \text{ for some } m \leq n) \to 0,$$

for with (d) and (c) this implies $\|S_{n,(n\cdot)} - \tilde{S}_{n,(n\cdot)}\| \to 0$ in probability. The proof of (7.2) constructs a Brownian motion with $\|\tilde{S}_{n,(n\cdot)} - B(\cdot)\| \to 0$, so the desired result follows from the triangle inequality and (6.9). Alternatively, you can use the result in Exercise 6.7.

To prove (e) we will use Lemma 3.5 of Dvoretsky (1972).

(f) If A_n is adapted to $\mathscr{G}_n$, then for any nonnegative $\delta \in \mathscr{G}_0$,

$$P\left(\bigcup_{m=1}^{n} A_m \Big| \mathscr{G}_0\right) \leq \delta + P\left(\sum_{m=1}^{n} P(A_m|\mathscr{G}_{m-1}) > \delta \Big| \mathscr{G}_0\right).$$

Proof of (f)   We proceed by induction. When $n = 1$ the conclusion says

$$P(A_1|\mathscr{G}_0) \leq \delta + P(P(A_1|\mathscr{G}_0) > \delta|\mathscr{G}_0).$$

This is obviously true on $\Omega_- \equiv \{P(A_1|\mathscr{G}_0) \leq \delta\}$ and also on $\Omega_+ \equiv \{P(A_1|\mathscr{G}_0) > \delta\} \in \mathscr{G}_0$, since on Ω_+,

$$P(P(A_1|\mathscr{G}_0) > \delta|\mathscr{G}_0) = 1 \geq P(A_1|\mathscr{G}_0).$$

To prove the result for n sets, observe that by the last argument the inequality is trivial on Ω_+. Let $B_m = A_m \cap \Omega_-$. Since $\Omega_- \in \mathscr{G}_0 \subset \mathscr{G}_{m-1}$, $P(B_m|\mathscr{G}_{m-1}) = P(A_m|\mathscr{G}_{m-1})$ on Ω_-. (See Exercise 1.1 in Chapter 4.) Applying the result for $n - 1$ sets with $\gamma = \delta - P(B_1|\mathscr{G}_0) \geq 0$,

$$P\left(\bigcup_{m=2}^{n} B_m \Big| \mathscr{G}_1\right) \leq \gamma + P\left(\sum_{m=2}^{n} P(B_m|\mathscr{G}_{m-1}) > \gamma \Big| \mathscr{G}_1\right).$$

Taking conditional expectation w.r.t. $\mathscr{G}_0$ and noting $\gamma \in \mathscr{G}_0$, $\gamma = \delta - P(B_1|\mathscr{G}_0)$

$$P\left(\bigcup_{m=2}^{n} B_m \Big| \mathscr{G}_0\right) \leq \gamma + P\left(\sum_{m=1}^{n} P(B_m|\mathscr{G}_{m-1}) > \delta \Big| \mathscr{G}_0\right).$$

$\bigcup_{2\le m\le n} B_m = (\bigcup_{2\le m\le n} A_m) \cap \Omega_-$ and $\sum_{1\le m\le n} P(B_m|\mathcal{G}_{m-1}) = \sum_{1\le m\le n} P(A_m|\mathcal{G}_{m-1})$ on Ω_-, so on Ω_-

$$P\left(\bigcup_{m=2}^{n} A_m \,\Big|\, \mathcal{G}_0\right) \le \delta - P(A_1|\mathcal{G}_0) + P\left(\sum_{m=1}^{n} P(A_m|\mathcal{G}_{m-1}) > \delta \,\Big|\, \mathcal{G}_0\right).$$

The result now follows from

$$P\left(\bigcup_{m=1}^{n} A_m \,\Big|\, \mathcal{G}_0\right) \le P(A_1|\mathcal{G}_0) + P\left(\bigcup_{m=2}^{n} A_m \,\Big|\, \mathcal{G}_0\right).$$

Let $C = \bigcup_{2\le m\le n} A_m$, observe that $1_{A_1\cup C} \le 1_{A_1} + 1_C$, and use the monotonicity of conditional expectations.

Proof of (e)   Let $A_m = \{|X_{n,m}| > \varepsilon_n\}$, $\mathcal{G}_m = \mathcal{F}_{n,m}$, and δ be a positive number. (f) implies

$$P(|X_{n,m}| > \varepsilon_n \text{ for some } m \le n) \le \delta + P\left(\sum_{m=1}^{n} P(|X_{n,m}| > \varepsilon_n|\mathcal{F}_{n,m-1}) > \delta\right).$$

To estimate the right-hand side we observe "Chebyshev's inequality" (Exercise 1.7 in Chapter 4) implies

$$\sum_{m=1}^{n} P(|X_{n,m}| > \varepsilon_n|\mathcal{F}_{n,m-1}) \le \varepsilon_n^{-2} \sum_{m=1}^{n} E(\hat{X}_{n,m}^2|\mathcal{F}_{n,m-1}) \to 0,$$

so

$$\limsup_{n\to\infty} P(|X_{n,m}| > \varepsilon_n \text{ for some } m \le n) \le \delta.$$

Since δ is arbitrary the proof of (e) and hence of (7.3) is complete.   □

For applications it is useful to have a result for a single sequence.

(7.4) Martingale Central Limit Theorem   Suppose X_n, $\mathcal{F}_n$, $n \ge 1$, is a martingale difference sequence and let $V_k = \sum_{1\le n\le k} E(X_n^2|\mathcal{F}_{n-1})$. If (i) $V_k/k \to \sigma^2 > 0$ in probability and (ii) $n^{-1} \sum_{m\le n} E(X_m^2 1_{(|X_m|>\varepsilon\sqrt{n})}) \to 0$, then $S_{(n\cdot)}/\sqrt{n} \Rightarrow \sigma B(\cdot)$.

Proof   Let $X_{n,m} = X_m/\sigma\sqrt{n}$, $\mathcal{F}_{n,m} = \mathcal{F}_m$. Changing notation and letting $k = nt$, our first assumption becomes (i) of (7.3). To check (ii) of (7.3) observe that

$$E \sum_{m=1}^{n} E(X_{n,m}^2 1_{(|X_{n,m}|>\varepsilon)}|\mathcal{F}_{n,m-1}) = \sigma^{-2} n^{-1} \sum_{m=1}^{n} E(X_m^2 1_{(|X_m|>\varepsilon\sqrt{n})}) \to 0.$$ □

Exercise 7.2   Use Doob's inequality to show that if $V_k/k \to 0$ in (7.4), then

$$\left(\max_{m\le n} S_m\right)\Big/\sqrt{n} \Rightarrow 0 \quad \text{and hence } S_{(n\cdot)}/\sqrt{n} \Rightarrow 0.$$

Exercise 7.3   *Chain-dependent random variables* (O'Brien (1974)). Let ζ_n be a Markov chain with state space, transition probability $p(i, j)$, and stationary distribution $\{X_n; n \ge 1\}$ are said to be chain dependent if for $\mathcal{G}_n = $ we have

$$P(\zeta_{n+1} = j, X_{n+1} \le x|\mathcal{G}_n) = p(\zeta_n, j)H_{\zeta_n}(x)$$

where $H_1, \ldots, H_n$ are given distribution functions. Intuitively which may be in one of a finite number of states, and the distri variable we observe depends on the value of ζ_n. Show tha $\int x^2 \, dH_i = \sigma_i^2 < \infty$, then X_n is a martingale difference sequence

$$n^{-1/2} S_{(n\cdot)} \Rightarrow \sigma B(\cdot) \quad \text{where } \sigma^2 = \sum_{m=1}^{M} \pi_m \sigma_m^2.$$

The unappetizing assumption $\int x \, dH_i = 0$ for all i will be dropped and 7.5.

To get from our result about martingales to our result abou sequences we begin by considering the intersection of the two cases.

(7.5) THEOREM   Suppose X_n, $n \in \mathbb{Z}$, is an ergodic stationary sequence integrable martingale differences; that is, $EX_n^2 < \infty$, and if $\mathcal{F}_n = \sigma(X_m, m \le $ $E(X_n|\mathcal{F}_{n-1}) = 0$. Let $S_n = X_1 + \cdots + X_n$. Then $S_{(n\cdot)}/n^{1/2} \Rightarrow \sigma B(\cdot)$.

Remark   This result we discovered independently by Billingsley (1961) and Ibrag (1963).

Proof $u_n \equiv E(X_n^2|\mathcal{F}_{n-1})$ can be written as $\varphi(X_{n-1}, X_{n-2}, \ldots)$, so (1.3) in Chapter implies u_n is stationary and ergodic, and the ergodic theorem implies

$$n^{-1} \sum_{m=1}^{n} u_m \to Eu_0 = EX_0^2 \quad \text{a.s.}$$

The last conclusion shows that (i) of (7.4) holds. To verify (ii), we observe

$$n^{-1} \sum_{m=1}^{n} E(X_m^2 1_{(|X_m|>\varepsilon\sqrt{n})}) = E(X_0^2 1_{(|X_0|>\varepsilon\sqrt{n})}) \to 0$$

by the dominated convergence theorem.   □

We come now to the promised central limit theorem for stationary sequences. Our proof is based on Scott (1973), but the key is an idea of Gordin (1969).

(7.6) THEOREM   Suppose X_n, $n \in \mathbb{Z}$, is an ergodic stationary sequence with $EX_n = 0$ and

To solve (∗) formally, we let

$$Z_0 = \sum_{j=0}^{\infty} E(X_j|\mathcal{F}_{-1})$$

$$\theta Z_0 = \sum_{j=0}^{\infty} E(X_{j+1}|\mathcal{F}_0)$$

$$Y_0 = \sum_{j=0}^{\infty} \{E(X_j|\mathcal{F}_0) - E(X_j|\mathcal{F}_{-1})\}$$

and check that

$$Y_0 + Z_0 - \theta Z_0 = E(X_0|\mathcal{F}_0) = X_0.$$

To justify the last calculation we need to know that the series in the definitions of Z_0 and Y_0 converge. Our assumption and shift invariance imply

$$\sum_{j=0}^{\infty} \|E(X_j|\mathcal{F}_{-1})\|_2 < \infty,$$

so the triangle inequality implies that the series for Z_0 converges in L^2. Since

$$\|E(X_j|\mathcal{F}_0) - E(X_j|\mathcal{F}_{-1})\|_2 \le \|E(X_j|\mathcal{F}_0)\|_2$$

(the left-hand side is the projection onto $K_0 \subset H_0$), the series for Y_0 also converges in L^2.

Putting the pieces together we have shown (7.6) with $\sigma^2 = EY_0^2$. To get the indicated formula for σ^2, observe that conditioning and using Cauchy–Schwarz,

$$|EX_0X_n| = |E(X_0E(X_n|\mathcal{F}_0))| \le \|X_0\|_2\|E(X_n|\mathcal{F}_0)\|_2.$$

Shift invariance implies $\|E(X_n|\mathcal{F}_0)\|_2 = \|E(X_0|\mathcal{F}_{-n})\|_2$, so the series converges absolutely.

$$ES_n^2 = \sum_{j=1}^{n}\sum_{k=1}^{n} EX_jX_k = nEX_0^2 + 2\sum_{m=1}^{n-1}(n-m)EX_0X_m.$$

From this it follows easily that

$$n^{-1}ES_n^2 \to EX_0^2 + 2\sum_{m=1}^{\infty} EX_0X_m.$$

To finish the proof let $T_n = \sum_{m=1}^{n} \theta^m Y_0$, observe $\sigma^2 = EY_0^2$, and

$$n^{-1}E(S_n - T_n)^2 = n^{-1}E(\theta Z_0 - \theta^{n+1}Z_0)^2 \le 4EZ_0^2/n \to 0,$$

since $(a - b)^2 \le (2a)^2 + (2b)^2$.   □

We turn now to examples. In the first one it is trivial to check the hypothesis of (7.6).

Example 7.1   *M-dependent sequences.* Let X_n, $n \in \mathbb{Z}$, be a stationary sequence with $EX_n = 0$, $EX_n^2 < \infty$, and suppose $\{X_j; j \le 0\}$ and $\{X_k: k \ge M\}$ are independent. In this case $E(X_0|\mathcal{F}_{-n}) = 0$ for $n \ge M$, so (7.6) implies $S_{n,(n\cdot)}/\sqrt{n} \Rightarrow \sigma B(\cdot)$ where

$$\sigma^2 = EX_0^2 + 2 \sum_{m=1}^{M-1} EX_0 X_m.$$

Exercise 7.4   Assume X_n and σ^2 are as in Example 7.1. Let $m = [n^{2/3}]$, let $Y_{n,k} = (X_{(k-1)m+1} + \cdots + X_{km-M})$ for $k \le K = [n^{1/3}]$, let $T_n = Y_{n,1} + \cdots + Y_{n,K}$, and $S_n = (X_1 + \cdots + X_n)$. (i) Use the Lindeberg–Feller theorem ((4.6) in Chapter 2) to conclude that $T_n/\sqrt{n} \Rightarrow \sigma\chi$. (ii) Show $(S_n - T_n)/\sqrt{n} \Rightarrow 0$ by computing second moments and use Exercise 2.8 in Chapter 2 to conclude $S_n/\sqrt{n} \Rightarrow \sigma\chi$.

Challenge   Use this outline with the Lindeberg–Feller theorem for martingales (7.3) to conclude $S_{(n\cdot)}/\sqrt{n} \Rightarrow \sigma B(\cdot)$.

Exercise 7.5   Consider the special case of Example 7.1 in which ξ_i, $i \in \mathbb{Z}$, are i.i.d. and take values H and T with equal probability. Let $X_n = f(\xi_n, \xi_{n+1})$ where $f(H, T) = 1$ and $f(i, j) = 0$ otherwise. $S_n = X_1 + \cdots + X_n$ counts the number of head runs in $\xi_1, \ldots, \xi_{n+1}$. Find constants μ and σ so that $(S_n - n\mu)/\sigma n^{1/2} \Rightarrow \chi$. What is the random variable Y_0 constructed in the proof of (7.6) in this case?

Example 7.2   *Markov chains.* Let ζ_n, $n \in \mathbb{Z}$, be an irreducible Markov chain on a countable state space S in which each ζ_n has the stationary distribution π. Let $X_n = f(\zeta_n)$ where $\sum f(x)\pi(x) = 0$ and $\sum f(x)^2\pi(x) < \infty$. In this case if $\mathcal{F}_{-n} = \sigma(\zeta_m, m \le -n)$, then

$$E(X_0|\mathcal{F}_{-n}) = \sum_y p^n(\zeta_{-n}, y)f(y)$$

where $p^n(x, y)$ is the n-step transition probability. Since $\sum f(x)\pi(x) = 0$,

$$\left| \sum_y p^n(x, y)f(y) \right| \le \sum_y |p^n(x, y) - \pi(y)||f(y)|$$

$$\le \|f\|_\infty \|p^n(x, \cdot) - \pi(\cdot)\|$$

where $\|f\|_\infty = \sup|f(x)|$ and $\|\cdot\|$ is the total variation norm. When S is finite, all f are bounded. If the chain is aperiodic, Exercise 5.7 in Chapter 5 implies that

$$\sup_x \|p^n(x, \cdot) - \pi(\cdot)\| \le Ce^{-\varepsilon n},$$

and the hypothesis of (7.6) is satisfied. To see that the limiting variance σ^2 may be 0 in this case, consider the chain with $S = \{1, 2, 3\}$ and

$$p(1, 2) = 1, \qquad p(2, 1) = p(2, 3) = 1/2, \qquad p(3, 1) = 1,$$

and take $f(1) = 1, f(2) = -1, f(3) = 0$.

Exercise 7.6   Consider the setup of Exercise 7.3 and allow $\int x \, dH_i = \mu_i$ to be different from 0 but assume (without loss of generality) that $\sum \mu_i \pi_i = 0$. Use (7.6) to conclude that $S_{(n \cdot)}/\sqrt{n} \Rightarrow \sigma B(\cdot)$.

Checking the hypotheses of (7.6) when S is infinite is harder but not too difficult for nice examples.

Exercise 7.7   Suppose $S = \{0, 1, 2, \ldots\}$, $p(x, x + 1) = p < 1/2$, $p(x, x - 1) = 1 - p$ for $x \geq 1$, and $p(0, 0) = 1 - p$. Let X_n be a Markov chain with this transition probability in which X_0 has the stationary distribution. Let $S_n = X_1 + \cdots + X_n$. (i) Show that there are constants μ and σ^2 so that $(S_n - n\mu)/\sqrt{n} \Rightarrow \sigma\chi$. (ii) Use the proof of (5.5) in Chapter 5 to conclude that the last conclusion holds for any initial distribution.

Our last example is simple to treat directly but will help us evaluate the strength of the conditions in (7.6) and in later theorems.

Example 7.3   *Moving average process.* Suppose

$$X_m = \sum_{k \geq 0} c_k \xi_{m-k} \quad \text{where} \quad \sum_{k \geq 0} c_k^2 < \infty$$

and the ξ_i, $i \in \mathbb{Z}$, are i.i.d. with $E\xi_i = 0$ and $E\xi_i^2 = 1$. If $\mathscr{F}_{-n} = \sigma(\xi_m; m \leq -n)$, then

$$\|E(X_0 | \mathscr{F}_{-n})\|_2 = \left\| \sum_{k \geq n} c_k \xi_{-k} \right\|_2 = \left(\sum_{k \geq n} c_k^2 \right)^{1/2}.$$

If, for example, $c_k = (1 + k)^{-p}$, $\|E(X_0 | \mathscr{F}_{-n})\|_2 \sim n^{(1/2)-p}$, and (7.6) applies if $p > 3/2$.

Exercise 7.8   Let Y_0, Z_0, T_n, and S_n be as in the proof of (7.6). Suppose

$$\sum_{j \geq 0} \|E(X_j | \mathscr{F}_0) - E(X_j | \mathscr{F}_{-1})\|_2 < \infty,$$

so that $EY_0^2 < \infty$. The triangle inequality and shift invariance imply $\|E(X_n | \mathscr{F}_0)\|_2 \to 0$ as $n \to \infty$. Use this to conclude that $n^{-1}E(S_n - T_n)^2 \to 0$, and $S_n/\sqrt{n} \Rightarrow \sigma\chi$ where $\sigma^2 = EY_0$. This shows that $\sum |c_k| < \infty$ is sufficient for the central limit theorem in Example 7.3.

Exercise 7.9   The condition in the last exercise is close to the best possible. Suppose the ξ_i take values 1 and -1 with equal probability and let $c_k = (1 + k)^{-p}$ where $1/2 < p < 1$. Show that $S_n/n^{1-p/2} \Rightarrow \sigma\chi$.

The last three examples show that in many cases it is easy to verify the hypothesis of (7.6) directly. To connect (7.6) with other results in the literature, we will introduce two sufficient conditions phrased in terms of "mixing" properties. In each case we will first give an estimate on covariances and then state a central limit theorem. Let

$$\alpha(\mathcal{G}, \mathcal{H}) = \sup\{|P(A \cap B) - P(A)P(B)| : A \in \mathcal{G}, B \in \mathcal{H}\}.$$

If $\alpha = 0$, $\mathcal{G}$ and $\mathcal{H}$ are independent, so α measures the dependence between the σ-fields.

(7.7) LEMMA   Let $p, q, r \in (1, \infty]$ with $1/p + 1/q + 1/r = 1$, and suppose $X \in \mathcal{G}$, $Y \in \mathcal{H}$ have $E|X|^p$, $E|Y|^q < \infty$. Then

$$|EXY - EXEY| \le 8\|X\|_p\|Y\|_q(\alpha(\mathcal{G}, \mathcal{H}))^{1/r}.$$

Here, we interpret $x^0 = 1$ for $x > 0$ and $0^0 = 0$.

Proof   If $\alpha = 0$, X and Y are independent and the result is true, so we can suppose $\alpha > 0$. We build up to the result in three steps, starting with the case $r = \infty$.

(a) $|EXY - EXEY| \le 2\|X\|_p\|Y\|_q$.

Proof of (a)   Hölder's inequality ((5.3) in the Appendix) implies $|EXY| \le \|X\|_p\|Y\|_q$, and Jensen's inequality implies $\|X\|_p\|Y\|_q \ge |E|X|E|Y|| \ge |EXEY|$, so the result follows from the triangle inequality.   □

(b) $|EXY - EXEY| \le 4\|X\|_\infty\|Y\|_\infty\alpha(\mathcal{G}, \mathcal{H})$.

Proof of (b)   Let $\eta = \text{sgn}\{E(Y|\mathcal{G}) - EY\} \in \mathcal{G}$. $EXY = E(XE(Y|\mathcal{G}))$, so

$$|EXY - EXEY| = |E(X\{E(Y|\mathcal{G}) - EY\})| \le \|X\|_\infty E|E(Y|\mathcal{G}) - EY|$$

$$= \|X\|_\infty E(\eta\{E(Y|\mathcal{G}) - EY\}) = \|X\|_\infty\{E(\eta Y) - E\eta EY\}.$$

Applying the last result with $X = Y$ and $Y = \eta$ gives

$$|E(Y\eta) - EYE\eta| \le \|Y\|_\infty|E(\zeta\eta) - E\zeta E\eta|$$

where $\zeta = \text{sgn}\{E(\eta|\mathcal{H}) - E\eta\}$. Now $\eta = 1_A - 1_B$ and $\zeta = 1_C - 1_D$, so

$$|E(\zeta\eta) - E\zeta E\eta| = |P(A \cap C) - P(B \cap C) - P(A \cap D) + P(B \cap D)$$

$$- P(A)P(C) + P(B)P(C) + P(A)P(D) - P(B)P(D)|$$

$$\le 4\alpha(\mathcal{G}, \mathcal{H}).$$

Combining the last three displays gives the desired result.   □

(c) $|EXY - EXEY| \le 6\|X\|_p \|Y\|_\infty \alpha(\mathcal{G}, \mathcal{H})^{1-1/p}$.

Proof of (c)   Let $C = \alpha^{-1/p} \|X\|_p$, $X_1 = X 1_{(|X| \le C)}$, and $X_2 = X - X_1$.

$$|EXY - EXEY| \le |EX_1 Y - EX_1 EY| + |EX_2 Y - EX_2 EY|$$

$$\le 4\alpha C \|Y\|_\infty + 2\|Y\|_\infty E|X_2|$$

by (b) and (a). Now

$$E|X_2| \le C^{-p+1} E(|X|^p 1_{(|X|>C)}) \le C^{-p+1} E|X|^p.$$

Combining the last two inequalities and using the definition of C give

$$|EXY - EXEY| \le 4\alpha^{1-1/p} \|X\|_p \|Y\|_\infty + 2\|Y\|_\infty \alpha^{1-1/p} \|X\|_p^{-p+1+p}.   □$$

Finally to prove (9), let $C = \alpha^{-1/q} \|Y\|_q$, $Y_1 = Y 1_{(|Y| \le C)}$, and $Y_2 = Y - Y_1$.

$$|EXY - EXEY| \le |EXY_1 - EXEY_1| + |EXY_2 - EXEY_2|$$

$$\le 6C \|X\|_p \alpha^{1-1/p} + 2\|X\|_p \|Y_2\|_\theta$$

where $\theta = (1 - 1/p)^{-1}$ by (c) and (a). Now

$$E|Y_2|^\theta \le C^{-q+\theta} E(|Y|^q 1_{(|Y|>C)}) \le C^{-q+\theta} E|Y|^q.$$

Combining the last two inequalities and using the definition of C give

$$|EXY - EXEY| \le 6\alpha^{-1/q} \|Y\|_q \|X\|_p \alpha^{1-1/p} + 2\|X\|_p \alpha^{1/\theta-1/q} \|Y\|_q,$$

proving (7.7).   □

Remark   The last proof is from Appendix III of Hall and Heyde (1980). They attribute (b) to Ibragimov (1962) and (c) and (7.7) to Davydov (1968).

Combining (7.6) and (7.7) gives:

(7.8) THEOREM   Suppose X_n, $n \in \mathbb{Z}$, is an ergodic stationary sequence with $EX_n = 0$, $E|X_0|^{2+\delta} < \infty$. Let $\alpha(n) = \alpha(\mathcal{F}_{-n}, \sigma(X_0))$ where $\mathcal{F}_{-n} = \sigma(X_m : m \le -n)$ and suppose

$$\sum_{n=1}^{\infty} \alpha(n)^{\delta/2(2+\delta)} < \infty.$$

If $S_n = X_1 + \cdots + X_n$, then $S_{(n \cdot)}/\sqrt{n} \Rightarrow \sigma B(\cdot)$ where $\sigma^2 = EX_0^2 + 2\sum_{n=1}^{\infty} EX_0 X_n$.

Remark Let $\bar{\alpha}(n) = \alpha(\mathscr{F}_{-n}, \mathscr{F}_0')$ where $\mathscr{F}_0' = \sigma(X_k, k \geq 0)$. When $\bar{\alpha}(n) \downarrow 0$, the sequence is called *strong mixing*. Rosenblatt (1956) introduced the concept as a condition under which the central limit theorem for stationary sequences could be obtained. Ibragimov (1962) proved $S_n/\sqrt{n} \Rightarrow \sigma\chi$ where $\sigma^2 = \lim_{n\to\infty} ES_n^2/n$ under the assumption

$$\sum_{n=1}^{\infty} \bar{\alpha}(n)^{\delta/(2+\delta)} < \infty.$$

See Ibragimov and Linnik (1971), Theorem 18.5.3, or Hall and Heyde (1980), Corollary 5.1, for a proof.

Proof To use (7.7) to estimate the quantity in (7.6) we observe that the Cauchy–Schwarz inequality implies that for Z, $Y \in \mathscr{F}$, $|E(ZY)| \leq \|Z\|_2 \|Y\|_2$, so taking $Z = E(X|\mathscr{F})$ gives

$$\|E(X|\mathscr{F})\|_2 = \sup\{E(XY): Y \in \mathscr{F}, \|Y\|_2 = 1\}.$$

Letting $p = 2 + \delta$ and $q = 2$ in (7.7) and recalling $EX_0 = 0$ show that if $Y \in \mathscr{F}_{-n}$,

$$|EX_0 Y| \leq 8\|X_0\|_{2+\delta}\|Y\|_2 \alpha(n)^{\delta/2(2+\delta)}.$$

Combining this with the first observation gives

$$\|E(X_0|\mathscr{F}_{-n})\|_2 \leq 8\|X_0\|_{2+\delta}\alpha(n)^{\delta/2(2+\delta)},$$

and it follows that the hypotheses of (7.6) are satisfied.   □

In the M-dependent case (Example 7.1), $\alpha(n) = 0$ for $n > M$, so (7.8) applies. As for Markov chains (Example 7.2), in this case

$$\alpha(n) = \sup_{A,B} |P(X_{-n} \in A, X_0 \in B) - \pi(A)\pi(B)|$$

$$\leq \sum_x \pi(x)\|p^n(x, \cdot) - \pi(\cdot)\|,$$

so the hypothesis of (7.8) can be checked if we know enough about the rate of convergence to equilibrium. If you could not do Exercise 7.7 before, try it again. It should be easier now.

Finally, to see how good the conditions in (7.8) are, we consider the special case of Example 7.3 in which the ξ_i are i.i.d. standard normals. Let

$$\rho(\mathscr{G}, \mathscr{H}) = \sup\{\text{corr}(X, Y): X \in \mathscr{G}, Y \in \mathscr{H}\}$$

where

$$\text{corr}(X, Y) = (EXY - EXEY)/(\|X - EX\|_2 \|Y - EY\|_2).$$

Clearly $\alpha(\mathcal{G}, \mathcal{H}) \leq \rho(\mathcal{G}, \mathcal{H})$. Kolmogorov and Rozanov (1964) have shown that when $\mathcal{G}$ and $\mathcal{H}$ are generated by Gaussian random variables, then $\rho(\mathcal{G}, \mathcal{H}) \leq 2\pi\alpha(\mathcal{G}, \mathcal{H})$. They proved this by showing that $\rho(\mathcal{G}, \mathcal{H})$ is the angle between $L^2(\mathcal{G})$ and $L^2(\mathcal{H})$. Using the geometric interpretation, we see that if $|c_k|$ is decreasing, then $\alpha(n) \leq \bar{\alpha}(n) \leq |c_n|$, so (7.8) requires $\sum |c_n|^{1/2 - \varepsilon} \leq \infty$, but Ibragimov's result applies if $\sum |c_n|^{1 - \varepsilon} < \infty$. As Exercise 7.9 (or direct computation) shows, the central limit theorem is valid if $\sum |c_n| < \infty$.

Our second mixing concept is more restrictive (and asymmetric) because we divide by $P(B) < 1$. Let

$$\beta(\mathcal{G}, \mathcal{H}) = \sup\{P(A|B) - P(A) : A \in \mathcal{G}, B \in \mathcal{H} \text{ with } P(B) > 0\}.$$

Clearly $\beta(\mathcal{G}, \mathcal{H}) \geq \alpha(\mathcal{G}, \mathcal{H})$ and $\beta(\mathcal{G}, \mathcal{H}) = 0$ implies that $\mathcal{G}$ and $\mathcal{H}$ are independent. The analogue of (7.7) for β is:

(7.9) LEMMA   Let $p, q \in (1, \infty)$ with $1/p + 1/q = 1$. Suppose $X \in \mathcal{G}$ and $Y \in \mathcal{H}$ have $E|X|^p, E|Y|^q < \infty$. Then

$$|EXY - EXEY| \leq 2\beta(\mathcal{G}, \mathcal{H})^{1/p} \|X\|_p \|Y\|_q.$$

Proof   Hölder's inequality ((5.3) in the Appendix) implies

$$|E(X - X')(Y - Y')| \leq \|X - X'\|_p \|Y - Y'\|_q.$$

The triangle inequality and Jensen's inequality imply

$$|EXEY - EX'EY'| \leq |EY||EX - EX'| + |EX'||EY - EY'|$$

$$\leq |EY|\|X - X'\|_p + |EX'|\|EY - EY'\|_q,$$

and a second application of this reasoning shows

$$|EX'| \leq |EX| + \|X - X'\|_p.$$

Since any X with $\|X\|_p < \infty$ can be approximated by simple X' in such a way that $\|X - X'\|_p \to 0$, it suffices to prove the result when

$$X = \sum x_i 1_{A_i}, \qquad Y = \sum y_j 1_{B_j}$$

where $A_1, \dots, A_m$ and $B_1, \dots, B_n$ are partitions of Ω.

(a) $|EXY - EXEY| = \left| \sum_i \sum_j x_i y_j \{P(A_i \cap B_j) - P(A_i)P(B_j)\} \right|$

$$= \left| \sum_i (x_i P(A_i)^{1/p})((P(A_i))^{1/q} \sum_j y_j\{P(B_j|A_i) - P(B_j)\}) \right|$$

$$\leq \|X\|_p \left(\sum_i P(A_i) \left| \sum_j y_j\{P(B_j|A_i) - P(B_j)\} \right|^q \right)^{1/q}$$

by Hölder's inequality. Since $1 = 1/q + 1/p$, another use of Hölder's inequality gives

$$\left| \sum_j y_j\{P(B_j|A_i) - P(B_j)\}^1 \right|^q$$

$$\leq \left(\sum_j |y_j|^q |P(B_j|A_i) - P(B_j)| \right) \cdot \left(\sum_j |P(B_j|A_i) - P(B_j)| \right)^{q/p},$$

and it follows that

(b) $\sum_i P(A_i) \left| \sum_j y_j\{P(B_j|A_i) - P(B_j)\} \right|^q$

$$\leq \sum_i P(A_i) \left(\left(\sum_j |y_j|^q |P(B_j|A_i) - P(B_j)| \right) \cdot \left(\sum_j |P(B_j|A_i) - P(B_j)| \right)^{q/p} \right)$$

$$\leq \left(\sum_i P(A_i) \sum_j |y_j|^q |P(B_j|A_i) + P(B_j)| \right) \sup_i \left(\sum_j |P(B_j|A_i) - P(B_j)| \right)^{q/p}.$$

The first term is $2\|Y\|_q$. To evaluate the second let C_i^+ be the union of the B_j for which $P(B_j|A_i) > P(B_j)$ and let C_i^- be the complement; then

(c) $\sum_j |P(B_j|A_i) - P(B_j)| = [P(C_i^+|A_i) - P(C_i^+)] + |P(C_i^-|A_i) - P(C_i^-)|$

$$\leq 2\beta(\mathcal{G}, \mathcal{H}).$$

Combining (a)–(c) gives (7.9).   □

Combining (7.6) and (7.9) gives:

(7.10) THEOREM   Suppose $X_n, n \in \mathbb{Z}$, is an ergodic stationary sequence with $EX_n = 0$, $E|X_0|^2 < \infty$. Let $\beta(n) = \alpha(\mathcal{F}_{-n}, \sigma(X_0))$ where $\mathcal{F}_{-n} = \sigma(X_m: m \leq -n)$ and suppose

$$\sum_{n=1}^{\infty} \beta(n)^{1/2} < \infty.$$

If $S_n = X_1 + \cdots + X_n$, then $S_{n, (n \cdot)} \Rightarrow \sigma B(\cdot)$ where $\sigma^2 = EX_0^2 + \sum_{n=1}^{\infty} EX_0 X_n$.

Remark   Let $\bar{\beta}(n) = \beta(\mathcal{F}_{-n}, \mathcal{F}_0')$ where $\mathcal{F}_0' = \sigma(X_0, X_1, \ldots)$. If $\bar{\beta}(n) \downarrow 0$ as $n \to \infty$, the sequence X_n is said to be *uniformly mixing*. Billingsley (1968), see Theorem 20.1 on p. 174, gives this result under the assumption $\sum \bar{\beta}(n)^{1/2} < \infty$. Our proof of (7.9) can be found on pp. 170–171 of his book.

Proof   To use (7.9) to estimate the quantity in (7.6) we observe that for reasons indicated in the proof of (7.8),

$$\|E(X|\mathcal{F})\|_2 = \sup\{E(XY): Y \in \mathcal{F}, \|Y\|_2 = 1\}.$$

Letting $p = q = 2$ in (7.9) and recalling $EX_0 = 0$ show that if $Y \in \mathcal{F}_{-n}$,

$$|EX_0 Y| \le 2\|X_0\|_2 \|Y\|_2 \beta(n)^{1/2},$$

and it follows that the hypotheses of (7.6) are satisfied.   □

As before, in the M-dependent case (Example 7.1), $\beta(n) = 0$ for $n > M$, so (7.10) applies. Turning to Markov chains (Example 7.2), in this case

$$\beta(n) = \sup_{x} \|p^n(x, \cdot) - \pi(\cdot)\|.$$

The chain is clearly not uniformly mixing if $\beta(n) = 1$ for all n. Conversely, if $\beta(N) < 1 - \varepsilon$ for some N, iterating gives

$$\sup_{x} \|p^{kN+j}(x, \cdot) - \pi(\cdot)\| \le (1 - \varepsilon)^k \quad \text{for } 0 \le j < N.$$

Example 7.3 shows that our second mixing condition is much stronger than the first.

Exercise 7.10   *Moving average process.* Show that if $c_k \ne 0$ for all k and the distribution of ξ_k is unbounded, $\beta(n) = 1$ for all n.

The last result is a little discouraging but there is at least one interesting example.

Example 7.4   The *continued fraction transformation* (see Exercise 1.8 in Chapter 6) is uniformly mixing with $\bar{\beta}(n) = a\rho^n$ and $\rho < 1$. See Chapter 9 of Lévy (1937). For more on this example see Billingsley (1968), pp. 192–194.

8 Kolmogorov–Smirnov Statistics, Brownian Bridge

Let $X_1, X_2, \ldots$ be i.i.d. with distribution F. In Section 8 of Chapter 1 we showed that with probability 1 the empirical distribution

$$\hat{F}_n(x) = \frac{1}{n} |\{m \leq n : X_m \leq x\}|$$

converges uniformly to $F(x)$. In this section we will investigate the rate of convergence when F is continuous. We impose this restriction so we can reduce to the case of a uniform distribution on $(0, 1)$ by setting $Y_n = F(X_n)$. Since $x \to F(x)$ is nondecreasing and continuous and no observations land in intervals of constancy of F, it is easy to see that if we let

$$\hat{G}_n(y) = \frac{1}{n} |\{m \leq n : Y_m \leq y\}|,$$

then

$$\sup_x |\hat{F}_n(x) - F(x)| = \sup_{0 < y < 1} |\hat{G}_n(y) - y|. \tag{8.1}$$

For the rest of the section then, we will assume $Y_1, Y_2, \ldots$ is i.i.d. uniform on $(0, 1)$. To be able to apply Donsker's theorem we will transform the problem. Put the observations $Y_1, Y_2, \ldots$ in increasing order: $U_1^n < U_2^n < \cdots < U_n^n$. I claim that

$$\sup_{0 < y < 1} \hat{G}_n(y) - y = \sup_{1 \leq m \leq n} m/n - U_m^n$$

$$\inf_{0 < y < 1} \hat{G}_n(y) - y = \inf_{1 \leq m \leq n} (m-1)/n - U_m^n, \tag{8.2}$$

since the sup occurs at a jump of $\hat{G}_n$ and the inf right before a jump. We will show that

$$D_n \equiv n^{1/2} \sup_{0 < y < 1} |\hat{G}_n(y) - y|$$

has a limit, so the extra $-1/n$ in the inf does not make any difference.

Our third and final manuever is to give a special construction of the order statistics $U_1^n < U_2^n < \cdots < U_n^n$. Let $W_1, W_2, \ldots$ be i.i.d. with $P(W_i > t) = e^{-t}$ and let $Z_n = W_1 + \cdots + W_n$.

(8.3) LEMMA $\{U_k^n : 1 \leq k \leq n\} \overset{d}{=} \{Z_k/Z_{n+1} : 1 \leq k \leq n\}$.

Proof   Using a little symbolic freedom, we can write the probability densities for the quantities in question as

$$P(Z_1 = t_1, \ldots, Z_{n+1} = t_{n+1}) = \prod_{m=1}^{n+1} \exp(-(t_m - t_{m-1})) = \exp(-t_{n+1})$$

if $0 = t_0 < t_1 < \cdots < t_{n+1}$. Integrating out $t_1, \ldots, t_n$ gives

$$P(Z_{n+1} = t_{n+1}) = \exp(-t_{n+1})t_{n+1}^n/n!,$$

so changing variables $u_i = t_i/t_{n+1}$, we find

$$P(Z_1/Z_{n+1} = u_1, \ldots, Z_n/Z_{n+1} = u_n | Z_{n+1} = t_{n+1}) = 1/n!,$$

the density function of the uniform distribution on $\{0 < u_1 < \cdots < u_n < 1\}$. This shows that even if we condition on the value of $Z_{n+1}, (Z_1/Z_{n+1}, \ldots, Z_n/Z_{n+1})$ has the desired distribution, and the proof of (8.3) is complete. □

We turn now to the limit law for D_n. As argued above, it suffices to consider

$$D_n' = n^{1/2} \max_{1 \le m \le n} \left| \frac{Z(m)}{Z(n+1)} - \frac{m}{n} \right| = \frac{n}{Z(n+1)} \max_{1 \le m \le n} \left| \frac{Z(m)}{n^{1/2}} - \frac{m}{n} \cdot \frac{Z(n+1)}{n^{1/2}} \right|$$

$$= \frac{n}{Z(n+1)} \max_{1 \le m \le n} \left| \frac{Z(m) - m}{n^{1/2}} - \frac{m}{n} \cdot \frac{Z(n+1) - n}{n^{1/2}} \right|.$$

If we let

$$B_n(t) = \begin{cases} (Z_m - m)/n^{1/2} & \text{if } t = m/n \\ \text{linear on each } [(m-1)/n, m/n], \end{cases}$$

then

$$D_n' = \frac{n}{Z(n+1)} \max_{0 \le t \le 1} \left| B_n(t) - t \left\{ B_n(1) + \frac{Z(n+1) - Z(n)}{n^{1/2}} \right\} \right|.$$

The strong law of large numbers implies $Z_{n+1}/n \to 1$ a.s., so the first factor will disappear in the limit. To find the limit of the second we observe that Donsker's theorem, (6.6), implies $B_n(\cdot) \Rightarrow B(\cdot)$, a Brownian motion, and computing second moments shows

$$(Z_{n+1} - Z_n)/n^{1/2} \to 0 \quad \text{in probability}.$$

$$\psi(\omega) = \max_{0 \le t \le 1} |\omega(t) - t\omega(1)|$$

is a continuous function from $C[0, 1]$ to $\mathbb{R}$, so it follows from Donsker's theorem that:

(8.4) THEOREM $D_n \Rightarrow \max_{0 \le t \le 1} |B_t - tB_1|$ where B_t is a Brownian motion starting at 0.

Remark   Doob (1949) suggested this approach to deriving results of Kolmogorov and Smirnov, which was later justified by Donsker (1952). Our proof follows Breiman (1968). The next exercise gives an alternative approach.

Exercise 8.1   M. Kac (1949). Let $\hat{G}_n(t) = |\{m \le n : Y_m \le t\}|$ where $Y_1, Y_2, \dots$ are i.i.d. uniform on $(0, 1)$. Let $N(n)$ have a Poisson distribution with mean n and let

$$\hat{H}_n(t) = \frac{1}{n}|\{m \le N(n) : Y_m \le t\}|.$$

We take $N(n)$ to have a Poisson distribution so that if (a_i, b_i) are disjoint intervals, Exercise 6.12 in Chapter 2 implies $\{m \le N(n) : Y_m \in (a_i, b_i)\}$ are independent Poissons. (i) Show that $\{\sqrt{n}(\hat{H}_n(t) - t), 0 \le t \le 1\} \Rightarrow \{B_t, 0 \le t \le 1\}$. (ii) Now $n\hat{H}_n(1) = N(n) \approx n + \chi n^{1/2}$ where χ has the standard normal distribution, and the Glivenko–Cantelli theorem implies that the extra $O(n^{1/2})$ particles are spread uniformly over $(0, 1)$. Let $\hat{J}_n(t) = \hat{G}_n(t) - \hat{H}_n(t)$. Show

$$n^{-1/2} \sup_{0 \le t \le 1} |n\hat{J}_n(t) - t(n\hat{H}_n(1) - n)| \to 0,$$

and it follows that

$$\sqrt{n} \sup_{0 \le t \le 1} |\hat{G}_n(t) - t| \Rightarrow \max_{0 \le t \le 1} |B_t - tB_1|.$$

To identify the distribution of the limit in (8.4), we will first prove

$$\{B_t - tB_1, 0 \le t \le 1\} \stackrel{\mathrm{d}}{=} \{B_t, 0 \le t \le 1 | B_1 = 0\}, \tag{8.5}$$

a process we will denote by B_t^0 and call the *Brownian bridge*. The event $B_1 = 0$ has probability 0, but it is easy to see what the conditional probability should mean. If $0 = t_0 < t_1 < \cdots < t_n < t_{n+1} = 1$, $x_0 = 0$, $x_{n+1} = 0$, and $x_1, \dots, x_n \in \mathbb{R}$, then

$$P(B(t_1) = x_1, \dots, B(t_n) = x_n | B(1) = 0) = \prod_{m=1}^{n+1} p_{t_m - t_{m-1}}(x_{m-1}, x_m)/p_1(0, 0) \tag{8.6}$$

where $p_t(x, y) = (2\pi t)^{-1/2} \exp(-(y - x)^2/2t)$.

Proof of (8.5)   Formula (8.6) shows that the f.d.d.'s of B_t^0 are multivariate normal and have mean 0. Since $B_t - tB_1$ also has this property, it suffices to show that the covariances are equal. We begin with the easy computation. If $s < t$,

$$E((B_s - sB_1)(B_t - tB_1)) = s - st - st + st = s(1 - t). \tag{8.7}$$

For the other process $P(B_s^0 = x, B_t^0 = y)$ is

$$\frac{\exp(-x^2/2s)}{(2\pi s)^{1/2}} \cdot \frac{\exp(-(y-x)^2/2(t-s))}{(2\pi(t-s))^{1/2}} \cdot \frac{\exp(-y^2/2(1-t))}{(2\pi(1-t))^{1/2}} \bigg/ (2\pi)^{-1/2}$$

$$= (2\pi)^{-1}(s(t-s)(1-t))^{-1/2} \exp(-(ax^2 + 2bxy + cy^2)/2)$$

where

$$a = \frac{1}{s} + \frac{1}{t-s} = \frac{t}{s(t-s)}, \qquad b = -\frac{1}{t-s},$$

$$c = \frac{1}{t-s} + \frac{1}{1-t} = \frac{1-s}{(t-s)(1-t)}.$$

Recalling the discussion at the end of Section 9 in Chapter 2 and noticing

$$\begin{bmatrix} \dfrac{t}{s(t-s)} & \dfrac{-1}{(t-s)} \\[2ex] \dfrac{-1}{(t-s)} & \dfrac{1-s}{(t-s)(1-t)} \end{bmatrix}^{-1} = \begin{bmatrix} s(1-s) & s(1-t) \\ s(1-t) & t(1-t) \end{bmatrix}$$

(multiply the matrices!) show (8.5) holds. □

Our final step in investigating the limit distribution of D_n is to compute the distribution of $\max_{0 \le t \le 1} B_t^0$. To do this we compute

$$P_x(T_a \wedge T_b > t, B_t \in A)$$

where $a < x < b$ and $A \subset (a, b)$. We begin by observing that

$$P_x(T_a \wedge T_b > t, B_t \in A) = P_x(B_t \in A) - P_x(T_a < T_b, T_a < t, B_t \in A) \qquad (*)$$

$$- P_x(T_b < T_a, T_b < t, B_t \in A).$$

If we let $\rho_a(y) = 2a - y$ be a reflection through a and observe that $\{T_a < T_b\} \in \mathscr{F}(T_a)$, then it follows from the proof of (3.9) that

$$P_x(T_a < T_b, T_a < t, B_t \in A) = P_x(T_a < T_b, B_t \in \rho_a A)$$

where $\rho_a A = \{\rho_a(y): y \in A\}$. To get rid of the $T_a < T_b$ we observe that

$$P_x(T_a < T_b, B_t \in \rho_a A) = P_x(B_t \in \rho_a A) - P_x(T_b < T_a, B_t \in \rho_a A).$$

Noticing that $B_t \in \rho_a A$ and $T_b < T_a$ imply $T_b < t$ and using the reflection principle again give

$$P_x(T_b < T_a, B_t \in \rho_a A) = P_x(T_b < T_a, B_t \in \rho_b \rho_a A)$$

$$= P_x(B_t \in \rho_b \rho_a A) - P_x(T_a < T_b, B_t \in \rho_b \rho_a A).$$

Repeating the last two calculations n more times gives

$$P_x(T_a < T_b, B_t \in \rho_a A) = \sum_{m=0}^{n} \{P_x(B_t \in \rho_a(\rho_b \rho_a)^m A) - P_x(B_t \in (\rho_b \rho_a)^{m+1} A)\}$$

$$+ P_x(T_a < T_b, B_t \in (\rho_b \rho_a)^{n+1} A).$$

Each pair of reflections pushes A farther away from 0, so letting $n \to \infty$ shows

$$P_x(T_a < T_b, B_t \in \rho_a A) = \sum_{m=0}^{\infty} P_x(B_t \in \rho_a(\rho_b \rho_a)^m A) - P_x(B_t \in (\rho_b \rho_a)^{m+1} A).$$

Interchanging the roles of a and b gives

$$P_x(T_b < T_a, B_t \in \rho_b A) = \sum_{m=0}^{\infty} P_x(B_t \in \rho_b(\rho_a \rho_b)^m A) - P_x(B_t \in (\rho_a \rho_b)^{m+1} A).$$

Combining the last two expressions with $(*)$ and using $\rho_c^{-1} = \rho_c$, $(\rho_a \rho_b)^{-1} = \rho_b^{-1} \rho_a^{-1}$ give

$$P_x(T_a \wedge T_b > t, B_t \in A) = \sum_{m=-\infty}^{\infty} P_x(B_t \in (\rho_b \rho_a)^m A) - P_x(B_t \in \rho_a(\rho_b \rho_a)^m A).$$

To prepare the last formula for applications let $A = (u, v)$ where $a < u < v < b$, notice that

$$\rho_b \rho_a(y) = y + 2(b - a),$$

and change variables in the second sum to get

$$P_x(T_a \wedge T_b > t, u < B_t < v) = \sum_{n=-\infty}^{\infty} \{P_x(u + 2n(b - a) < B_t < v + 2n(b - a)) \qquad (8.8)$$

$$- P_x(2b - v + 2n(b - a) < B_t < 2b - u + 2n(b - a))\}.$$

To check our arithmetic look back at Figure 7.3 and notice that if $x = a$ or b, the array of $+$ and $-$ signs is antisymmetric so the answer is 0. Setting $x = 0$, $a = -b$, $u = -\varepsilon$, and $v = \varepsilon$ gives

$$P_0\left(\max_{0 \le t \le 1} |B_t| < a, |B_1| < \varepsilon\right) = \sum_{n=-\infty}^{\infty} \{P_0(-\varepsilon + 4na < B_t < \varepsilon + 4na) \qquad (8.9)$$

$$- P_0(-\varepsilon + (4n + 2)a < B_1 < \varepsilon + (4n + 2)a)\}.$$

Figure 7.3

Dividing both sides by $P(|B_1| < \varepsilon)$ and letting $\varepsilon \to 0$ (leaving it to the reader to check that the dominated convergence theorem applies) give

$$P_0\left(\max_{0 \le t \le 1} |B_t^0| < a \right) = \sum_{m=-\infty}^{\infty} (-1)^m P_0(B_1 = 2ma)/P_0(B_1 = 0) \tag{8.10}$$

$$= 1 + 2 \sum_{m=1}^{\infty} (-1)^m \exp(-2m^2 a^2).$$

Exercise 8.2 Use Exercise 3.3 to conclude $P(\sup_{0 \le t \le 1} B_t^0 > a) = \exp(-2a^2)$.

Exercise 8.3 Use (8.8) to show that

$$P\left(\max_{0 \le t \le 1} B_t^0 - \min_{0 \le t \le 1} B_t^0 \ge x \right) = 2 \sum_{n=1}^{\infty} (4n^2 x^2 - 1) \exp(-2n^2 x^2).$$

Letting $\varepsilon = a$ in (8.9) gives

$$P_0\left(\max_{0 \le s \le t} |B_s| < a \right) = \sum_{n=-\infty}^{\infty} \{ P_x((4n - 1)a < B_t < (4n + 1)a) \tag{8.11}$$

$$- P_x((4n + 1)a < B_t < (4n + 3)a) \}.$$

The last expression can be rewritten as

$$\frac{4}{\pi} \sum_{m=0}^{\infty} \frac{(-1)^m}{2m + 1} \exp(-(2m + 1)^2 \pi^2 t / 8a^2); \tag{8.12}$$

see Feller, Vol. II (1971), p. 342 or p. 632. The next exercise gives a direct derivation.

Exercise 8.4 Let $u(t, x)$ denote the probability that Brownian motion starting at x stays in $(-a, a)$ up to time t. (i) Use the Markov property to show that if we let $\sigma = \inf\{t: B_t \notin (-a, a)\}$, then $u(t - s, B_{s \wedge \sigma})$ is a martingale. (ii) Observe $u(t, x) = E_x u(t - \varepsilon, B_{\varepsilon \wedge \sigma})$ and let $\varepsilon \to 0$ to conclude that $\partial u/\partial t = \partial^2 u/\partial x^2$ (see Exercise 5.7 for help). (iii) u satisfies the boundary conditions $u(0, x) = 1$ for $x \in (-a, a)$, $u(t, a) = u(t, -a) = 1$. Solving the differential equation by Fourier series leads to

$$u(t, x) = \frac{4}{\pi} \sum_{m=0}^{\infty} (2m + 1)^{-1} \exp(-(2m + 1)^2 \pi^2 t / 8a^2) \cos((2m + 1)\pi x / 2a).$$

Integrating from $-a$ to a gives (8.12).

9   Laws of the Iterated Logarithm

Our first goal is to show:

(9.1) *LIL for Brownian Motion* $\limsup_{t \to \infty} B_t/(2t \log \log t)^{1/2} = 1$ a.s.

Here LIL is short for "law of the iterated logarithm," a name that refers to the $\log \log t$ in the denominator. Once (9.1) is established we can use the Skorokhod representation to prove the analogous result for random walks with mean 0 and finite variance. The key to the proof of (9.1) is (3.8).

$$P_0\left(\max_{0 \leq s \leq 1} B_s > a\right) = P_0(T_a \leq 1) = 2P_0(B_1 \geq a). \tag{9.2}$$

To identify the asymptotic behavior of the right-hand side of (9.2) as $a \to \infty$, we use (1.3) from Chapter 1.

$$\int_x^\infty \exp(-y^2/2)\, dy \leq \frac{1}{x} \exp(-x^2/2) \tag{9.3}$$

$$\int_x^\infty \exp(-y^2/2)\, dy \sim \frac{1}{x} \exp(-x^2/2) \quad \text{as } x \to \infty \tag{9.4}$$

where $f(x) \sim g(x)$ means $f(x)/g(x) \to 1$ as $x \to \infty$.

The last result and Brownian scaling imply that

$$P_0(B_t > (tf(t))^{1/2}) \sim \kappa f(t)^{-1/2} \exp(-f(t)/2)$$

where $\kappa = (2\pi)^{-1/2}$ is a constant that we will try to ignore below. The last result implies

$$\sum_{n=1}^\infty P_0(B_n > (nf(n))^{1/2}) \begin{cases} < \infty & \text{when} \quad f(n) = (2 + \varepsilon) \log n \\ = \infty & \text{when} \quad f(n) = (2 - \varepsilon) \log n \end{cases}$$

and hence by the Borel–Cantelli lemma that

$$\limsup_{n \to \infty} B_n/(2n \log n)^{1/2} \leq 1 \quad \text{a.s.}$$

To replace $\log n$ by $\log \log n$ we have to look along exponentially growing sequences. Let $t_n = \alpha^n$ where $\alpha > 1$.

$$P_0\left(\max_{t_n \leq s \leq t_{n+1}} B_s > (t_n f(t_n))^{1/2}\right) \leq P_0\left(\max_{0 \leq s \leq t_{n+1}} B_s/(t_{n+1})^{1/2} > (f(t_n)/\alpha)^{1/2}\right)$$

$$\leq 2\kappa(f(t_n)/\alpha)^{-1/2} \exp(-f(t_n)/2\alpha)$$

by (9.2) and (9.3). If $f(t) = 2\alpha^2 \log \log t$, then

$$\log \log t_n = \log(n \log \alpha) = \log n + \log \log \alpha,$$

so

$$\exp(-f(t_n)/2\alpha) \leq C_\alpha n^{-\alpha}$$

where C_α is a constant that depends only on α, and hence

$$\sum_{n=1}^{\infty} P_0 \left(\max_{t_n \leq s \leq t_{n+1}} B_s > (t_n f(t_n))^{1/2} \right) < \infty.$$

Since $t \to (tf(t))^{1/2}$ is increasing and $\alpha > 1$ is arbitrary, it follows that

$$\limsup_{t \to \infty} B_t/(2t \log \log t)^{1/2} \leq 1. \tag{a}$$

To prove the other half of (9.1) we again let $t_n = \alpha^n$, but this time α will be large, since to get independent events we will we look at

$$P_0(B(t_{n+1}) - B(t_n) > (t_{n+1} f(t_{n+1}))^{1/2}) = P_0(B_1 > (\beta f(t_{n+1}))^{1/2})$$

where $\beta = t_{n+1}/(t_{n+1} - t_n) = \alpha/(\alpha - 1) > 1$. The last quantity is

$$\geq \frac{\kappa}{2}(\beta f(t_{n+1}))^{-1/2} \exp(-\beta f(t_{n+1})/2)$$

if n is large by (9.4). If $f(t) = (2/\beta^2) \log \log t$, then $\log \log t_n = \log n + \log \log \alpha$, so

$$\exp(-\beta f(t_{n+1})/2) \geq C_\alpha n^{-1/\beta}$$

where C_α is a constant that depends only on α, and hence

$$\sum_{n=1}^{\infty} P_0(B(t_{n+1}) - B(t_n) > (t_{n+1} f(t_{n+1}))^{1/2}) = \infty.$$

Since the events in question are independent, it follows from the second Borel–Cantelli lemma that

$$B(t_{n+1}) - B(t_n) > ((2/\beta^2) t_{n+1} \log \log t_{n+1})^{1/2} \quad \text{i.o.} \tag{b}$$

From (a) we get

$$\limsup_{n \to \infty} B(t_n)/(2t_n \log \log t_n)^{1/2} \leq 1. \tag{c}$$

Since $t_n = t_{n+1}/\alpha$ and $t \to \log \log t$ is increasing, combining (b) and (c) and recalling $\beta = (\alpha - 1)/\alpha$ give

$$\limsup_{n \to \infty} B(t_{n+1})/(2t_{n+1} \log \log t_{n+1})^{1/2} \geq (\alpha - 1)/\alpha - (1/\alpha)^{1/2}.$$

Letting $\alpha \to \infty$ now gives the desired lower bound and the proof of (9.1) is complete.

Exercise 9.1   Let $t_k = \exp(e^k)$. Find constants c_k so that $\limsup_{k \to \infty} B(t_k)/c_k = 1$ a.s.

Exercise 9.2   Use (3.9) and (9.1) to conclude $\limsup_{t \to 0} |B_t|/(2t \log \log(1/t))^{1/2} = 1$ a.s.

Blumenthal's 0–1 law (2.6) implies $P_0(B_t < h(t)$ for all t sufficiently small) $\in \{0, 1\}$. h is said to belong to the *upper class* if the probability is 1, the *lower class* if it is 0.

(9.5) *Kolmogorov's Test*   If $h(t)\uparrow$ and $t^{-1/2}h(t)\downarrow$, then h is upper or lower class according as

$$\int_0^1 t^{-3/2}h(t) \exp(-h^2(t)/2t)\, dt \quad \text{converges or diverges.}$$

The first proof of this was given by Petrovski (1935). Recalling (4.1), we see that the integrand is the probability of hitting $h(t)$ at time t. To see what (9.5) says, define $\lg_k(t) = \log(\lg_{k-1}(t))$ for $k \geq 2$ and $t > a_k = \exp(a_{k-1})$ where $\lg_1(t) = \log(t)$ and $a_1 = 0$. A little calculus shows that when $n \geq 4$,

$$h(t) = \left(2t \left\{ \lg_2(1/t) + \frac{3}{2}\lg_3(1/t) + \sum_{m=4}^{n-1} \lg_m(1/t) + (1 + \varepsilon)\, \lg_n(1/t) \right\} \right)^{1/2}$$

is upper or lower class according as $\varepsilon > 0$ or $\varepsilon \leq 0$.

Exercise 9.3   Approximate h by piecewise constant functions and show that if the integral in (9.5) converges, $h(t)$ is an upper class function. The proof of the other direction is much more difficult; see Motoo (1959) or Section 4.12 of Itô and McKean (1965).

Turning to random walk, we will prove a result due to Hartman and Wintner (1941):

(9.6) THEOREM   If $X_1, X_2, \ldots$ are i.i.d. with $EX_i = 0$ and $EX_i^2 = 1$, then

$$\limsup_{n \to \infty} S_n/(2n \log \log n)^{1/2} = 1.$$

Proof   By (6.3) we can write $S_n = B(T_n)$ with $T_n/n \to 1$ a.s. As in the proof of Donsker's theorem, this is all we will use in the argument below. (9.6) will follow from (9.1) once we show

$$(S_{[t]} - B_t)/(t \log \log t)^{1/2} \to 0 \quad \text{a.s.} \tag{a}$$

To do this we begin by observing that if $\varepsilon > 0$ and $t \geq t_0(\omega)$,

$$T_{[t]} \in [t/(1 + \varepsilon),\, t(1 + \varepsilon)]. \tag{b}$$

To estimate $S_{[t]} - B_t$ we let $M(t) = \sup\{|B(s) - B(t)| : t/(1 + \varepsilon) \leq s \leq t(1 + \varepsilon)\}$. To control the last quantity we let $t = (1 + \varepsilon)^k$ and notice that if $t_k \leq t \leq t_{k+1}$,

$$M(t) \leq \sup\{|B(s) - B(t)| : t_{k-1} \leq s \leq t_{k+2}\}$$

$$\leq 2 \sup\{|B(s) - B(t_{k-1})| : t_{k-1} \leq s \leq t_{k+2}\}.$$

Noticing $t_{k+2} - t_{k-1} = \delta t_{k-1}$ where $\delta = (1 + \varepsilon)^3 - 1$, scaling implies

$$P\left(\max_{t_{k-1} \leq s \leq t_{k+2}} |B(s) - B(t_{k-1})| > (3\delta t_{k-1} \log \log t_{k-1})^{1/2} \right)$$

$$= P\left(\max_{0 \leq r \leq 1} |B(r)| > (3 \log \log t_{k-1})^{1/2} \right)$$

$$\leq 2\kappa(3 \log \log t_{k-1})^{-1/2} \exp(-3 \log \log t_{k-1}/2)$$

by a now familiar application of (9.2) and (9.3). Summing over k and using (b) give

$$\limsup_{t \to \infty} (S_{[t]} - B_t)/(t \log \log t)^{1/2} \leq (3\delta)^{1/2}.$$

If we recall $\delta = (1 + \varepsilon)^3 - 1$ and let $\varepsilon \downarrow 0$, (a) follows and the proof of (9.6) is complete. □

Exercise 9.4   Show that if $E|X_i|^\alpha = \infty$ for some $\alpha < 2$, then $\limsup_{n \to \infty} |X_n|/n^{1/\alpha} = \infty$ a.s., so the law of the iterated logarithm fails. The last conclusion generalizes easily to situations in which

$$\sum_{m=1}^{\infty} P(|X_m| \geq (m \log \log m)^{1/2}) = \infty.$$

Strassen (1965) has shown an exact converse. If (9.6) holds, then $EX_i = 0$ and $EX_i^2 = 1$.

Exercise 9.5   Let X_n, $\mathscr{F}_n$ be a martingale difference sequence that satisfies the hypotheses of (7.5). Prove a law of the iterated logarithm for $S_n = X_1 + \cdots + X_n$.

Strassen's invariance principle   Let $X_1, X_2, \ldots$ be i.i.d. with $EX_i = 0$ and $EX_i^2 = 1$, let $S_n = X_1 + \cdots + X_n$, and let $S_{(n \cdot)}$ be the usual linear interpolation. Strassen (1964) proved:

(9.7) THEOREM   The limit set (i.e., the collection of limits of convergent subsequences) of

$$Z_n(\cdot) = (2n \log \log n)^{-1/2} S(n\cdot), \qquad n \geq 3$$

is $\mathcal{K} = \{f : f(x) = \int_0^x g(y) \, dy \text{ with } \int_0^1 g(y)^2 \, dy \leq 1\}$.

Jensen's inequality implies

$$f(1)^2 \leq \int_0^1 g(y)^2 \, dy \leq 1$$

with equality if and only if $f(t) = t$, so (9.7) contains (9.5) as a special case and provides some information about how the large values of S_n come about.

Exercise 9.6   Give a direct proof that under the hypotheses of (9.7), the limit set of $\{S_n/(2n \log \log n)^{1/2}\}$ is $[-1, 1]$.

History lesson   The LIL for sums of independent r.v.'s grew from the early efforts of Hausdorff (1913) and Hardy and Littlewood (1914) to determine the rate of convergence in Borel's theorem about normal numbers (see Example 2.4 in Chapter 6). The LIL for Bernoulli variables was reached in five steps:

Hausdorff (1913)	$S_n = o(n^{1/2+\varepsilon})$
Hardy and Littlewood (1914)	$S_n = O((n \log n)^{1/2})$
Steinhaus (1922)	$\limsup_{n \to \infty} S_n/(2n \log n)^{1/2} \leq 1$
Khintchine (1923)	$S_n = O((n \log \log n)^{1/2})$
Khintchine (1924)	$\limsup_{n \to \infty} S_n/((n/2) \log \log n)^{1/2} = 1$

We have $n/2$ instead of $2n$ in the last result, since Bernoulli variables have variance $1/4$. The LIL was proved for bounded independent r.v.'s by Kolmogorov (1929), who did not assume the summands were identically distributed. As mentioned earlier, (9.6) is due to Hartman and Wintner (1941). Strassen (1964) gave the proof (of (9.7) and hence (9.6)) using Skorokhod imbedding. The version above follows the treatment in Breiman (1968). For generalizations of (9.7) to martingales and processes with stationary increments, see Heyde and Scott (1973) and Hall and Heyde (1976, 1980).

There are versions of Kolmogorov's test for random walks. We say that c_n belongs to the upper class $\mathcal{U}$ if $P(S_n > c_n \text{ i.o.}) = 0$, and to the lower class $\mathcal{L}$ otherwise. For Bernoulli variables, P. Lévy (1937) showed

$$(2t \log \log t + a \log \log \log t)^{1/2} \begin{cases} \in \mathcal{U} & \text{if } a > 3 \\ \in \mathcal{L} & \text{if } a \leq 1. \end{cases}$$

Kolmogorov claimed and later Erdös (1942) proved $c_n \in \mathcal{U}$, $\mathcal{L}$ according as

$$\sum_{n=1}^{\infty} n^{-3/2} c_n \exp(-c_n^2/2n) < \infty, \; = \infty.$$

Feller (1943) gave a general version of the test for independent variables without assuming they all had the same distribution. His conditions are stronger than finite variance in the i.i.d. case.

Appendix:
The Essentials of Measure Theory

This appendix gives a complete treatment of the results from measure theory that we will need.

1 Lebesgue–Stieltjes Measures

To prove the existence of Lebesgue measure (and some related more general measures), we will use the Carathéodory extension theorem (1.1). To state that result we need several definitions in addition to the ones given in Section 1 of Chapter 1. A collection $\mathscr{A}$ of subsets of Ω is called an *algebra* (or *field*) if $A, B \in \mathscr{A}$ implies A^c and $A \cup B$ are in $\mathscr{F}$. Since $A \cap B = (A^c \cup B^c)^c$, it follows that $A \cap B \in \mathscr{F}$. Obviously a σ-algebra is an algebra. Two cases in which the converse is false are:

Example 1.1 $\Omega = \mathbb{Z} =$ the integers, $\mathscr{A} =$ the collection of $A \subset \mathbb{Z}$ so that A or A^c is finite.

Example 1.2 $\Omega = \mathbb{R}$, $\mathscr{A} =$ the collection of sets of the form

$$\bigcup_{i=1}^{k} (a_i, b_i] \quad \text{where } -\infty \le a_i < b_i \le \infty.$$

Exercise 1.1 Show that if $\mathscr{F}_1 \subset \mathscr{F}_2 \subset \cdots$ are σ-algebras, then $\bigcup_i \mathscr{F}_i$ is an algebra but need not be a σ-algebra.

Exercise 1.2 A set $A \subset \{1, 2, \ldots\}$ is said to have asymptotic density θ if

$$\lim_{n \to \infty} |A \cap \{1, 2, \ldots, n\}|/n = \theta.$$

Let $\mathscr{A}$ be the collection of sets for which the asymptotic density exists. Is $\mathscr{A}$ a σ-algebra? an algebra?

By a *measure on an algebra* $\mathscr{A}$ we mean a set function μ with

(i) $\mu(A) \geq \mu(\emptyset) = 0$ for all $A \in \mathcal{A}$, and

(ii) if $A_i \in \mathcal{A}$ are disjoint <u>and</u> <u>their</u> <u>union</u> <u>is</u> <u>in</u> $\mathcal{A}$, then $\mu(\bigcup_{i=1}^{\infty} A_i) = \sum_{i=1}^{\infty} \mu(A_i)$.

The underlined clause is unnecessary if $\mathcal{A}$ is a σ-algebra, so in that case the new definition coincides with the old one. The next exercise generalizes Exercise 1.1 in Chapter 1.

Exercise 1.3   We assume that all the sets mentioned are in $\mathcal{A}$.

(i) *Monotonicity:* If $A \subset B$, then $\mu(A) \leq \mu(B)$.

(ii) *Subadditivity:* If $A \subset \bigcup_i A_i$, then $\mu(A) \leq \sum_i \mu(A_i)$.

(iii) *Continuity from below:* If $A_i \uparrow A$ (i.e., $A_1 \subset A_2 \subset \cdots$ and $\bigcup_i A_i = A$), then $\mu(A_i) \uparrow \mu(A)$.

(iv) *Continuity from above:* If $A_i \downarrow A$ (i.e., $A_1 \supset A_2 \supset \cdots$ and $\bigcap_i A_i = A$) and $\mu(A_1) < \infty$, then $\mu(A_i) \downarrow \mu(A)$ as $i \uparrow \infty$.

μ is said to be *σ-finite* if there is a sequence of sets $A_n \in \mathcal{A}$, so that $\mu(A_n) < \infty$ and $\bigcup_n A_n = \Omega$. By using

$$A_n' = \bigcup_{m=1}^{n} A_m \quad \text{or} \quad A_n'' = A_n \cap \left(\bigcap_{m < n} A_m^c \right) \in \mathcal{A},$$

we can without loss of generality assume that $A_n \uparrow \Omega$ or the A_n are disjoint.

(1.1) Carathéodory's Extension Theorem   Let μ be a σ-finite measure on an algebra $\mathcal{A}$. Then μ has a unique extension to $\sigma(\mathcal{A}) =$ the smallest σ-algebra containing $\mathcal{A}$.

Exercise 1.4   Let $\mathbb{Z} =$ the integers and $\mathcal{A} =$ the collection of subsets so that A or A^c is finite. Let $\mu(A) = 0$ in the first case and $\mu(A) = 1$ in the second. Show that μ has no extension to $\sigma(\mathcal{A})$.

The next section is devoted to the proof of (1.1). To check the hypotheses of (1.1) for Lebesgue measure, we will prove a theorem, (1.3), that will be useful for other examples. To state that result we will need several definitions. A collection $\mathcal{S}$ of sets is said to be a *semialgebra* if (i) $S, T \in \mathcal{S}$ implies $S \cap T \in \mathcal{S}$, and (ii) if $S \in \mathcal{S}$, then S^c is a finite disjoint union of sets in $\mathcal{S}$. An important example of a semialgebra is $\mathcal{R}_o^d =$ the collection of sets of the form

$$(a_1, b_1] \times \cdots \times (a_d, b_d] \subset \mathbb{R}^d \quad \text{where } -\infty \leq a_i < b_i \leq \infty.$$

Exercise 1.5   Show that $\sigma(\mathcal{R}_o^d) = \mathcal{R}^d$, the Borel subsets of $\mathbb{R}^d$.

(1.2) LEMMA   If $\mathcal{S}$ is a semialgebra, then $\bar{\mathcal{S}} = \{$finite disjoint unions of sets in $\mathcal{S}\}$ is an algebra, called the *algebra generated by* $\mathcal{S}$.

Proof   Suppose $A = +_i S_i$ and $B = +_j T_j$ where $+$ denotes disjoint union and we assume the index sets are finite. Then $A \cup B = +_{i,j} S_i \cap T_j \in \bar{\mathcal{S}}$. As for complements, if $A = +_i S_i$, then $A^c = \bigcap_i S_i^c$. The definition of $\mathscr{S}$ implies $S_i^c \in \bar{\mathscr{S}}$. We have shown that $\bar{\mathscr{S}}$ is closed under intersection, so it follows by induction that $A^c \in \bar{\mathscr{S}}$. □

Let λ denote Lebesgue measure. The definition gives the values of λ on a semialgebra $\mathscr{S}$ (the half-open intervals). It is easy to see how to extend the definition to the algebra $\bar{\mathscr{S}}$ defined in (1.2). We let $\lambda(+_i(a_i, b_i]) = \sum_i (b_i - a_i)$. To assert that λ has an extension to $\sigma(\mathscr{S}) = \mathscr{R}$, we have to check that λ is a measure on $\mathscr{S}$; that is, if $A \in \mathscr{S}$ is a countable disjoint union of sets $A_i \in \mathscr{S}$, then $\lambda(A) = \sum_i \lambda(A_i)$. The next result simplifies that task somewhat.

(1.3) THEOREM   Let $\mathscr{S}$ be a semialgebra and let μ defined on $\mathscr{S}$ have $\mu(\varnothing) = 0$. Suppose (i) if $S \in \mathscr{S}$ is a finite disjoint union of sets $S_i \in \mathscr{S}$, then $\mu(S) = \sum_i \mu(S_i)$, and (ii) if $S \in \mathscr{S}$ is a countable disjoint union of sets $S_i \in \mathscr{S}$, then $\mu(S) \le \sum_i \mu(S_i)$. Then μ has a unique extension $\bar{\mu}$ that is a measure on $\bar{\mathscr{S}}$, the algebra generated by $\mathscr{S}$, and hence by (1.1) a unique extension ν that is a measure on $\sigma(\mathscr{S})$.

Proof   We define $\bar{\mu}$ on $\bar{\mathscr{S}}$ by $\bar{\mu}(A) = \sum_i \mu(S_i)$ whenever $A = +_i S_i$. To check that $\bar{\mu}$ is well defined suppose that $A = +_j T_j$ and observe $S_i = +_j(S_i \cap T_j)$ and $T_j = +_i(S_i \cap T_j)$, so (i) implies

$$\sum_i \mu(S_i) = \sum_{i,j} \mu(S_i \cap T_j) = \sum_j \mu(T_j).$$

To show that $\bar{\mu}$ is a measure on $\bar{\mathscr{S}}$ we observe that it follows from the definition that if $A = +_i B_i$ is a finite disjoint union of sets in $\bar{\mathscr{S}}$ and $B_i = +_j S_{i,j}$, then

$$\bar{\mu}(A) = \sum_{i,j} \mu(S_{i,j}) = \sum_i \bar{\mu}(B_i).$$

To extend the last conclusion to $A \in \bar{\mathscr{S}}$ that are countable disjoint unions $A = +_{i \ge 1} B_i$, we observe that each $B_i = +_j S_{i,j}$ with $S_{i,j} \in \mathscr{S}$ and $\sum_{i \ge 1} \bar{\mu}(B_i) = \sum_{i \ge 1, j} \mu(S_{i,j})$, so replacing the B_i's by $S_{i,j}$'s, we can without loss of generality suppose that the $B_i \in \mathscr{S}$. Now $A \in \bar{\mathscr{S}}$ implies $A = +_j T_j$ (a finite disjoint union) and $T_j = +_{i \ge 1} T_j \cap B_i$, so (ii) implies $\mu(T_j) \le \sum_{i \ge 1} \mu(T_j \cap B_i)$. Summing over j and observing that non-negative numbers can be summed in any order,

$$\bar{\mu}(A) = \sum_j \mu(T_j) \le \sum_{i \ge 1} \sum_j \mu(T_j \cap S_i) = \sum_{i \ge 1} \mu(S_i),$$

the last equality following from (i). To prove the opposite inequality, let $A_n = B_1 + \cdots + B_n$ and $C_n = A \cap A_n^c$. $C_n \in \bar{\mathscr{S}}$, since $\bar{\mathscr{S}}$ is an algebra, so finite additivity of $\bar{\mu}$ implies

$$\bar{\mu}(A) = \bar{\mu}(B_1) + \cdots + \bar{\mu}(B_n) + \bar{\mu}(C_n) \ge \bar{\mu}(B_1) + \cdots + \bar{\mu}(B_n),$$

and letting $n \to \infty$, $\bar{\mu}(A) \ge \sum_{i \ge 1} \bar{\mu}(B_i)$. □

Exercise 1.6   Let $\mathscr{S}$ be a semialgebra and μ a measure for which (i) in (1.3) holds. If $A, B_1, \ldots, B_n \in \mathscr{S}$ with $A \subset \bigcup_i B_i$, then $\mu(A) \leq \sum_i \mu(B_i)$.

With (1.3) established we are in a position to prove the existence of Lebesgue measure and a number of other measures on $(\mathbb{R}, \mathscr{R})$, where $\mathscr{R} =$ the Borel subsets of $\mathbb{R}$.

(1.4) THEOREM   Suppose F is (i) nondecreasing and (ii) right continuous; that is, $F(y) \downarrow F(x)$ when $y \downarrow x$. There is a unique measure μ on $(\mathbb{R}, \mathscr{R})$ with $\mu((a, b]) = F(b) - F(a)$ for all a, b.

Remark   A function F that has properties (i) and (ii) is called a *Stieltjes measure function*. To see the reasons for the two conditions observe that (a) if μ is a measure, then $F(b) - F(a) \geq 0$ and (b) part (iv) of Exercise 1.3 implies that if $\mu((a, y]) < \infty$ and $y \downarrow x > a$,

$$F(y) - F(a) = \mu((a, y]) \downarrow \mu((a, x]) = F(x) - F(a).$$

Conversely, if μ is a measure on $\mathbb{R}$ with $\mu((a, b]) < \infty$ when $-\infty < a < b < \infty$, then

$$F(x) = \begin{cases} c + \mu((0, x]) & \text{for } x \geq 0 \\ c - \mu((x, 0]) & \text{for } x < 0 \end{cases}$$

is a function F with $F(b) - F(a) = \mu((a, b])$, and any such function has this form with $c = F(0)$.

Proof   We begin by observing that

$$F(\infty) = \lim_{x \uparrow \infty} F(x) \quad \text{and} \quad F(-\infty) = \lim_{x \downarrow -\infty} F(x) \quad \text{exist},$$

and $\mu((a, b]) = F(b) - F(a)$ makes sense for all $-\infty \leq a < b \leq \infty$, since $F(\infty) > -\infty$ and $F(-\infty) < \infty$. It is clear that condition (i) in (1.3) holds, so we only have to check (ii). Suppose first that $-\infty < a < b < \infty$, and $(a, b] \subset \bigcup_{i \geq 1} (a_i, b_i]$ where (without loss of generality) $-\infty < a_i < b_i < \infty$. Pick $\delta > 0$ so that $F(a + \delta) < F(a) + \varepsilon$ and pick η_i so that

$$F(b_i + \eta_i) < F(b_i) + \varepsilon 2^{-i}.$$

The open intervals $(a_i, b_i + \eta_i)$ cover $[a + \delta, b]$ so there is a finite subcover (α_j, β_j), $1 \leq j \leq J$, with $\alpha_j < \beta_{j-1}$ for $2 \leq j \leq J$. (If a subcover is minimal—that is, no proper subcollection covers, it has this property.) From the monotonicity of F,

$$F(b) - F(a + \delta) \leq \sum_{j=1}^{J} F(\beta_j) - F(\alpha_j) \leq \sum_{i=1}^{\infty} F(b_i + \eta_i) - F(a_i),$$

so by the choice of δ and η_i,

$$F(b) - F(a) \le 2\varepsilon + \sum_{i=1}^{\infty} F(b_i) - F(a_i),$$

and since ε is arbitrary, we have proved the result in the case $-\infty < a < b < \infty$. To remove the last restriction observe that if $(a, b] \subset \bigcup_i (a_i, b_i]$ and $(A, B] \subset (a, b]$ has $-\infty < A < B < \infty$, then we have

$$F(B) - F(A) \le \sum_{i=1}^{\infty} F(b_i) - F(a_i).$$

The last result holds for any finite $(A, B] \subset (a, b]$, so the desired result follows. □

In Section 6 we will give a short proof of the existence of Lebesgue measure on $\mathbb{R}^d$, so we leave it to the reader to generalize the last proof to $\mathbb{R}^d$.

Exercise 1.7   Show that there is a unique measure λ on $(\mathbb{R}^d, \mathscr{R}^d)$ with

$$\lambda((a_1, b_1] \times \cdots \times (a_d, b_d]) = \prod_{i=1}^{d} (b_i - a_i) \quad \text{whenever } -\infty \le a_i < b_i \le \infty.$$

Sketch   The proof of (ii) in (1.3) is almost the same as before, but you might find Exercise 1.6 a useful substitute for the telescoping sum. To check (i), call $A = +_k B_k$ a *regular subdivision* of A if there are sequences $a_i = \alpha_{i,0} < \alpha_{i,1} < \cdots < \alpha_{i,n_i} = b_i$, so that each rectangle B_k has the form

$$(\alpha_{1,j_1-1}, \alpha_{1,j_1}] \times \cdots \times (\alpha_{d,j_d-1}, \alpha_{d,j_d}] \quad \text{where } 1 \le j_i \le n_i.$$

It is easy to see that for regular subdivisions $\lambda(A) = \sum_k \lambda(B_k)$. To extend this result to a general finite subdivision $A = +_j A_j$, subdivide further to get a regular one.

2   Carathéodory's Extension Theorem

This section is devoted to the proof of (1.1). The proof is slick but rather mysterious. The reader should not worry too much about the details but concentrate on the structure of the proof and the definitions introduced.

Uniqueness   We will prove that the extension is unique before tackling the more difficult problem of proving its existence. The key to our uniqueness proof is Dynkin's $\pi - \lambda$ theorem, a result that we will use many times below. As usual, we need a few definitions before we can state the result. $\mathscr{P}$ is said to be a *π-system* if it is closed under intersections. $\mathscr{L}$ is said to be a *λ-system* if it satisfies: (i) $\Omega \in \mathscr{L}$. (ii) If $A, B \in \mathscr{L}$ and $A \subset B$, then $B - A \in \mathscr{L}$. (iii) If $A_n \in \mathscr{L}$ and $A_n \uparrow A$, then $A \in \mathscr{L}$. The collection of rectangles $(a_1, b_1] \times \cdots \times (a_d, b_d]$ is a π-system. The reader will see in a moment that the next result is just what we need to prove uniqueness of the extension.

(2.1) $\pi - \lambda$ Theorem   If $\mathscr{P}$ is a π-system and $\mathscr{L}$ is a λ-system that contains $\mathscr{P}$, then $\sigma(\mathscr{P}) \subset \mathscr{L}$.

Proof   We will show that if $l(\mathscr{P})$ is the smallest λ-system containing $\mathscr{P}$, then $l(\mathscr{P})$ is a σ-field so $\sigma(\mathscr{P}) \subset l(\mathscr{P}) \subset \mathscr{L}$. To do this we begin by observing that a λ-system that is closed under intersection is a σ-field, since:

$$\text{if} \quad A \in \mathscr{L}, \quad \text{then} \quad A^c = \Omega - A \in \mathscr{L}$$

$$A \cup B = (A^c \cap B^c)^c$$

$$\bigcup_{i=1}^{n} A_i \uparrow \bigcup_{i=1}^{\infty} A_i \quad \text{as } n \uparrow \infty.$$

It is enough then to show that $l(\mathscr{P})$ is closed under intersection. To do this let $\mathscr{G}_A$ be the class of sets B such that $A \cap B \in l(\mathscr{P})$. It is easy to see that if $A \in \mathscr{P}$, then $\mathscr{G}_A$ is a λ-system. To check this observe: (i) $\Omega \in \mathscr{G}_A$ since $A \in \mathscr{P}$, (ii) $A \cap (B - C) = (A \cap B) - (A \cap C)$, (iii) if $B_n \uparrow B$, then $(A \cap B_n) \uparrow (A \cap B)$. Since $\mathscr{P} \subset \mathscr{G}_A$ by hypothesis, it follows that $l(\mathscr{P}) \subset \mathscr{G}_A$. Thus $A \in \mathscr{P}$ implies $l(\mathscr{P}) \subset \mathscr{G}_A$ or, put another way: if $A \in \mathscr{P}$ and $B \in l(\mathscr{P})$, then $A \cap B \in l(\mathscr{P})$. The last result says that if $B \in l(\mathscr{P})$, then $\mathscr{G}_B \supset \mathscr{P}$ and hence $\mathscr{G}_B \supset l(\mathscr{P})$, but this is what we wanted to show: if $B, C \in l(\mathscr{P})$, then $B \cap C \in l(\mathscr{P})$.   □

To prove that the extension in (1.1) is unique, we will show:

(2.2) THEOREM   Let $\mathscr{P}$ be a π-system. If v_1 and v_2 are measures that agree on $\mathscr{P}$ and there is a sequence $A_n \in \mathscr{P}$ with $A_n \uparrow \Omega$ and $v_i(A_n) < \infty$, then v_1 and v_2 agree on $\sigma(\mathscr{P})$.

Proof   Let $A \in \mathscr{P}$ have $v_1(A) = v_2(A) < \infty$. Let

$$\mathscr{L} = \{B : v_1(A \cap B) = v_2(A \cap B)\}.$$

Clearly $\Omega \in \mathscr{L}$. If $B \subset C \in \mathscr{L}$, then

$$v_1(A \cap (B - C)) = v_1(A \cap B) - v_1(A \cap C)$$

$$= v_2(A \cap B) - v_2(A \cap C) = v_2(A \cap (B - C)).$$

Here we use the fact that $v_1(A) = v_2(A) < \infty$. Finally if $B_n \in \mathscr{L}$ and $B_n \uparrow B$, then part (iii) of Exercise 1.3 implies

$$v_1(A \cap B) = \lim_{n \to \infty} v_1(A \cap B_n) = \lim_{n \to \infty} v_2(A \cap B_n) = v_2(A \cap B).$$

Since $\mathscr{P}$ is closed under intersection, the $\pi - \lambda$ theorem implies $\mathscr{L} \supset \sigma(\mathscr{P})$; that is, if $A \in \mathscr{P}$ with $v_1(A) = v_2(A) < \infty$ and $B \in \sigma(\mathscr{P})$, then $v_1(A \cap B) = v_2(A \cap B)$. Letting

$A_n \in \mathscr{P}$ with $A_n \uparrow \Omega$, $v_1(A_n) = v_2(A_n) < \infty$, and using the last result and part (iii) of Exercise 1.3, we have the desired conclusion. □

Exercise 2.1 Give an example of two probability measures $\mu \neq v$ on $\mathscr{F} = $ all subsets of $\{1, 2, 3, 4\}$ that agree on a collection of sets $\mathscr{C}$ with $\sigma(\mathscr{C}) = \mathscr{F}$; that is, the smallest σ-algebra containing $\mathscr{C}$ is $\mathscr{F}$.

Existence Our next step is to show that a measure (not necessarily σ-finite) defined on an algebra $\mathscr{A}$ has an extension to the σ-algebra generated by $\mathscr{A}$. If $E \subset \Omega$, we let $\mu^*(E) = \inf \sum_i \mu(A_i)$ where the infimum is taken over all sequences from $\mathscr{A}$ so that $E \subset \bigcup_i A_i$. Intuitively, if v is a measure that agrees with μ on $\mathscr{A}$, then it follows from part (ii) of Exercise 1.1 that

$$v(E) \leq v\left(\bigcup_i A_i\right) \leq \sum_i v(A_i),$$

so $\mu^*(E)$ is an upper bound on the measure of E. Intuitively, the measurable sets are the ones for which the upper bound is tight. Formally, we say that E is *measurable* if

$$\mu^*(F) = \mu^*(F \cap E) + \mu^*(F \cap E^c) \quad \text{for all sets } F. \tag{$*$}$$

The last definition is not very intuitive, but we will see in the proofs below that it works very well. We begin by showing that our new definitions extend the old.

(2.3) LEMMA If $A \in \mathscr{A}$, then $\mu^*(A) = \mu(A)$ and A is measurable.

Proof The first conclusion follows from part (ii) of Exercise 1.3. To prove the second we observe that it is immediate from the definition that μ^* has the following properties:

(i) *Monotonicity*: If $E \subset F$, then $\mu^*(E) \leq \mu^*(F)$.
(ii) *Subadditivity*: If $F \subset \bigcup_i F_i$, a finite or countable union, then $\mu^*(F) \leq \sum_i \mu^*(F_i)$.

Any set function with $\mu^*(\emptyset) = 0$ that satisfies (i) and (ii) is called an *outer measure*. Using (ii) with $F_1 = F \cap E$ and $F_2 = F \cap E^c$ (and $F_i = \emptyset$ otherwise) we see that to prove a set is measurable it is enough to show

$$\mu^*(F) \geq \mu^*(F \cap E) + \mu^*(F \cap E^c). \tag{$*'$}$$

Since the last inequality is trivial when $\mu^*(F) = \infty$, we can without loss of generality assume $\mu^*(F) < \infty$. To prove that $(*')$ holds when $E = A \in \mathscr{A}$, we observe that since $\mu^*(F) < \infty$, there is a sequence $B_i \in \mathscr{A}$ so that $\bigcup_i B_i \supset F$ and

$$\sum_i \mu(B_i) \leq \mu^*(F) + \varepsilon.$$

Since μ is additive on $\mathscr{A}$ and $\mu = \mu^*$ on $\mathscr{A}$, we have

$$\mu(B_i) = \mu^*(B_i \cap A) + \mu^*(B_i \cap A^c).$$

Summing over i and using the subadditivity of μ^* give

$$\mu^*(F) + \varepsilon \geq \sum_i \mu^*(B_i \cap A) + \sum_i \mu^*(B_i \cap A^c) \geq \mu^*(F \cap A) + \mu^*(F^c \cap A),$$

which proves the desired result, since ε is arbitrary.   □

(2.4) LEMMA   The class $\mathscr{A}^*$ of measurable sets is a σ-field and the restriction of μ^* to $\mathscr{A}^*$ is a measure.

Remark   This result is true for any outer measure.

Proof   It is clear from the definition that:

(a) If E is measurable, then E^c is.

Our first nontrivial task is to prove:

(b) If E_1 and E_2 are measurable, then $E_1 \cup E_2$ and $E_1 \cap E_2$ are.

Proof of (b)   To prove the first conclusion we observe that subadditivity implies

$$\mu^*(G \cap (E_1 \cup E_2)) + \mu^*(G \cap (E_1^c \cap E_2^c))$$

$$\leq \mu^*(G \cap E_1) + \mu^*(G \cap E_1^c \cap E_2) + \mu^*(G \cap E_1^c \cap E_2^c)$$

$$= \mu^*(G \cap E_1) + \mu^*(G \cap E_1^c) = \mu^*(G),$$

the two equalities following from the measurability of E_2 (let $F = G \cap E_1^c$ in (*)) and E_1, respectively. To prove that $E_1 \cap E_2$ is measurable, we observe $E_1 \cap E_2 = (E_1^c \cup E_2^c)^c$ and use (a).   □

(c) Let G be any set and $E_1, \ldots, E_n$ be disjoint measurable sets. Then

$$\mu^*\left(G \cap \left(\bigcup_{i=1}^n E_i\right)\right) = \sum_{i=1}^n \mu^*(G \cap E_i).$$

Proof of (c)   Let $F_m = \bigcup_{i \leq m} E_i$. E_n is measurable, $F_n \supset E_n$, and $F_{n-1} \cap E_n = \varnothing$, so

$$\mu^*(G \cap F_n) = \mu^*(G \cap F_n \cap E_n) + \mu^*(G \cap F_n \cap E_n^c)$$

$$= \mu^*(G \cap E_n) + \mu^*(G \cap F_{n-1}).$$

The desired result follows from this by induction.   □

(d) If the sets E_i are measurable, then $E = \bigcup_i E_i$ is measurable.

Proof of (d)   Let $E'_i = E_i \cap (\bigcap_{j<i} E^c_j)$. (a) and (b) imply E'_i is measurable, so we can suppose without loss of generality that the E_i are pairwise disjoint. Let $F_n = E_1 \cup \cdots \cup E_n$. F_n is measurable by (b), so

$$\mu^*(G) = \mu^*(G \cap F_n) + \mu^*(G \cap F^c_n)$$

$$\geq \mu^*(G \cap F_n) + \mu^*(G \cap E^c)$$

$$= \sum_{i=1}^{n} \mu^*(G \cap E_i) + \mu^*(G \cap E^c)$$

by (c). Letting $n \to \infty$ and using subadditivity,

$$\mu^*(G) \geq \sum_{i=1}^{\infty} \mu^*(G \cap E_i) + \mu^*(G \cap E^c) \geq \mu^*(G \cap E) + \mu^*(G \cap E^c). \qquad \square$$

The last step in the proof of (2.4) is:

(e) If $E = \bigcup_i E_i$ where $E_1, E_2, \ldots$ are disjoint and measurable, then

$$\mu^*(E) = \sum_{i=1}^{\infty} \mu^*(E_i).$$

Proof of (e)   Let $F_n = E_1 \cup \cdots \cup E_n$. By monotonicity and (c),

$$\mu^*(E) \geq \mu^*(F_n) = \sum_{i=1}^{n} \mu^*(E_i).$$

Letting $n \to \infty$ now and using subadditivity give the desired conclusion. $\qquad \square$

Exercise 2.2   Show that if $\mu^*(\Omega) < \infty$, a necessary and sufficient condition for E to be measurable is $\mu^*(\Omega) = \mu^*(E) + \mu^*(E^c)$.

Exercise 2.3   An outer measure on the subsets of $\mathbb{R}^d$ is said to be a *metric outer measure* if

$$\mu^*(A \cup B) = \mu^*(A) + \mu^*(B)$$

whenever $\text{dist}(A, B) \equiv \inf\{|x - y| : x \in A, y \in B\} > 0$. Show that in this case all Borel sets are measurable.

Exercise 2.4   *Hausdorff measure.* Suppose $h(t)$ is a continuous increasing function defined on $(0, y)$ with $y > 0$ that has $h(x) \downarrow 0$ when $x \downarrow 0$. If Ω is any metric space, let

$$\mu^*_h(E) = \lim_{\delta \to 0} \left(\inf \sum_{i=1}^{\infty} h(\text{diameter}(C_i)) \right)$$

where the infimum is taken over all covers of E by sets C_i, each of which has diameter

$< \delta$. Show μ_h^* is an outer measure. If we use (*) to define the measurable sets, then the resulting measure is called the Hausdorff measure with respect to h.

Exercise 2.5 Letting $h(t) = t^\alpha$ in the last exercise gives an outer measure μ_α^* that we will call the α-*dimensional Hausdorff measure*. Let $\Omega = \mathbb{R}^2$ and $\alpha = 1$. Show that if $\varphi: [0, 1] \to \mathbb{R}^2$ is 1–1 and differentiable, then

$$\mu_1^*(A) = \int_0^1 |\varphi'(s)| \, ds$$

where $|y| = $ the length of the vector y. Can you prove that A is measurable?

Remark A α-dimensional Hausdorff measure on $\mathbb{R}^d$ with $\alpha < d$ is an example of a "nice" measure that is not σ-finite.

Exercise 2.6 Let μ_α^* be as in the last exercise. If $A \subset \mathbb{R}^d$, $\dim(A) = \sup\{\alpha: \mu_\alpha^*(A) = \infty\}$ is called the *Hausdorff dimension* of A. Let $C = $ the Cantor set defined by removing $(1/3, 2/3)$ from $[0, 1]$ and then removing the middle third of each interval that remains. Show that $\dim(C) \le \log 2/\log 3$.

Remark The upper bound for $\dim(C)$ is actually the right answer, but that is considerably more difficult to establish. The answer may look strange at first but can be derived from a simple heuristic: Multiplying the Cantor set by $m = 3$ produces $k = 2$ copies of it. To see why this means the dimension is $\log k/\log m$, observe that multiplying a d-dimensional cube by $m = 2$ results in $k = 2^d$ copies.

3 Completion, Etc.

The proof of (1.1) given in the last section defines an extension on $\mathcal{A}^* \supset \sigma(\mathcal{A})$. Our next goal is to describe the relationship between these two σ-algebras. Let $\mathcal{A}_\sigma$ denote the collection of countable unions of sets in $\mathcal{A}$, and $\mathcal{B}_\delta$ denote the collection of countable intersections of sets in $\mathcal{B}$.

(3.1) LEMMA Let E be any set E with $\mu^*(E) < \infty$. (i) For any $\varepsilon > 0$, there is an $A \in \mathcal{A}_\sigma$ with $A \supset E$ and $\mu^*(A) \le \mu^*(E) + \varepsilon$. (ii) There is a $B \in \mathcal{A}_{\sigma\delta}$ with $B \supset E$ and $\mu^*(B) = \mu^*(E)$.

Proof By the definition of μ^*, there is a sequence A_i so that $A \equiv \bigcup_i A_i \supset E$ and $\sum_i \mu(A_i) \le \mu^*(E) + \varepsilon$. The definition of μ^* implies $\mu^*(A) \le \sum_i \mu(A_i)$, establishing (i). For (ii) let $A_n \in \mathcal{A}_\sigma$ with $A_n \supset E$ and $\mu^*(A_n) \le \mu^*(E) + 1/n$, and let $B = \bigcap_n A_n$. Clearly $B \in \mathcal{A}_{\sigma\delta}$, $B \supset E$, and hence by monotonicity, $\mu^*(B) \ge \mu^*(E)$. To prove the other inequality notice that $B \subset A_n$ and hence $\mu^*(B) \le \mu^*(A_n) \le \mu^*(E) + 1/n$ for any n. □

Exercise 3.1 Let $\mathcal{A}$ be an algebra, μ a measure on $\sigma(\mathcal{A})$, and $B \in \sigma(\mathcal{A})$ with $\mu(B) < \infty$. For any $\varepsilon > 0$ there is an $A \in \mathcal{A}$ with $\mu(A \triangle B) < \varepsilon$, where $A \triangle B = (A - B) \cup (B - A)$.

(3.2) THEOREM Suppose μ is σ-finite on $\mathscr{A}$. $B \in \mathscr{A}^*$ if and only if there is an $A \in \mathscr{A}_{\sigma\delta}$ and a set N with $\mu^*(N) = 0$ so that $B = A - N \, (= A \cap N^c)$.

Proof It follows from (2.3) and (2.4) if $A \in \mathscr{A}_{\sigma\delta}$, then $A \in \mathscr{A}^*$. $(*')$ in Section 2 and monotonicity imply sets with $\mu^*(N) = 0$ are measurable, so using (2.4) again, it follows that $A \cap N^c \in \mathscr{A}^*$. To prove the other direction, let Ω_i be a disjoint collection of sets with $\mu(\Omega_i) < \infty$ and $\Omega = \bigcup_i \Omega_i$. Let $B_i = B \cap \Omega_i$ and use (3.1) to find $A_i^n \in \mathscr{A}_\sigma$ so that $A_i^n \supset B_i$ and $\mu(A_i^n) \leq \mu^*(E_i) + 1/n2^i$. Let $A_n = \bigcup_i A_i^n$. $B \subset A_n$ and

$$A_n - B \subset \sum_{i=1}^{\infty} (A_i^n - B_i),$$

so by subadditivity

$$\mu^*(A_n - B) \leq \sum_{i=1}^{\infty} \mu^*(A_i^n - B_i) \leq 1/n.$$

Since $A_n \in \mathscr{A}_\sigma$, the set $A = \bigcap_n A_n \in \mathscr{A}_{\sigma\delta}$. Clearly $A \supset B$. Since $N \equiv A - B \subset A_n - B$ for all n, monotonicity implies $\mu^*(N) = 0$ and the proof of (3.2) is complete. □

A measure space $(\Omega, \mathscr{F}, \mu)$ is said to be *complete* if $\mathscr{F}$ contains all subsets of sets of measure 0. In the proof of (3.2) we showed that $(\Omega, \mathscr{A}^*, \mu^*)$ is complete. Our next result shows that $(\Omega, \mathscr{A}^*, \mu^*)$ is the completion of $(\Omega, \sigma(\mathscr{A}), \mu^*)$.

(3.3) THEOREM If $(\Omega, \mathscr{F}, \mu)$ is a measure space, then there is a complete measure space $(\Omega, \bar{\mathscr{F}}, \bar{\mu})$, called the *completion* of $(\Omega, \mathscr{F}, \mu)$, so that: (i) $E \in \bar{\mathscr{F}}$ if and only if $E = A \cup B$ where $A \in \mathscr{F}$ and $B \subset N$ with $\mu(N) = 0$, and (ii) $\bar{\mu}$ agrees with μ on $\mathscr{F}$.

Proof The first step is to check that $\bar{\mathscr{F}}$ is a σ-algebra. If $E_i = A_i \cup B_i$ where $A_i \in \mathscr{F}$ and $B_i \subset N_i$ where $\mu(N_i) = 0$, then $\bigcup_i A_i \in \mathscr{F}$ and subadditivity implies $\mu(\bigcup_i N_i) \leq \sum_i \mu(N_i) = 0$, so $\bigcup_i E_i \in \bar{\mathscr{F}}$. As for complements, if $E = A \cup B$ and $B \subset N$, then

$$E^c = A^c \cap B^c = (A^c \cap N^c) \cup (A^c \cap B^c \cap N).$$

$A^c \cap N^c$ is in $\mathscr{F}$ and $A^c \cap B^c \cap N \subset N$, so $E^c \in \bar{\mathscr{F}}$.

We define $\bar{\mu}$ in the obvious way: If $E = A \cup B$ where $A \in \mathscr{F}$ and $B \subset N$ where $\mu(N) = 0$, then we let $\bar{\mu}(E) = \mu(A)$. The first thing to show is that $\bar{\mu}$ is well defined; that is, if $E = A_i \cup B_i$, $i = 1, 2$, are two decompositions, then $\mu(A_1) = \mu(A_2)$. Let $A_0 = A_1 \cap A_2$ and $B_0 = B_1 \cup B_2$. $E = A_0 \cup B_0$ is a third decomposition with $A_0 \in \mathscr{F}$ and $B_0 \subset N_1 \cup N_2$, and has the pleasant property that if $i = 1$ or 2,

$$\mu(A_0) \leq \mu(A_i) \leq \mu(A_0) + \mu(N_1 \cup N_2) = \mu(A_0).$$

The last detail is to check that $\bar{\mu}$ is a measure, but that is easy. If $E_i = A_i \cup B_i$ are disjoint, then $\bigcup_i E_i$ can be decomposed as $\bigcup_i A_i \cup (\bigcup_i B_i)$ and the $A_i \subset E_i$ are disjoint, so

$$\bar{\mu}\left(\bigcup_i E_i\right) = \mu\left(\bigcup_i A_i\right) = \sum_i \mu(A_i) = \sum_i \bar{\mu}(E_i). \qquad \square$$

Remark   When $\mathcal{F} = \mathcal{R}$, the Borel subsets of $\mathbb{R}$, the completion $\bar{\mathcal{R}}$ is much larger than $\mathcal{R}$, at least from the standpoint of set theory. Let c denote the cardinality of the real line. Consulting Hewitt and Stromberg (1965), pp. 133–134, we see that then there are only c sets in $\mathcal{R}$, but an easy argument shows that there are at least 2^c sets in $\bar{\mathcal{R}}$.

Exercise 3.2   Use the Cantor set to conclude that there are 2^c sets in $\bar{\mathcal{R}}$.

Exercise 1.7 allows us to construct Lebesgue measure λ on $(\mathbb{R}^d, \mathcal{R}^d)$. Using (3.3) we can extend λ to be a measure on $(\mathbb{R}, \bar{\mathcal{R}}^d)$ where $\bar{\mathcal{R}}^d$ is the completion of $\mathcal{R}^d$. Having done this it is natural (if somewhat optimistic) to ask: Are there any sets that are not in $\bar{\mathcal{R}}^d$? The answer is yes, and we will now give an example of a nonmeasurable B in $\mathbb{R}$.

A Nonmeasurable Subset of [0, 1)

The key to our construction is the observation that λ is translation invariant; that is, if $A \in \bar{\mathcal{R}}$ and $x + A = \{x + y : y \in A\}$, then $x + A \in \bar{\mathcal{R}}$ and $\lambda(A) = \lambda(x + A)$. We say that $x, y \in [0, 1)$ are equivalent and write $x \sim y$ if $x - y$ is a rational number. By the axiom of choice there is a set B that contains exactly one element from each equivalence class. B is our nonmeasurable set. To prove that $B \notin \bar{\mathcal{R}}$, we prove:

(3.4) LEMMA   If $E \subset [0, 1)$ is in $\bar{\mathcal{R}}$, $x \in (0, 1)$, and $x +' E = \{(x + y) \bmod 1 : y \in E\}$, then $\lambda(E) = \lambda(x +' E)$.

Proof   Let $A = E \cap [0, 1 - x)$ and $B = E \cap [1 - x, 1)$. Let $A' = x + A = \{x + y : y \in A\}$ and $B' = x - 1 + B$. $A, B \in \bar{\mathcal{R}}$, so by translation invariance $A', B' \in \bar{\mathcal{R}}$ and $\lambda(A) = \lambda(A'), \lambda(B) = \lambda(B')$. Since $A' \subset [x, 1)$ and $B' \subset [0, x)$ are disjoint,

$$\lambda(E) = \lambda(A) + \lambda(B) = \lambda(A') + \lambda(B') = \lambda(x +' E).$$

From (3.4), it follows easily that B is not measurable; if it were, then $q +' B$, $q \in \mathbb{Q} \cap [0, 1)$ would be a countable disjoint collection of measurable subsets of $[0, 1)$ all with the same measure α. If $\alpha > 0$, then $\lambda([0, 1)) = \infty$, and if $\alpha = 0$, then $\lambda([0, 1)) = 0$. Neither conclusion is compatible with the fact that $\lambda([0, 1)) = 1$, so $B \notin \bar{\mathcal{R}}$.   □

Exercise 3.3   Show that if $A \subset \mathbb{R}$ has $\lambda(A) > 0$, there is a nonmeasurable $S \subset A$.

Letting $B' = B \times [0, 1]^{d-1}$ where B is our nonmeasurable subset of $(0, 1)$, we get a nonmeasurable set in $d > 1$. In $d = 3$ there is a much more interesting example, but we need the reader to do some preliminary work.

Exercise 3.4   In Euclidean geometry, two subsets of $\mathbb{R}^d$ are said to be *congruent* if one set can be mapped onto the other by translations and rotations. Show that two congruent measurable sets must have the same Lebesgue measure.

Sketch   Show that the definition of $\mu^*(A)$ is unchanged if we cover A by balls instead of by rectangles.

Banach–Tarski Theorem

Banach and Tarski (1924) used the axiom of choice to show that it is possible to partition the sphere $\{x: |x| \leq 1\}$ in $\mathbb{R}^3$ into a finite number of sets $A_1, \ldots, A_n$ and find congruent sets $B_1, \ldots, B_n$ whose union is two disjoint spheres of radius 1! Since congruent sets have the same Lebesgue measure, at least one of the sets A_i must be nonmeasurable. The construction relies on the fact that the group generated by rotations in $\mathbb{R}^3$ is not abelian. Lindenbaum (1926) showed that this cannot be done with any bounded set in $\mathbb{R}^2$. For a popular account of the Banach–Tarski theorem, see French (1988).

Solovay's Theorem

The axiom of choice played an important role in the last two constructions of nonmeasurable sets. Solovay (1970) proved that its use is unavoidable. In his own words, "We show that the existence of a non-Lebesgue measurable set cannot be proved in Zermelo–Frankel set theory if the use of the axiom of choice is disallowed." This should convince the reader that all subsets of $\mathbb{R}^d$ that arise "in practice" are in $\bar{\mathcal{R}}^d$.

4 Integration

Let μ be a σ-finite measure on $(\Omega, \mathcal{F})$. In this section we will define $\int f \, d\mu$ for a class of measurable functions. This is a four-step procedure:

Step 1 φ is said to be a *simple function* if

$$\varphi(\omega) = \sum_{i=1}^{n} a_i 1_{A_i}$$

and A_i are disjoint sets with $\mu(A_i) < \infty$. If φ is a simple function, we let

$$\int \varphi \, d\mu = \sum_{i=1}^{n} a_i \mu(A_i).$$

The representation of φ is not unique, since we have not supposed that the a_i are distinct. However, it is easy to see that the last definition does not contradict itself.

We will prove the next three conclusions four times, but before we can state them for the first time we need a definition. $\varphi \geq \psi$ μ *almost everywhere* (or $\varphi \geq \psi$ μ a.e.) means $\mu(\{\omega: \varphi(\omega) < \psi(\omega)\}) = 0$. When there is no doubt about what measure we are referring to, we drop the μ.

(4.1) LEMMA   Let φ and ψ be simple functions.

(i) If $\varphi \geq 0$ a.e., then $\int \varphi \, d\mu \geq 0$.
(ii) $\int a\varphi \, d\mu = a \int \varphi \, d\mu$.
(iii) $\int \varphi + \psi \, d\mu = \int \varphi \, d\mu + \int \psi \, d\mu$.

Proof   (i) and (ii) are immediate consequences of the definition. To prove (iii) suppose

$$\varphi = \sum_{i=1}^{m} a_i 1_{A_i} \quad \text{and} \quad \psi = \sum_{j=1}^{n} b_j 1_{B_j}.$$

To make the supports of the two functions the same we let $A_0 = \bigcup_i B_i - \bigcup_i A_i$, let $B_0 = \bigcup_i A_i - \bigcup_i B_i$, and let $a_0 = b_0 = 0$. Now

$$\varphi + \psi = \sum_{i=0}^{m} \sum_{j=0}^{n} (a_i + b_j) 1_{(A_i \cap B_j)}$$

and the $A_i \cap B_j$ are pairwise disjoint, so

$$\int (\varphi + \psi) \, d\mu = \sum_{i=0}^{m} \sum_{j=0}^{n} (a_i + b_j) \mu(A_i \cap B_j)$$

$$= \sum_{i=0}^{m} \sum_{j=0}^{n} a_i \mu(A_i \cap B_j) + \sum_{j=0}^{n} \sum_{i=0}^{m} b_j \mu(A_i \cap B_j)$$

$$= \sum_{i=0}^{m} a_i \mu(A_i) + \sum_{j=0}^{n} b_j \mu(B_j) = \int \varphi \, d\mu + \int \psi \, d\mu.$$

In the next to last step we used the fact that $A_i = +_j(A_i \cap B_j)$ and $B_j = +_i(A_i \cap B_j)$ where $+$ denotes a disjoint union. It is for this reason that we introduced B_0 and A_0.   □

We will prove (i)–(iii) three more times as we generalize our integral. See (4.3), (4.5), and (4.7). As a consequence of (i)–(iii) we get three more useful properties. To keep from repeating their proofs, which do not change, we will prove:

(4.2) LEMMA   If (i)–(iii) hold, then we have the following:

(iv) If $\varphi \leq \psi$ a.e., then $\int \varphi \, d\mu \leq \int \psi \, d\mu$.
(v) If $\varphi = \psi$ a.e., then $\int \varphi \, d\mu = \int \psi \, d\mu$.
(vi) $|\int f \, d\mu| \leq \int |f| \, d\mu$.

Proof   By (iii), $\int \varphi \, d\mu = \int \psi \, d\mu + \int (\varphi - \psi) \, d\mu$ and the second integral is ≥ 0 by (i), so (iv) holds. $\varphi = \psi$ a.e. implies $\varphi \leq \psi$ a.e. and $\psi \leq \varphi$ a.e., so (v) follows from two applications of (iv). To prove (vi) now notice that $f \leq |f|$, so (iv) implies $\int f \, d\mu \leq \int |f| \, d\mu$. $-f \leq |f|$, so (iv) and (ii) imply $-\int f \, d\mu \leq \int |f| \, d\mu$. Since $|y| = \max(y, -y)$, the result follows.   □

Step 2   Let E be a set with $\mu(E) < \infty$ and let f be a bounded function that vanishes on E^c. To define the integral of f we observe that if φ, ψ are simple functions that have $\varphi \leq f \leq \psi$, then we want to have

$$\int \varphi \, d\mu \leq \int f \, d\mu \leq \int \psi \, d\mu,$$

so we let

$$\int f \, d\mu = \sup_{\varphi \leq f} \int \varphi \, d\mu = \inf_{\psi \geq f} \int \psi \, d\mu. \tag{*}$$

Here and for the rest of Step 2 we assume that φ and ψ vanish on E^c. To justify (*) we have to prove that the sup and inf are equal. It follows from (iv) that

$$\sup_{\varphi \leq f} \int \varphi \, d\mu \leq \inf_{\psi \geq f} \int \psi \, d\mu.$$

To prove the other inequality, suppose $|f| \leq M$ and let

$$E_k = \{x \in E : kM/n \geq f(x) > (k-1)M/n\} \quad \text{for } -n \leq k \leq n$$

$$\psi_n(x) = \sum_{k=-n}^{n} kM/n 1_{E_k}$$

$$\varphi_n(x) = \sum_{k=-n}^{n} (k-1)M/n 1_{E_k}.$$

By definition $\psi_n(x) - \varphi_n(x) = (M/n)1_E$, so

$$\int \psi_n(x) - \varphi_n(x) \, d\mu = (M/n)\mu(E).$$

Since $\varphi_n(x) \leq f(x) \leq \psi_n(x)$, it follows that

$$\sup_{\varphi \leq f} \int \varphi \, d\mu \geq \int \varphi_n \, d\mu = -(M/n)\mu(E) + \int \psi_n \, d\mu \geq -(M/n)\mu(E) + \inf_{\psi \geq f} \int \psi \, d\mu.$$

The last inequality holds for all n, so the proof is complete.

(4.3) LEMMA   Let E be a set with $\mu(E) < \infty$. If f and g are bounded functions that vanish on E^c, then:

(i) If $f \geq 0$ a.e., then $\int f \, d\mu \geq 0$.
(ii) $\int af \, d\mu = a \int f \, d\mu$.
(iii) $\int f + g \, d\mu = \int f \, d\mu + \int g \, d\mu$.

Proof   (i) is clear from the definition. To prove (ii) we observe that if $a > 0$, then $a\varphi \leq af$ if and only if $\varphi \leq f$, so

$$\int af \, d\mu = \sup_{\varphi \leq f} \int a\varphi \, d\mu = \sup_{\varphi \leq f} a \int \varphi \, d\mu = a \sup_{\varphi \leq f} \int \varphi \, d\mu = a \int f \, d\mu.$$

For $a < 0$ we observe that $a\varphi \leq af$ if and only if $\varphi \geq f$, so

$$\int af \, d\mu = \sup_{\varphi \geq f} \int a\varphi \, d\mu = \sup_{\varphi \geq f} a \int \varphi \, d\mu = a \inf_{\varphi \geq f} \int \varphi \, d\mu = a \int f \, d\mu.$$

To prove (iii) we observe that if $\psi_1 \geq f$ and $\psi_2 \geq g$, then $\psi_1 + \psi_2 \geq f + g$, so

$$\inf_{\psi \geq f+g} \int \psi \, d\mu \geq \inf_{\psi_1 \geq f, \, \psi_2 \geq g} \int \psi_1 + \psi_2 \, d\mu.$$

Using linearity for simple functions it follows that

$$\int f + g \, d\mu = \inf_{\psi \geq f+g} \int \psi \, d\mu \geq \inf_{\psi_1 \geq f, \, \psi_2 \geq g} \int \psi_1 \, d\mu + \int \psi_2 \, d\mu = \int f \, d\mu + \int g \, d\mu.$$

To prove the other inequality replace sup by inf and $\geq$ by $\leq$ in the last proof, or observe that the last conclusion applied to $-f$ and $-g$ and (ii) imply

$$-\int f + g \, d\mu \geq -\int f \, d\mu - \int g \, d\mu. \qquad \square$$

Notation   An important special case of Step 2 is the integral of f over the set E:

$$\int_E f \, d\mu \equiv \int f \cdot 1_E \, d\mu.$$

Step 3   If $f \geq 0$, then we let

$$\int f \, d\mu = \sup \left\{ \int h \, d\mu : 0 \leq h \leq f, h \text{ is bounded and } \mu(\{x : h(x) > 0\}) < \infty \right\}.$$

The last definition is nice, since it is clear that this is well defined. The next result will help us compute the value of the integral.

(4.4) LEMMA   Let $E_n \uparrow \Omega$ have $\mu(E_n) < \infty$ and let $a \wedge b = \min(a, b)$. Then

$$\int_{E_n} f \wedge n \, d\mu \uparrow \int f \, d\mu \quad \text{as } n \uparrow \infty.$$

Proof   It is clear that the left-hand side increases as n does and each term is smaller than the integral on the right. To prove that the limit is $\int f \, d\mu$ observe that if $0 \leq h \leq f, h \leq M$, and $\mu(\{x: h(x) > 0\}) < \infty$, then for $n \geq M$,

$$\int_{E_n} f \wedge n \, d\mu \geq \int_{E_n} h \, d\mu = \int h \, d\mu - \int_{E_n^c} h \, d\mu$$

and

$$0 \leq \int_{E_n^c} h \, d\mu \leq M\mu(E_n^c \cap \{x: h(x) > 0\}) \to 0 \quad \text{as } n \to \infty. \qquad \square$$

(4.5) LEMMA   Suppose $f, g \geq 0$.

(i) $\int f \, d\mu \geq 0$.
(ii) If $a > 0$, then $\int af \, d\mu = a \int f \, d\mu$.
(iii) $\int f + g \, d\mu = \int f \, d\mu + \int g \, d\mu$.

Proof   (i) is trivial. (ii) is clear, since when $a > 0$, $ah \leq af$ if and only if $h \leq f$. For (iii) we observe that if $f \geq h$ and $g \geq k$, then $f + g \geq h + k$, so taking the sup over h and k in the defining class gives

$$\int f + g \, d\mu \geq \int f \, d\mu + \int g \, d\mu.$$

To prove the other direction we observe $(a + b) \wedge n \leq a \wedge n + b \wedge n$, so (iv) from (4.2) and (iii) from (4.3) imply

$$\int_{E_n} (f + g) \wedge n \, d\mu \leq \int_{E_n} f \wedge n \, d\mu + \int_{E_n} g \wedge n \, d\mu.$$

Letting $n \to \infty$ and using (4.4) give the desired result. $\qquad \square$

Exercise 4.1   Show that if $h \geq 0$ and $\int h \, d\mu = 0$, then $h = 0$ a.e.

Exercise 4.2   Let $f \geq 0$ and $E_{n,m} = \{x: m/2^n \leq x < (m + 1)/2^n\}$. As $n \uparrow \infty$,

$$\sum_{m=1}^{\infty} m/2^n \mu(E_{n,m}) \uparrow \int f \, d\mu.$$

Step 4   We say f is integrable if $\int |f| \, d\mu < \infty$. Let

$$f^+(x) = f(x) \vee 0 \quad \text{and} \quad f^-(x) = (-f(x)) \vee 0$$

where $a \vee b = \max(a, b)$. Clearly,

$$f(x) = f^+(x) - f^-(x) \quad \text{and} \quad |f(x)| = f^+(x) + f^-(x).$$

We define the integral of f by

$$\int f \, d\mu = \int f^+ \, d\mu - \int f^- \, d\mu.$$

The right-hand side is well defined, since $f^+, f^- \leq |f|$. For the final time we will prove our three properties. To do this it is useful to know:

(4.6) LEMMA   If $f = f_1 - f_2$ where $f_1, f_2 \geq 0$, then $\int f \, d\mu = \int f_1 \, d\mu - \int f_2 \, d\mu$.

Proof $f_1 + f^- = f_2 + f^+$ and all four functions are ≥ 0, so by (iii) of (4.5),

$$\int f_1 \, d\mu + \int f^- \, d\mu = \int f_1 + f^- \, d\mu = \int f_2 + f^+ \, d\mu = \int f_2 \, d\mu + \int f^+ \, d\mu.$$

Rearranging gives the desired conclusion.   □

(4.7) THEOREM   Suppose f and g are integrable.

(i) If $f \geq 0$ a.e., then $\int f \, d\mu \geq 0$.
(ii) $\int af \, d\mu = a \int f \, d\mu$.
(iii) $\int f + g \, d\mu = \int f \, d\mu + \int g \, d\mu$.

Proof   (i) is trivial. (ii) is clear, since if $a > 0$, $(af)^+ = a(f^+)$, and so on. To prove (iii) observe that $f + g = (f^+ + g^+) - (f^- + g^-)$, so using (4.6) and (4.5),

$$\int f + g \, d\mu = \int f^+ + g^+ \, d\mu - \int f^- + g^- \, d\mu$$

$$= \int f^+ \, d\mu + \int g^+ \, d\mu - \int f^- \, d\mu - \int g^- \, d\mu.$$ □

Notation for special cases

(a) If $(\Omega, \mathscr{F}, P)$ is a probability space and X is a random variable with $\int |X| \, dP < \infty$, then we define its *expected value* to be $EX = \int X \, dP$. EX is often called the *mean* of X and denoted by μ.
(b) When $(\Omega, \mathscr{F}, \mu) = (\mathbb{R}^d, \mathscr{R}^d, \lambda)$, we write $\int f(x) \, dx$ for $\int f \, d\lambda$.
(c) When $(\Omega, \mathscr{F}, \mu) = (\mathbb{R}, \mathscr{R}, \lambda)$ and $E = [a, b]$, we write $\int_a^b f(x) \, dx$ for $\int_E f \, d\lambda$.
(d) When $(\Omega, \mathscr{F}, \mu) = (\mathbb{R}, \mathscr{R}, \mu)$ with $\mu((a, b]) = G(b) - G(a)$, we write $\int f(x) \, dG(x)$ for $\int f \, d\mu$.
(e) When Ω is a countable set, $\mathscr{F} = $ all subsets of Ω, and μ is counting measure, we write

$$\sum_{i \in \Omega} f(i) \quad \text{for} \quad \int f \, d\mu.$$

We mention this example primarily to indicate that results for sums follow from those for integrals.

For the rest of this section we will consider the case $(\Omega, \mathcal{F}, \mu) = (\mathbb{R}, \mathcal{R}, \lambda)$.

Littlewood's Principles

Speaking of the theory of functions of a real variable, Littlewood (1944) said, "The extent of knowledge required is nothing like so great as is sometimes supposed. There are three principles, roughly expressible in the following terms:

1. Every measurable set is roughly a finite union of intervals.
2. Every measurable function is almost continuous.
3. Every convergent sequence of measurable functions is almost uniformly convergent.

Most of the results of the theory are fairly intuitive applications of the theory and the student armed with them should be equal to most occasions when real variable theory is called for."

Exercise 3.1 above and Exercise 2.8 in Chapter 1 give versions of the first and third principles. The next two exercises develop a version of the second.

Exercise 4.3   Let g be an integrable function on $\mathbb{R}$ and $\varepsilon > 0$. (i) Use the definition of the integral to conclude there is a simple function $\varphi = \sum_k b_k 1_{A_k}$ with $\int |g - \varphi| \, dx < \varepsilon$. (ii) Use Exercise 3.1 to approximate the A_k by finite unions of intervals to get a *step function*

$$h = \sum_{m=1}^{n} c_m 1_{(a_{m-1}, a_m)}$$

with $a_0 < a_1 < \cdots < a_n$, so that $\int |\varphi - h| < \varepsilon$. (iii) Round the corners of h to get a continuous function k so that $\int |h - k| \, dx < \varepsilon$.

Remark   See Exercise 5.4 of Chapter 4 for another approach.

Exercise 4.4   Prove *Lusin's theorem*. Let $\delta > 0$. If f is a measurable function, we can find a continuous $j(x)$ with $\lambda(\{x : f(x) \neq j(x)\}) < \delta$.

Sketch   Begin by supposing $f = 0$ on $[0, 1]^c$. Pick a continuous function k_n that vanishes on $(-1, 2)^c$ and has $\lambda(\{x : |f(x) - k_n(x)| > \eta/2^n\}) < \eta/2^n$. Observe that $K = \{x : |k_n(x) - k_{n+1}(x)| \leq 3\eta/2^{n+1}$ for all $n \geq 1\}$ is a closed set, $k_n \to k_\infty$ uniformly on K, and a function that is continuous on a closed set K^c can be extended to be continuous on $\mathbb{R}$ by making it linear on the bounded intervals that make up K^c.

Exercise 4.5   Prove the *Riemann–Lebesgue lemma*. If g is integrable, then

$$\lim_{n \to \infty} \int g(x) \cos nx \, dx = 0.$$

Hint   If g is a step function, this is easy. Now use Exercise 4.3.

Riemann Integration

Our treatment of the Lebesgue integral would not be complete if we did not prove the classic theorem of Lebesgue that identifies the functions for which the Riemann integral exists. Let $-\infty < a < b < \infty$. A subdivision σ of $[a, b]$ is a finite sequence $a = x_0 < x_1 < \cdots < x_n = b$. Given a subdivision σ, we define:

Upper Riemann sum: $U(\sigma) = \sum_{i=1}^{n} (x_{i+1} - x_i) \sup\{f(y): y \in [x_{i-1}, x_i]\}$.
Lower Riemann sum: $L(\sigma) = \sum_{i=1}^{n} (x_{i+1} - x_i) \inf\{f(y): y \in [x_{i-1}, x_i]\}$.

We say that f is *Riemann integrable on* $[a, b]$ *in the liberal sense* if

$$\inf_{\sigma} U(\sigma) = \sup_{\sigma} L(\sigma).$$

The function $q(x)$ that is 1 if x is irrational and 0 if x is rational is the classic example of a function that is not Riemann integrable on $[0, 1]$ in the liberal sense but is Lebesgue integrable on $[0, 1]$. ($q 1_{[0, 1]}$ is a simple function!) The next result gives a necessary condition for Riemann integrability.

(4.8) THEOREM   If f is Riemann integrable on $[a, b]$ in the liberal sense, then f is bounded and continuous a.e. on $[a, b]$.

Proof   If f is unbounded above, then $U(\sigma) = \infty$ for all subdivisions. Likewise if f is unbounded below, $L(\sigma) = -\infty$ for all subdivisions. Let

$$u_n(x) = \sup\{f(y): |x - y| < 2^{-n} \text{ and } y \in [a, b]\}$$

$$v_n(x) = \inf\{f(y): |x - y| < 2^{-n} \text{ and } y \in [a, b]\}.$$

Exercise 2.4 in Chapter 1 implies u_n and v_n are measurable. Let

$$f^0 = \lim_{n \to \infty} u_n \quad \text{and} \quad f_0 = \lim_{n \to \infty} v_n.$$

$f^0(x) \geq f_0(x)$ with equality if and only if f is continuous at x. Given a subdivision σ,

$$\sup\{f(y): y \in [x_{i-1}, x_i]\} \geq f^0(x) \quad \text{for } x \in (x_{i-1}, x_i),$$

so $U(\sigma) \geq \int_{[a,b]} f^0 \, dx$, the integral existing since f^0 is bounded and measurable. Similar reasoning shows that any lower Riemann sum has $\int_{[a,b]} f_0 \, dx \geq L(\sigma)$, so if f is Riemann integrable in the liberal sense, $\int_{[a,b]} f^0 - f_0 \, dx = 0$ and it follows from Exercise 4.1 that $f^0 = f_0$ a.e.   □

To state a converse to (4.8) we need two definitions. The mesh of a subdivision $= \sup x_i - x_{i-1}$. f is said to be *Riemann integrable on* $[a, b]$ *in the strict sense* if for any sequence of subdivisions with mesh $\to 0$, $U(\sigma_n) - L(\sigma_n) \to 0$. An argument by contradiction shows that when this occurs, $U(\sigma_n)$ and $L(\sigma_n)$ have limits that are independent of the sequence of subdivisions chosen.

(4.9) THEOREM   If f is bounded and continuous a.e. on $[a, b]$, then f is Riemann integrable on $[a, b]$ in the strict sense.

Proof   We need a little more theory before we can give a simple proof of this. See Exercise 5.6. □

Exercise 4.6   Give examples to show that for a function f defined on $\mathbb{R}$ neither statement implies the other. (i) f is continuous a.e. (ii) There is a continuous function g so that $f = g$ a.e.

Exercise 4.7   Let $(\Omega, \mathscr{F}, \mu)$ be a finite measure space and let f be a function with $|f| < M$. Given a subdivision $-M = x_0 < x_1 < \cdots < x_n = M$, define:

Upper Lebesgue sum: $\bar{U}(\sigma) = \sum_{m=1}^{n} x_m \mu(\{\omega: f(\omega) \in [x_{m-1}, x_m)\})$.
Lower Lebesgue sum: $\bar{L}(\sigma) = \sum_{m=1}^{n} x_{m-1} \mu(\{\omega: f(\omega) \in [x_{m-1}, x_m)\})$.

Show that if mesh $(\sigma_n) \to 0$, then $\bar{U}(\sigma_n), \bar{L}(\sigma_n) \to \int f\, d\mu$. In short, in Riemann integration we subdivide the domain and in Lebesgue integration we subdivide the range.

5 Properties of the Integral

In this section we will develop properties of the integral defined in the last section. From (4.7) and (4.2), we have:

(5.1) THEOREM   Suppose f and g are integrable.

 (i) If $f \geq 0$ a.e., then $\int f\, d\mu \geq 0$.
 (ii) $\int af\, d\mu = a \int f\, d\mu$.
 (iii) $\int f + g\, d\mu = \int f\, d\mu + \int g\, d\mu$.
 (iv) If $f \geq g$ a.e., then $\int f\, d\mu \geq \int g\, d\mu$.
 (v) If $f = g$ a.e., then $\int f\, d\mu = \int g\, d\mu$.
 (vi) $|\int f\, d\mu| \leq \int |f|\, d\mu$.

Our first result generalizes (vi).

(5.2) *Jensen's Inequality*   Suppose φ is convex; that is,

$$\lambda\varphi(x) + (1 - \lambda)\varphi(y) \geq \varphi(\lambda x + (1 - \lambda)y)$$

for all $\lambda \in (0, 1)$ and $x, y \in \mathbb{R}$. If f and $\varphi(f)$ are integrable, then

$$\varphi\left(\int f\, d\mu \right) \leq \int \varphi(f)\, d\mu.$$

Proof   Let $c = \int f\, d\mu$ and let $l(x) = ax + b$ be a linear function that has $l(c) = \varphi(c)$ and $\varphi(x) \geq l(x)$. To see that such a function exists recall that convexity implies

$$\lim_{h \downarrow 0} (\varphi(y) - \varphi(y - h))/h \leq \lim_{h \downarrow 0} (\varphi(y + h) - \varphi(y))/h.$$

(The limits exist, since the sequences are monotone.) If we let a be any number between the two limits and let $l(x) = a(x - c) + \varphi(c)$, then l has the desired properties. With the existence of l established, the rest is easy:

$$\int \varphi(f) \, d\mu \geq \int (af + b) \, d\mu = a \int f \, d\mu + b = l\left(\int f \, d\mu\right) = \varphi\left(\int f \, d\mu\right). \qquad \square$$

Let $\|f\|_p = (\int |f| \, d\mu)^{1/p}$ for $1 \leq p < \infty$, and notice $\|cf\|_p = |c| \|f\|_p$ for $c \in \mathbb{R}$.

(5.3) Hölder's Inequality   If $p, q \in (1, \infty)$ with $1/p + 1/q = 1$, then

$$\int |fg| \, d\mu \leq \|f\|_p \|g\|_q.$$

Proof   If $\|f\|_p$ or $\|g\|_q = 0$, then $|fg| = 0$ a.e., so it suffices to prove the result when $\|f\|_p$ and $\|g\|_q > 0$ or by dividing both sides by $\|f\|_p \|g\|_q$, when $\|f\|_p = \|g\|_q = 1$. Fix $y \geq 0$ and let

$$\varphi(x) = x^p/p + y^q/q - xy \quad \text{for } x \geq 0.$$

$$\varphi'(x) = x^{p-1} - y \quad \text{and} \quad \varphi''(x) = (p - 1)x^{p-2},$$

so φ has a minimum at $x_o = y^{1/(p-1)}$. $x_o^p = y^{p/(p-1)} = y^q$, so

$$\varphi(x_o) = y^q(1/p + 1/q) - y^{1/(p-1)}y = 0,$$

and it follows that $xy \leq x^p/p + y^q/q$. Letting $x = |f|, y = |g|$, and integrating,

$$\int |fg| \, d\mu \leq \frac{1}{p} + \frac{1}{q} = 1. \qquad \square$$

Exercise 5.1   The special case $p = q = 2$ is called the *Cauchy–Schwarz inequality*. Give a direct proof of the result in this case by observing that for any θ,

$$0 \leq \int (f + \theta g)^2 \, d\mu = \int f^2 \, d\mu + \theta\left(2 \int fg \, d\mu\right) + \theta^2\left(\int g^2 \, d\mu\right),$$

so the quadratic on the right-hand side has at most one real root.

Exercise 5.2   Let $\|f\|_\infty = \inf\{M : \mu(\{x : |f(x)| > M\}) = 0\}$. Prove that

$$\int |fg| \, d\mu \leq \|f\|_1 \|g\|_\infty.$$

Exercise 5.3   Show that $\|f\|_\infty = \lim_{p\to\infty} \|f\|_p$.

Exercise 5.4   *Minkowski's inequality.* (i) Suppose $p \in (1, \infty)$. The inequality $|f + g|^p \le 2^p(|f|^p + |g|^p)$ shows that if $\|f\|_p$ and $\|g\|_p$ are $< \infty$, then $\|f + g\|_p < \infty$. Apply Hölder's inequality to $|f||f + g|^{p-1}$ and $|g||f + g|^{p-1}$ to show $\|f + g\|_p \le \|f\|_p + \|g\|_p$. (ii) Show that the last result remains true when $p = 1$ or $p = \infty$.

Our next goal is to give conditions that guarantee

$$\lim_{n\to\infty} \int f_n \, d\mu = \int \left(\lim_{n\to\infty} f_n \right) d\mu.$$

(5.4) Bounded Convergence Theorem   Let E be a set with $\mu(E) < \infty$. Suppose f_n vanishes on E^c, $|f_n(x)| \le M$, and $f_n \to f$ a.e. Then

$$\int f \, d\mu = \lim_{n\to\infty} \int f_n \, d\mu.$$

Example 5.1   The functions $f_n = 1_{[n,n+1)}(x)$ on $\mathbb{R}$ show that the conclusion of (5.4) does not hold when $\mu(E) = \infty$. (Here and in similar situations that follow we assume that $\mathbb{R}$ is equipped with the Borel sets $\mathcal{R}$ and Lebesgue measure λ.)

Proof   Let $\varepsilon > 0$, $G_n = \{x : |f_n(x) - f(x)| < \varepsilon\}$, and $B_n = E - G_n$.

$$\left| \int f \, d\mu - \int f_n \, d\mu \right| = \left| \int (f - f_n) \, d\mu \right| \le \int |f - f_n| \, d\mu$$

$$= \int_{G_n} |f - f_n| \, d\mu + \int_{B_n} |f - f_n| \, d\mu \le \varepsilon\mu(E) + 2M\mu(B_n).$$

Let $C_n = \{x : |f_m(x) - f(x)| \ge \varepsilon \text{ for some } m \ge n\} = \bigcup_{m\ge n} B_m$. $f_n \to f$ a.e. implies $\mu(B_n) \le \mu(C_n) \to 0$. $\varepsilon > 0$ is arbitrary so the proof is complete.

Exercise 5.5   For the bounded convergence theorem, the condition $f_n \to f$ a.e. can be replaced by $f_n \to f$ *in measure*; that is, for all $\varepsilon > 0$, $\mu(\{x : |f_n(x) - f(x)| > \varepsilon\}) \to 0$.

Exercise 5.6   Use the last exercise to prove (4.9).

Hint   Given a subdivision σ, let

$$f^\sigma(x) = \sup\{f(y) : y \in [x_{i-1}, x_i]\} \quad \text{for } x \in (x_{i-1}, x_i).$$

Notice that $\int_{[a,b]} f^\sigma(x) \, dx = U(\sigma)$, and show that $f^\sigma \to f$ in measure.

(5.5) Fatou's Lemma   If $f_n \ge 0$, then $\liminf_{n\to\infty} \int f_n \, d\mu \ge \int (\liminf_{n\to\infty} f_n) \, d\mu$.

Example 5.2   Example 5.1 shows that we may have strict inequality in (5.5). The functions $f_n = n1_{(0,\,1/n]}(x)$ on $(0, 1)$ show that this can happen on a space of finite measure.

Proof   Let $g_n(x) = \inf_{m \geq n} f_m(x)$. $f_n(x) \geq g_n(x)$ and as $n \uparrow \infty$,

$$g_n(x) \uparrow g(x) = \liminf_{n \to \infty} f_n(x).$$

Since $\int f_n \, d\mu \geq \int g_n \, d\mu$, it suffices to show that $\liminf_{n \to \infty} \int g_n \, d\mu \geq \int g \, d\mu$. Let $E_m \uparrow \Omega$ be sets of finite measure. $g_n \geq 0$ and for fixed m,

$$(g_n \wedge m) \cdot 1_{E_m} \to (g \wedge m) \cdot 1_{E_m} \quad \text{a.e.,}$$

so the bounded convergence theorem (5.4) implies

$$\liminf_{n \to \infty} \int g_n \, d\mu \geq \int_{E_m} g_n \wedge m \, d\mu \to \int_{E_m} g \wedge m \, d\mu.$$

Taking the sup over m and using (4.4) give the desired result.   □

(5.6) Monotone Convergence Theorem   If $f_n \geq 0$ and $f_n \uparrow f$, then $\int f_n \, d\mu \uparrow \int f \, d\mu$.

Proof   Fatou's lemma (5.5) implies $\liminf \int f_n \, d\mu \geq \int f \, d\mu$. On the other hand, $f_n \leq f$ implies $\limsup \int f_n \, d\mu \leq \int f \, d\mu$.   □

Exercise 5.7   If $g_n \uparrow g$ and $\int |g_1| \, d\mu < \infty$, then $\int g_n \, d\mu \uparrow \int g \, d\mu$.

Exercise 5.8   If $g_n \geq 0$, then $\int \sum_{n=0}^{\infty} g_n \, d\mu = \sum_{n=0}^{\infty} \int g_n \, d\mu$.

Exercise 5.9   Let $f \geq 0$. (i) Show that $\int f \wedge n \, d\mu \uparrow \int f \, d\mu$ as $n \to \infty$. (ii) Use (i) to conclude that if g is integrable and $\varepsilon > 0$, then we can pick $\delta > 0$ so that $\mu(A) < \delta$ implies $\int_A |g| \, d\mu < \varepsilon$.

(5.7) Dominated Convergence Theorem   If $f_n \to f$ a.e., $|f_n| \leq g$ for all n, and g is integrable, then $\int f_n \, d\mu \to \int f \, d\mu$.

Proof $f_n + g \geq 0$, so Fatou's lemma implies

$$\liminf_{n \to \infty} \int f_n + g \, d\mu \geq \int f + g \, d\mu.$$

Subtracting $\int g \, d\mu$ from both sides gives

$$\liminf_{n \to \infty} \int f_n \, d\mu \geq \int f \, d\mu.$$

Applying the last result to $-f_n$, we get

$$\limsup_{n \to \infty} \int f_n \, d\mu \le \int f \, d\mu,$$

and the proof is complete.   □

Exercise 5.10   If f is integrable and E_n are disjoint sets with union E, then

$$\sum_{n=0}^{\infty} \int_{E_n} f \, d\mu = \int_E f \, d\mu.$$

So if $f \ge 0$, then $v(E) = \int_E f \, d\mu$ defines a measure.

Exercise 5.11   Show that $\int_1^\infty e^{-t} \, dt = \lim_{n \to \infty} \int_1^{\theta n} (1 - t/n)^n \, dt$ if $\theta \le 1$ but not if $\theta > 2$.

Exercise 5.12   Show that if f is integrable on $[a, b]$, $g(x) = \int_{[a, x]} f(y) \, dy$ is continuous.

Exercise 5.13   Show that if f has $\|f\|_p = (\int |f|^p \, d\mu)^{1/p} < \infty$, then there are simple functions f_n so that $\|f_n - f\|_p \to 0$.

6 Product Measures, Fubini's Theorem

Let $(X, \mathscr{A}, \mu_1)$ and $(Y, \mathscr{B}, \mu_2)$ be two σ-finite measure spaces. Let

$$\Omega = X \times Y = \{(x, y): x \in X, y \in Y\}.$$

$$\mathscr{S} = \text{the sets of the form } A \times B \text{ where } A \in \mathscr{A} \text{ and } B \in \mathscr{B}.$$

Sets in $\mathscr{S}$ are called *rectangles*. It is easy to see that $\mathscr{S}$ is a semialgebra:

$$(A \times B) \cap (C \times D) = (A \cap C) \times (B \cap D)$$

$$(A \times B)^c = (A^c \times B) \cup (A \times B^c) \cup (A^c \times B^c).$$

Let $\mathscr{F} = \mathscr{A} \times \mathscr{B}$ be the σ-algebra generated by $\mathscr{S}$.

(6.1) THEOREM   There is a unique measure μ on $\mathscr{F}$ with $\mu(A \times B) = \mu_1(A)\mu_2(B)$.

Notation μ is often denoted by $\mu_1 \times \mu_2$.

Proof   By (1.3) it is enough to show that if $A \times B = +_i(A_i \times B_i)$, then

$$\mu(A \times B) = \sum_i \mu(A_i \times B_i).$$

For each $x \in A$, let $I(x) = \{i : x \in A_i\}$. $B = +_{i \in I(x)} B_i$, so

$$1_A(x)\mu_2(B) = \sum_i 1_{A_i}(x)\mu_2(B_i).$$

Integrating with respect to μ_1 and using Exercise 6.8 give

$$\mu_1(A)\mu_2(B) = \sum_i \mu_1(A_i)\mu_2(B_i),$$

which proves (6.1). $\square$

Exercise 6.1   Let $\mathscr{A}_o \subset \mathscr{A}$ and $\mathscr{B}_o \subset \mathscr{B}$ be semialgebras with $\sigma(\mathscr{A}_o) = \mathscr{A}$ and $\sigma(\mathscr{B}_o) = \mathscr{B}$. Then there is a unique measure μ on $\mathscr{A} \times \mathscr{B}$ that has $\mu(A \times B) = \mu_1(A)\mu_2(B)$ for $A \in \mathscr{A}_o$ and $B \in \mathscr{B}_o$. The point of this exercise is that we can define Lebesgue measure on $\mathbb{R}^2$ by the requirement that $\lambda((a, b] \times (c, d]) = (b - a)(d - c)$.

Using (6.1) and induction it follows that if $(\Omega_i, \mathscr{F}_i, \mu_i)$, $i = 1, \ldots, n$, are σ-finite measure spaces, and $\Omega = \Omega_1 \times \cdots \times \Omega_n$, there is a unique measure μ on the σ-algebra $\mathscr{F}$ generated by sets of the form $A_1 \times \cdots \times A_n$, $A_i \in \mathscr{F}_i$, that has

$$\mu(A_1 \times \cdots \times A_n) = \prod_{m=1}^{n} \mu_i(A_i).$$

When $(\Omega_i, \mathscr{F}_i, \mu_i) = (\mathbb{R}, \mathscr{R}, \lambda)$ for all i, the result is Lebesgue measure on the Borel subsets of n-dimensional Euclidean space $\mathbb{R}^n$.

Returning to the case in which $(\Omega, \mathscr{F}, \mu)$ is the product of two measure spaces, our next goal is to prove:

(6.2) Fubini's Theorem   If $f \geq 0$ or $\int |f| \, d\mu < \infty$, then

$$\int_X \int_Y f(x, y)\mu_2(dy)\mu_1(dx) = \int_{X \times Y} f \, d\mu = \int_Y \int_X f(x, y)\mu_1(dx)\mu_2(dy). \qquad (*)$$

Two technical things that need to be proved before we can assert that the first integral makes sense are:

For each x, $y \to f(x, y)$ is $\mathscr{B}$-measurable.
$x \to \int_Y f(x, y)\mu_2(dy)$ is $\mathscr{A}$-measurable.

We begin with the case $f = 1_E$. Let $E_x = \{y : (x, y) \in E\}$ be the *cross section* at x.

(6.3) LEMMA   If $E \in \mathscr{F}$, then $E_x \in \mathscr{B}$.

Proof $(E^c)_x = (E_x)^c$ and $(\bigcup_i E_i)_x = \bigcup_i (E_i)_x$, so if $\mathscr{E}$ is the collection of sets E for which $E_x \in \mathscr{B}$, then $\mathscr{E}$ is a σ-algebra. Since $\mathscr{E}$ contains the rectangles, the result follows. $\square$

(6.4) LEMMA   If $E \in \mathscr{F}$, then $g(x) \equiv \mu_2(E_x)$ is $\mathscr{A}$-measurable and $\int_X g \, d\mu_1 = \mu(E)$.

Notice that it is not obvious that the collection of sets for which the conclusion is true is a σ-algebra, since $\mu(E_1 \cup E_2) = \mu(E_1) + \mu(E_2) - \mu(E_1 \cap E_2)$. Dynkin's $\pi - \lambda$ theorem (2.1) was tailor-made for situations like this.

Proof of (6.4)   If conclusions hold for E_n and $E_n \uparrow E$, then (2.5) in Chapter 1 and the monotone convergence theorem imply that they hold for E. Since μ_1 and μ_2 are σ-finite, it is enough then to prove the result for $E \subset A \times B$ with $\mu_1(A) < \infty$ and $\mu_2(B) < \infty$, or taking $\Omega = A \times B$ we can suppose without loss of generality that $\mu(\Omega) < \infty$. Let $\mathscr{L}$ be the collection of sets E for which the conclusions hold. We will now check that $\mathscr{L}$ is a λ-system. Property (i) of a λ-system is trivial. (iii) follows from the first sentence in the proof. To check (ii) we observe that

$$\mu_2((A - B)_x) = \mu_2(A_x - B_x) = \mu_2(A_x) - \mu_2(B_x),$$

and integrating over x gives the second conclusion. Since $\mathscr{L}$ contains the rectangles, a π-system that generates $\mathscr{F}$, the desired result follows from the $\pi - \lambda$ theorem. $\square$

Proof of (6.2)   We prove the result by verifying it in four increasingly general special cases.

Case 1:   If $E \in \mathscr{F}$ and $f = 1_E$, then $(*)$ follows from (6.4).

Case 2:   Since each integral is linear in f, it follows that $(*)$ holds for simple functions.

Case 3:   Now if $f \geq 0$ and we let $f_n(x) = ([2^n f(x)]/2^n) \wedge n$ where $[x] = $ the largest integer $\leq x$, then the f_n are simple and $f_n \uparrow f$, so it follows from the monotone convergence theorem that $(*)$ holds for all $f \geq 0$.

Case 4:   The general case now follows by writing $f(x) = f(x)^+ - f(x)^-$ and applying Case 3 to f^+, f^-, and $|f|$. The condition $\int |f| \, d\mu < \infty$ guarantees that all the integrals are finite. $\square$

To illustrate why the various hypotheses of (6.2) are needed, we will now give some examples where the conclusion fails.

Example 6.1   Let $X = Y = \{1, 2, \ldots\}$ with $\mathscr{A} = \mathscr{B} = $ all subsets and $\mu_1 = \mu_2 = $ counting measure. For $m \geq 1$ let $f(m, m) = 1$ and $f(m + 1, m) = -1$, and let $f(m, n) = 0$ otherwise. We claim that

$$\sum_m \sum_n f(m, n) = 1, \quad \text{but} \quad \sum_n \sum_m f(m, n) = 0.$$

A picture is worth several dozen words:

$\vdots$	$\vdots$	$\vdots$	$\vdots$	$\vdots$	
0	0	0	1	-1	$\cdots$
0	0	1	-1	0	$\cdots$
0	1	-1	0	0	$\cdots$
1	-1	0	0	0	$\cdots$

Example 6.2   Let $X = (0, 1)$, $Y = (1, \infty)$, both equipped with the Borel sets and Lebesgue measure. Let $f(x, y) = e^{-xy} - 2e^{-2xy}$.

$\int_0^1 \int_1^\infty f(x, y)\, dy\, dx = \int_0^1 x^{-1}(e^{-x} - e^{-2x})\, dx$ exists and is > 0.

$\int_1^\infty \int_0^1 f(x, y)\, dx\, dy = \int_1^\infty y^{-1}(e^{-2y} - e^{-y})\, dy$ exists and is < 0.

A similar example, but with more arduous calculus, is provided by

$$g(x, y) = \frac{x^2 - y^2}{(x^2 + y^2)^2} \quad \text{for } 0 < x, y < 1.$$

In this case

$$\int_0^1 \int_0^1 g(x, y)\, dy\, dx = \pi/4 = -\int_0^1 \int_0^1 g(x, y)\, dx\, dy.$$

The next example indicates why μ_1 and μ_2 must be σ-finite.

Example 6.3   Let $X = (0, 1)$ with $\mathscr{A} =$ the Borel sets and $\mu_1 =$ Lebesgue measure. Let $Y = (0, 1)$ with $\mathscr{B} =$ all subsets and $\mu_2 =$ counting measure. Let $f(x, y) = 1$ if $x = y$ and 0 otherwise.

$\int_Y f(x, y)\mu_2(dy) = 1$ for all x, so $\int_X \int_Y f(x, y)\mu_2(dy)\mu_1(dx) = 1$.

$\int_X f(x, y)\mu_1(dx) = 0$ for all y, so $\int_Y \int_X f(x, y)\mu_1(dy)\mu_2(dx) = 0$.

Our last example shows that measurability is important, or maybe that some of the axioms of set theory are not as innocent as they seem.

Example 6.4   By the axiom of choice and the continuum hypothesis one can define an order relation $<'$ on $(0, 1)$ so that $\{x: x <' y\}$ is countable for each y. Let $X = Y = (0, 1)$, $\mathscr{A} = \mathscr{B} =$ the Borel sets, and $\mu_1 = \mu_2 =$ Lebesgue measure. Let $f(x, y) = 1$ if $x <' y$, $= 0$ otherwise. Since $\{x: x <' y\}$ is countable,

$$\int_X f(x, y)\mu_1(dx) = 0 \quad \text{for all } y$$

and

$$\int_Y f(x, y)\mu_2(dy) = 1 \quad \text{for all } x.$$

We turn now to applications of (6.2).

Exercise 6.2   If $\int_X \int_Y |f(x, y)|\mu_2(dy)\mu_1(dx) < \infty$, then

$$\int_X \int_Y f(x, y)\mu_2(dy)\mu_1(dx) = \int_{X \times Y} f\, d(\mu_1 \times \mu_2) = \int_Y \int_X f(x, y)\mu_1(dx)\mu_2(dy).$$

A sample application of the last result is:

Exercise 6.3 Let $X = \{1, 2, \ldots\}$, $\mathcal{A} = $ all subsets of X, and $\mu_1 = $ counting measure. If $\sum_n \int |f_n| \, d\mu < \infty$, then $\sum_n \int f_n \, d\mu = \int \sum_n f_n \, d\mu$. Show that this conclusion is also a consequence of the convergence theorems in Section 5.

Exercise 6.4 Let $g \geq 0$ be a measurable function on $(X, \mathcal{A}, \mu)$. Let $Y = [0, \infty)$, $\mathcal{B} = $ Borel subsets, and $\lambda = $ Lebesgue measure. Use (6.2) to conclude that

$$\int_X g \, d\mu = \mu \times \lambda(\{(x, y): 0 \leq y < g(x)\}) = \int_0^\infty \mu(\{x: g(x) > y\}) \, dy.$$

Hint $g(x) = \int_{[0, \infty)} 1_{(y < g(x))} \, dy.$

Exercise 6.5 Let F, G be Stieltjes measure functions and let μ, ν be the corresponding measures on $(\mathbb{R}, \mathcal{R})$. Show that

(i) $\int_{(a, b]} F(y) - F(a) \, dG(y) = \mu \times \nu(\{(x, y): a < x \leq y \leq b\})$
(ii) $\int_{(a, b]} F(y) \, dG(y) + \int_{(a, b]} G(y) \, dF(y) = F(b)G(b) - F(a)G(a) + \sum_{x \in (a, b]} \mu(\{x\})\nu(\{x\}).$

To see the second term is needed, let $F(x) = G(x) = 1_{[0, \infty)}(x)$ and $a < 0 < b$.

(iii) If $F = G$ is continuous, then $\int_{(a, b]} F(y) \, dF(y) = (F^2(b) - F^2(a))/2.$

Exercise 6.6 Let μ be a finite measure on $\mathbb{R}$ and $F(x) = \mu((-\infty, x])$. Show that

$$\int F(x + c) - F(x) \, dx = c\mu(\Omega).$$

Exercise 6.7 Show that $e^{-xy} \sin x$ is integrable in the strip $0 < x < a$, $0 < y$. Perform the double integral in the two orders to get:

$$\int_0^a \frac{\sin x}{x} \, dx = \frac{\pi}{2} - (\cos a) \int_0^\infty \frac{e^{-ay}}{1 + y^2} \, dy - (\sin a) \int_0^\infty \frac{ye^{-ay}}{1 + y^2} \, dy,$$

and replace $1 + y^2$ by 1 to conclude

$$\left| \int_0^a \frac{\sin x}{x} \, dx - \frac{\pi}{2} \right| \leq \frac{2}{a}.$$

Exercise 6.8 Let $(X, \mathcal{A}, \mu_1)$ and $(Y, \mathcal{B}, \mu_2)$ be σ-finite measure spaces. Suppose $f \geq 0$ is measurable with respect to the completion of $\mathcal{A} \times \mathcal{B}$. Then:

(i) For μ_1 a.e. x, $y \to f(x, y)$ is $\mathcal{B}$-measurable.
(ii) $x \to \int_Y f(x, y)\mu_2(dy)$ is $\mathcal{A}$-measurable.
(iii) $\int_X \int_Y f(x, y)\mu_2(dy)\mu_1(dx) = \int_{X \times Y} f \, d(\mu_1 \times \mu_2).$

Hint Start with $f = 1_E$ and use (3.1) to split the proof into two cases $E \in \mathcal{A} \times \mathcal{B}$ and $\mu_1 \times \mu_2(E) = 0.$

7 Kolmogorov's Extension Theorem

To construct some of the basic objects of study in probability theory we will need an existence theorem for measures on infinite product spaces. Let $\mathbb{N} = \{1, 2, \ldots\}$ and

$$\mathbb{R}^{\mathbb{N}} = \{(\omega_1, \omega_2, \ldots) : \omega_i \in \mathbb{R}\}.$$

We equip $\mathbb{R}^{\mathbb{N}}$ with the product σ-algebra $\mathscr{R}^{\mathbb{N}}$, which is generated by the *finite-dimensional rectangles* = sets of the form $\{\omega : \omega_i \in (a_i, b_i]$ for $i = 1, \ldots, n\}$.

Exercise 7.1   Show that the σ-algebra just defined is the same as the one generated by

$$\{\omega : \omega_i \in A_i \text{ for } i = 1, \ldots, n\} \quad \text{where } A_i \in \mathscr{R}.$$

(7.1) *Kolmogorov's Extension Theorem*   Suppose we are given probability measures μ_n on $(\mathbb{R}^n, \mathscr{R}^n)$ that are consistent; that is,

$$\mu_{n+1}((a_1, b_1] \times \cdots \times (a_n, b_n] \times \mathbb{R}) = \mu_n((a_1, b_1] \times \cdots \times (a_n, b_n]).$$

Then there is a unique probability measure P on $(\mathbb{R}^{\mathbb{N}}, \mathscr{R}^{\mathbb{N}})$ with

$$P(\omega : \omega_i \in (a_i, b_i], 1 \le i \le n) = \mu_n((a_1, b_1] \times \cdots \times (a_n, b_n]).$$

An important example of a consistent sequence of measures is given in:

Example 7.1   Let $F_1, F_2, \ldots$ be distribution functions and let μ_n be the measure on $\mathbb{R}^n$ with

$$\mu_n((a_1, b_1] \times \cdots \times (a_n, b_n]) = \prod_{m=1}^{n} (F_m(b_m) - F_m(a_m)).$$

Proof   Let $\mathscr{S}$ be the sets of the form $\{\omega : \omega_i \in (a_i, b_i], 1 \le i \le n\}$, and use the last equality in (7.1) to define P on $\mathscr{S}$. $\mathscr{S}$ is a semialgebra, so by (1.3) it is enough to show that if $A \in \mathscr{S}$ is a disjoint union of $A_i \in \mathscr{S}$, then $P(A) = \sum_i P(A_i)$. If the union is finite, then all the A_i are determined by the values of a finite number of coordinates and the conclusion follows from results in Section 6.

Suppose now that the union is infinite. Let $\mathscr{A} = \{$finite disjoint unions of sets in $\mathscr{S}\}$ be the algebra generated by $\mathscr{S}$. Since $\mathscr{A}$ is an algebra,

$$B_n \equiv A - \bigcup_{i=1}^{n} A_i$$

is a finite disjoint union of rectangles, and by the result for finite unions,

$$P(A) = \sum_{i=1}^{n} P(A_i) + P(B_n).$$

It suffices then to show:

(7.2) LEMMA   If $B_n \in \mathcal{A}$ and $B_n \downarrow \varnothing$, then $P(B_n) \downarrow 0$.

Proof   Suppose $P(B_n) \downarrow \delta > 0$. By repeating sets in the sequence we can suppose

$$B_n = \bigcup_{k=1}^{K_n} \{\omega: \omega_i \in (a_i^k, b_i^k], 1 \le i \le n\} \quad \text{where } -\infty \le a_i^i < b_i^i \le \infty.$$

The strategy of proof is to approximate the B_n from within by compact rectangles with almost the same probability and then use a diagonal argument to show that $\bigcap_n B_n \ne \varnothing$. There is a set $C_n \subset B_n$ of the form

$$C_n = \bigcup_{k=1}^{K_n} \{\omega: \omega_i \in [\bar{a}_i^k, \bar{b}_i^k], 1 \le i \le n\} \quad \text{with } -\infty < \bar{a}_k^i < \bar{b}_k^i < \infty$$

that has $P(B_n - C_n) \le \delta/2^{n+1}$. Let $D_n = \bigcap_{m=1}^{n} C_m$.

$$P(B_n - D_n) \le \sum_{m=1}^{n} P(B_m - C_m) \le \delta/2,$$

so $P(D_n) \downarrow$ a limit $\ge \delta/2$. Now there are sets $C_n^*, D_n^* \subset \mathbb{R}^n$ so that

$$C_n = \{\omega: (\omega_1, \ldots, \omega_n) \in C_n^*\} \quad \text{and} \quad D_n = \{\omega: (\omega_1, \ldots, \omega_n) \in D_n^*\}.$$

C_n^* is a finite union of rectangles, so

$$D_n^* = C_n^* \cap \bigcap_{m=1}^{n-1} (C_m^* \times \mathbb{R}^{n-m})$$

is a compact set. For each m let $\omega_m \in D_m$. $D_m \subset D_1$, so $\omega_{m,1}$ (i.e., the first coordinate of ω_m) is in D_1^*. Since D_1^* is compact, we can pick a subsequence $m(1, j) \ge j$ so that as $j \to \infty$,

$$\omega_{m(1,j),1} \to \text{a limit } \theta_1.$$

For $m \ge 2$, $D_m \subset D_2$ and hence $(\omega_{m,1}, \omega_{m,2}) \in D_2^*$. Since D_2^* is compact, we can pick a subsequence of the previous subsequence (i.e., $m(2, j) = m(1, i_j)$ with $i_j \ge j$) so that as $j \to \infty$,

$$\omega_{m(2,j),2} \to \text{a limit } \theta_2.$$

Continuing in this way we define $m(k, j)$ as a subsequence of $m(k - 1, j)$ so that as

$j \to \infty$,

$$\omega_{m(k,j),k} \to \text{a limit } \theta_k.$$

Let $\omega_i' = \omega_{m(i,i)}$. ω_i' is a subsequence of all the subsequences, so $\omega_{i,k}' \to \theta_k$ for all k. Now $\omega_{i,1}' \in D_1^*$ for all $i \geq 1$ and D_1^* is closed, so $\theta_1 \in D_1^*$. Turning to the second set, $(\omega_{i,1}', \omega_{i,2}') \in D_2^*$ for $i \geq 2$, since

$$\omega_i' = \omega_{m(i,i)} = \omega_{j_i} \quad \text{with } j_i \geq i.$$

Since D_2^* is closed, we conclude $(\theta_1, \theta_2) \in D_2^*$. Repeating the last argument we conclude that $(\theta_1, \ldots, \theta_k) \in D_k^*$ for all k. This shows that $\theta = (\theta_1, \theta_2, \ldots) \in D_k$ for all k, so

$$\varnothing \neq \bigcap_k D_k \subset \bigcap_k B_k, \quad \text{a contradiction.} \qquad \square$$

8   Radon–Nikodym Theorem

In this section we prove the Radon–Nikodym theorem. To develop that result we begin with a topic that at first may appear to be unrelated. Let $(\Omega, \mathscr{F})$ be a measurable space. α is said to be a *signed measure* on $(\Omega, \mathscr{F})$ if: (i) α takes values in $(-\infty, \infty]$, (ii) $\alpha(\varnothing) = 0$, and (iii) if $E = +_i E_i$ is a disjoint union, then $\alpha(E) = \sum_i \alpha(E_i)$, in the following sense:

If $\alpha(E) < \infty$, the sum converges absolutely and $= \alpha(E)$.
If $\alpha(E) = \infty$, then $\sum_i \alpha(E_i)^- < \infty$ and $\sum_i \alpha(E_i)^+ = \infty$.

Clearly, a signed measure cannot be allowed to take both the values ∞ and $-\infty$, since then $\alpha(A) + \alpha(B)$ might not make sense. In most formulations a signed measure is allowed to take values in either $(-\infty, \infty]$ or $[-\infty, \infty)$. We will ignore the second possibility to simplify statements later. As usual, we turn to examples to help explain the definition.

Example 8.1   Let μ be a measure, f be a function with $\int f^- \, d\mu < \infty$, and $\alpha(A) = \int_A f \, d\mu$.

Example 8.2   Let μ_1 and μ_2 be measures with $\mu_2(\Omega) < \infty$, and let $\alpha(A) = \mu_1(A) - \mu_2(A)$.

The Jordan decomposition ((8.4) below) will show that Example 8.2 is the general case. To derive that result we begin with two definitions. A set A is *positive* if every measurable $B \subset A$ has $\alpha(B) \geq 0$. A set A is *negative* if every measurable $B \subset A$ has $\alpha(B) \leq 0$.

Exercise 8.1   In Example 8.1, A is positive if and only if $\mu(A \cap \{x: f(x) < 0\}) = 0$.

(8.1) LEMMA   (i) Every measurable subset of a positive set is positive. (ii) If the sets A_n are positive, then $A = \bigcup_n A_n$ is also positive.

Proof   (i) is trivial. To prove (ii) observe that

$$B_n = A_n \cap \left(\bigcap_{m=1}^{n} A_m^c \right) \subset A_n$$

are positive, disjoint, and $\bigcup_n B_n = \bigcup_n A_n$. Let $E \subset A$ be measurable, and let $E_n = E \cap B_n$. $\alpha(E_n) \geq 0$, since B_n is positive, so $\alpha(E) = \sum_n \alpha(E_n) \geq 0$.   □

The conclusions in (8.1) remain valid if *positive* is replaced by *negative*. The next result is the key to the proof of (8.3).

(8.2) LEMMA   Let E be a measurable set with $\alpha(E) < 0$. Then there is a negative set $F \subset E$ with $\alpha(F) < 0$.

Proof   If E is negative, this is true. If not, let n_1 be the smallest positive integer so that there is an $E_1 \subset E$ with $\alpha(E_1) \geq 1/n_1$. Let $k \geq 2$. If $F_k = E - (E_1 \cup \cdots \cup E_{k-1})$ is negative, we are done. If not, let n_k be the smallest positive integer so that there is an $E_k \subset F_k$ with $\alpha(E_k) \geq 1/n_k$. If the construction does not stop for any $k < \infty$, let

$$F = \bigcap_k F_k = E - \left(\bigcup_k E_k \right).$$

Since $0 > \alpha(E) > -\infty$ and $\alpha(E_k) \geq 0$, it follows from the definition of signed measure that

$$\alpha(E) = \alpha(F) + \sum_{k=1}^{\infty} \alpha(E_k),$$

$\alpha(F) \leq \alpha(E) < 0$, and the sum is finite. From the last observation and the construction it follows that F can have no subset G with $\mu(G) > 0$, for then $\mu(G) \geq 1/N$ for some N and we would have a contradiction.   □

(8.3) *Hahn Decomposition*   Let α be a signed measure. Then there is a positive set A and a negative set B so that $\Omega = A \cup B$ and $A \cap B = \emptyset$.

Proof   Let $c = \inf\{\alpha(B) : B \text{ is negative}\} \leq 0$. Let B_i be negative sets with $\alpha(B_i) \downarrow c$. Let $B = \bigcup_i B_i$. By (8.1) B is negative, so by the definition of c, $\alpha(B) \geq c$. To prove $\alpha(B) \leq c$ we observe that $\alpha(B) = \alpha(B_i) + \alpha(B - B_i) \leq \alpha(B_i)$, since B is negative, and let $i \to \infty$. The last two inequalities show that $\alpha(B) = c$ and it follows from our definition of a signed measure that $c > -\infty$. Let $A = B^c$. To show A is positive, observe that if A contains a set with $\alpha(E) < 0$, then by (8.2) it contains a negative set F with $v(F) < 0$, but then $B \cup F$ would be a negative set with $\alpha(B \cup F) = \alpha(B) + \alpha(F) < c$, a contradiction.   □

The Hahn decomposition is not unique. In Example 8.1, A can be any set with

$$\{x: f(x) > 0\} \subset A \subset \{x: f(x) \geq 0\} \quad \text{a.e.}$$

where $B \subset C$ a.e. means $\mu(B \cap C^c) = 0$. The last example is typical of the general situation. Suppose $\Omega = A_1 \cup B_1 = A_2 \cup B_2$ are two Hahn decompositions. $A_2 \cap B_1$ is positive and negative so it is a *null set*; all its subsets have measure 0. Similarly $A_1 \cap B_2$ is a null set.

Two measures μ_1 and μ_2 are said to be *mutually singular* if there is a set A with $\mu_1(A) = 0$ and $\mu_2(A^c) = 0$. In this case, we also say μ_1 is *singular with respect to* μ_2 and write $\mu_1 \perp \mu_2$.

Exercise 8.2   Show that the uniform distribution on the Cantor set (Example 1.8 in Chapter 1) is singular with respect to Lebesgue measure.

(8.4) *Jordan Decomposition*   Let α be a signed measure. There are mutually singular measures α_+ and α_- so that $\alpha = \alpha_+ - \alpha_-$. Moreover, there is only one such pair.

Proof   Let $\Omega = A \cup B$ be a Hahn decomposition. Let $\alpha_+(E) = \alpha(E \cap A)$ and $\alpha_-(E) = -\alpha(E \cap B)$. Since A is positive and B is negative, α_+ and α_- are measures. $\alpha_+(A^c) = 0$ and $\alpha_-(A) = 0$, so they are mutually singular. To prove uniqueness suppose $\alpha = \nu_1 - \nu_2$ and D is a set with $\nu_1(D) = 0$ and $\nu_2(D^c) = 0$. If we set $C = D^c$, then $\Omega = C \cup D$ is a Hahn decomposition, and it follows from the choice of D that

$$\nu_1(E) = \nu(C \cap E) \quad \text{and} \quad \nu_2(E) = -\nu(D \cap E).$$

Our uniqueness result for the Hahn decomposition shows that $A \cap D = A \cap C^c$ and $B \cap C = A^c \cap C$ are null sets, so $\alpha(E \cap C) = \alpha(E \cap (A \cup C)) = \alpha(E \cap A)$ and $\nu_1 = \alpha_+$.

$\square$

Exercise 8.3   Show that $\alpha^+(A) = \sup\{\alpha(B): B \subset A\}$.

Remark   Let α be a *finite signed measure* (one that does not take the value ∞) on $(\mathbb{R}, \mathscr{R})$. Let $\alpha = \alpha_+ - \alpha_-$ be its Jordan decomposition. Let $A(x) = \alpha((-\infty, x])$, $F(x) = \alpha_+((-\infty, x])$, and $G(x) = \alpha_-((-\infty, x])$. $A(x) = F(x) - G(x)$, so the distribution function for a finite signed measure can be written as a difference of two bounded increasing functions. It follows from Exercise 8.2 that the converse is also true. Let $|\alpha| = \alpha^+ + \alpha^-$. $|\alpha|$ is called the *total variation of* α, since in this example $|\alpha|((a, b])$ is the total variation of A over $(a, b]$ as defined in analysis textbooks. See, for example, Royden (1988), p. 103. We exclude the left endpoint of the interval, since a jump there makes no contribution to the total variation on $[a, b]$ but it does appear in $|\alpha|$.

Our third and final decomposition is:

(8.5) *Lebesgue Decomposition*   Let μ, ν be σ-finite measures. ν can be written as $\nu_r + \nu_s$ where ν_s is singular with respect to μ and

$$v_r(E) = \int_E g \, d\mu.$$

Proof By decomposing $\Omega = +_i \Omega_i$ we can suppose without loss of generality that μ and v are finite measures. Let $\mathscr{G}$ be the set of $g \geq 0$, so that $\int_E g \, d\mu \leq v(E)$ for all E.

(a) If $g, h \in \mathscr{G}$, then $g \vee h \in \mathscr{G}$.

Proof of (a)   Let $A = \{g > h\}$, $B = \{g \leq h\}$.

$$\int_E g \vee h \, d\mu = \int_{E \cap A} g \, d\mu + \int_{E \cap B} h \, d\mu \leq v(E \cap A) + v(E \cap B) = v(E).$$

Let $\kappa = \sup\{\int g \, d\mu : g \in \mathscr{G}\} \leq v(\Omega) < \infty$. Pick g_n so that $\int g_n \, d\mu > \kappa - 1/n$ and let $h_n = g_1 \vee \cdots \vee g_n$. By (a), $h_n \in \mathscr{G}$. As $n \uparrow \infty$, $h_n \uparrow h$. The definition of κ, the monotone convergence theorem, and the choice of g_n imply that

$$\kappa \geq \int h \, d\mu = \lim_{n \to \infty} \int h_n \, d\mu \geq \lim_{n \to \infty} \int g_n \, d\mu = \kappa. \qquad \square$$

Let $v_r(E) = \int_E h \, d\mu$ and $v_s(E) = v(E) - v_r(E)$. The last detail is to show:

(b) v_s is singular with respect to μ.

Proof of (b)   Let $\varepsilon > 0$ and let $\Omega = A_\varepsilon \cup B_\varepsilon$ be a Hahn decomposition for $v_s - \varepsilon\mu$. Using the definition of v_r and then the fact that A_ε is positive for $v_s - \varepsilon\mu$,

$$\int_E h + \varepsilon 1_{A(\varepsilon)} \, d\mu = v_r(E) + \varepsilon\mu(A_\varepsilon) \leq v(E)$$

for all E, so $k = h + \varepsilon 1_{A(\varepsilon)} \in \mathscr{G}$. It follows that $\mu(A_\varepsilon) = 0$, for if not, then $\int k \, d\mu > \kappa$, a contradiction. Letting $A = \bigcup_n A_{1/n}$, we have $\mu(A) = 0$. To see that $v_s(A^c) = 0$ observe that if $v_s(A^c) > 0$, then $(v_s - \varepsilon\mu)(A^c) > 0$ for small ε, a contradiction since $A^c \subset B_\varepsilon$, a negative set. $\qquad \square$

Exercise 8.4   Prove that the Lebesgue decomposition is unique. Note that you can suppose without loss of generality that μ and v are finite.

We are finally ready for the main business of the section. We say a measure v is *absolutely continuous with respect to* μ (and write $v \ll \mu$) if $\mu(A) = 0$ implies that $v(A) = 0$.

Exercise 8.5   If $\mu_1 \ll \mu_2$ and $\mu_2 \perp v$, then $\mu_1 \perp v$.

(8.6) *Radon-Nikodym Theorem*   If μ, v are σ-finite measures and v is absolutely continuous with respect to μ, then there is a $g \geq 0$ so that $v(E) = \int_E g \, d\mu$. If h is another such function, then $g = h \ \mu$ a.e.

Proof Let $v = v_r + v_s$ be any Lebesgue decomposition. Let A be chosen so that $v_s(A^c) = 0$ and $\mu(A) = 0$. Since $v \ll \mu$, $0 = v(A) \geq v_s(A)$, and $v_s \equiv 0$. To prove uniqueness observe that if $\int_E g\, d\mu = \int_E h\, d\mu$ for all E, then letting $E \subset \{g > h, g \leq n\}$ be any subset of finite measure, we conclude $\mu(g > h, g \leq n) = 0$ for all n, so $\mu(g > h) = 0$ and similarly $\mu(g < h) = 0$. □

Example 8.3 (8.6) may fail if μ is not σ-finite. Let $(\Omega, \mathscr{F}) = (\mathbb{R}, \mathscr{R})$, $\mu = $ counting measure, and $v = $ Lebesgue measure. The assumption that v is σ-finite can be dropped. See Exercise 8.12 below.

The function g whose existence is proved in (8.6) is often denoted $d\mu/dv$. This notation suggests the following properties, whose proofs are left to the reader.

Exercise 8.6   If v_1 and $v_2 \ll \mu$, then $(d(v_1 + v_2)/d\mu) = dv_1/d\mu + dv_2/d\mu$.

Exercise 8.7   If $v \ll \mu$ and $f \geq 0$, then $\int f\, dv = \int f(dv/d\mu)\, d\mu$.

Exercise 8.8   If $\pi \ll v \ll \mu$, then $d\mu/d\pi = d\mu/dv \cdot dv/d\pi$.

Exercise 8.9   If $v \ll \mu$ and $\mu \ll v$, then $d\mu/dv = (dv/d\mu)^{-1}$. (Take $\pi = \mu$ in the last exercise.)

Exercise 8.10   Let F be the set defined by taking $[0, 1]$ and successively removing an open interval of length a centered at $1/2$, open intervals of length a^2 centered at $1/4$ and $3/4$, and of length a^3 centered at $1/8, 3/8, 5/8, 7/8, \ldots$. When $a = 1/3$, the result is the Cantor set. When $a < 1/3$, the result is a fat Cantor set with positive Lebesgue measure. Define a distribution F by the two requirements (i) $F(x) = x$ when $x = m/2^n$ and (ii) the corresponding measure has $\mu(F^c) = 0$. Prove that μ is absolutely continuous for $a < 1/3$ by writing down its density function.

Exercise 8.11   Suppose $v \ll \mu$ and $\mu(E) < \infty$ imply $v(E) < \infty$. For any $\varepsilon > 0$ there is a $\delta > 0$ so that $\mu(E) < \delta$ implies $v(E) < \varepsilon$. Prove this without using (8.6). Give an example $v \ll \mu$ for which the conclusion fails.

Exercise 8.12   *Another proof of* (8.6). Let $(\Omega, \mathscr{F}, \mu)$ be a finite measure space and suppose $v \ll \mu$. Prove that there is a $g \geq 0$ so that $\int_E g\, d\mu = v(E)$ for all $E \in \mathscr{F}$ by carrying out the following steps. (i) If $c > 0$, let $\Omega = A_c \cup B_c$ be a Hahn decomposition for $v - c\mu$. Let $A_0 = \Omega$, $B_0 = \varnothing$. Think of A_c, B_c as the sets where g is Above, Below c. If $b < c$, then $\mu(B_b - B_c) = 0$. (ii) Let $N = \bigcup_{q < r} B_q - B_r$ where the union is over rationals q and r, and let $B_q' = B_q \cup N$. Show that $g(x) = \inf\{q \in \mathbb{Q} : x \in B_q'\}$, with inf $\varnothing = \infty$, defines a measurable function. (iii) Let $E_k = E \cap (B_{(k+1)/n} - B_{k/n})$ and $E_\infty = E - \bigcup_q B_q$. Show that $v(E_k)$ and $\int_{E_k} g\, d\mu$ are each between $(k/n)\mu(E_k)$ and $(k + 1)/n\mu(E_k)$ and use this to reach the desired conclusion.

Exercise 8.13   (8.6) *implies* (8.5). Suppose μ and v are σ-finite on $(\Omega, \mathscr{F})$. Let $\pi = \mu + v$, let $f = d\mu/d\pi$ and $g = dv/d\pi = 1 - f$. Let $A = \{\omega : f(\omega) > 0\}$ and $B = \{\omega : f(\omega) = 0\}$. Show that $v_r(E) = v(E \cap A)$, $v_s(E) = v(E \cap B)$ give the desired decomposition.

Facts about Some Important Distributions

In this section we collect some facts that are scattered throughout Chapter 1. For the characteristic functions of these distributions see Section 3 of Chapter 2.

The **Bernoulli distribution** has $P(X = 1) = p$, $P(X = 0) = 1 - p$. In Example 5.3 of Chapter 1 we showed $EX = p$, $\text{var}(X) = p(1 - p)$, and if $X_1, X_2, \ldots$ are i.i.d. Bernoulli random variables, then $S_n = X_1 + \cdots + X_n$ has a **binomial distribution**

$$P(S_n = m) = \binom{n}{m} p^m (1 - p)^{n-m} \quad \text{where} \quad \binom{n}{m} = n!/(m!(n - m)!).$$

The **exponential distribution** with parameter λ, or exponential(λ) for short, has density function

$$P(X = x) = \lambda e^{-\lambda x} \text{ for } x \geq 0, \quad = 0 \text{ for } x < 0.$$

In Example 3.1. in Chapter 1 we computed $EX = 1/\lambda$ and $\text{var}(X) = 1/\lambda^2$.

The **gamma distribution** with parameters α and λ, or gamma(α, λ) for short, has density function

$$P(X = x) = \lambda^\alpha x^{\alpha-1} e^{-\lambda x}/\Gamma(\alpha) \text{ for } x \geq 0, \quad = 0 \text{ for } x < 0$$

where

$$\Gamma(\alpha) = \int_0^\infty x^{\alpha-1} e^{-x} \, dx.$$

In Exercise 4.10 of Chapter 1 we showed that if $X = $ gamma(α, λ) and $Y = $ gamma(β, λ) are independent, then $X + Y = $ gamma($\alpha + \beta, \lambda$). Since gamma($1, \lambda$) $=$ exponential(λ), it follows that if $X_1, X_2, \ldots$ are i.i.d. exponential(λ), then $X_1 + \cdots + X_n = $ gamma(n, λ). In view of the last fact it should not be surprising that if $X = $ gamma(α, λ), then $EX = \alpha/\lambda$.

The **geometric distribution** has $P(X = k) = p^{k-1}(1 - p)$ for $k \geq 1$. In Exercise 5.12 of Chapter 1 we computed $EX = 1/p$, $\text{var}(X) = (1 - p)/p^2$.

435

References

O. Adelman (1985) Brownian motion never increases: a new proof of a result of Dvoretsky, Erdös, and Kakutani. Israel J. Math. 50, 189–192

D. Aldous and P. Diaconis (1986) Shuffling cards and stopping times. Amer. Math. Monthly 93, 333–348

P. H. Algoet and T. M. Cover (1988) A sandwich proof of the Shannon McMillan Breiman theorem. Ann. Probab. 16, 899–909

E. S. Andersen and B. Jessen (1948) On the introduction of measures in infinite product sets. Danske Vid. Selsk. Mat. Fys. Medd. 25, No. 4

D. V. Anosov (1963) Ergodic properties of geodesic flows on closed Riemannian manifolds of negative curvature. Soviet Math. Doklady 4, 1153–1156

D. V. Anosov (1967) Geodesic flows on compact Riemannian manifolds of negative curvature. Proceedings of the Steklov. Inst. of Math., No. 90

K. Athreya, D. McDonald, and P. Ney (1978) Coupling and the renewal theorem. Amer. Math. Monthly 85, 809–814

K. Athreya and P. Ney (1972) *Branching processes.* Springer–Verlag, New York

K. Athreya and P. Ney (1978) A new approach to the limit theory of recurrent Markov chains. Trans. AMS 245, 493–501

L. Bachelier (1900) Théorie de la spéculation. Ann. Sci. École Norm. Sup. 17, 21–86

R. Ballerini and S. Resnick (1985) Records from improving populations. J. Appl. Probab. 22, 487–502

R. Ballerini and S. Resnick (1987) Records in the presence of a linear trend. Adv. Appl. Probab. 19, 801–828

S. Banach and A. Tarski (1924) Sur la décomposition des ensembles de points en parties respectivements congruent. Fund. Math. 6, 244–277

O. Barndorff-Nielsen (1961) On the rate of growth of partial maxima of a sequence of independent and identically distributed random variables. Math. Scand. 9, 383–394

L. E. Baum and P. Billingsley (1966) Asymptotic distributions for the coupon collector's problem. Ann. Math. Statist. 36, 1835–1839

F. Benford (1938) The law of anomalous numbers. Proc. Amer. Phil. Soc. 78, 552–572

J. D. Biggins (1977) Chernoff's theorem in branching random walk. J. Appl. Probab. 14, 630–636

J. D. Biggins (1978) The asymptotic shape of branching random walk. Adv. Appl. Probab. 10, 62–84

J. D. Biggins (1979) Growth rates in branching random walk. Z. Warsch. verw. Gebiete 48, 17–34

P. Billingsley (1961) The Lindeberg–Lévy theorem for martingales. Proc. AMS 12, 788–792

P. Billingsley (1968) *Weak convergence of probability measures*. John Wiley & Sons, New York

P. Billingsley (1979) *Probability and measure*. John Wiley & Sons, New York

G. D. Birkhoff (1931) Proof of the ergodic theorem. Proc. Nat. Acad. Sci. 17, 656–660

D. Blackwell and D. Freedman (1964) The tail σ-field of a Markov chain and a theorem of Orey. Ann. Math. Statist. 35, 1291–1295

R. M. Blumenthal and R. K. Getoor (1968) *Markov processes and their potential theory*. Academic Press, New York

E. Borel (1909) Les probabilités dénombrables et leurs applications arithmétiques. Rend. Circ. Mat. Palermo 27, 247–271

L. Breiman (1957) The individual ergodic theorem of ergodic theory. Ann. Math. Statist. 28, 809–811, Correction 31, 809–810

L. Breiman (1968) *Probability*. Addison–Wesley, Reading, MA

K. Burdzy (1989) On the non-increase of Brownian motion, preprint

Y. S. Chow and H. Teicher (1988) *Probability theory: independence, interchangeability, martingales*, second edition. Springer–Verlag, New York

K. L. Chung (1961) A note on the ergodic theorem of information theory. Ann. Math. Statist. 32, 612–614

K. L. Chung (1974) *A course in probability theory*, second edition. Academic Press, New York

K. L. Chung, P. Erdös, and T. Sirao (1959) On the Lipshitz's condition for Brownian motion. J. Math. Soc. Japan 11, 263–274

K. L. Chung and W. H. J. Fuchs (1951) On the distribution of values of sums of independent random variables. Memoirs of the AMS, No. 6

V. Chvátal and D. Sankoff (1975) Longest common subsequences of two random sequences. J. Appl. Probab. 12, 306–315

Z. Ciesielski (1961) Hölder condition for realizations of Gaussian processes. Trans. AMS 99, 403–413

J. Cohen, H. Kesten, and C. Newman (1985) *Random matrices and their applications*. AMS Contemporary Math. 50, Providence, RI

J. T. Cox and R. Durrett (1981) Limit theorems for percolation processes with necessary and sufficient conditions. Ann. Probab. 9, 583–603

B. Davis (1983) On Brownian slow points. Z. Warsch. verw. Gebiete 64, 359–367

Y. A. Davydov (1968) Convergence of distributions generated by stationary stochastic processes. Theory Probab. Appl. 13, 691–696

Y. Derriennic (1983) Une théorème ergodique presque sous additif. Ann. Probab. 11, 669–677

P. Diaconis and D. Freedman (1980) Finite exchangeable sequences. Ann. Probab. 8, 745–764

J. Dieudonné (1948) Sur la théorème de Lebesgue–Nikodym, II. Ann. Univ. Grenoble 23, 25–53

M. Donsker (1951) An invariance principle for certain probability limit theorems. Memoirs of the AMS, No. 6

M. Donsker (1952) Justification and extension of Doob's heuristic approach to the Kolmogorov–Smirnov theorems. Ann. Math. Statist. 23, 277–281

J. L. Doob (1949) A heuristic approach to the Kolmogorov–Smirnov theorems. Ann. Math. Statist. 20, 393–403

J. L. Doob (1953) *Stochastic processes*. John Wiley & Sons, New York

P. Doyle and L. Snell (1984) *Random walks and electrical networks*. Carus Monograph, Math. Assoc. of America

L. E. Dubins (1968) On a theorem of Skorokhod. Ann. Math. Statist. 39, 2094–2097

L. E. Dubins and D. A. Freedman (1965) A sharper form of the Borel–Cantelli lemma and the strong law. Ann. Math. Statist. 36, 800–807

L. E. Dubins and D. A. Freedman (1979) Exchangeable processes need not be distributed mixtures of independent and identically distributed random variables. Z. Warsch. verw. Gebiete 48, 115–132

R. Durrett (1984) *Brownian motion and martingales in analysis.* Wadsworth, Belmont, CA

R. Durrett and S. Resnick (1978) Functional limit theorems for dependent random variables. Ann. Probab. 6, 829–846

A. Dvoretsky (1972) Asymptotic normality for sums of dependent random variables. Proc. 6th Berkeley Symp., Vol. II, 513–535

A. Dvoretsky and P. Erdös (1950) Some problems on random walk in space. Proc. 2nd Berkeley Symp. 353–367

A. Dvoretsky, P. Erdös, and S. Kakutani (1961) Nonincrease everywhere of the Brownian motion process. Proc. 4th Berkeley Symp., Vol. II, 103–116

E. B. Dynkin (1955) Continuous one-dimensional Markov processes. Dokl. Akad. Nauk. SSR 105, 405–408

E. B. Dynkin (1961) Some limit theorems for sums of independent random variables with infinite mathematical expectation. Selected Translations in Math. Stat. and Prob., Vol. 1, 171–189

E. B. Dynkin (1965) *Markov processes.* Springer–Verlag, New York

P. Erdös (1942) On the law of the iterated logarithm. Ann. Math. 43, 419–436

P. Erdös and M. Kac (1946) On certain limit theorems of the theory of probability. Bull. AMS 52, 292–302

P. Erdös and M. Kac (1947) On the number of positive sums of independent random variables. Bull. AMS 53, 1011–1020

P. Erdös and B. Szerkeres (1935) A combinatorial problem in geometry. Compositio Math. 2, 463–470

N. Etemadi (1981) An elementary proof of the strong law of large numbers. Z. Warsch. verw. Gebiete 55, 119–122

A. M. Faden (1985) The existence of regular conditional probabilities: necessary and sufficient conditions. Ann. Probab. 13, 288–298

M. Fekete (1923) Über die Verteilung der Wurzeln bei gewissen algebraischen Gleichungen mit ganzzahligen Koeffizienten. Math. Z. 17, 228–249

W. Feller (1943) The general form of the so-called law of the iterated logarithm. Trans. AMS 54, 373–402

W. Feller (1946) A limit theorem for random variables with infinite moments. Amer. J. Math. 68, 257–262

W. Feller (1961) A simple proof of renewal theorems. Comm. Pure Appl. Math. 14, 285–293

W. Feller (1968) *An introduction to probability theory and its applications*, Vol. I, third edition. John Wiley & Sons, New York

W. Feller (1971) *An introduction to probability theory and its applications*, Vol. II, second edition. John Wiley & Sons, New York

S. R. Foguel (1969) *The ergodic theory of Markov processes.* Van Nostrand, New York

D. Freedman (1965) Bernard Friedman's urn. Ann. Math. Statist. 36, 956–970

D. Freedman (1971a) *Brownian motion and diffusion.* Originally published by Holden-Day, San Francisco. Second edition by Springer–Verlag, New York

D. Freedman (1971b) *Markov chains.* Originally published by Holden-Day, San Francisco. Second edition by Springer–Verlag, New York

D. Freedman (1980) A mixture of independent and identically distributed random variables need not admit a regular conditional probability given the exchangeable σ-field. Z. Warsch. verw. Gebiete 51, 239–248

R. M. French (1988) The Banach–Tarski theorem. Math. Intelligencer 10, No. 4, 21–28

B. Friedman (1949) A simple urn model. Comm. Pure Appl. Math. 2, 59–70

H. Furstenberg and H. Kesten (1960) Products of random matrices. Ann. Math. Statist. 31, 451–469

A. Garsia (1965) A simple proof of E. Hopf's maximal ergodic theorem. J. Math. Mech. 14, 381–382

M. L. Glasser and I. J. Zucker (1977) Extended Watson integrals for the cubic lattice. Proc. Nat. Acad. Sci. 74, 1800–1801

B. V. Gnedenko (1943) Sur la distribution limité du terme maximum d'une série aléatoire. Ann. Math. 44, 423–453

B. V. Gnedenko and A. V. Kolmogorov (1954) *Limit distributions for sums of independent random variables.* Addison–Wesley, Reading, MA

M. I. Gordin (1969) The central limit theorem for stationary processes. Soviet Math. Doklady 10, 1174–1176

D. R. Grey (1989) Persistent random walks may have arbitrarily large tails. Adv. Appl. Probab. 21, 229–230

D. Griffeath and T. M. Liggett (1982) Critical phenomena for Spitzer's reversible nearest particle systems. Ann. Probab. 10, 881–895

P. Hall (1982) *Rates of convergence in the central limit theorem.* Pitman, Boston

P. Hall and C. C. Heyde (1976) On a unified approach to the law of the iterated logarithm for martingales. Bull. Austral. Math. Soc. 14, 435–447

P. Hall and C. C. Heyde (1980) *Martingale limit theory and its application.* Academic Press, New York

P. R. Halmos (1950) *Measure theory.* Van Nostrand, New York

P. R. Halmos (1956) *Lectures on ergodic theory.* Chelsea, New York

J. M. Hammersley (1970) A few seedlings of research. Proc. 6th Berkeley Symp., Vol. I, 345–394

G. H. Hardy and J. E. Littlewood (1914) Some problems of Diophantine approximation. Acta Math. 37, 155–239

G. H. Hardy and E. M. Wright (1959) *An introduction to the theory of numbers,* fourth edition. Oxford Univ. Press, London

T. E. Harris (1956) The existence of stationary measures for certain Markov processes. Proc. 3rd Berkeley Symp., Vol. II, 113–124

P. Hartman and A. Wintner (1941) On the law of the iterated logarithm. Amer. J. Math. 63, 169–176

F. Hausdorff (1913) *Grundzüge der Mengenlehre.* Viet, Leipzig

E. Hewitt and L. J. Savage (1956) Symmetric measures on Cartesian products. Trans. AMS 80, 470–501

E. Hewitt and K. Stromberg (1965) *Real and abstract analysis.* Springer–Verlag, New York

C. C. Heyde (1963) On a property of the lognormal distribution. J. Royal. Stat. Soc. B 29, 392–393

C. C. Heyde (1967) On the influence of moments on the rate of convergence to the normal distribution. Z. Warsch. verw. Gebiete 8, 12–18

C. C. Heyde and D. J. Scott (1973) Invariance principles for the law of the iterated logarithm for martingales and for processes with stationary increments. Ann. Probab. 1, 428–436

T. Hida (1980) *Brownian motion.* Springer–Verlag, New York

J. L. Hodges, Jr., and L. LeCam (1960) The Poisson approximation to the binomial distribution. Ann. Math. Statist. 31, 737–740

P. Hoel, S. Port, and C. Stone (1972) *Introduction to stochastic processes.* Houghton Mifflin, Boston

G. Hunt (1956) Some theorems concerning Brownian motion. Trans. AMS 81, 294–319

I. A. Ibragimov (1962) Some limit theorems for stationary processes. Theory Probab. Appl. 7, 349–382

I. A. Ibragimov (1963) A central limit theorem for a class of dependent random variables. Theory Probab. Appl. 8, 83–89

I. A. Ibragimov and Y. V. Linnik (1971) *Independent and stationary sequences of random variables.* Wolters–Noordhoff, Groningen

H. Ishitani (1977) A central limit theorem for the subadditive process and its application to products of random matrices. RIMS, Kyoto 12, 565–575

K. Itô and H. P. McKean (1965) *Diffusion processes and their sample paths.* Springer–Verlag, New York

N. C. Jain and S. Orey (1969) On the range of random walk. Israel J. Math. 6, 373–380

N. C. Jain and W. E. Pruitt (1971) The range of random walk. Proc. 6th Berkeley Symp., Vol. III, 31–50

M. Kac (1947a) Brownian motion and the theory of random walk. Amer. Math. Monthly 54, 369–391

M. Kac (1947b) On the notion of recurrence in discrete stochastic processes. Bull. AMS 53, 1002–1010

M. Kac (1949) On deviations between theoretical and empirical distribution functions. Proc. Nat. Acad. Sci. 35, 252–257

M. Kac (1959) *Statistical independence in probability, analysis, and number theory.* Carus Monographs, Math. Assoc. of America

S. Karlin and H. M. Taylor (1975) *A first course in stochastic processes,* second edition. Academic Press, New York

Y. Katznelson and B. Weiss (1982) A simple proof of some ergodic theorems. Israel J. Math. 42, 291–296

E. Keeler and J. Spencer (1975) Optimal doubling in backgammon. Operations Research 23, 1063–1071

H. Kesten (1986) Aspects of first passage percolation. In *École d'été de probabilités de Saint–Flour XIV.* Lecture Notes in Math 1180, Springer–Verlag, New York

H. Kesten (1987) Percolation theory and first passage percolation. Ann. Probab. 15, 1231–1271

A. Khintchine (1923) Über dyadische Brüche. Math. Z. 18, 109–116

A. Khintchine (1924) Über einen Satz der Wahrscheinlichkeitsrechnung. Fund. Math. 6, 9–20

J. Kielson and D. M. G. Wishart (1964) A central limit theorem for processes defined on a Markov chain. Proc. Camb. Phil. Soc. 60, 547–567

J. F. C. Kingman (1968) The ergodic theory of subadditive stochastic processes. J. Roy. Stat. Soc. B. 30, 499–510

J. F. C. Kingman (1973) Subadditive ergodic theory. Ann. Probab. 1, 883–909

J. F. C. Kingman (1975) The first birth problem for age dependent branching processes. Ann. Probab. 3, 790–801

A. N. Kolmogorov (1929) Über das Gesetz des iterierten Logarithmus. Math. Ann. 101, 126–135

A. N. Kolmogorov and Y. A. Rozanov (1964) On strong mixing conditions for stationary Gaussian processes. Theory Probab. Appl. 5, 204–208

K. Kondo and T. Hara (1987) Critical exponent of susceptibility for a general class of ferromagnets in $d > 4$ dimensions. J. Math. Phys. 28, 1206–1208

U. Krengel (1985) *Ergodic theorems.* deGruyter, New York

A. Lasota and M. MacKay (1985) *Probabilistic properties of deterministic systems.* Cambridge Univ. Press, London

J. F. LeGall (1985) Un théorème central limite pour le nombre de points visités par une marche aléatoire plane récurrente. C. R. Acad. Sci. Paris 300, 505–508

J. F. LeGall (1986a) Propriétés d'intersection des marches aléatoires, I. Convergence vers le temps local d'intersection. Comm. Math. Phys. 104, 471–507

J. F. LeGall (1986b) Propriétés d'intersection des marches aléatoires, II. Étude de cas critiques. Comm. Math. Phys. 104, 509–528

R. Leipnik (1981) The lognormal law and strong non-uniqueness of the moment problem. Theory Probab. Appl. 26, 850–852

S. Leventhal (1988) A proof of Liggett's version of the subadditive ergodic theorem. Proc. AMS 102, 169–173

P. Lévy (1931) Sur un théorème de M. Khintchine. Bull. Sci. Math. (2), 55, 145–160

P. Lévy (1937) *Théorie de l'addition des variables aléatoires.* Gauthier–Villars, Paris

P. Lévy (1939) Sur certains processus stochastiques homogènes. Compositio Math. 7, 283–339

P. Lévy (1948) *Processus stochastiques et mouvement Brownien.* Gauthier–Villars, Paris

T. M. Liggett (1985) An improved subadditive ergodic theorem. Ann. Probab. 13, 1279–1285

A. Lindenbaum (1926) Contributions à l'étude de l'espace metrique. Fund. Math. 8, 209–222

T. Lindvall (1977) A probabilistic proof of Blackwell's renewal theorem. Ann. Probab. 5, 482–485

T. Lindvall (1979) On coupling of discrete renewal processes. Z. Warsch. verw. Gebiete 48, 57–70

J. E. Littlewood (1944) *Lectures on the theory of functions.* Oxford Univ. Press, London

B. F. Logan and L. A. Shepp (1977) A variational problem for random Young tableaux. Adv. Math. 26, 206–222

T. Lyons (1983) A simple criterion for transience of a Markov chain. Ann. Probab. 11, 393–402

H. P. McKean (1969) *Stochastic integrals.* Academic Press, New York

B. McMillan (1953) The basic theorems of information theory. Ann. Math. Statist. 24, 196–219

M. Motoo (1959) Proof of the law of the iterated logarithm through diffusion equation. Ann. Inst. Stat. Math. 10, 21–28

S. Newcomb (1881) Note on the frequency of use of the different digits in natural numbers. Amer. J. Math. 4, 39–40

J. Neveu (1965) *Mathematical foundations of the calculus of probabilities.* Holden–Day, San Francisco

J. Neveu (1975) *Discrete parameter martingales.* North-Holland, Amsterdam

G. O'Brien (1974) Limit theorems for sums of chain dependent processes. J. Appl. Probab. 11, 582–587

D. Ornstein (1969) Random walks. Trans. AMS 138, 1–60

V. I. Oseledec (1968) A multiplicative ergodic theorem. Lyapunov characteristic numbers for synmaical systems. Trans. Moscow Math. Soc. 19, 197–231

R. E. A. C. Paley and N. Wiener (1934) *Fourier transforms in the complex domain.* Amer. Math. Soc. Colloq. Pub. XIX

R. E. A. C. Paley, N. Wiener, and A. Zygmund (1933) Notes on random functions. Math. Z. 37, 647–668

I. Petrovski (1935) Zur ersten Randwertaufgabe der Wärmeleitungsgleichung. Compositio Math. 1, 383–419

E. J. G. Pitman (1956) On derivatives of characteristic functions at the origin. Ann. Math. Statist. 27, 1156–1160

S. C. Port and C. J. Stone (1969) Potential theory of random walks on abelian groups. Acta Math. 122, 19–114

M. S. Ragunathan (1979) A proof of Oseledĕc's multiplicative ergodic theorem. Israel J. Math. 32, 356–362

R. Raimi (1976) The first digit problem. Amer. Math. Monthly 83, 521–538

S. Resnick (1987) *Extreme values, regular variation, and point processes.* Springer–Verlag, New York

D. Revuz (1984) *Markov chains*, second edition. North-Holland, Amsterdam

D. H. Root (1969) The existence of certain stopping times on Brownian motion. Ann. Math. Statist. 40, 715–718

M. Rosenblatt (1956) A central limit theorem and a strong mixing condition. Proc. Nat. Acad. Sci. 42, 43–47

H. Royden (1988) *Real analysis*, third edition. McMillan, New York

D. Ruelle (1979) Ergodic theory of differentiable dynamical systems. IHES Pub. Math. 50, 275–306

C. Ryll–Nardzewski (1951) On the ergodic theorems, II. Studia Math. 12, 74–79

L. J. Savage (1972) *The foundations of statistics*, second edition. Dover, New York

D. J. Scott (1973) Central limit theorems for martingales and for processes with stationary independent increments using a Skorokhod representation approach. Adv. Appl. Probab. 5, 119–137

C. Shannon (1948) A mathematical theory of communication. Bell Systems Tech. J. 27, 379–423

L. A. Shepp (1964) Recurrent random walks may take arbitrarily large steps. Bull. AMS 70, 540–542

S. Sheu (1986) Representing a distribution by stopping Brownian motion: Root's construction. Bull. Austral. Math. Soc. 34, 427–431

A. Skorokhod (1965) *Studies in the theory of random processes.* Originally published by Addison Wesley, Reading, MA. Second edition (1982), Dover, New York

N. V. Smirnov (1949) Limit distributions for the terms of a variational series. AMS Transl. Series 1, No. 67

R. Smythe and J. C. Wierman (1978) *First passage percolation on the square lattice.* Lecture Notes in Math 671, Springer–Verlag, New York

R. M. Solovay (1970) A model of set theory in which every set of reals is Lebesgue measurable. Ann. Math. 92, 1–56

E. Sparre–Andersen and B. Jessen (1984) On the introduction of measures in infinite product spaces. Danske Vid. Selsk. Mat.–Fys. Medd. 25, No. 4

F. Spitzer (1964) *Principles of random walk.* Van Nostrand, Princeton, NJ

J. M. Steele (1989) Kingman's subadditive ergodic theorem. Ann. Inst. H. Poincaré 25, 93–98

C. Stein (1987) *Approximate computation of expectations.* IMS Lecture Notes, Vol. 7

H. Steinhaus (1922) Les probabilités denombrables et leur rapport à la théorie de la mesure. Fund. Math. 4, 286–310

C. J. Stone (1969) On the potential operator for one dimensional recurrent random walks. Trans. AMS 136, 427–445

J. Stoyanov (1987) *Counterexamples in probability.* John Wiley & Sons, New York

V. Strassen (1964) An invariance principle for the law of the iterated logarithm. Z. Warsch. verw. Gebiete 3, 211–226

V. Strassen (1965) A converse to the law of the iterated logarithm. Z. Warsch. verw. Gebiete 4, 265–268

V. Strassen (1967) Almost sure behavior of the sums of independent random variables and martingales. Proc. 5th Berkeley Symp., Vol. II, 315–343

H. Thorisson (1987) A complete coupling proof of Blackwell's renewal theorem. Stoch. Proc. Appl. 26, 87–97

H. Trotter (1958) A property of Brownian motion paths. Illinois J. Math. 2, 425–433

P. van Beek (1972) An application of Fourier methods to the problem of sharpening the Berry–Esseen inequality. Z. Warsch. verw. Gebiete 23, 187–196

A. M. Vershik and S. V. Kerov (1977) Asymptotic behavior of the Plancherel measure of the symmetric group and the limit form of random Young tableau. Dokl. Akad. Nauk SSR 233, 1024–1027

H. Wegner (1973) On consistency of probability measures. Z. Warsch. verw. Gebiete 27, 335–338

L. Weiss (1955) The stochastic convergence of a function of sample successive differences. Ann. Math. Statist. 26, 532–536

N. Wiener (1923) Differential space. J. Math. Phys. 2, 131–174

K. Yosida and S. Kakutani (1939) Birkhoff's ergodic theorem and the maximal ergodic theorem. Proc. Imp. Acad. Tokyo 15, 165–168

A. Zygmund (1947) *Trigonometric series*. Cambridge Univ. Press, London

Index

Notation

$\mathbb{N}$	natural numbers 1, 2, ...
$\mathbb{Z}$	integers
$\mathbb{Q}$	rational numbers
$\mathbb{R}$	real numbers
$\mathbb{C}$	complex numbers

Real numbers

$[x]$	integer part of x, the largest integer $n \leq x$
$x \wedge y$	minimum of x and y
$x \vee y$	maximum of x and y
x^+	positive part, $x \vee 0$
x^-	negative part, $(-x) \vee 0$
$\text{sgn}(x)$	the sign of x, 1 if $x > 0$, -1 if $x < 0$, 0 if $x = 0$
$x_n \to x$	$\lim_{n \to \infty} x_n = x$
$a_n \uparrow$	$a_1 \leq a_2 \leq \dots$
$a_n \downarrow a$	$a_1 \geq a_2 \geq \dots$ and $a_n \to a$
$\sim$	asymptotically, $a_n \sim b_n$ means $a_n/b_n \to 1$ as $n \to \infty$
O	$f(t)$ is $O(t^2)$ as $t \to 0$ means $\limsup_{t \to 0} |f(t)|/t^2 < \infty$
o	$f(t)$ is $o(t)$ as $t \to 0$ means $f(t)/t \to 0$ as $t \to 0$

Complex numbers, $z = a + bi$

$\bar{z}$	complex conjugate, $= a - bi$
$|z|$	modulus, $= (a^2 + b^2)^{1/2}$
$\text{Re } z$	real part of z, $= a$
$\text{Im } z$	imaginary part of z, $= b$

Vectors $(x_1, \ldots, x_d) \in \mathbb{R}^d$

$\lvert x \rvert$	the length of x, $(x_1^2 + \ldots + x_d^2)^{1/2}$
$\lVert x \rVert_1$	the L^1 norm, $\lvert x_1 \rvert + \ldots + \lvert x_d \rvert$
$x \cdot y$	dot product, $x_1 y_1 + \ldots + x_d y_d$
$\bar{A}$	closure of A
A^o	interior of A
∂A	boundary of A, $= \bar{A} - A^o$
$B(x, r)$	ball of radius r with center at x, $\{y \colon \lvert x - y \rvert < r\}$
$\partial B(x, r)$	boundary of $B(x, r)$, $\{y \colon \lvert x - y \rvert = r\}$
$x + A$	$\{x + y \colon y \in A\}$
rA	$\{rx \colon x \in A\}$

Set Theory

$\varnothing$	empty set
$A \cup B$	the union of A and B
$A \cap B$	the intersection of A and B
$A + B$	disjoint union, i.e., $A + B = A \cup B$ and indicates that $A \cap B = \varnothing$
A^c	the complement of A
$A - B$	difference, $A \cap (B^c)$
$A \,\Delta\, B$	symmetric difference, $(A - B) \cup (B - A)$
$\limsup A_n$	$\bigcap_{m \geq 1}(\bigcup_{n \geq m} A_n) = $ points in infinitely many A_n
$\liminf A_n$	$\bigcup_{m \geq 1}(\bigcap_{n \geq m} A_n) = $ points in all but finitely many A_n
$A_n \downarrow A$	$A_1 \supset A_2 \supset \ldots$ and $\bigcap_n A_n = A$
$A_n \uparrow A$	$A_1 \subset A_2 \subset \ldots$ and $\bigcup_n A_n = A$

Probability

$\mathscr{R}^d$	Borel subsets of $\mathbb{R}^d$
$\lvert A \rvert$	Lebesgue measure of A
1_A	indicator function, $= 1$ on A, $= 0$ on A^c
$X \in \mathscr{F}$	X is measurable with respect to $\mathscr{F}$; see Section 2 of Chapter 1
$\sigma(\underline{\quad})$	σ-field generated by $\underline{\quad}$
$\sigma(\mathscr{C})$	the smallest σ-field containing all the sets in $\mathscr{C}$
$\sigma(X)$	the smallest σ-field $\mathscr{G}$ so that $X \in \mathscr{G}$
$\mathscr{F}_n \uparrow \mathscr{F}_\infty$	$\mathscr{F}_1 \subset \mathscr{F}_2 \subset \ldots,\ \sigma(\bigcup \mathscr{F}_n) = \mathscr{F}_\infty$
$\mathscr{F}_n \downarrow \mathscr{F}_\infty$	$\mathscr{F}_1 \supset \mathscr{F}_2 \supset \ldots,\ \bigcap \mathscr{F}_n = \mathscr{F}_\infty$
EX	expected value of X; see Section 3 of Chapter 1
$\lVert X \rVert_p$	$(E\lvert X \rvert^p)^{1/p}$
$\mathrm{var}(X)$	the variance of X, $= E(X - EX)^2$
$L^2(\mathscr{F})$	$\{X \colon X \in \mathscr{F},\ EX^2 < \infty\}$
$E(X \mid \mathscr{F})$	conditional expectation of X given $\mathscr{F}$; see Section 1 of Chapter 4

$P(A\vert\mathscr{F})$	$E(X\vert\mathscr{F})$ when $X = 1_A$
χ	random variable with a standard normal distribution
$\mathscr{N}(x)$	$P(\chi \le x)$, normal distribution function
$X \overset{d}{=} Y$	X and Y have the same distribution
$\Rightarrow$	converges weakly; see Section 2 of Chapter 2

Abbreviations

a.e.	almost everywhere
a.s.	almost surely
ch.f.	characteristic function
CLT	central limit theorem
d.f.	distribution function
f.d.d.'s	finite-dimensional distributions
g.c.d.	greatest common divisor
i.i.d.	independent and identically distributed
i.o.	infinitely often
LIL	law of the iterated logarithm
l.s.c.	lower semicontinuous
MCT	monotone class theorem
r.c.d.	regular conditional distribution
u.s.c.	upper semicontinuous
r.v.	random variable
w.r.t.	with respect to